Paleoenvironments and paleohydrology of the Mojave and southern Great Basin Deserts

Edited by

Yehouda Enzel
Institute of Earth Sciences and Department of Geography
The Hebrew University of Jerusalem
Jerusalem 91904
Israel

and

Stephen G. Wells
Desert Research Institute
2215 Raggio Parkway
Reno, Nevada 89512-1095
USA

and

Nicholas Lancaster
Desert Research Institute
2215 Raggio Parkway
Reno, Nevada 89512-1095
USA

THE
GEOLOGICAL
SOCIETY
OF AMERICA

Special Paper 368

3300 Penrose Place, P.O. Box 9140 ▪ Boulder, Colorado 80301-9140 USA

2003

Copyright © 2003, The Geological Society of America, Inc. (GSA). All rights reserved. GSA grants permission to individual scientists to make unlimited photocopies of one or more items from this volume for noncommercial purposes advancing science or education, including classroom use. For permission to make photocopies of any item in this volume for other noncommercial, nonprofit purposes, contact the Geological Society of America. Written permission is required from GSA for all other forms of capture or reproduction of any item in the volume including, but not limited to, all types of electronic or digital scanning or other digital or manual transformation of articles or any portion thereof, such as abstracts, into computer-readable and/or transmittable form for personal or corporate use, either noncommercial or commercial, for-profit or otherwise. Send permission requests to GSA Copyrights Permissions, 3300 Penrose Place, P.O. Box 9140, Boulder, Colorado 80301-9140, USA.

Copyright is not claimed on any material prepared wholly by government employees within the scope of their employment.

Published by The Geological Society of America, Inc.
3300 Penrose Place, P.O. Box 9140, Boulder, Colorado 80301
www.geosociety.org

Printed in U.S.A.

GSA Books Science Editor: Abhijit Basu

Library of Congress Cataloging-in-Publication Data

Paleoenvironments and paleohydrology of the Mojave and southern Great Basin deserts /
Edited by Yehouda Enzel, Stephen G. Wells, Nicholas Lancaster.
    p. cm. — (Special paper ; 368)
  Includes bibliographical references.
  ISBN 0-8137-2368-X (pbk.)
    1. Paleohydrology—California--Mojave Desert. I. Enzel, Yehouda, 1955- II. Wells, Stephen G.
III. Lancaster, Nicholas, 1948- IV. Special papers (Geological Society of America) ; 368.

QE39.5.P27 P355 2003
560'.45'09794—dc21

                                    2002192780

Cover: Photograph of oil-on-hardboard painting by Mark Hamlin (Tecopa, California) done in 1980. View in painting is approximately 4 miles from Baker, California, southeastward across the Soda Lake playa toward the Cowhole and Old Dad Mountains. Providence Mountains form the skyline.

10 9 8 7 6 5 4 3 2 1

Contents

Acknowledgments v

Dedication vii

Preface ix

1. *Pliocene and Pleistocene evolution of the Mojave River, and associated tectonic development of the Transverse Ranges and Mojave Desert, based on borehole stratigraphy studies and mapping of landforms and sediments near Victorville, California* 1
   B.F. Cox, J.W. Hillhouse, and L.A. Owen

2. *Stratigraphy and paleontology of the middle to late Pleistocene Manix Formation, and paleoenvironments of the central Mojave River, southern California* 43
   George T. Jefferson

3. *Late Pleistocene lakes along the Mojave River, southeast California* 61
   Yehouda Enzel, Stephen G. Wells, and Nicholas Lancaster

4. *Late Quaternary geology and paleohydrology of pluvial Lake Mojave, southern California* 79
   Stephen G. Wells, William J. Brown, Yehouda Enzel, Roger Y. Anderson, and Leslie D. McFadden

5. *Latest Pleistocene lake highstands in Death Valley, California* 115
   Diana E. Anderson and Stephen G. Wells

6. *Latest Quaternary paleohydrology of Silurian Lake and Salt Spring basin, Silurian Valley, California* 129
   Kirk C. Anderson and Stephen G. Wells

7. *Isotopic and geochemical evidence for Holocene-age groundwater in regional flow systems of south-central Nevada* 143
   Timothy P. Rose and M. Lee Davisson

8. *Late Quaternary paleohydrologic and paleotemperature change in southern Nevada* 165
   Jay Quade, Richard M. Forester, and Joseph F. Whelan

9. *Regional response of alluvial fans to the Pleistocene-Holocene climatic transition, Mojave Desert, California* 189
   Eric V. McDonald, Leslie D. McFadden, and Stephen G. Wells

10. ***Late Quaternary variations in alluvial fan sedimentologic and geomorphic processes,
    Soda Lake basin, eastern Mojave Desert, California*** 207
    Adrian M. Harvey and Stephen G. Wells

11. ***Late Quaternary eolian dynamics, Mojave Desert, California*** 231
    Nicholas Lancaster and Vatche P. Tchakerian

Acknowledgements

The editors thank the following people for their attention to detail and insightful reviews, without which this Special Paper could never have been assembled: K. Adams, B. Albright, T. Bullard, M. Dettinger, A.R. Gillespie, A. Harvey, B. Harrison, T. Lowenstein, M. Mifflin, J.R. Miller, D. Morton, D.R. Muhs, B.D. Newman, J. Oviatt, M. Reheis, J. Ritter, P. Sadler, S. Sharpe, G.I. Smith, J.M. Thomas, J. Tinsley, D.P. Whistler, I.J. Winograd, M.O. Woodburne, and J. Zimbelman.

Dedication

Robert P. Sharp

This monograph on the paleoenvironments and paleohydrology of the Mojave and southern Great Basin Deserts is fittingly dedicated to Robert P. Sharp, who for more than four decades has been one of the premier geomorphologists in the world. Bob Sharp used the Mojave Desert in teaching in the natural laboratories of eolian, fluvial, lacustrine environments, and as a backyard for his research into the way Earth's surface was formed and changes in response to the many forces at work on it. From Caltech, he guided numerous class field trips to many of the areas mentioned in this volume. Bob's approach distilled the key elements of a complex situation into an understandable, plausible and testable model. Much of Bob's work in the deserts of the American southwest focused on active processes: erosion by blowing sand and the behavior of wind ripples, the shifting of dune fields from year to year, and the movement of the enigmatic "sliding stones" on Racetrack Playa. He also was one of the first to take a hard look at the formation of rock varnish, a dark material that commonly coats rocks in arid environments.

Bob's research is insightful and wide-ranging, and his studies of permafrost, mountain glaciations, and the relative dating of moraines and fans are as widely known as his desert studies. Farther afield, Bob pioneered the application of terrestrial geomorphology to other planets, notably Mars. His contributions on Martian geomorphology include studies of fretted terrain, glaciers, south polar cap, and mass wasting. It is also important to recall that, in the 1950s, Bob anticipated the role isotope geochemistry would play in revolutionizing geology and, together with colleague Sam Epstein at Caltech, pioneered the use of stable isotopes in ice cores to study climate change.

Bob's long career in Quaternary geology and geomorphology serves as a beacon to those who follow: this is the way that a successful career and a successful life can be conducted. Bob's intellectual contributions, his rich insights, and his gentle and constructive criticism have enriched the lives of those fortunate enough to have known and worked with him, and the stamp of his intellect will remain in his fields of study for generations.

Alan Gillespie
University of Washington, Seattle
November 10, 2002

Preface

The Mojave Desert of southern California and southern Nevada occupies an area of approximately 140,000 square kilometers, making it the smallest division of the North American arid zone. Within this relatively small area occurs a great diversity of landforms and deposits resulting from the complex tectonic and climatic history of the region during the late Cenozoic.

The southern boundary of the Mojave Desert is defined by the Transverse Ranges and the San Andreas fault system. Its northern boundary is in part provided by the Garlock fault but is mainly transitional to the southern Great Basin Desert. To the east, the Mojave Desert is considered to end at the Colorado River, although ecologically related areas occur in northwestern Arizona and extreme southwestern Utah.

The Mojave Desert is part of the Basin and Range physiographic province. Landforms are therefore characterized by broad valleys that separate mountain blocks bounded by faults. Relief in many parts of the Mojave Desert is generally high and range crests average 500 m above adjacent valleys but with a general decrease toward its southeastern margin near the Colorado River. Drainage is mostly endoreic and dominated by ephemeral streams that route water and sediment from mountain watersheds to terminal basins. The largest drainage of the region is the Mojave River, which heads in the Transverse Ranges and flows north and east toward Death Valley, presently terminating in Silver Lake playa.

Vegetation in the Mojave Desert is dominated by two shrub species, creosote bush (*Larrea tridentata*) and bursage (*Franseria dumosa*), which together make up 70% of the vegetation cover in many areas. The vegetation changes with elevation and from west to east. Compared to the Sonoran Desert, the Mojave Desert has a lower diversity of perennial species, but a higher diversity of annuals and ephemerals. The Mojave is generally poor in succulent plants.

The Mojave Desert is located at 34° to 37° N, within the area of descending air of the Hadley and Ferrel circulations that form the zone of the global subtropical high pressure cells. Its location in the rain shadow of the Transverse Ranges, which limit the inland penetration of moisture from the Pacific Ocean accentuates the arid conditions. In winter, Pacific storms penetrate the area from the west and northwest. In summer, the southwest North America "monsoon" brings rainfall from the south and east. These two sources of rain produce a gradient from winter-rainfall domination (>90% of the rainfall) in the western Mojave Desert to the summer rainfall–dominated southeastern Mojave Desert where it is in transition to the Sonoran Desert.

The average annual rainfall in the Mojave Desert ranges from about 50 mm at the lowest elevations to approximately 300 mm at its highest elevations. Most of the area experiences 100–200 mm/year. Mean annual lake evaporation is 1400–2000 with pan evaporation reaching 3000 mm/year. Mean monthly temperatures during July and August exceed 30 °C and are ~10 °C in December–February. Evidence of wetter conditions that affected the area during various episodes in the Quaternary, in contrast to the present-day harsh climatic conditions, were the motivation of many studies, including many of those in this volume.

The diversity of relief and drainage provides a great variety of landforms and deposits within a small geographical area. Landforms include desert mountains, alluvial fans and bajadas, pediments and volcanic landforms, playas, and eolian dunes. The close geographic proximity of many of these systems has stimulated study of the response of the arid land geomorphology and hydrology to Quaternary climatic changes, and such studies have provided many of the dominant paradigms of modern arid lands geomorphology (e.g., Bull, 1991; Kocurek and Lancaster, 1999; McFadden et al., 1987; Wells et al., 1987).

The physiographic diversity of the Mojave Desert, together with its proximity to the major metropolitan areas of southern California with noted universities and research institutions, have engendered a long history of seminal research by such individuals as Elliott Blackwelder, Robert Sharp, Charles Hunt, and their students. One of the earliest volumes that was published on the Mojave Desert (Thompson, 1929) still serves as a resource to modern researchers. This area also attracted intensive archaeological research that involved geoarchaeological studies that provided a large volume of paleoenvironmental data. The Mojave Desert has also been the location of many pioneering applications of remote sensing to understanding surface processes and landforms, as well as analog testing sites for planetary geology. In recent years, Mojave Desert research has been facilitated by the California State University Desert Studies Center at Zzyzx and the University of California Granite Mountains Preserve.

The close proximity of the Mojave Desert to expanding urban populations in the Los Angeles basin and Las Vegas, together with major military installations and the Nevada Test Site, have focused attention on the human impacts on desert landforms and hydrology via civilian and military off-road vehicle use (Webb and Wilshire, 1983), the nuclear weapons and nuclear waste disposal programs, and intensive use of water resources. These conflicts have been addressed in part by creation of the Mojave National Preserve and the Death Valley National Park, but urban encroachment and off-road vehicle use continue to be threats to the landforms of the Mojave Desert.

In this volume, we bring together a series of papers that summarize recent and ongoing research on major aspects of the paleohydrology and paleoenvironments of the Mojave Desert and its response to Quaternary tectonic and climatic changes. The contributions can be grouped into four main categories: (1) Late Tertiary and middle to late Pleistocene evolution of the Mojave Desert and its drainage systems, with papers by Cox et al. and Jefferson; (2) the paleohydrologic record of the late Pleistocene lakes of the Mojave River and Death Valley Basins, with papers by Enzel et al., Wells et al., D. Anderson and Wells, and K. Anderson and Wells; (3) paleohydrologic records from groundwater systems in southern Nevada with contributions from Rose and Davisson and Quade et al.; and (4) the response of alluvial fan and eolian systems to regional changes in climate, with papers from McDonald et al., Harvey and Wells, and Lancaster and Tchakerian.

Yehouda Enzel
Stephen G. Wells
Nicholas Lancaster

REFERENCES CITED

Bull, W.B., 1991, Geomorphic responses to climatic change: New York, Oxford University Press, 326 p.

Kocurek, G., and Lancaster, N., 1999, Aeolian sediment states: Theory and Mojave Desert Kelso Dunefield example: Sedimentology, v. 46, no. 3, p. 505–516.

McFadden, L.D., Wells, S.G., and Jercinovich, M.J., 1987, Influences of eolian and pedogenic processes on the origin and evolution of desert pavements: Geology, v. 15, p. 504–508.

Thompson, D.G., 1929, The Mojave Desert region, California: U.S. Geological Survey Water-Supply Paper 578, p. 759 p.

Webb, R.H., and Wilshire, H.G., 1983, Environmental effects of off-road vehicles: Impacts and management in arid regions: New York, Springer Verlag, p. 534.

Wells, S.G., McFadden, L.D., and Dohrenwend, J.C., 1987, Influence of late Quaternary climatic changes on geomorphic and pedogenic processes on a desert piedmont, eastern Mojave Desert, California: Quaternary Research, v. 27, p. 130–146.

Pliocene and Pleistocene evolution of the Mojave River, and associated tectonic development of the Transverse Ranges and Mojave Desert, based on borehole stratigraphy studies and mapping of landforms and sediments near Victorville, California

B.F. Cox
J.W. Hillhouse
U.S. Geological Survey, MS 975, 345 Middlefield Road, Menlo Park, California 94025, USA

L.A. Owen
Department of Earth Sciences, University of California, Riverside, California 92521, USA

ABSTRACT

Pliocene and Pleistocene continental sediments near Victorville, California, record the downstream growth of the early Mojave River, which was controlled partly by uplift of the central Transverse Ranges along the San Andreas fault, and also by coeval arching of the southern Mojave Desert. Boreholes at former George Air Force Base penetrate three conformable stratigraphic units with a total thickness of ~160 m. The basal unit comprises >50 m of lithic-arkose sand and polymictic gravel deposited by a southward-flowing braided stream and tributary alluvial fans. The middle unit consists of 8–25 m of clay and silt deposited in a shallow lake or wetland. The uppermost unit consists of 55–110 m of arkosic sand, silt, and gravel deposited by the northward-flowing ancestral Mojave River. The three units are herein termed the lower alluvial unit, middle lacustrine unit, and upper fluvial unit, respectively. The upper fluvial unit forms a northwest-tapering clastic wedge, and the middle lacustrine unit rises to the northwest along its base. The clastic wedge evidently is an orogenic deposit derived from the San Bernardino Mountains, and the underlying lakebeds accumulated at the tip of the wedge as it expanded to the northwest against a southward-facing paleoslope.

Magnetostratigraphic analysis of borehole cores shows that the stratigraphic succession at the air base ranges in age from ca. 4.2 Ma to <0.78 Ma. The diachronous middle lacustrine unit accumulated between 2.55 and 1.95 Ma near the southeast corner of the base, and between 1.18 and 1.05 Ma near the northern end of the base. The top of the upper fluvial unit is dated at ca. 60–70 ka by optically stimulated luminescence methods. The sediments accumulated most rapidly between 1.2 and 1.0 Ma, during an episode of humid regional climate that is documented at Searles Lake, California. Using rates of sediment accumulation determined at the air base, we estimate that a thick (244 m) southeastern extension of the upper fluvial unit beneath southernmost Victorville began accumulating ca. 3.3 Ma. Therefore, the birth of the ancestral Mojave River probably predates significant uplift along the northern front of the San Bernardino Mountains, which began after 2–3 Ma according to previous studies. The

Cox, B.F., Hillhouse, J.W., and Owen, L.A., 2003, Pliocene and Pleistocene evolution of the Mojave River, and associated tectonic development of the Transverse Ranges and Mojave Desert, based on borehole stratigraphy studies and mappping of landforms and sediments near Victorville, California, in Enzel, Y., Wells, S.G., and Lancaster, N., eds., Paleoenvironments and paleohydrology of the Mojave and southern Great Basin Deserts: Boulder, Colorado, Geological Society of America Special Paper 368, p. 1–42. © 2003 Geological Society of America.

initial advancement of the river into the Mojave Desert possibly was induced by early uplift of San Gorgonio Mountain beside the San Andreas fault.

Our borehole investigation and associated geologic mapping studies disclosed northwest-trending syndepositional faults near the southwest corner of the air base and a low, southward-facing anticline directly north of the base. These structures were produced by dextral shearing and north-south contraction of the southern Mojave Desert during the uplift of the Transverse Ranges. The anticline apparently represents the western end of a regional basement arch that extends west-northwest ~250 km across the southern Mojave Desert. The growth of this arch controlled the advancement of the Mojave River northward beyond the Victorville region. The middle lacustrine unit pinches out on its south flank, and the steady northward tapering of the upper fluvial unit ends at its crest. Relatively thin (8–25 m) deposits of the upper fluvial unit cap the lower alluvial unit unconformably on the broad northern limb of the arch—a stratigraphic pattern that persists northward at least 25 km to Iron Mountain. A southward-flowing stream deposited much of the lower alluvial unit across this stretch of the Mojave Desert. Therefore, the deposition of the upper fluvial unit by the northward-flowing ancestral Mojave River denotes a reversal of the regional paleoslope. This reversal was probably produced by gentle tilting on the northern limb of the arch.

Arching of the southern Mojave Desert occurred mainly during the Pliocene and early Pleistocene. However, the reversal of regional drainage was not completed until the middle Pleistocene, ca. 575–475 ka, when the ancestral Mojave River finally aggraded its bed to a broad saddle at the crest of the arch. This saddle marked the abandoned canyon of the former southward-flowing stream. After rapidly descending the northern limb of the arch, the river may have been confined briefly to Harper Lake basin. However, it soon advanced eastward into the Lake Manix basin, where previous studies date its arrival at ca. 500 ka. The optically stimulated luminescence dates from the top of the upper fluvial unit suggest that the Mojave River began incising its modern canyon between Victorville and Barstow ca. 60–70 ka. Downcutting was likely induced by renewed regional contraction that broadly arched the southern Mojave Desert. The large volume of sediment eroded from the canyon probably filled in a significant area of the Lake Manix basin. This in turn must have increased the overflow of pluvial lakes, thereby accelerating the erosion that breached the eastern sill of the basin ca. 18 ka.

INTRODUCTION

During the late Cenozoic, oblique convergence of tectonic plates along strike-slip faults of the San Andreas system uplifted extensive mountain belts in southern and western California. These transpressional uplifts transformed regional patterns of drainage and sedimentation. Thus, a record of the evolving landscape is preserved in the deposits, erosional surfaces, and areal configuration of major stream systems. In this paper, we explore the depositional history and longitudinal growth of the early Mojave River and consider how the evolution of the river reflects the tectonic development of the central Transverse Ranges and neighboring Mojave Desert of southern California.

The Mojave River occupies the largest drainage basin in the Mojave Desert region (Fig. 1). Its headwaters debouch from the northwest front of the San Bernardino Mountains, converging in an intermittent river channel that sweeps northward and eastward 200 km to a terminal sink at Silver Lake playa (Thompson, 1929; Martin, 1994; Enzel and Wells, 1997; Enzel et al., this volume). Given the broad extent of its drainage basin, the deposits and geomorphic surfaces along the Mojave River contain an important record of regional tectonic and paleoenvironmental history.

The development of the Mojave River culminated a sequence of regional paleogeographic and paleotectonic events. During the middle Miocene, 18–10 Ma, the Mojave Desert drained to the southwest near Cajon Pass (Fig. 1) (Woodburne and Golz, 1972; Woodburne, 1975; Foster, 1980). At the end of this period, the landscape of the southern Mojave Desert and central Transverse Ranges apparently consisted of a deeply weathered surface with minimal topographic relief (Baker, 1911; Oberlander, 1972). In the late Miocene, 10–5 Ma, the ancestral Transverse Ranges were uplifted along the San Gabriel fault, and the resulting mountain barrier presumably diverted streams flowing southward from the Mojave Desert (Meisling and Weldon, 1989). During the Pliocene, 5–2 Ma, fluviolacustrine basins developed immediately north and south of the uplifted region. The Old Woman Sandstone of Shreve (1968) and the Phelan Peak Formation of Weldon et al. (1993) accumulated in an east-west–trending basin along the north edge of the ancestral San Bernardino Mountains (Shreve, 1968; Sadler, 1982b; May and

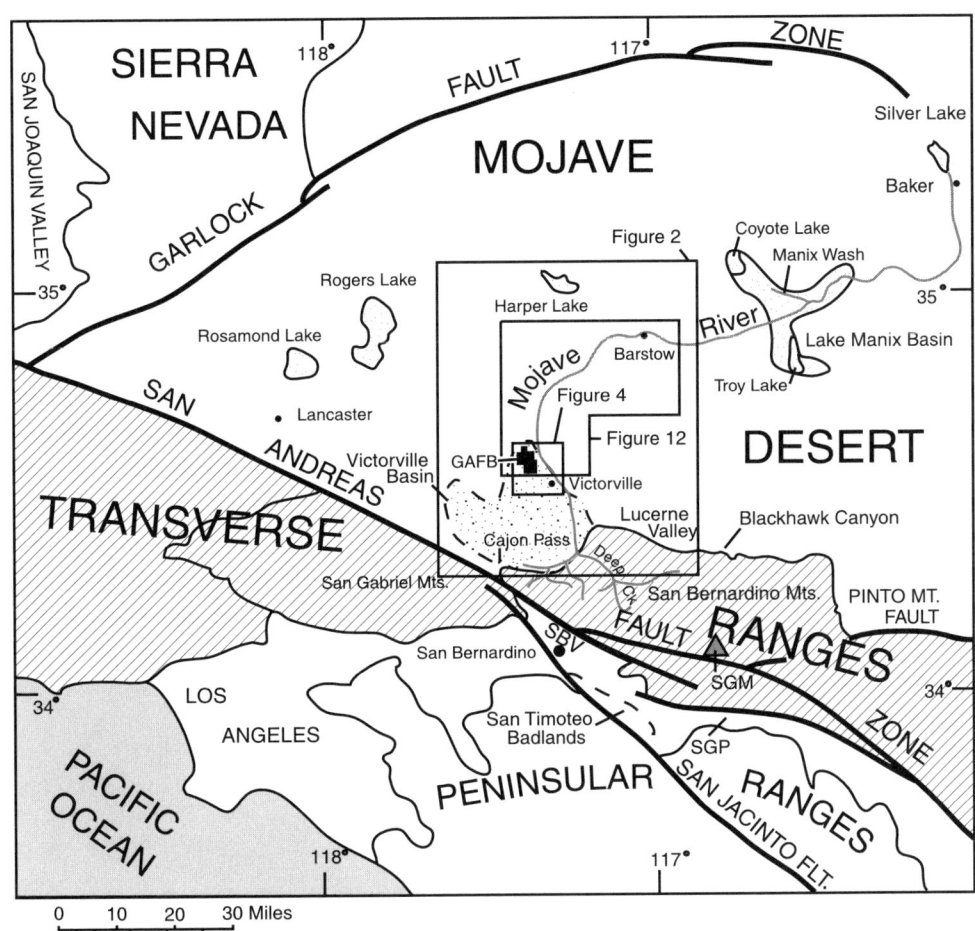

Figure 1. Index map showing location of Mojave River drainage basin, Victorville structural basin, and former George Air Force Base (GAFB) in relation to geologic provinces and major faults of southern California. Diagonal ruling denotes Transverse Ranges. Inset boxes show location of Figures 2, 4, and 12. SBV—San Bernardino Valley; SGM—San Gorgonio Mountain; SGP—San Gorgonio Pass. Outline of Lake Manix basin represents approximate late Wisconsin highstand of Lake Manix.

Repenning, 1982; Meisling, 1984; Meisling and Weldon, 1989; Powell and Matti, 1998). Simultaneously, streams draining the south slope of the ancestral San Gabriel Mountains deposited a southward-prograding wedge of sand and gravel in the San Timoteo Badlands (Fig. 1) (Matti and Morton, 1975, 1993; Albright, 1999). These episodes of basin development and sedimentation suggest that tectonic uplift of the Transverse Ranges continued or resumed during the Pliocene (Sadler and Reeder, 1983; Meisling and Weldon, 1989; Albright, 1999).

In an insightful early study of the Lake Manix basin east of Barstow (Fig. 1), Buwalda (1914) concluded that the basin was originally occupied by a dry playa, and that perennial lakes first developed during the Pleistocene with the arrival of the ancestral Mojave River. He surmised that the early river advanced across the Mojave Desert either in response to tectonic uplift of the San Bernardino Mountains or to a general increase in precipitation during times of Pleistocene glaciation. He also suggested that contemporaneous tectonism within the desert region might have directed the river toward the Lake Manix basin.

More recently, Meisling (1984) and Meisling and Weldon (1989) reported that the ancestral Mojave River emerged at the northwest front of the San Bernardino Mountains in the latest Pliocene to early Pleistocene, ca. 2.0–1.5 Ma, with the onset of rapid uplift along the northern flank of the range. Their chief evidence was the "Ord River deposits," a deeply eroded mass of fluvial sand and gravel near the mouth of Deep Creek (Figs. 1 and 2). Yet, despite its early appearance at the mountain front, the Mojave River probably did not venture east of Barstow until the middle Pleistocene, ca. 0.5 Ma, when fluvial and lacustrine sediments began accumulating in the Lake Manix basin (Jefferson, 1985, 1994, 1999, this volume). Thus, combining the stratigraphic evidence from the upper and lower regions of its drainage basin, the ancestral Mojave River apparently advanced from the San Bernardino Mountains to Lake Manix basin over an extended interval of ~1.0–1.5 m.y.

Nagy and Murray (1996) envisioned more rapid, climatically controlled growth of the early Mojave River. They speculated that the river would have progressed from the San Bernardino Mountains to the Lake Manix basin during a single Pleistocene pluvial cycle, thus within ~100,000 yr. Although appealing in its simplicity, this hypothesis is inconsistent with the stratigraphic evidence cited above. If valid, it would imply that the Mojave River either emerged at the mountain front much later, or arrived at Lake Manix basin much earlier, than indicated by the studies of

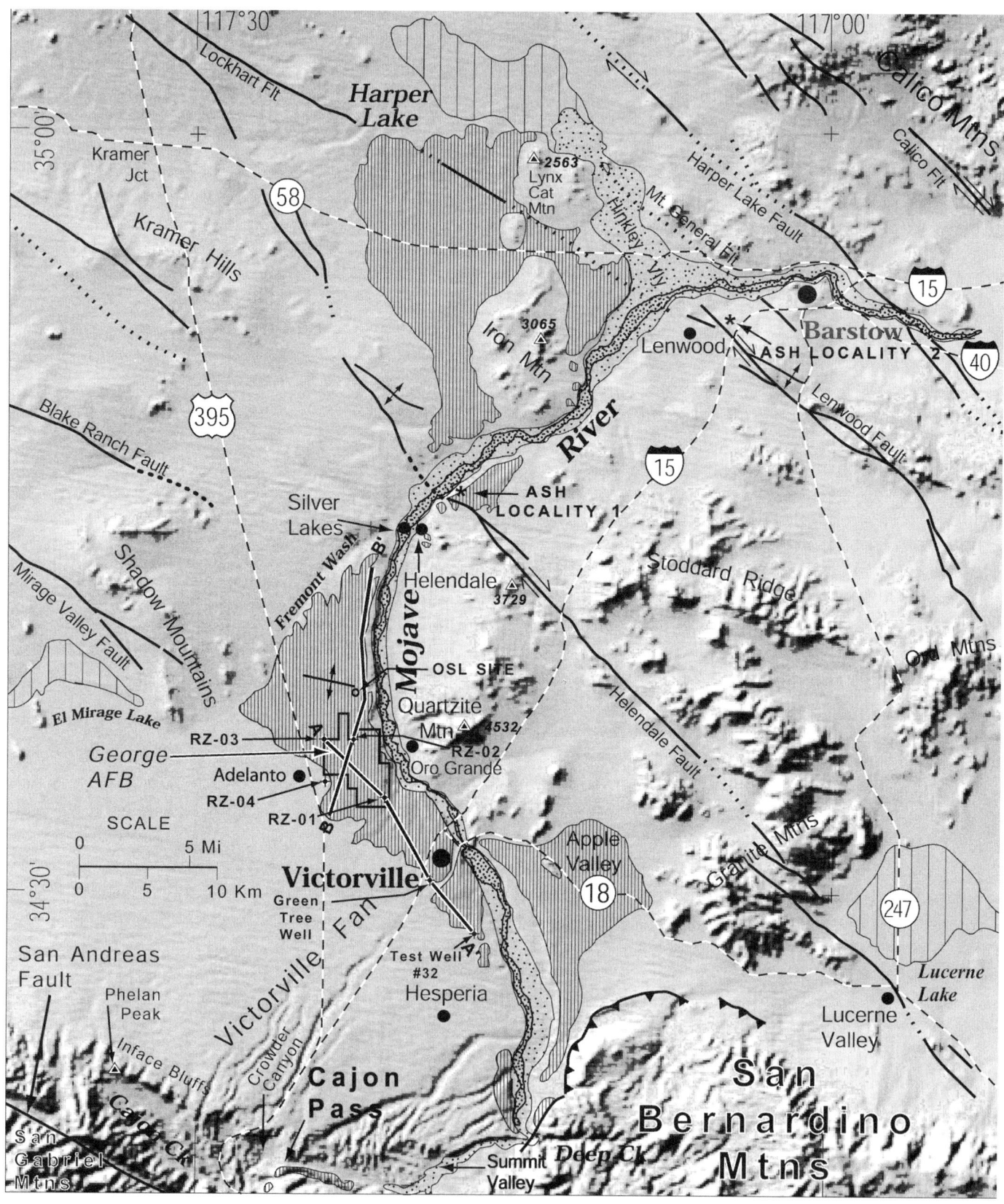

Figure 2. Shaded-relief map showing distribution of ancestral Mojave River deposits (narrow ruling) and location of boreholes and other study sites and sample localities between Cajon Pass and Barstow, California. Also shows active channel (dense stipples) and Holocene floodplain (sparse stipples) of the modern Mojave River, and dry lakebeds (broad ruling). Modified after Cox and Tinsley (1999). Lines A–A' and B–B' indicate location of cross sections shown in Figures 6 and 13. Elevation in feet above sea level is indicated for several peaks between Victorville and Harper Lake (triangles). Structural symbols are explained on Figure 12.

Meisling (1984), Meisling and Weldon (1989), and Jefferson (1985, 1994, 1999, this volume). The hypothesis also assumes that the river did not encounter any challenging topographic obstacles as it forged a route across the desert.

The latter assumption is questionable, as the modern Mojave River follows a circuitous path through mountainous terrain of the southern Mojave Desert (Fig. 2). Furthermore, in any given location, most of the topographic relief seems to predate the oldest deposits of the river. The sequence of regional paleogeographic and paleotectonic events cited above also suggests that the Mojave River probably developed across irregular or sloping terrain. Access toward the north may have been blocked by a southward-inclined regional paleoslope inherited from the middle Miocene landscape (Woodburne, 1975; Foster, 1980; Meisling and Weldon, 1989), or by younger uplifts and basins generated by north-south contraction during the late Miocene to Holocene (Bartley et al., 1990). If such barriers were present, then the river's pioneering journey across the Mojave Desert could easily have taken at least 1.0–1.5 m.y. to complete.

Previous investigations revealed deposits of the ancestral Mojave River in the Victorville region, particularly in the subsurface at former George Air Force Base (GAFB) (IT Corporation, 1992; Sibbett, 1996, 1999; Chrisley, 1997), and in the bluffs of the Mojave River between Victorville and Helendale (Bowen, 1954). These deposits, which received little attention until recently, are key to reconciling the apparent differences in drainage history between the upper and lower regions of the Mojave River basin. In the present study, we analyzed sediment cores from four deep boreholes at GAFB to establish the physical stratigraphy, depositional environments, and age of the ancestral Mojave River deposits and associated sedimentary units. This work included a detailed magnetostratigraphic investigation of each borehole. Supplementary paleomagnetic measurements were performed on oriented samples collected from outcrops of fluvial strata near GAFB and from a layer of volcanic ash that underlies the ancestral Mojave River deposits near Helendale. Staff of the U.S. Geological Survey Tephrochronology Project at Menlo Park, California, performed microprobe analyses on the Helendale ash bed and two additional ash layers near Lenwood and determined likely geochemical correlations.

The magnetostratigraphic studies enabled us to date and correlate the stratigraphic section at GAFB through reference to the modern geomagnetic polarity time scale (Fig. 3). We also employed the modern time scale to adjust paleomagnetic ages from previous studies in the Cajon Pass region (e.g., Meisling and Weldon, 1989; Weldon et al., 1993). Optically stimulated luminescence methods were applied to determine the age of the youngest fluvial deposits at the top of the stratigraphic succession. Drilling records from other boreholes were examined to extend stratigraphic interpretations southeastward beneath the city of Victorville. We also analyzed widespread stratigraphic and structural evidence, including previously unpublished findings, that we and our colleagues observed along the Mojave River between Victorville and Barstow (Cox and Wilshire, 1993, 1994; Densmore et al., 1997; Stamos et al., 2001). Finally, we consulted published accounts of several late Cenozoic sedimentary basins around the central Transverse Ranges and in the central Mojave Desert.

We address the following key questions: (1) When did the Mojave River originate and what was the nature of preexisting regional drainage and topography? (2) How and when did the river develop its integrated drainage basin across the Mojave Desert? (3) How does the depositional and erosional history of the Mojave River correlate with the growth and abandonment of the Victorville Fan? (4) What is the age and significance of the deeply incised modern channel of the Mojave River between Cajon Pass and Barstow? (5) What does the drainage history of the Mojave River imply about the regional tectonic history of the Mojave Desert and central Transverse Ranges?

Some initial results and interpretations were summarized by Cox et al. (1998), Cox and Tinsley (1999), and Reynolds and Cox (1999), and a preliminary version of the present paper was released as an open-file report (Cox and Hillhouse, 2000).

GEOLOGIC SETTING

Our study focuses on the Victorville region and neighboring former George Air Force Base (GAFB) (now Southern California

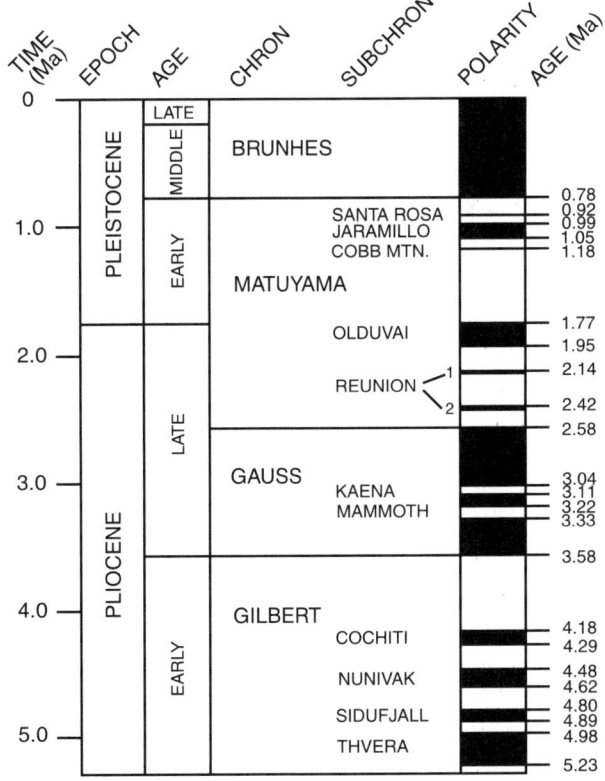

Figure 3. Correlation of Pliocene and Pleistocene epochs and ages with the geomagnetic polarity time scale (Berggren et al., 1995). Polarity zones (black—normal polarity; white—reversed polarity) are calibrated according to Singer et al. (1999) and Cande and Kent (1995).

Logistics Airport), which lie in the southern Mojave Desert ~100 km northeast of Los Angeles (Fig. 1). The geologic history of the study area is tied to the evolution of the nearby San Andreas fault, which is the main structural break between the North American and Pacific tectonic plates in southern California. The San Andreas fault slices obliquely across the central Transverse Ranges near Cajon Pass, ~30 km southwest of Victorville, separating the San Gabriel Mountains on the west from the San Bernardino Mountains on the east. These two mountain ranges were uplifted between the Mojave Desert and the Peninsular Ranges by transpressional stresses along a major restraining bend of the fault (Dibblee, 1975a; Sadler, 1982a, 1982b; Sadler and Reeder, 1983; Meisling and Weldon, 1989; Matti and Morton, 1993; Morton and Matti, 1993a; Weldon et al., 1993; Spotila and Sieh, 2000).

Regional gravity surveys indicate that the region between Cajon Pass and GAFB is underlain by a structural depression 40 km wide that is filled with Cenozoic sediments up to 1300 m thick (Fig. 1) (Mabey, 1960; Biehler et al., 1988; Subsurface Surveys, 1990; Spotila and Sieh, 2000). This feature has been termed the Cajon basin (Dibblee, 1967, Figure 71; 1975b), but here we refer to it as the Victorville basin to maintain a clear distinction from the drainage basin of Cajon Creek, which is located nearby to the south (Fig. 2).

The San Bernardino Mountains border the Victorville basin to the southeast (Fig. 1). The area of the range lying north of the San Andreas fault comprises two main geomorphic domains. Much of the area consists of a broad northern plateau standing 1–1.5 km above adjacent lowlands of the southern Mojave Desert. South of this plateau the range rises another 1–1.5 km to the summit of San Gorgonio Mountain (elevation 3505 m), which is the highest point in southern California. The San Bernardino Mountains consist of granitic rocks and gneiss that contain sparse pendants of siliceous and calcareous metasedimentary rocks. The western end of the range, which includes the drainage basin of Deep Creek, adjoins the Victorville basin along an arcuate, northeast-trending front produced by tectonic uplift in the Pleistocene and late Pliocene (Meisling and Weldon, 1989). Near the mouth of Deep Creek, southeast-dipping thrust and reverse faults place granitic rocks atop fluvial deposits of the basin (Fig. 2) (Meisling, 1984; Meisling and Weldon, 1989). To the west, near Cajon Pass, anticlinal and monoclinal folds have arched up the granitic basement rocks and overlying sediments, thereby producing the narrow western "wing" of the San Bernardino Mountains (Dibblee, 1975b; Meisling and Weldon, 1989; Kenney and Weldon, 1999). High peaks of the eastern San Gabriel Mountains lie nearby across the San Andreas fault, framing the southwest side of the Victorville basin. These mountains contain extensive bodies of schistose metasediments and metabasalt (Pelona Schist), in addition to granitic and gneissic rocks.

The basin is bordered to the north by local peaks and ridges of pre-Cenozoic basement rocks (Bowen, 1954; Dibblee, 1960b, 1967; Miller, 1981; Bortugno and Spittler, 1986). The Mojave River exits the north end of the basin through an alluviated gap between Quartzite Mountain and the southeastern Shadow Mountains (Fig. 2). The uplands on either side of this gap contain large pendants of siliceous and calcareous metasedimentary rocks that are intruded by Mesozoic granitic rocks. Silicic metavolcanic rocks crop out extensively in neighboring ridges north and east of Quartzite Mountain. Complex faulting and warping probably produced the irregular northern, western, and eastern margins of the basin. However, the upper levels of the basin fill largely conceal the associated deformational structures, which are therefore poorly understood.

A deep stratigraphic cross section of Cenozoic sediments and underlying basement rocks is exposed near Cajon Pass, where the headwaters of Cajon Creek breach the southern edge of the Victorville basin (Fig. 2). This section contains a thick succession of Miocene to Pleistocene continental sediments. The cumulative thickness is ~3150 m (Meisling and Weldon, 1989), but only ~1200 m of the succession appears to be associated physically and genetically with the Victorville basin.

The lower part of the succession comprises two very thick and largely coeval units of arkosic sandstone, siltstone, and conglomerate—the Cajon Formation of Meisling and Weldon (1989) (2440 m) and the Crowder Formation (980 m). These formations both accumulated during the Miocene, 18–13.5 Ma and 17–9.5 Ma, respectively (Woodburne and Golz, 1972; Weldon, 1985; Reynolds, 1991). They were deposited in separate basins by streams draining southward from the Mojave Desert (Woodburne and Golz, 1972; Foster, 1980). The basins were juxtaposed by southwest-directed thrusting between 9.5 and 5 Ma, during the uplift of the ancestral Transverse Ranges (Meisling and Weldon, 1989; Weldon et al., 1993). The Cajon Formation is confined to a narrow fault-bounded block near the San Andreas fault and evidently does not extend northward into the Victorville basin. However, the Crowder Formation, lying in the upper plate of the old thrust fault, apparently is unrestricted to the north and probably constitutes much of the deep basin fill between Cajon Pass and Victorville (Bortugno and Spittler, 1986, section A–A′).

The Cajon and Crowder Formations are overlain unconformably by the Pliocene to early Pleistocene Phelan Peak Formation of Weldon et al. (1993), which consists of an upward-coarsening succession of sand, silt, and gravel. The type section near Phelan Peak (Fig. 2) is ~500 m thick and was deposited between 4.4 and 1.4 Ma (ages revised from Weldon et al., 1993, Figure 8). The lower half of the section consists of low-energy fluvial and lacustrine sediments that probably were derived from both the north and south sides of the basin, whereas the upper half, deposited after 2.5 Ma, consists of coarse alluvial-fan deposits derived from an elevated region nearby to the south (Foster, 1980). Weldon et al. (1993) proposed that the upland source area was a remnant of the late Miocene ancestral Transverse Ranges that was transported northwestward past the Cajon Pass region by the San Andreas fault.

The Victorville Fan deposits of Meisling and Weldon (1989) conformably overlie the Phelan Peak Formation between Phelan Peak and Cajon Pass (Kenney and Weldon, 1999). The Victorville

Fan deposits are an upward-coarsening succession of weakly consolidated sand and gravel ~200 m thick that crops out spectacularly in the Inface Bluffs (Fig. 2). In ascending order, the succession comprises the Harold Formation, the Shoemaker Gravel, and the older alluvium of Noble (1954). Based on distinctive rock detritus, including debris of the Pelona Schist and Lowe Granodiorite, all three units were derived from basement rocks of the San Gabriel Mountains lying southwest of the San Andreas fault (Foster, 1980; Meisling and Weldon, 1989). Deposits at the head of Crowder Canyon (Fig. 2) accumulated during the early Pleistocene, between 1.7 and 0.78 Ma (ages revised from Meisling and Weldon, 1989; Weldon et al., 1993, Figure 4). Sedimentation evidently was induced by strike-slip movements of the San Andreas fault that transported the high, central and eastern parts of the San Gabriel Mountains alongside relatively subdued terrain of the southern Mojave Desert (Meisling, 1984; Weldon, 1986; Meisling and Weldon, 1989; Weldon et al., 1993).

Deposits at the southeast side of the Victorville basin record a different source area and drainage system associated with the San Bernardino Mountains. The Ord River deposits of Meisling and Weldon (1989), which are exposed in the walls of the Mojave River canyon along the northwest front of the range, consist of coarse-grained, well-stratified fluvial sand and gravel composed mainly of granitic debris. Meisling (1984) described these deposits and interpreted their paleogeographic significance. The younger parts of the unit contain clasts of metasedimentary and metavolcanic rocks recycled from the Crowder Formation, but Pelona Schist and other detritus indicative of the San Gabriel Mountains are absent throughout the unit. The main source area was the drainage basin of ancestral Deep Creek, which probably was the principal tributary of the ancestral Mojave River in the early Pleistocene. The Ord River deposits intertongue to the west with deposits of the Harold Formation and Shoemaker Gravel, which demonstrates that the ancestral Mojave River and neighboring Victorville Fan simultaneously aggraded the southern Victorville basin during the early Pleistocene (Meisling, 1984; Meisling and Weldon, 1989).

Drainage from the San Bernardino Mountains is also recorded in the stratigraphy of the northern Victorville basin near Victorville and GAFB. Borehole investigations indicate that the uppermost 160 m of the basin fill beneath GAFB consists of three main units of weakly consolidated sediments. These are a basal unit of compositionally heterogeneous sand, silt, and gravel, a middle unit of clay and silt, and an upper unit of granitic sand, silt, and gravel (IT Corporation, 1992; Montgomery Watson, 1995; Sibbett, 1996, 1999; Chrisley, 1997). Correlation within a dense array of boreholes at GAFB disclosed northwest-trending gravel-filled channels in the upper unit, which suggests it was deposited by the ancestral Mojave River (IT Corporation, 1992; Sibbett, 1996, 1999). The middle unit pinches out near the north end of GAFB and is thus confined to the Victorville basin, as delineated by gravity data (Montgomery Watson, 1995; Chrisley, 1997). However, the lower and upper units continue to the north, forming an attenuated, disconformable succession that crops out

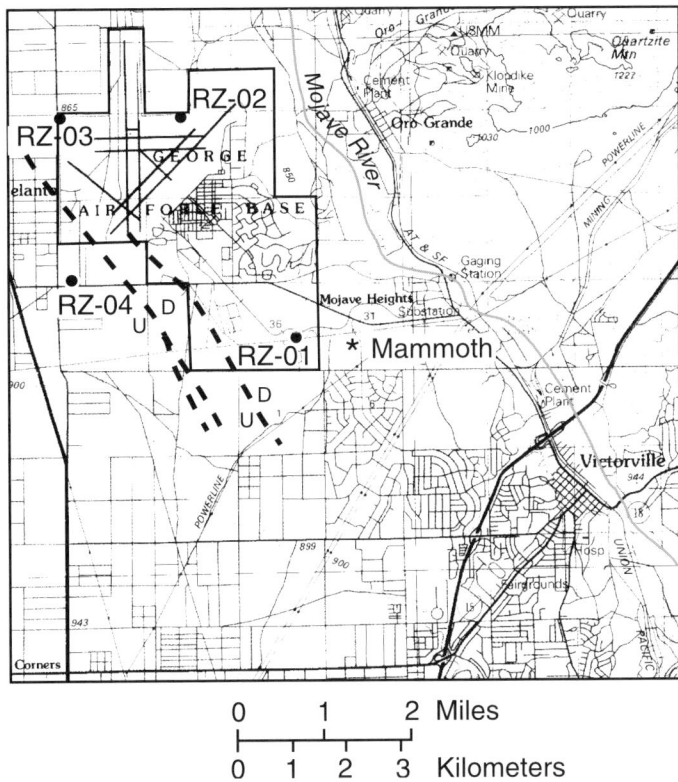

Figure 4. Map of former George Air Force Base and Victorville, California, showing location of the four principal boreholes investigated for this study (RZ-01, -02, -03, and -04). Star indicates location of fossil mammoth (*Mammuthus meridionalis*) reported by Scott et al. (1997). Heavy dashed lines represent faults deduced from aerial-photographic lineaments and offset stratigraphy in borehole RZ-04; D—downthrown side; U—upthrown side. Base from U.S. Geological Survey, Victorville 30 × 60 minute quadrangle, California, 1982.

downstream along the Mojave River to Iron Mountain (Cox et al., 1998; Cox and Tinsley, 1999; Reynolds and Cox, 1999).

Vertebrate fossils collected from the upper unit at various sites around Victorville are generally consistent with a middle Pleistocene (late Irvingtonian or earliest Rancholabrean) age (Jefferson, 1986; Reynolds, 1989; Reynolds and Reynolds, 1994a). This age interpretation was locally bolstered by the determination of normal magnetic polarity in the upper unit (Meisling, 1984, locality HRF; K. Meisling, 1984, personal commun., cited in Reynolds, 1989; K. Meisling, 1985, personal commun., cited in Reynolds and Reynolds, 1994a). However, a recent study reported possible early Pleistocene vertebrates in the upper part of the unit (Scott et al., 1997). Previous studies have yielded little information regarding the age and origin of the underlying two units.

The Quaternary surficial geology of the greater Victorville region is dominated by three large-scale geomorphic elements: the Victorville Fan, the George surface, and the Mojave River canyon. The first two features form the upper bounding surface of the Victorville basin, whereas the Mojave River canyon is incised into the basin fill. The Victorville Fan is a broad piedmont or bajada that descends northeastward from Cajon Pass to Victorville (Fig. 2). Early Pleistocene-age first-cycle fan gravels (Victorville Fan deposits of Meisling and Weldon, 1989) underlie the relatively steeply sloping head of the fan. The gentler lower slopes are blanketed by late Pleistocene-age alluvium and aeolian deposits composed of detritus recycled from the head of the fan (Bortugno and Spittler, 1986; Reynolds and Cox, 1999).

Erosion and sedimentary recycling prevailed on the Victorville Fan after ca. 0.5 Ma, when southwestern tributaries of the ancestral Mojave River eroded a series of channels into the southeast shoulder of the fan (Meisling, 1984; McFadden and Weldon, 1987; Reynolds and Cox, 1999). These channels are represented by an inset stream terrace in Summit Valley, and by several older ravines whose outlines are conspicuous on Figure 2. Runoff from the San Gabriel Mountains was funneled northeastward through the incised channels, which terminated the supply of fresh alluvium to the broad surface of the fan. The runoff was subsequently diverted to the southeast in the middle or late Pleistocene, when Cajon Creek eroded headward along the San Andreas fault and beheaded the Victorville Fan along the Inface Bluffs (Meisling, 1984).

The George surface is a deeply dissected, gently inclined fluvial platform at the northeast toe of the Victorville Fan (Cox and Tinsley, 1999). It is well preserved near Apple Valley, Victorville, and GAFB, and an accordant, apparently equivalent surface envelops Iron Mountain northeast of the Helendale fault (Fig. 2). The George surface is a late Pleistocene river terrace representing the broad floodplain of the ancestral Mojave River (IT Corporation, 1992; Sibbett, 1996, 1999; Cox et al., 1998; Cox and Tinsley, 1999). Prior to the entrenchment of the Mojave River canyon, the floodplain was 3–13 km wide and 75 km long, extending downstream from the mouth of Deep Creek to Harper Lake (Fig. 2).

The modern Mojave River occupies a canyon 1–2 km wide that extends from the northwest front of the San Bernardino Mountains downstream nearly to Barstow (Fig. 2). The gradient of the active river channel between Deep Creek and Victorville is nearly identical to that of the adjacent George surface, which supports the view that the ancestral Mojave River produced the latter feature. The floor of the river canyon typically lies ~60 m below the George surface between southern Apple Valley and the Helendale fault, but its depth locally reaches 75 m near GAFB. The canyon gradually shallows downstream from the Helendale fault, ending at Hinkley Valley.

LITHOSTRATIGRAPHY OF BOREHOLE CORES

Prior to the recent closure of George Air Force Base (GAFB), many wells and exploratory boreholes were drilled to determine the geohydrologic framework and subsurface distribution of contaminants at the air base (IT Corporation, 1992; Montgomery Watson, 1995). Through the courtesy of S.M. Chrisley of the Montgomery Watson company, we acquired nearly continuous, 2-inch-diameter mud-rotary cores from four drilling sites located around the perimeter of the facility (Fig. 4, sites RZ-01, -02, -03, and -04). We focused our investigation on the borehole sediment cores because our primary target—alluvium of the ancestral Mojave River—was reported to form a thick layer beneath GAFB (IT Corporation, 1992; Sibbett, 1996). Moreover, the boreholes penetrate a stratigraphic interval of ~160 m, whereas the thickness of alluvium exposed in the walls of the Mojave River canyon near Victorville and GAFB does not exceed 75 m.

The vertical distribution of lithofacies in each borehole is summarized on Figure 5. This diagram also shows inferred lateral stratigraphic relations between the holes. No core was available for the lowest quarter of borehole RZ-02 or the upper half of RZ-04. The lithology of these intervals is generalized from the original drilling logs (Montgomery Watson, 1995, appendix D). We extended the stratigraphic column at site RZ-01 14 m above the top of the borehole by measuring strata exposed on nearby walls of a ravine (Fig. 5).

Principal stratigraphic trends

Critical evidence pertaining to the evolution of the ancestral Mojave River is found in the detrital composition and texture, lateral variation in thickness, and weathering characteristics of deposits observed in the borehole cores from GAFB. Here we summarize features that support the major conclusions of this paper. The three main stratigraphic units recognized by our study are described individually in greater detail in the following sections.

In agreement with previous studies (IT Corporation, 1992; Montgomery Watson, 1995; Chrisley, 1997), our observations indicate that the boreholes intersect three conformable lithostratigraphic units, which we designate as the lower alluvial unit, middle lacustrine unit, and upper fluvial unit (Fig. 5). The lower alluvial and upper fluvial units each comprise thick deposits of sand, silt, and gravel, whereas the middle lacustrine unit comprises thinner deposits of ostracode-bearing calcareous clay and silt, interspersed with minor amounts of sand. The lower alluvial and upper fluvial units each coarsen vertically away from the middle lacustrine unit. The lower alluvial unit coarsens downward from poorly sorted silty sand into moderately sorted pebbly sand, and the upper fluvial unit coarsens upward from interlayered sand and silt, through relatively homogeneous sand, into interlayered sand and pebble-cobble gravel. The upper fluvial unit is thickest in borehole RZ-01, near the southeast corner of the air base, and is thinner in boreholes RZ-02, -03, and -04, which lie along the northern and western margins of the base. Therefore, this unit apparently forms a northwest-thinning wedge, and the middle lacustrine unit rises to the northwest along its base (Fig. 5).

Differences in detrital composition indicate that the lower and upper units were derived from distinct source areas. Deposits of the lower alluvial unit are polymictic, containing debris of granitic, metavolcanic, and metasedimentary rocks, and nonmetamorphosed volcanic rocks. Potential sources for these assorted rock types are present in bedrock terranes of the southern and central Mojave Desert. The upper fluvial unit is more homogeneous in composition, consisting mostly of granitic detritus. Its

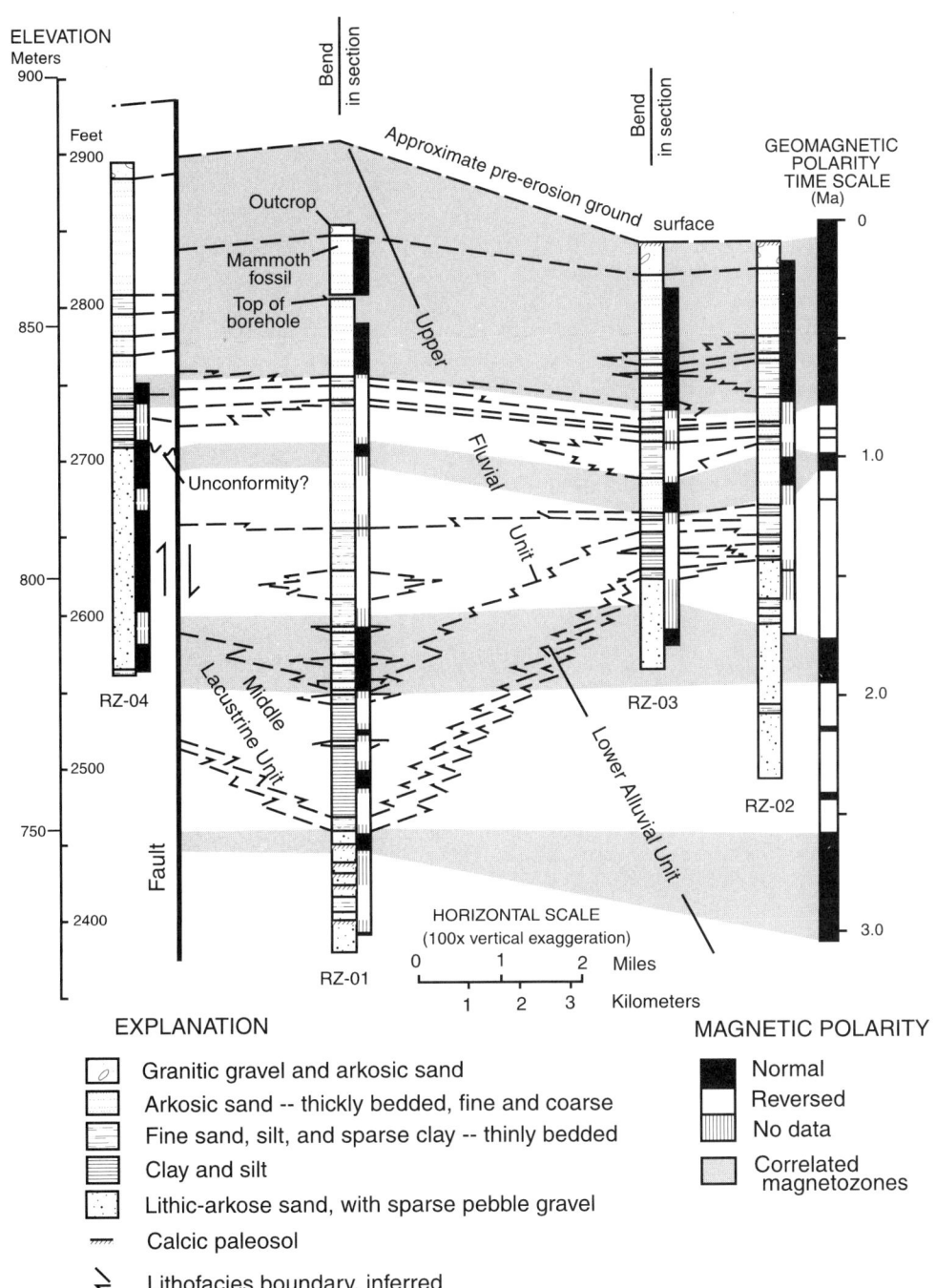

Figure 5. Lithostratigraphy and magnetostratigraphy of boreholes at former George Air Force Base, California. Vertical scale is 100 times horizontal scale. Borehole locations are shown on Figures 2, 4, and 12. There are two sharp bends in the section, at sites of boreholes RZ-01 and RZ-03. Critical subsurface relations are most readily visualized by imagining that view is toward the northwest, with borehole RZ-01 in the foreground and the other boreholes in the background. Approximate stratigraphic level of fossil mammoth locality is projected onto lithologic column of borehole RZ-01.

deposits closely resemble arkosic sediments of the modern Mojave River that are derived mainly from the northwestern San Bernardino Mountains. The middle lacustrine unit has a mixed provenance manifested by discrete layers of arkosic sand and lithic-arkosic sand. Except for rare concentrations of Pelona Schist in the upper fluvial unit, the entire succession lacks conspicuous rock detritus from the San Gabriel Mountains region.

The deposits are also distinguished by varying degrees of weathering, oxidation, diagenesis, and consolidation, all of which tend to be most advanced in the lower alluvial unit. A series of well-developed calcic paleosols near the bottom of borehole RZ-01 is particularly noteworthy (Fig. 5). Although the upper fluvial unit is largely unweathered and otherwise unaltered, it is capped by a paleosol containing mature argillic and calcic horizons. This paleosol is associated with a broad, deeply incised geomorphic platform that evidently represents the floodplain of the ancestral Mojave River (Sibbett, 1999); Cox and Tinsley (1999) termed this feature the "George surface."

Lower alluvial unit

The deepest deposits in each of the four boreholes comprise lithic-arkosic sand, silt, and polymictic gravel that we designate as the lower alluvial unit (Fig. 5). Although the boreholes penetrate as deeply as 50 m into this unit, none of them reaches its base, which therefore remains undefined. The lower alluvial unit is overlain by the middle lacustrine unit in each of the boreholes and in nearby bluffs of the Mojave River. However, the lakebeds pinch out near the northern and eastern margins of GAFB (Montgomery Watson, 1995; Chrisley, 1997). Beyond this pinch-out, the lower alluvial unit is capped unconformably by the upper fluvial unit, a relationship that extends northward 25 km along the Mojave River between GAFB and Iron Mountain (Cox and Tinsley, 1999).

Deposits of the lower alluvial unit characteristically are moderately weathered and oxidized and contain more conspicuous pedogenic and diagenetic features than the overlying units. They generally are slightly to moderately consolidated, are tinted yellowish brown to strong brown (Munsell hues 10YR–7.5YR), and commonly contain abundant authigenic calcium carbonate and clay minerals. Two thickly bedded sandy facies are predominant. One consists of silty sand, and the other of "clean" sand and pebble gravel. These facies locally are interlayered and jointly constitute ~90% of the total drilled thickness of the unit in the four boreholes. The abundance of the silty sand facies generally increases upward in the boreholes at the expense of clean sand and gravel. Sparse thick beds of silt and fine sand constitute a subordinate, third facies.

Deposits of the silty sand facies are poorly stratified and poorly sorted. Silt and very fine-grained to medium-grained sand are the dominant constituents, but coarse sand, granules, and small pebbles also are abundant, particularly in the lower parts of beds. The deposits of clean sand and gravel are prominently stratified and moderately sorted, typically consisting of medium-grained to very coarse-grained pebbly sand and subordinate sandy pebble gravel. Individual beds in both facies are 30–180 cm thick and are stacked to form compound layers up to 3 m thick.

The sandy and gravelly facies of the lower alluvial unit are characterized by diverse detrital textures and provenance. Sand-size grains are mainly subangular, but sparse rounded grains evidently record subsidiary aeolian transport or recycling from preexisting sedimentary rocks. The sand consists of slightly lithic arkose with ~1–5% lithic fragments and conspicuous accessory epidote (also see Chrisley, 1997). Biotite typically is sparse and strongly weathered. Gravel clasts consist of aplite and assorted felsic plutonic rocks; silicic metavolcanic rocks, including abundant ash-flow tuff; nonmetamorphosed hornblende-biotite dacite; and assorted metasedimentary rocks, including quartzite, marble, and calc-silicate rocks. The relative abundance of the different clast types varies considerably from layer to layer within each borehole. Clasts as large as 15–20 cm were observed in outcrops of clean sandy gravel directly east of GAFB. Most clasts are angular to subrounded. However, many of the dacite clasts are rounded, which suggests they were transported a greater distance or recycled from preexisting deposits of gravel or conglomerate.

The subordinate facies of silt and fine sand forms several beds 1–3 m thick (Fig. 5). There are two variants of this facies. One comprises structureless beds of consolidated, well-oxidized clayey silt and fine sand intercalated well below the top of the lower alluvial unit in boreholes RZ-01, -02, and -04. The other variant forms a thick layer of well-stratified friable silt and fine sand that caps the lower alluvial unit in boreholes RZ-01, -03, and -04. The latter deposit is laminated to thinly bedded and locally (in RZ-01) contains sediment-filled horizontal burrows ~15 mm in diameter. It grades upward from oxidized yellowish-brown silty sand into olive-gray sandy silt, thus effecting a gradual, seemingly conformable, transition between the lower alluvial unit and overlying middle lacustrine unit. We suspect this boundary truly is conformable in boreholes RZ-01 and RZ-03. However, evidence discussed in a later section ("Syndepositional faulting at GAFB") suggests there is a significant unconformity at the base of the transitional layer in borehole RZ-04. Interlayered alluvial-fan and lacustrine facies establish a conformable transition between the lower alluvial unit and middle lacustrine unit in borehole RZ-02 (Fig. 5).

Paleosols with mature argillic and calcic horizons are common in the lower alluvial unit. The calcic horizons mainly contain filamentous and nodular calcium carbonate (stage I and II carbonate morphology; Gile et al., 1966; Bachman and Machette, 1977), but several stage-III carbonate crusts were observed in borehole RZ-01 (Fig. 5). Throughout the unit, grains of plagioclase and dacite are partly altered to clay. This pervasive effect apparently is a product of intrastratal diagenesis.

The lower alluvial unit at GAFB evidently comprises an assemblage of braided-stream, alluvial-fan, and playa deposits. Deposits of clean sand and pebble gravel probably accumulated in channels of a major southward-flowing braided stream and possibly also in washes of large alluvial fans. Deposits of poorly sorted silty sand are ascribed to sheet floods in low-gradient fluvial or alluvial-fan environments, and the intercalated thick beds of well-oxidized clayey silt and fine sand probably formed in playas. Finally, the transitional layer of well-stratified silt and fine sand that caps the unit probably is a fluviodeltaic deposit. In general, the greater degree of weathering and soil development as compared to deposits of the upper fluvial unit suggests that the lower alluvial unit either accumulated more slowly or formed under a different climatic regime.

Middle lacustrine unit

Each of the four cored boreholes at GAFB contains a fine-grained medial unit that consists of calcareous clay and silt, with lesser amounts of interlayered and disseminated sand. This stratigraphic interval is designated as the middle lacustrine unit. Its thickness varies from 8 to 25 m in the boreholes (Fig. 5). As with the overlying upper fluvial unit, the maximum thickness is found in borehole RZ-01, near the southeast corner of GAFB. The lacustrine unit also crops out locally in the Mojave River bluffs at the east edge of the base. In the outcrops it is no more than 3 m thick and gradually pinches out to the north, terminating 2040 m S65°E from borehole RZ-02. Based on borehole data, the unit also pinches out in the subsurface 2.6 km north of borehole RZ-02 (Fig. 6) (Montgomery Watson, 1995; Chrisley, 1997).

Several lithologic features distinguish the middle lacustrine unit from the other deposits in the local stratigraphic succession. First, it contains thick intervals of structureless to prominently laminated clay and clayey silt. Some of these intervals are nearly devoid of sand and coarse silt (particularly in boreholes RZ-01 and RZ-03). By contrast, the underlying and overlying units contain only sparse thin layers of clay intercalated amongst beds of sand and coarse silt. Second, the lacustrine unit consists of interlayered greenish (olive-gray—Munsell hues 5Y-10Y-2.5GY) and brownish (10YR and 2.5Y) sediments, whereas the units below and above are composed exclusively of brownish sediments. Third, the lacustrine mud contains abundant aquatic microfossils, particularly ostracodes, diatoms, and seeds and stems of water plants; fish scales and gastropods are occasionally observed (R. Forester, 1999, written commun.). The mud typically effervesces in dilute hydrochloric acid, indicating the presence of calcareous microfossils, matrix, or cement. Much of the calcium carbonate consists of very fine-grained material of probable inorganic origin that is disseminated within marly clay. Muddy sediments in the other units mostly lack aquatic microfossils and generally are noncalcareous. Finally, there is virtually no evidence of soil development in the lacustrine unit, whereas the lower alluvial unit contains incipient to mature argillic and calcic soil horizons and the upper fluvial unit contains incipient argillic horizons.

The abundance and character of intercalated sand varies vertically and laterally in the middle lacustrine unit. The lowest three-quarters of the 25-m-thick lacustrine succession in borehole RZ-01 consists of homogeneous clayey mud with only sparse disseminated sand grains and rare thin layers of sand. However, in the uppermost quarter of this succession, and throughout the entire lacustrine interval in the other boreholes, clayey mud is interlayered with coarse silt and fine sand, and locally with sparse beds of medium and coarse sand. Some of the arenaceous deposits consist of olive-gray ostracodal sand and silt, while others consist of oxidized, yellowish-brown sand and silt that lack ostracodes but contain mud cracks and root traces. The abundant fine sand intercalated near the top of the lacustrine unit in borehole RZ-01 produces a transitional, conformable boundary with the overlying upper fluvial unit. This boundary is more abrupt, yet still demonstrably conformable, in the other three boreholes. The lacustrine section in borehole RZ-02 contains thicker sandy intervals than the laterally equivalent section in borehole RZ-03 (Fig. 5), which indicates that the depositional environment varied significantly between these two sites.

Sandy strata in the lacustrine unit comprise two distinct compositional varieties. One consists of essentially pure, biotite-rich arkose that closely resembles deposits of the upper fluvial unit. The other consists of sparsely micaceous, slightly lithic

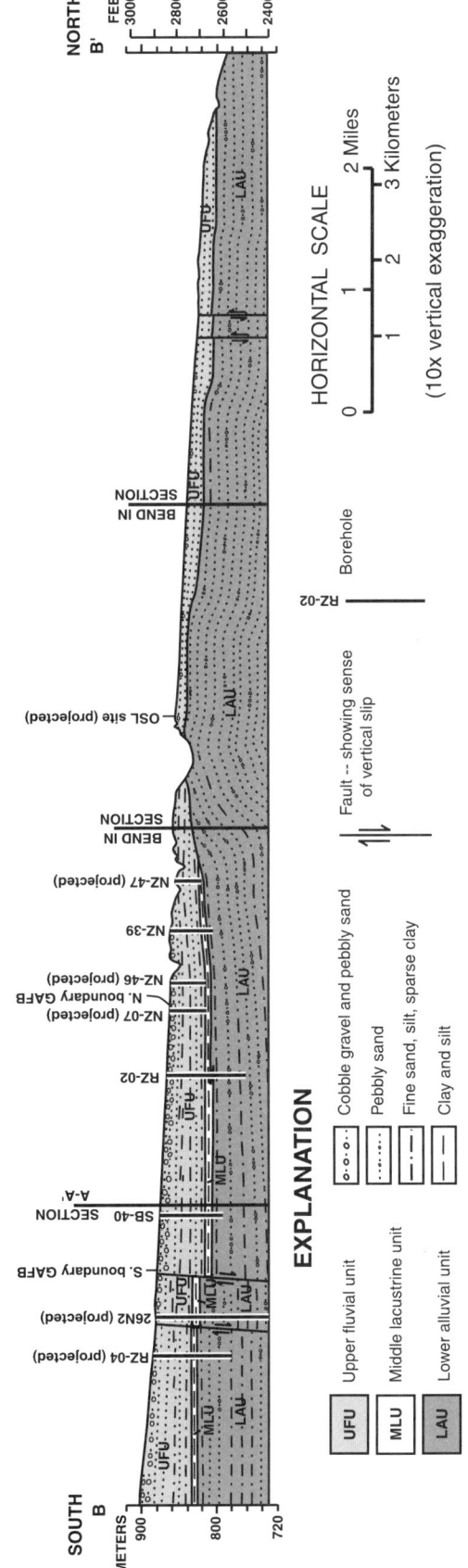

Figure 6. North-south cross section between former George Air Force Base and Silver Lakes, California, showing faults and folds that deform the Pliocene-Pleistocene succession. Line of section is shown on Figure 2. Vertical scale is 10 times horizontal scale. Southern half of section is constructed mainly from borehole data; northern half is derived from outcrops in bluffs of the Mojave River. Interpretation of boreholes RZ-02 and RZ-04 is based on this study; other boreholes are from Montgomery Watson (1995).

arkose like that in the lower alluvial unit. The lithic-arkose sand tends to be concentrated near the base of the unit, whereas the pure arkose is concentrated near the top.

The interlayered sand and fine mud, alternating green and brown colors, local root casts and mud cracks, and assorted aquatic organisms indicate that the middle lacustrine unit accumulated in a shallow lake or wetland bordered by fluviodeltaic mudflats. Lower parts of the lacustrine succession in boreholes RZ-01 and RZ-03 contain abundant veins and nodules of calcium carbonate. The carbonate material probably precipitated from upwelling groundwater, which implies the lake or wetland was recharged in part by springs.

Upper fluvial unit and capping paleosol

The uppermost deposits in the boreholes comprise a texturally heterogeneous, but compositionally homogeneous, succession of granitic sand, silt, and gravel that is here designated as the upper fluvial unit. These deposits directly underlie the broad geomorphic platform that surrounds GAFB, and thick sections through the unit are exposed in neighboring bluffs of the Mojave River canyon. The deposits are part of a belt of dissected fluvial sediments that extends 75 km along the Mojave River, from Deep Creek to Harper Lake (Fig. 2; Cox and Tinsley, 1999). After compensating for surficial erosion, the unit is ~110 m thick near the southeast corner of GAFB (borehole RZ-01), but it thins markedly to the north and west (Fig. 5). The thickness decreases to ~55 m near the western and northern margins of the air base (boreholes RZ-02, -03, and -04) and to ~25 m in outcrops 3.2 km north of borehole RZ-02.

The upper fluvial unit contains four distinct lithofacies, each consisting of loose to weakly consolidated yellowish-brown and olive-brown sediments (Munsell hues 10YR and 2.5 Y). Two facies of thickly bedded sand are abundant and intimately interlayered throughout the unit. One of these consists of silty fine sand and the other of pebbly coarse sand. A third facies consists of thinly bedded fine sand, silt, and sparse clay. The thinly bedded, fine-grained sediments constitute much of the lowest third of the upper fluvial unit in borehole RZ-01, thus largely accounting for the greater thickness of the unit beneath the southeastern part of GAFB (Fig. 5). The same lithologic assemblage is found at higher stratigraphic levels in boreholes RZ-02 and RZ-03 near the north end of the air base, where it extends upward to ~20 m below the George surface. A fourth facies consists of thick beds of pebble and cobble gravel, which are interlayered with thickly bedded fine and coarse sand near the top of the unit. Overall, the unit coarsens irregularly upward, mainly due to the concentration of silt-rich and gravel-rich sediments at opposite ends of the stratigraphic column.

Beds within the gravel facies and the two thickly bedded sand facies are 30–200 cm thick. Multiple beds of either sand facies commonly are stacked to form compound layers as much as 4–5 m thick. The thickly bedded fine sand is poorly sorted, consisting mainly of fine to very fine sand, combined with abundant silt, moderate amounts of medium to very coarse sand, and sparse granules and small pebbles. This material forms nearly structureless beds that characteristically are capped gradationally with a thin layer of sandy silt.

By contrast, the pebbly coarse sand typically is moderately sorted, consisting mostly of medium to very coarse sand with only sparse interstitial silt. The pebbles range up to 25 mm in diameter. Textural and heavy-mineral laminations commonly are visible in the borehole cores. Outcrops in the Mojave River bluffs reveal other features including lenticular channels, trough and low-angle cross-stratification and "rip-up" clasts of sandy silt. The outcrops also display intertonguing and intergrading between laterally adjacent thick beds of pebbly coarse sand and silty fine sand, which indicates that these two facies accumulated simultaneously in directly adjoining environments. The gravel beds concentrated near the top of the section generally resemble the deposits of pebbly coarse sand and contain a similar matrix of medium to very coarse sand. However, they differ by containing more abundant and larger clasts which, in the outcrops, range to at least 20 cm in diameter.

In the two thickly bedded sand facies and in the gravel, sand grains are mostly subangular and are derived almost exclusively from felsic plutonic rocks and gneiss. Flakes of fresh biotite are conspicuous. Clasts are mostly subangular to subrounded, although many of the larger cobble-size clasts are rounded. The clasts consist of felsic plutonic rocks and granitic gneiss, accompanied by minor amounts of quartzite and silicic metavolcanic rocks. Pebbles of schistose metasediments derived from the Pelona Schist of the San Gabriel Mountains were locally observed in cobble gravel near the top of the unit (e.g., Reynolds and Cox, 1999, field trip stop 8). However, fragments of the schist were not evident at deeper stratigraphic levels in any of the four borehole cores.

Deposits of thinly bedded sand, silt, and clay range from 1 to 8 m thick. Individual layers within these fine-grained intervals are mostly 2–20 cm thick, although beds of fine sand occasionally are as thick as 60 cm. The layering locally is disrupted by desiccation cracks, root traces, and rare burrows. The sand consists of silty, fine to very fine-grained, micaceous (biotitic) arkose. Ostracodes are very rarely observed in layers of silt and clay. In each borehole, the basal deposit of the upper fluvial unit consists of a thin (10–200 cm) interval of variegated, olive-gray to yellowish-brown, micaceous fine sand with ripple-drift cross-lamination, heavy-mineral laminations, and intercalated thin layers of silt. This interval effects a conformable transition between the highest lacustrine muds of the middle unit and the lowest coarse fluvial sands of the upper unit.

The deposits of pebbly coarse sand and pebble-cobble gravel closely resemble sediments in the nearby active channel of the Mojave River, and by analogy we infer that they were deposited by turbulent floodwaters in shallow braided channels of the northward-flowing ancestral Mojave River. The intimately associated deposits of thickly bedded silty fine sand probably represent floodplain splays that accumulated alongside the channels when floods overtopped their banks. Thinly bedded fine sand, silt, and clay apparently are paludal sediments that accumulated in ephemeral ponds and marshes on a delta plain or river floodplain.

The George surface at the top of the upper fluvial unit is capped by a paleosol containing well-developed argillic and calcic horizons. This feature is exposed in road cuts and other shallow excavations around GAFB, and it crops out sporadically at the crest of the Mojave River bluffs north and east of the base. In most areas, the paleosol is partly eroded and covered by a veneer of Holocene sheet wash and eolian sand. Several good exposures of the paleosol were encountered in a pipeline trench on the east side of Helendale Road, 3.2–6.5 km north of borehole RZ-02. There the soil profile is ~0.75 m thick and consists of an eroded argillic horizon ~45 cm thick and an underlying calcic horizon ~30 cm thick. The argillic horizon consists of clayey sand with coarse prismatic peds; damp material is strong brown (Munsell soil color 7.5YR 4/6). The calcic horizon consists of structureless, chalky calcium carbonate that impregnates the sandy parent material, completely obscuring primary detrital texture and stratification (stage III to III+ carbonate morphology; Gile et al., 1966; Bachman and Machette, 1977).

Beneath the capping paleosol, most deposits of the upper fluvial unit are relatively unaltered and unweathered, containing only incipient argillic soil horizons. Thus, the deposits generally are less weathered than deposits of the lower alluvial unit, and the pedologic evidence does not reveal any major hiatuses in the accumulation of the unit.

MAGNETOSTRATIGRAPHY

Paleomagnetic methods

Samples for paleomagnetic analysis from George Air Force Base (GAFB) were collected from the dried cores RZ-01, -02, -03, and -04 at intervals ranging from 0.1 to 3 m apart, wherever consolidated fine-grained sediments were recovered. The magnetic polarities and approximate stratigraphic positions of the samples are shown on Figures 7–9, and the detailed analytical results are presented in Table DR1[1]. The larger gaps in the sampling occurred in sections of gravel and loose sand not suitable for paleomagnetic analysis. To prepare standard cylindrical specimens (10.5 cc), we sawed each sample into a 2.3-cm-thick disk, then cored the disk with a diamond drill cooled by compressed air. A few samples, too sandy to allow drilling by this dry method, were shaped with hand tools. In addition to the borehole samples, oriented blocks were collected from four outcrops of consolidated fine sand near the southeast corner of GAFB (Table DR1 [see footnote 1]; outcrops 1–4), and from a layer of volcanic ash near the Helendale fault (Table DR1, outcrop 5).

All measurements of remanent magnetization were made with a cryogenic magnetometer at laboratories of the U.S. Geological Survey in Menlo Park. Demagnetization treatments employed a commercial alternating-field apparatus (400 Hz, 100 mT maximum field) equipped with a specimen tumbler. Thermal demagnetization was carried out in air in a magnetically shielded furnace (<5 nT).

Determinations of magnetic inclination required a dual demagnetization procedure consisting of first applying alternating fields of 5 mT and 10 mT, followed by progressive heating from 150 to 590 °C in the low-field furnace. This procedure stripped away spurious magnetic remanence caused by the original coring, the recent geomagnetic field, and oxidation of iron-bearing minerals. The alternating-field steps were critical in determining whether a sample had been inadvertently overturned in the core box. If the magnetization removed by the low alternating field was directed downward, as would be expected from the magnetic effect imparted by the recent geomagnetic field or the drill-stem, then we were assured that the core had not been inverted during handling. The few misoriented specimens detected by this method were confirmed by detailed visual inspection of the core, and then corrections were applied to the analytical results. Principal-component analysis (Kirschvink, 1980) was next used to calculate the best-fit magnetic direction to the thermal demagnetization data in all but a few cases. The exceptions were friable specimens that had to be encased in plastic boxes; these samples were given alternating-field treatment only.

Paleomagnetic results

The higher-quality determinations of magnetic directions were obtained by heatings of 300–580 °C (Fig. 10, Table DR1). All but a few specimens yielded a stable magnetization direction after such treatment. A measure of quality is given by the Maximum Angular Deviation (MAD) derived from a least-squares fit to the thermal demagnetization data (Kirschvink, 1980). Seventy percent of the specimens gave MAD values of <5°, indicating determinations of very high quality. Low-quality results with MAD >15° were flagged, and we weighed this group, comprising 5% of the total collection, lightly in the interpretation of polarity. For the specimens given alternating-field treatments only, it was generally found that polarity was adequately determined by applying fields of 10 mT to 40 mT. Two specimens from core RZ-03 had magnetizations that failed to stabilize after demagnetization treatment, so inclinations could not be determined. The cored strata were assumed to be flat lying and the boreholes were assumed to be vertical, so no tilt corrections were made. The declination results are not listed, because no attempt was made to reorient the cores with regard to azimuth.

The geomagnetic polarity was defined from the inclination data as follows: normal polarity, > +20°; reversed polarity, < –20°; intermediate polarity, inclination 20° or shallower. These definitions take into account the natural variation of the geomagnetic field about the long-term axial dipole direction. Intermediate determinations might reflect transition of the field between polarity states, a brief excursion of the geomagnetic field, or partial removal of secondary magnetization components.

---

[1]GSA Data Repository item 2003068, analytical data (Tables DR1–DR4) and site information (Fig. DR1), is available on request from Documents Secretary, GSA, P.O. Box 9140, Boulder, CO 80301-9140, USA, editing@geosociety.org, or at www.geosociety.org/pubs/ft2003.htm.

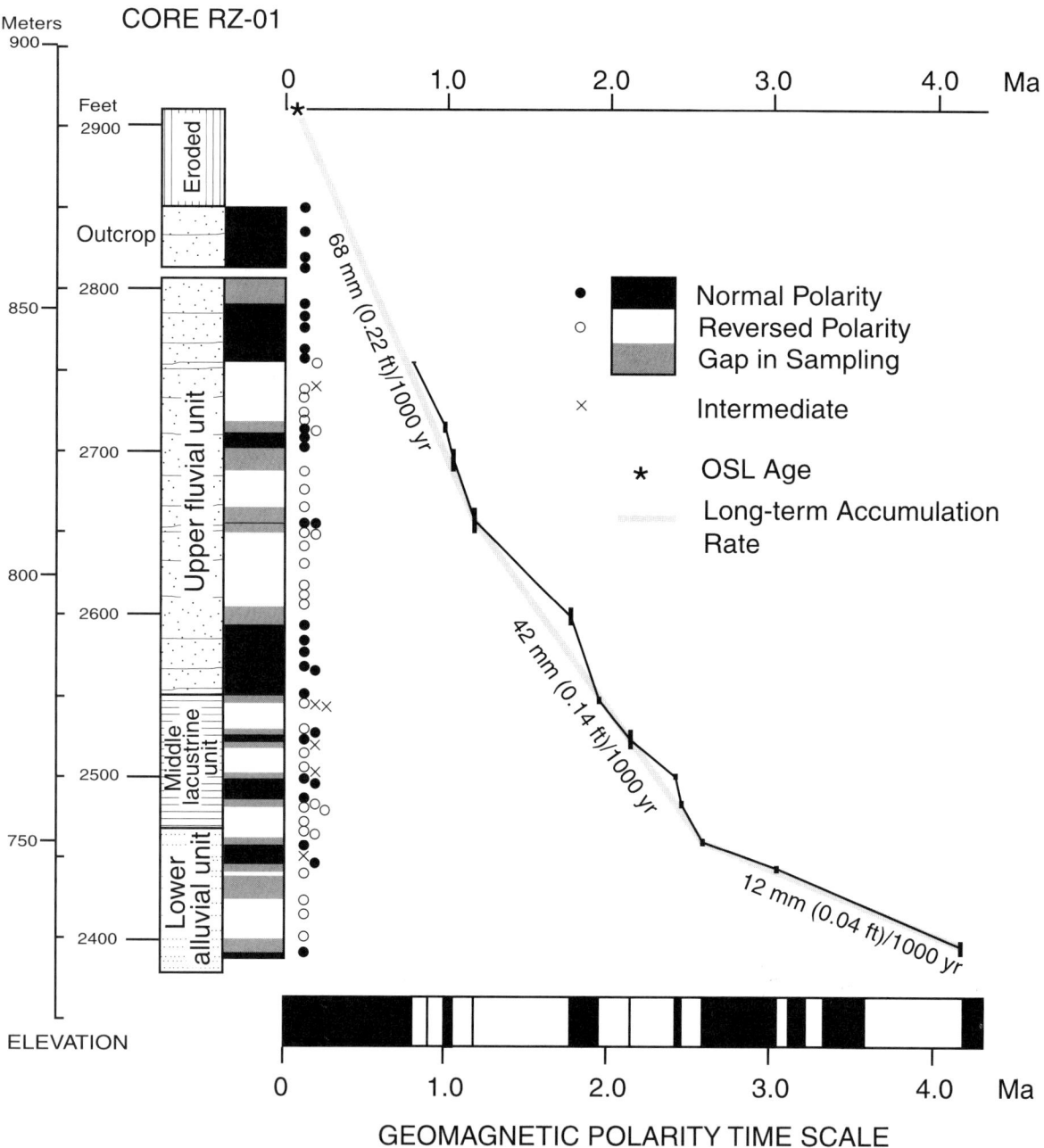

Figure 7. Age vs. stratigraphic position in borehole RZ-01 and nearby outcrops at former George Air Force Base, as correlated with the geomagnetic polarity time scale. Vertical bars indicate uncertainty in stratigraphic position of the polarity transitions. Small circles and crosses show polarities determined for individual sediment samples, and adjacent column shows inferred magnetozones. Shaded line illustrates major changes in mean rate of sediment accumulation. The optically stimulated luminescence (OSL) age at top of section is projected from sample site north of the air base (site location is shown on Figures 2 and 12).

Boreholes RZ-01 and RZ-03 are well represented by samples from top to bottom. The lowest quarter of borehole RZ-02 and upper half of borehole RZ-04 were not cored, so the sampling of these holes is incomplete. From the inclination data, a sequence of magnetozones was defined for each borehole (Figs. 7, 8, 9). Uncertainties in the exact stratigraphic positions of the magnetozone boundaries are caused by gaps in the sampling that are denoted on the figures.

When the magnetozones of RZ-01, RZ-02, and RZ-03 are compared, there is a consistent pattern with a thick normal-polarity zone at the top, generally followed lower in the section by dominantly reversed polarity. All three cores show two or more thin

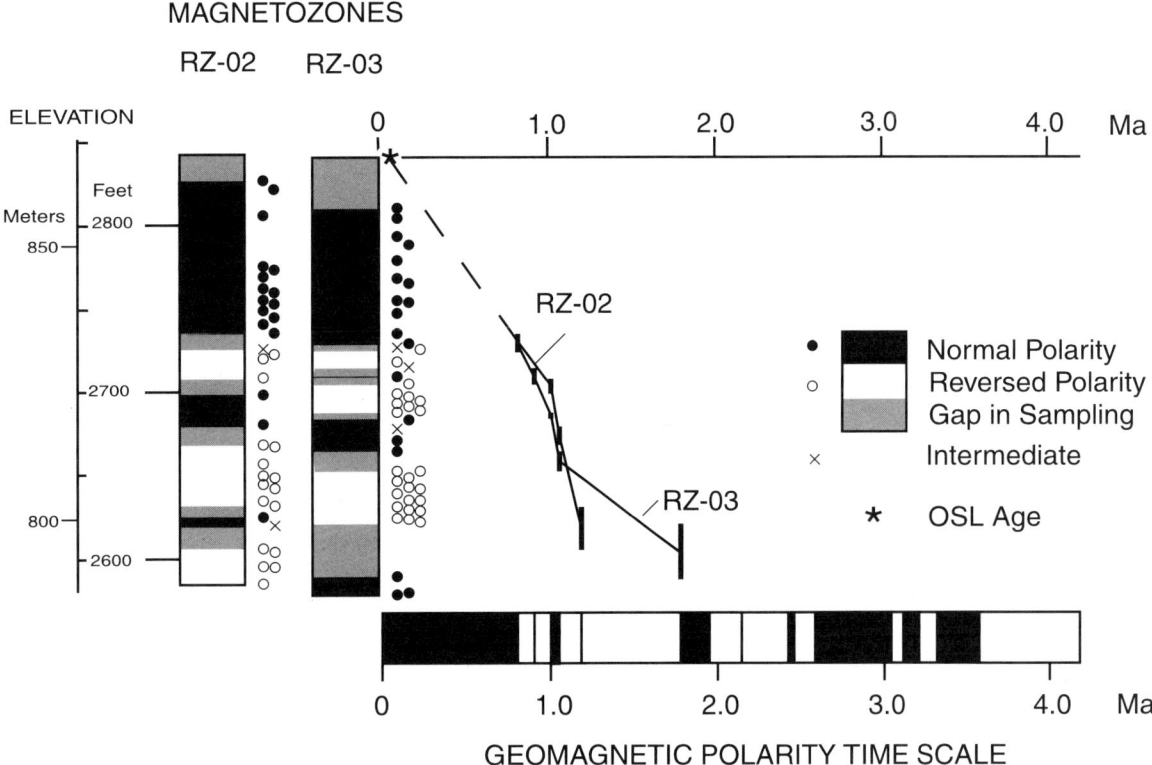

Figure 8. Age vs. stratigraphic position in boreholes RZ-02 and RZ-03 at former George Air Force Base, as correlated with the geomagnetic polarity time scale. Vertical bars indicate uncertainty in stratigraphic position of the polarity transitions. Small circles and crosses show polarities determined for individual sediment samples, and adjacent columns show inferred magnetozones. The optically stimulated luminescence (OSL) age at top of section is projected from sample site north of the air base (site location is shown on Figures 2 and 12).

zones of normal polarity within the dominantly reversed lower interval. In contrast, RZ-04 shows three reversed-polarity horizons intercalated within a broad interval of normally polarized strata.

Correlation with the geomagnetic polarity time scale

Figure 7 shows our preferred correlation of the RZ-01 magnetozones with the polarity time scale as calibrated by Singer et al. (1999) for the period younger than 1.2 Ma and by Cande and Kent (1995) for the interval 1.2–5.0 Ma. This correlation assumes that deposition was not broken by any lengthy hiatuses, and thereby yields the youngest possible ages for the stratigraphic section beneath GAFB. Other interpretations that would assign older ages to the magnetozones are permissible. However, such interpretations are less likely to be accurate, because the thick normal-polarity zone, as defined at the top of RZ-01 and in the outcrop above the wellhead, is best correlated with the lengthy Brunhes Chron. The uppermost polarity transition ~16 m below the wellhead is most likely the Brunhes/Matuyama boundary dated at 0.78 Ma. The next two thin normal-polarity zones are attributed to the Jaramillo (0.986–1.053 Ma) and Cobb Mountain (1.18 Ma) Subchrons. The well-defined transition near the top of the middle lacustrine unit is most likely the beginning of the Olduvai Subchron at 1.95 Ma. The transition directly beneath the lakebeds is tentatively correlated with the beginning of the Matuyama Chron at 2.58 Ma.

The sections penetrated in RZ-02 and RZ-03 overlap stratigraphically the upper two-thirds of RZ-01. The Brunhes/Matuyama boundary, Jaramillo Subchron, and the end of the Olduvai Subchron are correlated at similar elevations among these three boreholes (Figs. 5 and 8). Correlation of RZ-04 with the time scale and the other boreholes is problematical (Figs. 5 and 9). In contrast to the other boreholes, RZ-04 exhibits much thinner reversed-polarity zones between elevations of 790–840 m. From mapping of lineaments and an offset geomorphic surface near the RZ-04 wellhead, we infer that a fault has raised the RZ-04 stratigraphic section relative to the sections in the other boreholes (see "Syndepositional faulting at GAFB").

CHRONOLOGY AND STRATIGRAPHIC CORRELATION

Our studies indicate that the four-cored boreholes at George Air Force Base (GAFB) penetrate a conformable succession of Pliocene and Pleistocene strata (Fig. 11). Continuity of sedimentation in the lacustrine unit and overlying upper fluvial unit is implied

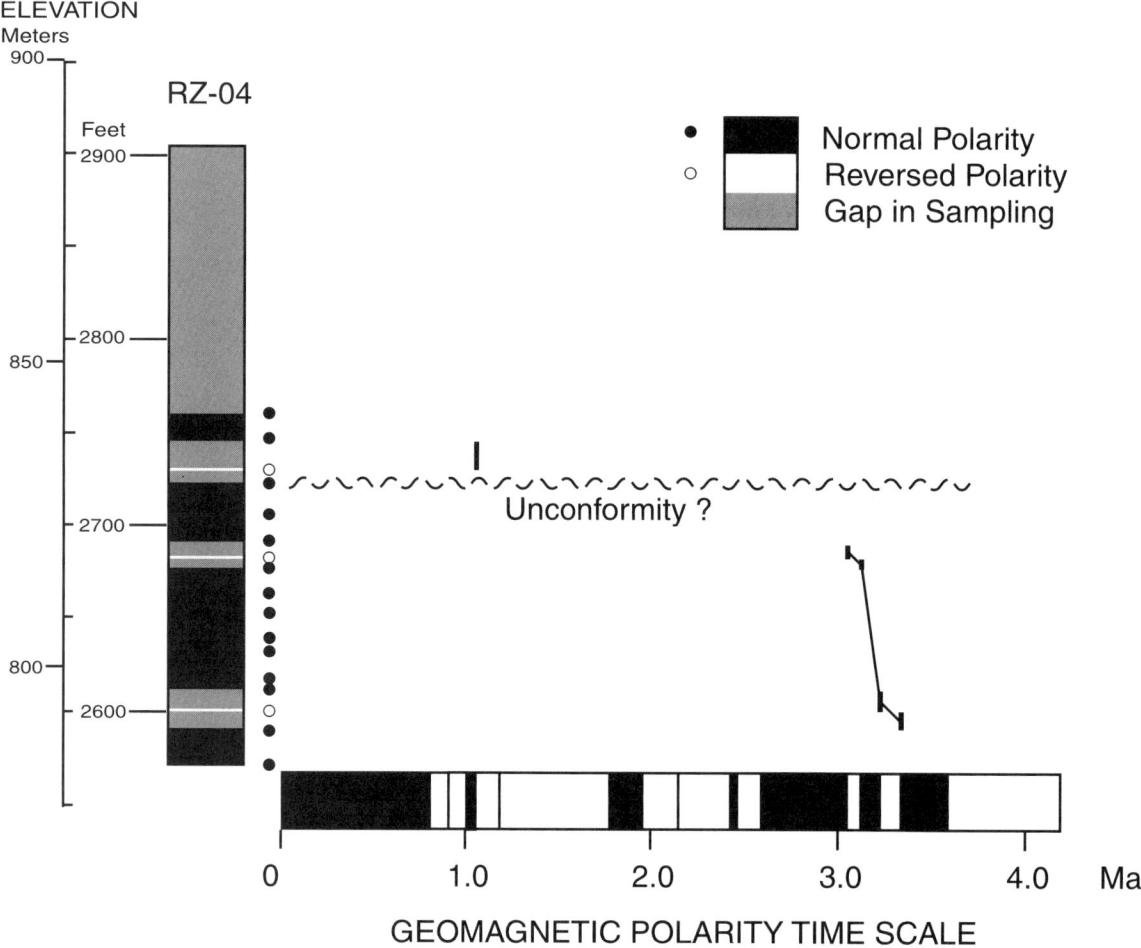

Figure 9. Age vs. stratigraphic position in borehole RZ-04 at former George Air Force Base, as correlated with the geomagnetic polarity time scale. Vertical bars indicate uncertainty in stratigraphic position of the polarity transitions. Small circles show polarities determined for individual sediment samples, and adjacent column shows inferred magnetozones.

by the transitional contact observed between these units in each of the boreholes, and by the absence of mature paleosols or obvious erosional hiatuses. The lower alluvial unit is also transitional with the lacustrine unit in boreholes RZ-01, -02, and -03, but mature soil horizons in the lower unit in borehole RZ-01 may indicate gaps in sedimentation as great as 10,000–100,000 yr. There also appears to be an unconformity at the top of the lower alluvial unit in borehole RZ-04. The results of our magnetostratigraphic investigation and optically stimulated luminescence study allow us to date and correlate most of the stratigraphic succession, especially in boreholes RZ-01, -02, and -03 (Fig. 5). The correlation of tectonically dislocated strata in borehole RZ-04 is addressed in a following section ("Syndepositional faulting at GAFB").

Age of the George surface

The George surface or pre-erosion ground surface at the top of the upper fluvial unit was targeted for dating because it records the final filling of the Victorville basin and predates the incision of the Mojave River canyon. Our correlation with the geomagnetic polarity time scale implies that the George surface is younger than the Brunhes/Matuyama boundary, dated at 0.78 Ma (Figs. 3 and 5). The thickness and internal character of the capping paleosol suggests that the surface is late Pleistocene in age. About 40 km south of GAFB (Fig. 2), a soil with a comparably developed argillic horizon caps the oldest southward-draining stream terrace of Cajon Creek (unit Qoa-d of Weldon and Sieh, 1985; McFadden and Weldon, 1987). The soil is dated at 55 ± 12 ka based on lateral separation of 1.3 km across the San Andreas fault (McFadden and Weldon, 1987).

We applied the optically stimulated luminescence (OSL) method to date deposits associated with the George surface near GAFB. The OSL method relies on the interaction of ionizing radiation with electrons in semiconducting crystals, such as quartz and feldspar (Huntley et al., 1985; Aitken, 1998). This interaction produces a cumulative metastable charge within

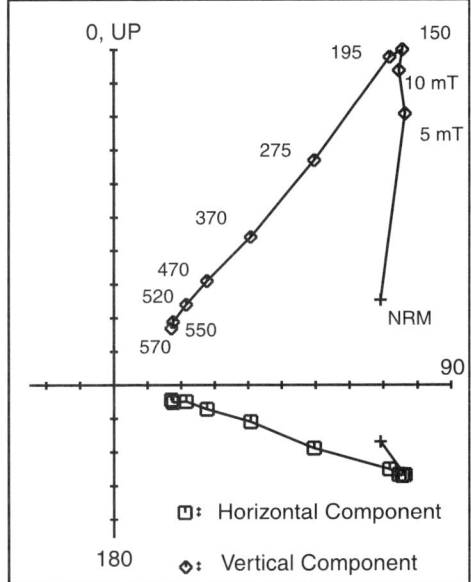

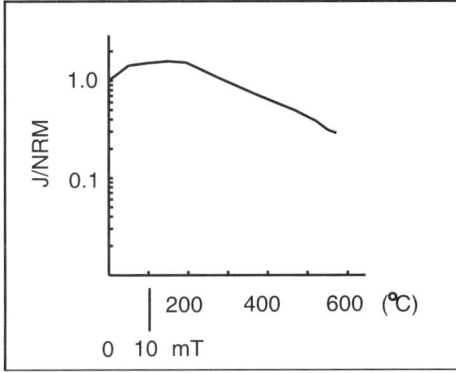

Figure 10. Plot showing the demagnetization behavior of specimen 7J078 from borehole RZ-01. Upper diagram shows the projection of the magnetization components as the natural remanent magnetization (NRM) is reduced by alternating-field demagnetization followed by heating in a low-field furnace. Temperature is in degrees C; mT, milliTesla. Lower diagram shows reduction of the normalized intensity of magnetization (J/NRM) after the alternating-field and thermal treatments.

buried detrital mineral grains. The charge is released with the emission of photons (luminescence) when the grains are exposed to sunlight or illuminated in the lab. The natural radiation flux responsible for an observed quantity of luminescence is determined by measuring the luminescence released from a series of artificially irradiated subsamples. By this procedure, the equivalent dose (D_E), or artificial radiation flux required to produce the observed luminescence, is determined. The in situ dose rate of radiation is measured directly in the field, or is estimated by measuring the concentration of radionuclides in the sediment. The age of sediment burial is then determined using the equation: Age = D_E/dose rate.

To obtain material for OSL dating, four sediment cores were collected from the walls of a pipeline trench 3.2 km due north of borehole RZ-02 (Fig. 2). Near the top of the trench, a veneer of unweathered fine-grained eolian sand and sandy sheet wash unconformably overlies the upper fluvial unit and its capping paleosol along a channeled erosional contact. Two of the cores were collected from the material above the unconformity. The second pair of cores was extracted from coarse-grained sand of the upper fluvial unit ~20 cm below the base of the paleosol. Laboratory preparation of the samples and OSL measurements were undertaken in the Luminescence Laboratory at the University of California, Riverside. Analytical data (Tables DR2–DR4) and site information (Fig. DR1) are available in the GSA Data Repository (see footnote 1).

The preliminary results of the OSL measurements on quartz are 4.3 ± 1.5 ka and 6.2 ± 1.0 ka for the deposits above the unconformity, and 60.8 ± 5.4 ka and 71.7 ± 13.3 ka for the fluvial sediments beneath the paleosol. These data imply that the maximum age of the George surface is ca. 60–70 ka. We suspect that this also approximates the *true* age of the surface, because the paleosol seems to have developed on the essentially uneroded top of the upper fluvial unit. The OSL ages are similar to the age of 55 ± 12 ka that was reported for the Qoa-d stream terrace at Cajon Creek (McFadden and Weldon, 1987). Thus, they support the tentative correlation between that terrace and the George surface that is suggested by pedologic criteria.

Age of the upper fluvial unit and associated vertebrate fauna

The foregoing results of OSL dating allow us to extend plots of age versus stratigraphic position upward to the ground surface (Figs. 7 and 8). The internal consistency of the extended plots in turn suggests that 60–70 ka, or ca. 65 ka, is a reasonable age for the top of the upper fluvial unit, as the resulting post-Matuyama sedimentation rate is comparable to preceding rates determined solely from magnetostratigraphic data. Thus, the upper fluvial unit apparently was deposited during the Pleistocene and latest Pliocene, between 1.95 and 0.065 Ma (Figs. 3, 5, and 11). The base of the unit becomes younger northwestward from 1.95 Ma in borehole RZ-01, to ca. 1.05 Ma in boreholes RZ-02 and RZ-03 (Fig. 5). The older age of the basal horizon in RZ-01 is largely accounted for by a thick interval of thinly bedded fine sand, silt, and clay deposited between 1.95 and 1.18 Ma. Although these deposits are largely restricted to RZ-01, a similar collection of fine-grained fluvial sediments accumulated between ca. 1.0 and 0.5 Ma at the sites of boreholes RZ-02 and RZ-03 (Fig. 5).

Vertebrate fossils collected at numerous sites around Victorville and GAFB have elicited conflicting interpretations regarding the age of the upper fluvial unit. Most of the local fossil localities lie within the belt of ancestral Mojave River deposits located west of the Mojave River between northern Hesperia and GAFB (Fig. 2). Several brief reports (Jefferson, 1986; Reynolds, 1989; Reynolds and Reynolds, 1994a) proposed that the vertebrates are chiefly middle Pleistocene (late Irvingtonian), or ca.

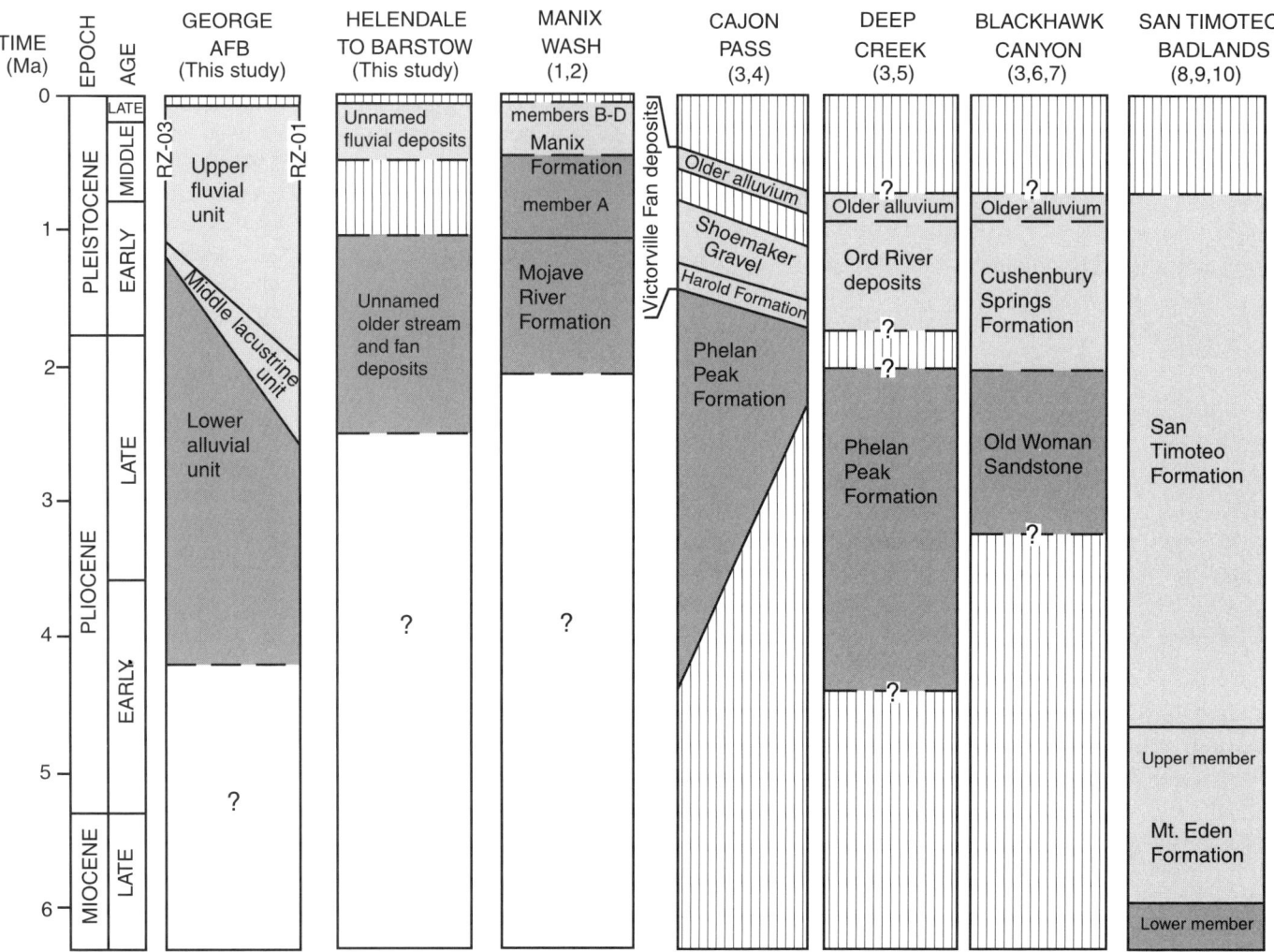

Figure 11. Correlation diagram showing age relations of upper Cenozoic strata at former George Air Force Base as compared to other sites north and south of the central Transverse Ranges (geographic locations shown on Figures 1 and 2). Vertical ruling denotes depositional or erosional hiatus; column is unpatterned where base of section is concealed. Dark shading indicates detrital source is outside the Transverse Ranges; light shading, source within Transverse Ranges. Columns for the air base and Cajon Pass are composite and show lateral variations in age; left and right margins of Cajon Pass column correspond to Phelan Peak and Crowder Canyon sections of Weldon et al. (1993), respectively. Sources of data: (1) Jefferson (1994); (2) Nagy and Murray (1996); (3) Meisling and Weldon (1989); (4) Weldon et al. (1993); (5) Meisling (1984); (6) Shreve (1968); (7) May and Repenning (1982); (8) Fraser (1931); (9) Matti and Morton (1993); (10) Albright (1999).

0.8–0.5 Ma. However, two taxa collected near Victorville ostensibly are older than middle Pleistocene, based on their minimum ages reported from other regions. These comprise a Pliocene(?) cotton rat, *Sigmodon* cf. *S. minor* (Reynolds and Reynolds, 1994a), and an early(?) Pleistocene mammoth, *Mammuthus meridionalis* (Scott et al., 1997).

The results of our investigation and paleomagnetic evidence from previous studies (Meisling, 1984, locality HRF; K. Meisling, 1984, personal commun., cited in Reynolds, 1989; K. Meisling, 1985, personal commun., cited in Reynolds and Reynolds, 1994a) suggest that the vertebrate fossil assemblages near Victorville are predominantly middle Pleistocene and younger, as was originally concluded by Jefferson (1986). The combined magnetostratigraphic evidence shows that a thick zone of normal-polarity strata extends 33–45 m below the George surface at Victorville and GAFB. Most of the vertebrate fossils apparently were collected from this zone. If the normal-polarity zone is accurately correlated with the Brunhes Chron, then the fossils are younger than 0.78 Ma.

Mammuthus meridionalis is particularly relevant to our study, because a well-preserved specimen was found near GAFB (Figs. 4 and 5). The specimen consists of a skull, mandible, pelvis, and several ribs discovered ~1145 m east of borehole RZ-01 at an approximate elevation of 869 m (E. Scott, 1997, written commun.). The bones were unearthed from a distinctive interval of sand and gravel that we traced westward to the site of the borehole by mapping a laterally extensive, thick bed of cobble gravel whose base lies at ~858 m elevation. This lithostratigraphic correlation associates the mammoth with normally polarized sediments of the upper fluvial unit 9–12 m above the top of borehole RZ-01 (Fig. 5; Table DR1, outcrops 3 and 4). This stratigraphic level lies well above the Brunhes/Matuyama boundary, as interpreted by our study.

To test the stratigraphic correlation, we performed paleomagnetic measurements on sediments from two outcrops near the fossil site (Table DR1, outcrops 1 and 2). The outcrops lie stratigraphically below and above the fossil horizon, at approximate elevations of 858 and 875 m. Both samples yielded a normal polarity, which is consistent with a stratigraphic position above the Brunhes/Matuyama boundary. This suggests that the specimen of *M. meridionalis* is younger than 0.78 Ma and thus is middle Pleistocene or younger, rather than early Pleistocene. This finding is unusual but is not unprecedented. Occurrences of *M. meridionalis* considered to be 0.6 Ma or younger have been reported from several localities in southeastern California and bordering parts of Arizona, including three sites along the Colorado River (Agenbroad et al., 1992) and a site in the Tecopa beds near Shoshone, California (McDaniel and Jefferson, 2002).

Assuming, once again, that our magnetostratigraphic data are accurately correlated with the geomagnetic polarity time scale, then the age of the Victorville mammoth can be estimated more precisely from its stratigraphic position 20 m below the George surface and 26 m above the Brunhes/Matuyama boundary. By interpolation, the apparent age is 375 ka, or late-middle Pleistocene. The accuracy of this estimate is uncertain, owing to the broad spacing between the dated stratigraphic horizons.

Age and local correlation of the middle lacustrine unit

The lithostratigraphic correlation of the middle lacustrine unit at GAFB was uncertain at the outset of our study because the deposits are anomalously thick and deep in borehole RZ-01 (Fig. 5). This stratigraphic arrangement persuaded previous investigators that the lakebeds in RZ-01 are unrelated to those in the other boreholes (Montgomery Watson, 1995). However, a distinctive assemblage of lithologic features, and the consistent intercalation of the lakebeds between lithic-arkose sand of the lower alluvial unit and nearly pure arkosic sand of the upper fluvial unit (Chrisley, 1997), help to confirm the original continuity of the unit. The greater depth of the lakebeds in borehole RZ-01 accordingly seems to reflect either tectonic displacement or time-transgressive sedimentation.

Intertonguing between the lacustrine unit and lower alluvial unit is evident in boreholes RZ-02 and RZ-03, which lie close enough together that thin intervals of strata can be traced fairly confidently from hole to hole (Fig. 5). Comparison of these boreholes indicates that the lake or wetland first became established at the site of RZ-03 before expanding to the site of RZ-02. Migration of the lacustrine environment across a greater distance may have produced the large stratigraphic misalignment relative to borehole RZ-01. If so, then the lacustrine unit should be significantly older in RZ-01. This is apparently confirmed by a distinct ostracode assemblage in borehole RZ-01 (R. Forester, 1999, written commun.) and is quantified by our magnetostratigraphic correlations, which indicate that the lakebeds were deposited between 2.55 and 1.95 Ma in borehole RZ-01, and between ca. 1.18 and 1.05 Ma in boreholes RZ-02 and RZ-03 (Fig. 5). Thus, the age of the middle lacustrine unit ranges from late Pliocene to early Pleistocene at GAFB, and its mean age and duration of accumulation both decrease to the northwest (Fig. 11). The lakebeds in borehole RZ-04 apparently are coeval with those in RZ-02 and RZ-03, because they contain the same ostracode assemblage found in RZ-03 (R. Forester, 1999, personal commun.) and are directly overlain by a normal-polarity magnetozone that we interpret to be the Jaramillo Subchron (Fig. 5).

Age of the lower alluvial unit

Our magnetostratigraphic interpretations suggest that the lower alluvial unit in boreholes RZ-01, -02, and -03 was deposited during the Pliocene and early Pleistocene, between ca. 4.2 and 1.18 Ma. The maximum age of the unit is derived from the lower magnetozones of RZ-01, which we have tentatively correlated with the Gauss and upper Gilbert Chrons (Figs. 3 and 7). The polarity transition near the bottom of the hole is tentatively correlated with the end of the Cochiti Subchron, dated at 4.18 Ma. Our proposed age correlation requires that the Mammoth and Kaena Subchrons were missed, either due to a disconformity or to a gap in sampling between elevations of 738–744 m. Our examination of the core from borehole RZ-01 revealed multiple disconformities in the lower alluvial unit, which are indicated by several stage-III pedogenic carbonate horizons between elevations of 730–747 m (Fig. 5). Comparably developed calcic soil horizons are not found elsewhere in the stratigraphic section, except within the capping paleosol on the George surface. Depositional hiatuses corresponding to these buried paleosols could account for the missing magnetozones at the base of the section.

Sediment accumulation rates at GAFB

Plots of age versus stratigraphic position (Figs. 7, 8, and 9) show significant changes in the rate of sediment accumulation (not corrected for postdepositional compaction) at GAFB during the Pliocene and Pleistocene. We have superimposed a generalized sedimentation-rate curve on the detailed record from borehole RZ-01 to illustrate some of the principal trends (Fig. 7). Two sharp inflections at the Matuyama/Gauss boundary (2.58 Ma) and the Cobb Mountain Subchron (1.18 Ma) divide the curve into three main segments. The pre-Matuyama segment indicates slow deposition of the lower alluvial unit at 12 mm/10^3 yr during the early to late Pliocene. The central segment represents deposition of the middle lacustrine unit and the thinly bedded, fine-grained basal part of the upper fluvial unit at an intermediate rate of

42 mm/10^3 yr during the late Pliocene to early Pleistocene. Finally, the post–Cobb Mountain segment records relatively rapid accumulation of thickly bedded sand and gravel in the main body of the upper fluvial unit at a mean rate of 68 mm/10^3 yr during the early to late Pleistocene. The basal part of this last interval, deposited between the Cobb Mountain Subchron and the end of the Jaramillo Subchron (1.18–0.99 Ma), evidently accumulated most rapidly, 94 mm/10^3 yr. The mean accumulation rate for the entire upper fluvial unit in borehole RZ-01 is 58 mm/10^3 yr.

The record for boreholes RZ-02 and RZ-03 (Fig. 8) again indicates that the sedimentation rate increased dramatically following the deposition of the lower alluvial unit. Although the data from RZ-02 or RZ-03 alone do not define the accumulation rate of the lower unit, the combined data projected onto borehole RZ-03 restricts the rate to <22 mm/10^3 yr between the Olduvai and Cobb Mountain Subchrons (1.77–1.18 Ma). Thus, the relatively slow accumulation of the lower alluvial unit that was determined for Pliocene-age deposits near the southeast corner of GAFB also applies to early Pleistocene-age deposits of this unit at the north end of the air base. By contrast, data from borehole RZ-02 indicate that early Pleistocene-age deposits of the middle lacustrine unit accumulated nearly five times faster, 100 mm/10^3 yr, between the Cobb Mountain and Jaramillo Subchrons (1.18–1.05 Ma). The overlying basal strata of the upper fluvial unit accumulated at a similar elevated pace during the Jaramillo Subchron (1.05–0.99 Ma). The accumulation rate slowed before the onset of the Brunhes Chron (0.78 Ma), after which the bulk of the upper fluvial unit in boreholes RZ-02 and RZ-03 was deposited at the modest rate of 44 mm/10^3 yr.

The foregoing observations show that between ca. 1.2 and 1.0 Ma a pulse of relatively rapid sedimentation affected much of GAFB (sites of boreholes RZ-01, -02, and -03) and presumably also extended into surrounding areas of the basin. This implies that deposition was affected by a significant paleotectonic or paleoenvironmental event. Two possible explanations include arching of the westernmost San Bernardino Mountains that began ca. 1.5 Ma (Meisling and Weldon, 1989; Kenney and Weldon, 1999), and an episode of humid regional climate between 1.3 and 1.0 Ma that is documented at Searles Lake basin east of the southern Sierra Nevada (Smith, 1984). Uplift of the source area and a more humid climate may have worked in concert to boost the volume of sediments that was delivered to the basin.

The magnetostratigraphic data also reveal that the rate of accumulation of the upper fluvial unit varied laterally across GAFB. Deposits of this unit accumulated significantly more rapidly near borehole RZ-01 as compared to boreholes RZ-02 and RZ-03. Deposition after the beginning of the Jaramillo Subchron (1.05 Ma) produced ~64 m of strata in borehole RZ-01 and ~53 m in each of boreholes RZ-02 and RZ-03. These thicknesses correspond to mean sedimentation rates of 65 mm/10^3 yr in RZ-01 and 54 mm/10^3 yr in RZ-02 and RZ-03. The deposits possibly accumulated more rapidly toward the southeast because this direction leads up-gradient toward the inferred source area in the San Bernardino Mountains. Later in this report, we exploit this lateral gradient in sedimentation rate in order to date the inception of the ancestral Mojave River.

Magnetostratigraphic and paleopedologic evidence suggests that the older, Pliocene-age, deposits of the lower alluvial unit also accumulated at a variable rate across GAFB. As was mentioned earlier, this unit evidently accumulated at 12 mm/10^3 yr near borehole RZ-01 (Fig. 7). However, late Pliocene-age deposits in borehole RZ-04, 4 km to the northwest, may have accumulated as rapidly as 90 mm/10^3 yr between ca. 3.2 and 3.1 Ma (Fig. 9). This finding is tentative, because the cited accumulation rates from both boreholes are based on provisional magnetostratigraphic correlations. However, paleopedologic evidence points to a similar conclusion. Carbonate crusts and other signs of advanced pedogenesis are common in the lower alluvial unit in borehole RZ-01 (Fig. 5), whereas comparable features are rare in borehole RZ-04. Thus, Pliocene sediments evidently accumulated more rapidly at the site of borehole RZ-04 than at borehole RZ-01.

Although the sediment accumulation rates estimated from GAFB cores vary significantly, even the fastest rates of ~100 mm/10^3 yr associated with the middle lacustrine unit and upper fluvial unit are sluggish compared to characteristic values of 300–700 mm/10^3 yr reported for closed Pleistocene basins in the western United States (Dohrenwend et al., 1991, p. 342). Paleogeographic and structural factors may account for this difference. Located near the north edge of the Victorville basin (Fig. 1), GAFB is fairly remote from major sediment sources in the Transverse Ranges, >30 km to the south, and is also removed from the structural axis and depocenter of the basin. Local evidence of syndepositional arching exists directly north of GAFB (see "Folding at northern margin of the Victorville basin"), and it is likely that a broader region straddling the air base was simultaneously gently uplifted relative to the basin axis. Furthermore, the Victorville basin drained externally during much of its history and possibly never developed deep topographic closure. It probably drained to the east periodically during the late Pliocene and early Pleistocene (see "Suspected former drainage connection between Victorville region and Lucerne Valley"), and unimpeded drainage to the north prevailed during the middle and late Pleistocene (see "Escape of the ancestral Mojave River from the Victorville basin"). We previously noted that the middle lacustrine unit apparently consists mostly of wetland or shallow-lacustrine sediments rather than deposits of deep perennial lakes. In combination, the distance from source area, through-flowing drainage, minimal topographic closure, and proximity to an uplifted basin margin may explain the slower sedimentation at GAFB, as compared to the closed Pleistocene basins.

Regional correlation with deposits north and east of GAFB

The stratigraphic succession in the subsurface at GAFB correlates with widespread deposits to the north and east in the southern and central Mojave Desert. The regional outcrop distribution of key Pliocene and Pleistocene sedimentary units identified by our field survey (Fig. 12) differs significantly from that mapped by previous workers (Dibblee, 1960a, 1960c, 1967; Bortugno and Spittler, 1986). Deposits of clay, silt, and fine sand that are equivalent to the middle lacustrine unit crop out locally in the Mojave River bluffs directly east of the air base (Fig. 12, older lake or wetland deposits) but are absent downstream to the north. Two unnamed major units of sand and gravel equivalent to the lower alluvial and upper fluvial units crop out extensively in vertical succession northward to Hinkley Valley (Fig. 12, older stream and fan deposits, and ancestral Mojave River deposits). The thickness of the upper unit varies from 8 to 25 m in the river bluffs between GAFB and Iron Mountain. Outcrops of the lower unit are as much as 100 m thick along this stretch of the Mojave River valley, and gravity data (Mabey, 1960; Biehler et al., 1988; Subsurface Surveys, 1990) suggest that a much greater thickness is concealed in the subsurface. Directly southeast of Barstow, geophysical and borehole studies suggest that deposits of the lower unit concealed beneath the Mojave River channel and bordering alluvial fans may be as thick as 600 m (Zohdy and Bisdorf, 1994; Densmore et al., 1997, Figure 4, unit QTof).

The two unnamed major units of sand and gravel mentioned above are generally separated by an erosional unconformity and are distinguished by contrasts in geologic structure, geomorphic

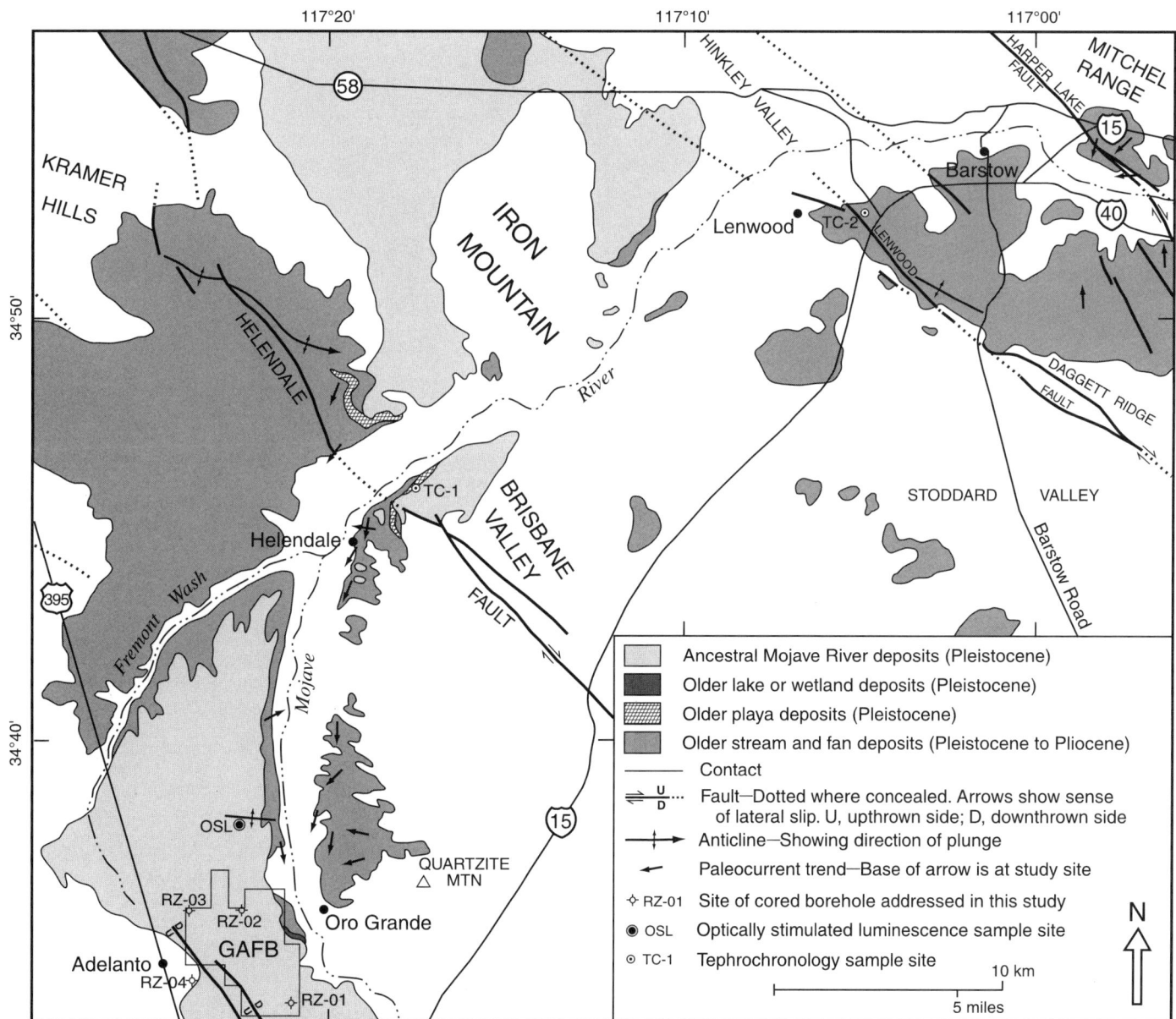

Figure 12. Outcrop distribution of key Pliocene and Pleistocene sedimentary units between former George Air Force Base and Barstow, California. Arrows indicate paleocurrent directions within unit of older stream and alluvial-fan deposits, as determined from imbricated clasts. Also shows tephrochronology sample sites (TC-1 and TC-2).

expression, sedimentary facies, and detrital provenance. The lower unit is prominently folded and faulted, and erosion has destroyed nearly all vestiges of its primary alluvial topography. It consists of alluvial-fan and fluvial deposits of lithic-arkose sand and polymictic gravel derived from assorted plutonic, volcanic, and metamorphic rocks of the central and southern Mojave Desert. Playa deposits of calcareous and gypsiferous clay, silt, fine sand, and tufa are intercalated within it near Helendale and Lenwood (Dibblee, 1960a, 1960c); the playa deposits near Helendale are shown on Figure 12. The upper unit, by contrast, is relatively undeformed and, although deeply incised by erosion, locally preserves broad remnants of a former depositional surface that apparently correlates with the George surface at GAFB. It mainly consists of fluvial deposits of arkosic sand and gravel. By analogy to compositionally similar sediments in the active channel of the Mojave River, the old fluvial deposits evidently were derived from plutonic and gneissic rocks of the northwestern San Bernardino Mountains.

To clarify the age and correlation of the lower stratigraphic unit between GAFB and Barstow, we investigated three intercalated layers of silicic volcanic ash near Helendale and Lenwood (Fig. 12, locs. TC-1, TC-2). Staff of the U.S. Geological Survey Tephrochronology Project at Menlo Park, California, performed microprobe analyses and evaluated potential geochemical correlations. Each layer consists of unweathered, white to light-gray, fine-grained vitric ash. Two beds of air-fall ash were sampled from an upward-coarsening sequence of playa and distal alluvial-fan deposits ~30 m thick located northeast of the Lenwood fault (Fig. 12, loc. TC-2). The lower ash bed, which is ~12 cm thick, is exposed in a fresh road cut on the southwest side of State Highway 58 (lat. 34°52.63′ N, long. 117°5.00′ W; elevation 710 m). This bed lies along the contact between a thick interval of gypsiferous silt and clay that forms the lower part of the road cut, and an overlying thick interval of poorly sorted muddy sand and silt. The upper ash bed is ~8 cm thick and crops out near the top of the interval of muddy sand and silt and ~470 m south of the lower ash bed (lat. 34°52.39′ N, long. 117°4.94′ W; elevation 738 m). Deposits of pebbly lithic-arkose sand and polymictic pebble gravel that crop out several meters above the upper ash bed cap the local stratigraphic succession. The upper ash bed is tentatively correlated geochemically with an ash layer at Honey Lake, California, that was erupted from a Cascadian source sometime between 2.06 and 2.5 Ma (ash bed CDGD-5; A. Sarna-Wojcicki, 2000, written commun.). The lower ash bed is chemically similar and may represent an earlier eruption from the same source.

A layer of stream-deposited ash as much as 50 cm thick crops out discontinuously on the southeast wall of the Mojave River canyon northeast of the Helendale fault (Fig. 12, loc. TC-1; lat. 34°45.97′ N, long. 117°17.70′ W; elevation 766 m). This ash bed underlies a sequence of arkosic sand and gravel 15–25 m thick that correlates lithologically with the upper fluvial unit at GAFB. It overlies playa deposits of calcareous sand, silt, and clay 0–6 m thick. The playa deposits in turn overlie and intertongue northwestward into deposits of poorly stratified lithic-arkose sand and polymictic pebble-cobble gravel whose thickness exposed in the canyon walls is ~30 m. The latter deposits resemble alluvial-fan facies observed in the lower alluvial unit at GAFB. The ash has an unusually shallow and westerly magnetic direction (Table DR1, outcrop 5). It is tentatively correlated geochemically with the PIC0-40A ash bed near Ventura, California, which is restricted between 0.9 and 1.2 Ma and has chemical affinity to ashes from the Long Valley region (A. Sarna-Wojcicki, 1998, written commun.).

Overall, the tephrochronologic evidence suggests that the unnamed lower unit of stream, alluvial-fan, and playa deposits between Helendale and Barstow, more specifically the upper parts of that unit that are exposed along the Mojave River, accumulated between ca. 2.5 Ma and 1 Ma. Thus, the deposits apparently are late Pliocene to early Pleistocene in age and correlate with the upper parts of the lower alluvial unit in the GAFB boreholes (Figs. 5 and 11). Owing to the diachronous nature of sedimentation at the air base, the lower unit between Helendale and Barstow also correlates temporally with parts of the middle lacustrine unit and upper fluvial unit. The playa deposits near the Helendale fault superficially resemble the middle lacustrine unit at GAFB and probably are similar in age, but they evidently accumulated in a separate, relatively saline basin that mainly lay northeast of the Helendale fault. The upper unit of fluvial deposits between Helendale and Barstow is poorly dated, but it apparently is equivalent to the upper part of the upper fluvial unit at GAFB, based on comparable stratigraphic position, detrital texture, provenance, and geomorphic expression. Its basal deposits are younger than 0.9–1.2 Ma, as constrained by the underlying ash bed near Helendale. The OSL age of ca. 60–70 ka, which was determined for fluvial deposits directly beneath the George surface near GAFB, may approximately date the top of the unit, assuming that fluvial sedimentation ended more or less synchronously between GAFB and Barstow.

Other correlative deposits crop out in the Lake Manix basin (Fig. 1), ~90 km northeast of GAFB. Previous workers distinguished two superposed major stratigraphic units near the intersection of the Mojave River and Manix Wash (Fig. 11). In the basal unit, the Mojave River Formation of Nagy and Murray (1991, 1996), sand and sparse pebble-cobble gravel of distal alluvial-fan origin overlie and intertongue with gypsiferous silt and clay deposited in a playa. The thickness of the outcropping strata is ~95 m, and additional deposits are concealed in the subsurface. The alluvial-fan sediments consist of angular, locally derived rock detritus, including abundant volcanic debris (Buwalda, 1914; Nagy and Murray, 1991). A similar assemblage of alluvial-fan and playa deposits occupies the southern margin and subsurface of Mojave Valley east of Barstow (Dibblee, 1970; Cox and Wilshire, 1994; Densmore et al., 1997, Figures 3 and 5, section B–B′, unit QTof). The upward-coarsening succession of playa and distal alluvial-fan deposits at Manix Wash also closely resembles the ash-bearing deposits northeast of the Lenwood fault (Fig. 12, loc. TC-2). Tephrochronologic and magnetostratigraphic evidence suggests that the Mojave River Formation accumulated approximately 2.1 to 1 Ma (Nagy and Murray, 1991, 1996; Pluhar et al., 1991). Thus, the formation apparently correlates with the unnamed lower unit between Helendale and Barstow, and also with parts of the lower alluvial, middle lacustrine, and upper fluvial units at GAFB (Fig. 11).

The upper stratigraphic unit at Manix Wash consists of the Manix Formation of Jefferson (1985, 1994, this volume), which has a maximum exposed thickness of 39 m and comprises four lithostratigraphic units (Fig. 11, members A–D). Member A constitutes basal deposits of sand and boulder-bearing cobble gravel that accumulated on alluvial fans. These deposits consist of poorly sorted, subangular lithic sand and polymictic gravel derived from local bedrock sources. They contain abundant volcanic detritus and thus share a common provenance with the distinctly finer grained alluvial-fan deposits of the underlying Mojave River Formation. Overlying members B–D comprise intertonguing deposits of silt, clay, sand, and granule to cobble gravel that accumulated in fluviodeltaic and lacustrine environments. By contrast to member A, the sand and gravel in members B and D typically are moderately to well sorted and contain abundant subrounded detritus of felsic plutonic rocks including granodiorite (Jefferson, 1985). The upper three members also contain fairly abundant microfossils and fossil vertebrates (Jefferson, this volume).

Isotopic, paleontologic, and tephrochronologic data suggest that the Manix Formation accumulated between 1 Ma and 18 ka (Jefferson, 1985, 1994, this volume; Meek, 1989, 1999; Nagy and Murray, 1996). Thus, it apparently is largely coeval with sections of the upper fluvial unit drilled near the north end of GAFB (Fig. 5, boreholes RZ-02 and RZ-03). However, its youngest deposits were laid down as much as 50,000 yr after the upper fluvial unit ceased accumulating at GAFB. The abrupt change in detrital provenance near the base of member B, which is analogous to the compositional change between the lower alluvial unit and upper fluvial unit at GAFB, evidently occurred ca. 500 ka, based on tentative correlation with marine oxygen isotope stages (Jefferson, 1985, 1994, 1999, this volume). Member D, which consists of relatively coarse sand and gravel of fluviodeltaic origin, is ca. 60 ka at its base, as judged by U/Th-series ages of fossil bone fragments (Jefferson, this volume). This age is similar to the optically stimulated luminescence ages of ca. 60–70 ka that we obtained from the top of the upper fluvial unit near GAFB.

Regional correlation with deposits south of GAFB

The strata drilled at GAFB are broadly coeval with Pliocene and Pleistocene sediments that crop out along the north flank of the San Bernardino Mountains and eastern San Gabriel Mountains (Fig. 11). In particular, the upper fluvial unit at GAFB is contemporaneous with the Victorville Fan deposits and Ord River deposits of Meisling and Weldon (1989), and probably also with the poorly dated Cushenbury Springs Formation of Shreve (1968) at Blackhawk Canyon. The upper fluvial unit and Ord River deposits share the same provenance, each consisting mostly of granitic and gneissic detritus derived from the northwestern San Bernardino Mountains. However, the upper fluvial unit is predominantly sandy, rather than gravelly, and therefore appears to be the distal facies equivalent of the Ord River deposits (Cox et al., 1998). The upper fluvial unit also appears to range to a significantly younger age than the Ord River deposits (Fig. 11). The upper fluvial unit and Ord River deposits both correlate temporally, but not lithologically, with the Victorville Fan deposits. The latter deposits are dissimilar because they contain abundant detritus of Pelona Schist, Lowe Granodiorite, and other rocks of the San Gabriel Mountains (Foster, 1980; Meisling and Weldon, 1989).

The upper fluvial unit and Victorville Fan deposits each are time-transgressive, but the upper fluvial unit spans a broader interval of time (Fig. 11). The oldest Victorville Fan deposits probably are ca. 1.7 Ma (base of the Harold Formation at Crowder Canyon) (age revised from Weldon et al., 1993). By contrast, our work suggests that the upper fluvial unit ranges back to 1.95 Ma at GAFB, and its southeastern extension beneath Victorville may be considerably older, as is explained in the following section. The accumulation of the Victorville Fan deposits apparently ended during the middle Pleistocene. The uppermost unit in the succession (older alluvium of Noble, 1954) contains the Brunhes/Matuyama boundary (0.78 Ma) and is cut by an eastward-sloping inset stream terrace dated at ca. 0.5 Ma (Meisling, 1984; Weldon, 1986; McFadden and Weldon, 1987; Weldon et al., 1993). The age of the stream terrace (and by implication, that of the older alluvium as well) is estimated in part from a thick (15 m) argillic soil horizon that caps the incised alluvial fill (McFadden and Weldon, 1987). By contrast, the much thinner (0.75 m) paleosol on the George surface, the optically stimulated luminescence ages of ca. 60–70 ka determined for underlying fluvial deposits, and the considerable depth of the Brunhes/Matuyama polarity boundary, all suggest that sedimentation continued into the late Pleistocene near GAFB.

In discussing the age of the George surface, we noted that the associated paleosol and OSL ages suggest a correlation with the Qoa-d stream terrace of McFadden and Weldon (1987) at Cajon Creek, which they dated at 55 ± 12 ka. This interpretation conflicts with a correlation proposed by Meisling (1984). He inferred that the Qoa-d terrace at Cajon Creek is equivalent to a stream terrace that is inset well below the level of the George surface at Summit Valley and Apple Valley. However, the inset terrace is capped by a relatively poorly developed soil (Williamson and Wells, 1994). Thus, the George surface seems to be a better match for the Qoa-d terrace.

Pliocene to lowest Pleistocene strata at GAFB correlate temporally with the Phelan Peak Formation of Weldon et al. (1993) near Cajon Pass and Deep Creek, and with the Old Woman Sandstone of Shreve (1968) at Blackhawk Canyon and neighboring areas on the north flank of the central San Bernardino Mountains (Fig. 11). As a result of time-transgressive sedimentation at GAFB, the lower alluvial unit, middle lacustrine unit, and upper fluvial unit all overlap in age with the Phelan Peak Formation, but the long-ranging lower alluvial unit is the closest match. The lower alluvial unit and the Phelan Peak also are physically comparable, to the extent that they each contain abundant alluvial-fan deposits with interspersed paleosols and calcite-cemented horizons (Meisling and Weldon, 1989). However, the detrital composition of the two units differs significantly, which indicates they were derived from separate source areas. Green clay reported in the lower part of the Phelan Peak Formation (Foster, 1980; a subfacies within his "western facies of Crowder Formation") is reminiscent of argillaceous sediments in the middle lacustrine unit at GAFB but is significantly older, ca. 3.8 Ma, based on magnetostratigraphic and radiometric data (Weldon et al., 1993).

The stratigraphic succession at GAFB is comparable in age to the San Timoteo Formation of Fraser (1931), which lies on the opposite side of the central Transverse Ranges southeast of San Bernardino (Figs. 1 and 11; San Timoteo Badlands). The San Timoteo Formation comprises an upward-coarsening succession of Pliocene and Pleistocene sand and gravel deposited by braided streams and associated alluvial fans that drained southward from the ancestral Transverse Ranges (Matti and Morton, 1993; Albright, 1999). Although they are broadly coeval with the San Timoteo Formation, the strata at GAFB and other sites north of the San Bernardino Mountains differ in one significant respect— their Pliocene-age deposits were mainly derived from sources lying *outside* the area of the Transverse Ranges (Fig. 11).

SOUTHEASTERN EXTENSION OF THE CLASTIC WEDGE

Our stratigraphic analysis of the upper fluvial unit at George Air Force Base (GAFB) suggests that it is a diachronous, wedge-shaped body deposited by the ancestral Mojave River (Fig. 5). The bottom of the unit deepens and increases in age to the southeast beneath the air base, which suggests it may be even deeper and older beneath Victorville and Hesperia. To clarify the geometry and paleogeographic significance of this clastic wedge, we attempted to determine its subsurface configuration beneath a larger area of the Victorville basin.

Because of its relatively fine-grained clayey texture, the middle lacustrine unit at GAFB is readily distinguished from adjacent sandy units on borehole electrical logs. Drilling records from boreholes RZ-01, -02, -03, and -04 at GAFB (S. Chrisley, 1997, written commun.) indicate that the resistivity (16-inch normal) of the middle lacustrine unit is uniformly low, mostly 5–10 Ωm (ohm-meters), excluding deposits in borehole RZ-02, where abundant sand layers yield higher resistivity values. By contrast, the resistivity of the adjacent units is typically much higher, ranging from 5 to 65 Ωm for the lower alluvial unit and 15 to 100 Ωm for the upper fluvial unit.

Using the resistivity logs as a correlation tool, we found that the stratigraphic succession at GAFB can be traced southeastward beneath central and southern Victorville (Figs. 2 and 13; Green Tree Well, Test Well #32). Several wells drilled at the municipal golf course in central Victorville encountered two major units: an upper unit of interlayered sand, silt, and minor clay 155–180 m thick, and a lower unit of greenish to brownish clay and silt at least 21 m thick (S. Dickey, 1997, personal commun.). In each of the wells, drilling was terminated before the base of the lower unit was reached. This cluster of wells is represented on Figure 13 by the Green Tree well, which intersected the top of the clay-rich unit at an elevation of ~733 m. Resistivity logs from the wells indicate values of 15–70 Ωm for the upper unit and 2–5 Ωm for the lower unit (S. Dickey, 1998, written commun.). Based on their lithologic and geophysical characteristics, the upper and lower units at the Victorville golf course seem to correlate with the upper fluvial unit and middle lacustrine unit at GAFB.

Test Well #32 in southern Victorville penetrated three major units. An upper unit 232 m thick with resistivity of 10–75 Ωm seems to correspond to the upper fluvial unit at GAFB. A middle unit ~37 m thick with resistivity of 2–5 Ωm is correlated with the middle lacustrine unit. Finally, a lower unit ~58 m thick with resistivity of 5–15 Ωm is correlated with the lower alluvial unit.

The results of the foregoing analysis suggest that the upper fluvial unit thins steadily to the northwest between southernmost Victorville and the northern margin of GAFB (Fig. 13). Thus, the northwestward thinning that was initially diagnosed at GAFB (Fig. 5) evidently is symptomatic of a broader stratigraphic pattern. The wedge-shaped configuration of the upper unit implies that the fluvial tract of the ancestral Mojave River prograded to the northwest and may have reached southern Victorville long before GAFB. Therefore, in order to document the earliest history of the river, chronologic evidence is needed from the deeper, more southerly deposits of the wedge.

The deepest, and presumably oldest, deposits along the extended transect of the clastic wedge illustrated in Figure 13 are encountered in Test Well #32. Although we have no direct means of dating these deposits, the age of the base of the wedge can be estimated by applying the sediment accumulation rates determined at GAFB. In a previous section, we noted that the mean accumulation rate of the upper fluvial unit increases southeastward across GAFB. Sediments deposited since the beginning of the Jaramillo Subchron (1.05 Ma) accumulated at 54 mm/10^3 yr in boreholes RZ-02 and RZ-03, and 65 mm/10^3 yr in borehole RZ-01. Boreholes RZ-01 and RZ-03 are ~6.0 km apart. Thus, the rate increases southeastward at ~2.4 mm/10^3 yr/km.

The complete thickness of the upper fluvial unit in borehole RZ-01 was deposited at a mean rate of 58 mm/10^3 yr. If this rate increased southeastward across Victorville at 2.4 mm/10^3 yr/km, then the fluvial succession at the site of Test Well #32, 11 km to the southeast, would have accrued at 76 mm/10^3 yr. Before the George surface was eroded, the upper fluvial unit was ~244 m thick at the site of Test Well #32. If the unit accumulated at 76 mm/10^3 yr until 65 ka, then the age of its base would be ca. 3.3 Ma.

Stratigraphic relations at the northwest front of the San Bernardino Mountains further define the age and geographic extent of the fluvial wedge. Meisling (1984), and Meisling and

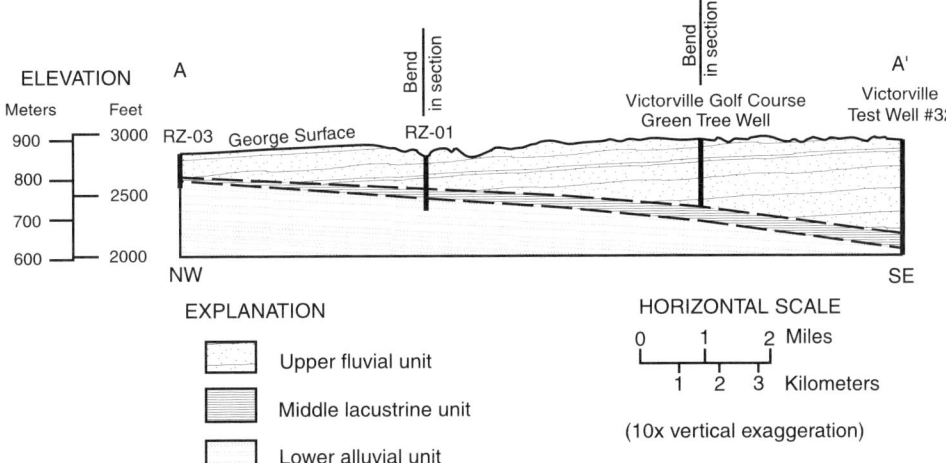

Figure 13. Northwest-southeast cross section between former George Air Force Base and southern Victorville, California, showing southeastward descent of middle lacustrine unit beneath wedge-shaped mass of upper fluvial unit. Line of section is shown on Figure 2. Vertical scale is 10 times horizontal scale.

Weldon (1989) proposed that their Ord River deposits—more specifically, a deeply dissected subunit of fluvial sand and gravel near the mouth of Deep Creek (Deep Creek facies of the Ord River gravel; Meisling, 1984)—contain the earliest deposits of the ancestral Mojave River. These sediments have reversed paleomagnetic polarity, which implies they are early Pleistocene or older (Meisling, 1984). Judging from their antiquity and granitic provenance, they probably are the coarse-grained proximal counterpart of the upper fluvial unit at GAFB (Cox et al., 1998). Thus, the Ord River deposits apparently represent the southeastern end of the clastic wedge.

Nearby in eastern Summit Valley (Fig. 2), the Ord River deposits overlie fine-grained volcaniclastic sandstone, siltstone, and claystone of the Phelan Peak Formation (Meisling, 1984, his "volcanogenic eastern facies of the Crowder Formation"). A layer of volcanic ash in the latter unit yielded an apatite fission-track age of 3.8 ± 0.4 Ma (Meisling and Weldon, 1989). Together, this radiometric age and our estimated age for the base of the upper fluvial unit in Test Well #32 seem to restrict the basal deposits at the proximal end of the clastic wedge to between 3.8 and 3.3 Ma. The outcrops near Deep Creek and in eastern Summit Valley are located 13–16 km southeast of Test Well #32 (Fig. 2). The expansion of the clastic wedge across this considerable distance possibly consumed much of the half million years between 3.8 and 3.3 Ma. Thus, the oldest river deposits may have accumulated near the beginning of this time interval.

If the foregoing interpretation is correct, then the Phelan Peak Formation and Ord River deposits probably are not separated by a major depositional or erosional hiatus. Meisling and Weldon (1989, Figure 4) demonstrated from magnetostratigraphy and intertonguing relations that the Ord River deposits and Victorville Fan deposits are coeval, and they suggested that both units unconformably overlie the Phelan Peak Formation. Based on subsequent field studies, however, Kenney and Weldon (1999) concluded that the Victorville Fan deposits actually rest conformably on the Phelan Peak Formation near Phelan Peak and Cajon Pass. This latter finding implies that the alleged unconformity between the Ord River deposits and Phelan Peak Formation also warrants reevaluation. Meisling (1984, p. 48) observed that the predominantly volcaniclastic Phelan Peak Formation in eastern Summit Valley contains intercalated beds of granitic gravel derived from the San Bernardino Mountains region. These beds evidently are similar in composition to the Ord River deposits and thus might constitute even older deposits of the ancestral Mojave River. Under this scenario, the river may have begun constructing an embryonic clastic wedge as early as 3.8 Ma, while the Phelan Peak Formation was still accumulating near the future site of Summit Valley.

DEFORMATIONAL STRUCTURES AND ASSOCIATED CRUSTAL MOVEMENTS IN THE SOUTHERN MOJAVE DESERT

Our field observations and borehole studies disclosed new geomorphic, stratigraphic, and structural evidence of vertical crustal movements along the Mojave River between Victorville and Iron Mountain. Here we summarize some of the key evidence demonstrating that the early Mojave River evolved within a dynamic tectonic setting, and that vertical movements north of George Air Force Base (GAFB) helped to control the river's advancement across the southern Mojave Desert.

Folding at northern margin of the Victorville basin

The Victorville basin as defined by gravity data and basement topography ends directly north of GAFB, where the upper levels of the basin fill extend northward between Quartzite Mountain and the southeastern Shadow Mountains (Fig. 2). Several noteworthy stratigraphic and structural features are localized in this area. The main item is a low, east-west–trending, southward-facing anticline that intersects the west wall of the Mojave River canyon (Fig. 6). Deformation is most pronounced in the lower alluvial unit on the south limb of this fold, where strata dip as steeply as 8° and probably descend as much as 60 m toward the south. Higher on the same limb, the base of the upper fluvial unit dips nearly 1° and descends ~30 m to the south, and the capping George surface drops ~14 m to the south. Strata on the broad north limb of the fold dip northward very gently, generally <1°. This limb is modified by two minor north-facing monoclinal flexures that deform the base of the upper fluvial unit without deflecting the overlying George surface.

Stratigraphic and structural relations imply that the folding north of GAFB was contemporaneous with sedimentation and affected patterns of sediment accumulation. Uplift resulting from the folding evidently physically restricted the deposition of the middle lacustrine unit, which pinches out low on the south limb of the fold (Fig. 6). This suggests that folding began before ca. 1.18–1.05 Ma, which is the estimated depositional age of the lacustrine unit in nearby borehole RZ-02. Directly north of the pinch-out, the upper fluvial unit overlies the lower alluvial unit with pronounced angular unconformity. The upper unit thins northward from ~45 m at the base of the southern limb, to 14 m near the crest of the fold. This abrupt variation in thickness probably resulted in large part from the deposition of the upper fluvial unit against a preexisting fold-generated slope. Thus, the uplift of the basin margin caused by folding may have delayed the advancement of the ancestral Mojave River northward from the Victorville region. The offset in the George surface indicates that the most recent episode of folding occurred after ca. 60–70 ka.

Syndepositional faulting at GAFB

Evidence from boreholes indicates that the folding directly north of GAFB was synchronous with faulting near the southwest corner of the air base (Figs. 5 and 6). Strata in borehole RZ-04 are consistently elevated relative to equivalent strata in the other three boreholes (Fig. 5). The magnitude of the stratigraphic offsets increases down the hole, which implies that some of the deformation was contemporaneous with sedimentation. Strata near the middle and top of the upper fluvial unit appear to be uplifted ~15 m relative to equivalent strata in borehole RZ-01. At deeper levels, the middle lacustrine unit lies ~20–25 m higher in borehole RZ-04 than in RZ-02 and RZ-03. We previously noted (see "Age and local correlation of the middle lacustrine unit") that the wetland or shallow lake deposits sampled in the latter three boreholes are apparently very similar in age. This temporal equivalence, combined with spatial proximity and comparable thickness, suggests that the lakebeds in these boreholes were originally deposited at approximately the same elevation. Therefore, tectonic displacement, rather than paleotopography or diachronous sedimentation, probably best explains the anomalous elevation of the lakebeds in borehole RZ-04.

The magnetostratigraphy of the lower alluvial unit in borehole RZ-04 is dominated by thick normal-polarity zones (Figs. 5 and 9). We tentatively correlate this polarity sequence with the Gauss Chron, which ranges from 3.58 to 2.58 Ma (Fig. 3). If this correlation is accurate, and if the middle lacustrine unit began accumulating ca. 1.18 Ma at this site, then the lower alluvial unit and middle lacustrine unit evidently are separated by a hiatus of at least 1.4 m.y. in this borehole (2.58–1.18 Ma). At least 30 m of strata seems to be missing owing to erosion or nondeposition. The corresponding unconformity probably is located at the base of a thin interval of fine sand and silt that directly underlies the deposits of lacustrine clay and silt in borehole RZ-04 (Fig. 5). The fine sand in this interval consists of unweathered, biotite-bearing arkose that is unlike sand of the lower alluvial unit but indistinguishable from sand of the upper fluvial unit.

Together, the prelacustrine hiatus and the offset strata within the middle lacustrine and upper fluvial units suggest that the site of borehole RZ-04 was cumulatively uplifted at least 50 m relative to the other boreholes. The corresponding structural discontinuity that separates borehole RZ-04 from the other boreholes is represented schematically by a single vertical fault on Figure 5. In reality, at least two faults apparently are involved. The existence of faults is supported by our identification of two northwest-trending aerial-photographic lineaments between boreholes RZ-01 and RZ-04 (Figs. 4 and 6). These lineaments are parallel to the main strand of the Mirage Valley fault zone and project toward a northern splay of the zone that crosses the Shadow Mountains ~13 km northwest of borehole RZ-04 (Fig. 2).

Geomorphic evidence of uplift is locally evident at a site southwest of the lineaments and ~2.2 km south of borehole RZ-04 (Figs. 2 and 6, southern end of section B–B′). An elevated topographic bench at this site is underlain by granitic gravel and arkosic sand of the upper fluvial unit. The bench apparently is an uplifted outlier of the George surface. The elevation of the bench as estimated from topographic contours is ~903 m. By contrast, adjacent remnants of the George surface directly east of the lineaments lie at ~891 m. Thus, the bench apparently was uplifted ~12 m by faults that coincide with the lineaments. This is comparable to the vertical displacement of 15 m that we have estimated from offset strata in the upper half of borehole RZ-04.

The vertical movements ascribed to faulting between boreholes RZ-01 and RZ-04 compare closely in magnitude and timing to the uplift produced by folding directly north of GAFB. In either area, the estimated net displacement is within the range of 50–60 m, and in each case the bulk of the displacement occurred before 1.18 Ma. Furthermore, in both areas the most recent vertical movements have offset the George surface 12–15 m, probably during the past 70,000 yr. From the consistency of these relations, it appears that two distinct tectonic pulses, during the early Pleistocene and latest Pleistocene, deformed the region surrounding GAFB.

There may be additional faults that escaped detection by our limited survey of four boreholes at GAFB. However, we found no evidence of significant vertical displacement between boreholes RZ-01, -02, and -03. For example, our data do not support a previous interpretation that ascribed the anomalous depth and thickness of the lakebeds in borehole RZ-01 to syndepositional faulting or folding (Chrisley, 1997, Figure 40). Although such an interpretation plausibly explains the lateral variation in the depth and thickness of the middle lacustrine unit, it does not account for the contrast in the age of this unit that is indicated by magnetostratigraphy (Fig. 5) and disparate ostracode assemblages (R. Forester, 1999, written commun.).

Uplift by monoclinal folding along the Helendale fault

Our field studies also revealed evidence of recurrent vertical displacement along the Helendale fault near its intersection with the Mojave River. The Helendale fault bounds the northeast side of a large basement ridge ~15 km northeast of GAFB (Fig. 2). This ridge terminates directly east of the Mojave River, where its northwest tip and the bordering fault zone are buried by a succession of continental sediments that probably first accumulated during the late Pliocene. Vertical movement on the fault zone at depth has deformed the overlying sediments in this area into a gently dipping, northeast-facing monoclinal flexure. The geometry of the monocline suggests that the associated concealed strand of the Helendale fault might be a southwest-dipping reverse fault. Other evidence of crustal contraction at a high angle to the Helendale fault is found nearby on the west side of the Mojave River, where the crest of a large southeast-plunging anticline locally nearly parallels the trace of the fault (Figs. 2 and 12) (Bowen, 1954; Dibblee, 1960a, 1967).

Deposits exposed in the monocline and in neighboring flat-lying sections east of the Mojave River comprise three superposed units: a lower unit of lithic-arkose sand and polymictic pebble-cobble gravel, a middle unit of calcareous sand, silt, and clay, and an upper unit of arkosic sand and gravel. These units are described in a preceding section of this paper ("Regional correlation with deposits north and east of GAFB") and their outcrop distribution is shown on Figure 12. Stratigraphic and structural relations within the monocline indicate that the southwest side of the Helendale fault was uplifted repeatedly before and after the arrival of the ancestral Mojave River. The cumulative uplift recorded by the deformed strata is at least 36 m. Tilted beds of the lower and middle units in the core of the monocline dip 10°–15° to the northeast and are ~18 m thick. They are unconformably overlain by deposits of the upper unit that dip northeastward at 2°–3°. Thus, the southwest side of the fault zone evidently was uplifted ~18 m between the deposition of the middle and upper units. There also appears to be an unconformity between the lower and middle units in the monocline, which suggests that an earlier episode of uplift occurred prior to the deposition of the middle unit. Based on the layer of volcanic ash that locally overlies the middle unit northeast of the monocline, this earlier movement apparently occurred before ca. 0.9–1.2 Ma. Uplift of the land surface southwest of the Helendale fault may have blocked the southward-flowing stream system that previously deposited the lower unit, thereby producing a local playa basin in which the middle unit accumulated.

The top of the upper fluvial unit (George surface) is deflected upward ~9 m as it crosses to the southwest side of the monocline, and the thickness of this unit decreases southwestward from 24 m to 15 m across the structure. Thus, the southwest side of the Helendale fault evidently was uplifted an additional 18 m after the ancestral Mojave River arrived in this region. Half of this displacement occurred after the Mojave River abandoned the George surface at ca. 60–70 ka. Overall, the magnitude and timing of the vertical displacements recorded by the monocline are comparable to those determined for the previously discussed folds and faults near GAFB.

Reversal of regional paleoslope north of GAFB

A combination of stratigraphic and structural features suggests that a broad crustal panel extending at least from GAFB to Iron Mountain was tilted gently to the north during the early Pleistocene, thus facilitating the advancement of the early Mojave River beyond the Victorville region. Throughout this region, sediments of the northward-flowing ancestral Mojave River overlie a composite unit deposited by a southward-flowing trunk stream and tributary alluvial fans (Fig. 12) (Cox et al., 1998). The old Mojave River deposits correlate with the uppermost deposits of the upper fluvial unit at GAFB, and the underlying unit correlates with the lower alluvial unit at the air base.

Deposits of the ancient southward-flowing stream crop out extensively east of the Mojave River between Oro Grande and Helendale, and they are locally exposed west of the river near GAFB and north of Helendale. Some of the best exposures are in road cuts of National Trails Highway near Helendale. The stream deposits consist of well-stratified, moderately sorted, coarse sand and gravel with a compositionally diverse assemblage of slightly to strongly abraded clasts. Imbricated clasts within the stream deposits indicate southward-flowing paleocurrents, whereas deposits of tributary alluvial fans in the same region are characterized by westward- and eastward-flowing paleocurrents (Fig. 12).

In contrast to the conformable succession observed in most of the boreholes at GAFB (excluding borehole RZ-04), deposits of the ancestral Mojave River crop out unconformably atop the older stream and fan deposits between GAFB and Iron Mountain. The contact is a disconformity or low-angle unconformity that seems to exhibit significant erosional relief. On the west wall of the Mojave River canyon between Oro Grande and Helendale, the contact lies 30–60 m below deposits of the older south-flowing stream system that crop out east of the river. Part of this difference in elevation can be ascribed to postdepositional deformation. However, after compensating for down-to-west displacement on several north-trending and north-northeast–trending faults east of the river, a residual difference in elevation of at least 30 m remains.

Between GAFB and Iron Mountain, the thickness of the ancestral Mojave River deposits fluctuates between 8 m and 25 m (e.g., Figure 6). As the unit neither thins to the north nor intertongues with underlying deposits, it apparently was not deposited on a southward-facing slope. Instead, the land surface probably was horizontal or sloped gently to the north. This implies that the former southward-inclined paleoslope was flattened or reversed by northward tilting before the ancestral Mojave River entered this region.

A very slight amount of tilting probably would have been sufficient to reverse the paleoslope. Excluding alluvial-fan systems and mountain streams, the gradients of large fluvial channels typically are a fraction of 1°. For example, the mean gradient of the modern Mojave River between GAFB and Iron Mountain is 3.3 m/km, or 0.2°. Imperceptibly slight tilting would reverse such a slope.

Subtle tilting on a large scale, if present, should be reflected by variation in the elevation of laterally extensive stratigraphic or geomorphic reference planes. Two features descend to the north in agreement with broad northward tilting. One of these is the upper boundary of sediments laid down by the ancient southward-flowing stream. East of the Mojave River between Oro Grande and Helendale, deposits of sand and gravel laid down by the old fluvial system are overlain by westward-sloping Pleistocene alluvial fans. The elevation of the intervening contact, as measured ~2 km east of the modern Mojave River floodplain, drops 73 m in 9.1 km between northwest ¼ sec. 5, T. 6 N., R. 4 W., and southwest ¼ sec. 4, T. 7 N., R. 4 W. This corresponds to a gradient of 8.0 m/km or a dip of 0.5° to the north. This evidence is consistent with northward tilting, but does not prove it; the structural relief along the contact might reflect erosion or displacement along unidentified faults.

The topography of modern basement ridges along the Mojave River is consistent with northward tilting of a crustal block extending 42 km from GAFB to Harper Lake (Fig. 2). The elevation of the highest summits decreases steadily to the north between Quartzite Mountain and Lynx Cat Mountain. A line between these two peaks dips northward at 14.5 m/km, or 0.8°. It seems unlikely that this series of peaks would have descended consistently to the north across this considerable distance during the Miocene and Pliocene, when the southern Mojave Desert region drained toward the south. Therefore, we suspect that the major summits along this north-south transect originally were more or less accordant, and that the present topographic pattern reflects differential uplift by broad northward tilting, supplemented by vertical displacement on structural discontinuities such as the Helendale fault. The net topographic relief along the transect implies that Quartzite Mountain and neighboring areas may have been uplifted as much as 600 m relative to Lynx Cat Mountain and Harper Lake.

The timing of the postulated tilting event can be deduced from stratigraphic relations near GAFB. The latest major episode of tilting probably postdates vigorous flow of the southward-draining stream system and predates the advancement of the ancestral Mojave River beyond the Victorville basin. In borehole RZ-02, thick deposits of moderately sorted coarse sand and gravel that we suspect were deposited by the southward-draining trunk stream system lie at least 69 m below the present land surface. Shallower deposits of the lower alluvial unit generally consist of poorly sorted, silty sand that more likely was deposited by local alluvial-fan drainages. The transition between these two sedimentary facies lies between the Olduvai and Cobb Mountain Subchrons, which suggests that vigorous southward streamflow ceased sometime between 1.77 and 1.18 Ma. In the next section, we propose that the Mojave River escaped from the Victorville basin and flowed north to Harper Lake at ca. 475–575 ka. Thus, significant northward tilting apparently occurred during the early to middle Pleistocene, between 1.8 and 0.5 Ma; however, we suspect that the complete episode of tilting spanned a much broader period between the early Pliocene and late Pleistocene.

PALEOGEOGRAPHIC EVOLUTION OF THE MOJAVE RIVER DRAINAGE BASIN

Origin and initial northward growth of the ancestral Mojave River

During the Pliocene and Pleistocene, a sedimentary basin near Victorville was filled by separate depositional systems that converged from the southeast, southwest, and north sides of the basin. Granitic detritus eroded from the San Bernardino Mountains was delivered to the southeast side of the basin by ancestral Deep Creek, which was the principal tributary of the ancestral Mojave River. The present-day Deep Creek drains much of the northwestern San Bernardino Mountains, and its late Pliocene to early Pleistocene precursor probably drained an even larger area (Sadler and Reeder, 1983). Coarse fluvial gravel deposited near the mountain front formed the Ord River deposits of Meisling and Weldon (1989), whereas distal deposits of gravel, sand, silt, and clay formed the middle lacustrine and upper fluvial units at George Air Force Base (GAFB).

Concentrated streamflow probably emerged at the northwest front of the San Bernardino Mountains before 3.3 Ma. The onset of the fluvial system as early as 3.8 Ma may be recorded by lenses of granitic gravel that Meisling (1984) observed amid fine-grained volcaniclastic sediments of the Phelan Peak Formation in eastern Summit Valley. Concurrently, ca. 4.4–2.5 Ma, fine-grained fluvial and lacustrine deposits composed of granitic and metamorphic detritus were accumulating near Phelan Peak (ages revised from Weldon et al., 1993, Figure 8).

Thus, the headwaters of the ancestral Mojave River evidently initially drained into the eastern part of the "Phelan Peak basin," which was a precursor of the Victorville basin. This east-west–trending trough was bordered to the southeast by the emerging San Bernardino Mountains, to the southwest by an area uplifted along the San Andreas fault, and to the north by a broad paleoslope down which streams drained southward from the Mojave Desert region (Meisling and Weldon, 1989). The upland to the southwest apparently was very low, because coarse clastic debris did not enter from this side of the basin before 2.5 Ma (Foster, 1980; Weldon et al., 1993).

The initial development of the ancestral Mojave River must have been influenced by the configuration of the Phelan Peak basin, including the position of its topographic axis. The structural axis of the modern Victorville basin underlies Hesperia (Fig. 2), where the sedimentary fill reaches its maximum thickness of ~1300 m (Mabey, 1960; Biehler et al., 1988; Subsurface Surveys, 1990; Spotila and Sieh, 2000). The topographic axis and depocenter of the Phelan Peak basin probably occupied this same region. Test Well #32 lies a short distance (6.5 km) north of the inferred center of the Phelan Peak basin, whereas GAFB lies well up on its northern flank. Our analysis of the clastic wedge of the ancestral Mojave River suggests that the fluvial tract prograded relatively rapidly across the south flank and center of the basin, advancing 16 km between eastern Summit Valley and Test Well #32 between 3.8 and 3.3 Ma (mean rate of 32 m/10^3 yr). Subsequently, the expanding wedge probably was restrained by the southward-inclined paleoslope on the north flank of the basin. According to our estimates, the fluvial tract spanned the 11-km gap between Test Well #32 and borehole RZ-01 between 3.3 and 1.95 Ma (8 m/10^3 yr), and it advanced 6 km between boreholes RZ-01 and RZ-03 between 1.95 and 1.05 Ma (7 m/10^3 yr).

The individual rates of fluvial progradation estimated for the two segments south of GAFB may be inaccurate, as they depend on extrapolated sedimentation rates and other assumptions (see "Southeastern Extension of the Clastic Wedge"). Nevertheless, the mean rate for these two segments combined is ~15 m/10^3 yr, based on dated strata in eastern Summit Valley and in borehole RZ-01. This is about twice the rate determined for the segment between boreholes RZ-01 and RZ-03. As was previously discussed, the stratigraphic relations in Summit Valley indicate that 3.8 Ma is the *maximum* possible age for the inception of the fluvial wedge. If the wedge originated at a later time, then it would have advanced from Summit Valley to GAFB at a mean rate exceeding 15 m/10^3 yr. Thus, the northward-expansion of the fluvial tract apparently decelerated somewhere between Summit Valley and GAFB. The opposing paleoslope on the north flank of the basin probably contributed to the slowing of the wedge by shifting the balance between lateral and vertical accretion of sediments. If, as we suspect, the early fluvial tract also broadened downstream in plan view, then the resulting radial dispersal of sediments would have tended to further retard the northward advancement of the fluvial tract.

The southward-inclined paleoslope on the north flank of the basin probably was inherited from the middle Miocene landscape, but the gradient may have been steepened locally during the Pliocene and Pleistocene by north-south tectonic contraction of the basin and by isostatic subsidence of the basin fill. The ancestral Mojave River gradually overcame this obstacle by depositing a northwest-thinning clastic wedge (Fig. 13). Several stratigraphic and sedimentologic features reflect the northward progradation of the fluvial tract during the growth the clastic wedge. These include (1) northwest-younging contacts at the base and top of the middle lacustrine unit (Fig. 5); (2) northwestward thinning of the upper fluvial unit (Figs. 5 and 13); (3) upward coarsening of the middle lacustrine and upper fluvial units (Fig. 5; most evident in borehole RZ-01); and (4) a sharp increase in the sediment accumulation rate at the base of the middle lacustrine unit (Figs. 7 and 8).

Depositional model and key analogs

The land surface produced by the growing clastic wedge descended toward the northwest, away from the detrital source area in the San Bernardino Mountains. The middle lacustrine unit evidently accumulated at the intersection between this fluvial slope and an opposing southward-inclined depositional slope on the north side of the sedimentary basin. The influx of debris from the actively rising San Bernardino Mountains apparently shifted the topographic axis of the basin progressively northward, such that the middle lacustrine unit lapped northwestward across the lower alluvial unit and was in turn overlapped from the southeast by the upper fluvial unit. This lateral migration of facies belts produced the diachronous stratigraphic succession observed in the boreholes at GAFB. A comparable northward-overlapping succession of Pliocene and Pleistocene alluvial-fan, lake, and stream deposits is found nearby to the east in Lucerne Valley (Figs. 1, 2) (R. Powell and J. Matti, 1999, written commun.).

Similar asymmetric basin architecture also exists in the southwestern Mojave Desert, ~70 km west of Victorville (Fig. 1). Regional gravity studies revealed a very deep (1500–3000 m) continental basin near Lancaster (Mabey, 1960, Figure 30). Evidence from boreholes indicates that this "east Antelope basin" (Dibblee, 1967, Figure 71) contains a diachronous body of lacustrine clay, silt, and fine sand 15–120 m thick, which is sandwiched between much thicker bodies of sand and gravel (Dutcher and Worts, 1963, Figure 3; Londquist et al., 1993, Figure 3). The lacustrine unit lies ~245 m beneath the land surface near the northern front of the Transverse Ranges. It rises steadily away from the mountains, eventually intersecting the land surface 24–32 km to the north, at the playas of Rosamond Lake and Rogers Lake. The deposits of sand and gravel above the lacustrine unit form a northward-tapering wedge beneath the piedmont of the western San Gabriel Mountains. The deeper deposits of sand and gravel that underlie the lacustrine stratum have not been studied in detail, but they presumably were derived from sources in the western Mojave Desert. The fill of the east Antelope basin is poorly dated, but the upper two units probably are mainly Pleistocene and Holocene in age. The accumulation of these units is still under way on the modern piedmont and playas.

An older possible analog lies *south* of the central Transverse Ranges, in the San Timoteo Badlands (Figs. 1 and 11). Geologic and geochronologic studies of outcrops have identified a conformable succession consisting of three main units (Matti and Morton, 1975, 1993; Albright, 1999). A lower unit as much as 550 m thick consists of fluvial sandstone deposited between 6.3 and 6.0 Ma. A medial unit ~40 m thick consists of lacustrine claystone, siltstone, and fine-grained sandstone deposited between 6.0 and 4.6 Ma. An upper unit that may be as thick as 2000 m consists of stream-deposited sandstone and conglomerate deposited between 4.6 and 0.5 Ma. The lower and medial units constitute the Mount Eden Formation of Fraser (1931), and the upper unit constitutes the San Timoteo Formation of Fraser (1931) (Fig. 11). Like the upper fluvial unit at GAFB, the San Timoteo Formation gradually coarsens upward, ranging from ripple-laminated fine-grained sandstone at its base, through interbedded fine- and coarse-grained sandstone, into capping deposits of interbedded sandstone and conglomerate.

The sandy lower member of the Mount Eden Formation was deposited as local valley fill by streams emerging at the northern margin of the Peninsular Ranges. The San Timoteo Formation and fine-grained upper member of the Mount Eden Formation, by contrast, were derived from more distant sources in the central Transverse Ranges (Matti and Morton, 1975, 1993). The lower and middle parts of the San Timoteo Formation were derived from the San Gabriel Mountains and San Bernardino Valley regions south of the San Andreas fault, whereas the upper part, deposited after 1.5 Ma, was derived from the headwaters of the Santa Ana River north of the fault (Morton and Matti, 1993b; Albright, 1999). The San Timoteo Formation was deposited by a large alluvial-fan/braided-stream complex that prograded southward from the Transverse Ranges (Matti and Morton, 1993), and we here propose that the underlying upper member of the Mount Eden Formation probably accumulated in shallow lakes or wetlands at the distal end of the prograding clastic wedge.

Thus, orogenic clastic wedges derived from the central Transverse Ranges invaded late Cenozoic continental basins on either side of the mountains. In the southern Mojave Desert, lakes and wetlands developed where northward-sloping orogenic deposits intersected a southward-facing regional slope occupied by active streams and alluvial fans. A southward-dipping, diachronous lacustrine stratum was deposited as the abundant detritus shed from the Transverse Ranges forced this depositional interface to the north. In the Victorville basin and neighboring Lucerne Valley, the northward flux of sediments was initiated by uplift of the San Bernardino Mountains in the late Pliocene. In the east Antelope basin, deposition occurred mainly during the Pleistocene in response to northwestward translation of the previously uplifted San Gabriel Mountains along the San Andreas fault.

By analogy with the southern Mojave Desert region, the lacustrine deposits in the upper member of the Mount Eden Formation may have accumulated diachronously where a clastic wedge prograding southward from the Transverse Ranges intersected an opposing fluvial slope on the north flank of the Peninsular Ranges. The southward-prograding system apparently records early uplift of the southeastern San Gabriel Mountains and neighboring San Bernardino Valley region in the late Miocene. If the lacustrine member truly was deposited at the tip of an expanding orogenic wedge, then this sector of the Transverse Ranges apparently began rising by 6 Ma. This interpretation is consistent with apatite fission-track data from crystalline rocks of the eastern San Gabriel Mountains, which record an episode of rapid cooling beginning ca. 7 Ma (Blythe et al., 2000).

Suspected former drainage connection between Victorville region and Lucerne Valley

A low drainage divide within alluvial deposits southwest of the Granite Mountains now separates neighboring lowlands near Victorville and Lucerne Valley (Fig. 2). This geomorphic feature and independent stratigraphic and structural evidence suggest that the continental basins in these two areas were interconnected during the Pliocene and Pleistocene. Meisling (1984) and Meisling and Weldon (1989) described a small outcrop of fine-grained lacustrine or fluvial sediments that overlies granitic basement on the broad salient of the northwestern San Bernardino Mountains north of Deep Creek (their "Rock Springs Road deposits"). These sediments are paleomagnetically reversed and thus are early Pleistocene or older (Meisling, 1984). Meisling and Weldon (1989) suggested that the outcrop is a western outlier of the Old Woman Sandstone. If correct, this implies that the Pliocene fluviolacustrine basin in Lucerne Valley originally extended southwest of the Granite Mountains and merged with the Phelan Peak basin (Meisling and Weldon, 1989, Figure 12).

The present configuration of the basement surface near Victorville, Lucerne Valley, and adjoining areas of the northwestern San Bernardino Mountains likewise suggests that a single sedimentary basin once extended across this region. Structure contour maps of the basement surface constructed from outcrop, borehole, and gravity data (Spotila and Sieh, 2000, Figures 4, 6, and 7) seem to indicate that the deep basin centered beneath Hesperia—the Victorville basin of this report—once merged with a smaller basin centered 6 km west of the town of Lucerne Valley. During the Pliocene, these basins probably formed a single large structural depression—the Phelan Peak basin—that extended eastward nearly to the Helendale fault. The Rock Springs Road deposits of Meisling (1984) probably accumulated on the floor of this depression before it was disrupted by the upthrusting of the northwestern San Bernardino Mountains.

If the Victorville basin retained an eastward drainage route inherited from the Phelan Peak basin, then water and sediments transported by the early Mojave River potentially were distributed across a broad area extending far beyond the Victorville region. The expanded surface area of this larger basin could have accommodated a greater volume of runoff through ponding, infiltration, and evaporation. It is also possible that a drainage outlet existed east of Lucerne Valley. If so, then fluvial transport of sediments through such an eastern outlet might help explain the relatively slow accumulation rates at GAFB during the latest Pliocene and early Pleistocene, as compared to closed Pleistocene basins in the western United States (Dohrenwend et al., 1991, p. 342) (see "Sediment accumulation rates at GAFB"). Drainage toward the east also may have prevented the ancestral Mojave River from producing a deep perennial lake in the Victorville region. However, thrust faulting, upwarping, and the growth of local alluvial fans associated with uplift of the northern San Bernardino Mountains must have progressively restricted the fluvial passage south of the Granite Mountains. The resulting topographic obstructions probably ultimately helped to determine the northward course of the modern Mojave River.

Alluvial-fan systems at southwest margins of the Phelan Peak and Victorville basins

Coarse alluvial-fan gravels derived from elevated ground south of the San Andreas fault form the upper half of the Phelan Peak Formation near Phelan Peak. Based on magnetostratigraphy, these sediments were deposited between 2.5 and 1.4 Ma (ages revised from Weldon et al., 1993, Figure 8). The sources of the granitic, gneissic, and marble clasts in these deposits originally lay nearby to the southwest (Foster, 1980), but they probably were subsequently displaced 65–100 km to the northwest by the San Andreas fault (Weldon et al., 1993). The alluvial fans grew northward into the Phelan Peak basin and its successor, the Victorville basin, contemporaneously with the clastic wedge of the ancestral Mojave River. However, the fans evidently originated later than the river and in response to a different tectonic stimulus. Whereas the fluvial wedge resulted from in situ uplift of the San Bernardino Mountains north of the San Andreas fault, the fans evidently were shed from a mobile source terrain that was transported past the Cajon Pass region by the San Andreas fault (Weldon et al., 1993).

The Phelan Peak Formation is overlain conformably (Kenney and Weldon, 1999) by the Victorville Fan deposits of Meisling and Weldon (1989), which comprise the Harold Formation, Shoemaker Gravel, and the older alluvium of Noble (1954)—all deposited on northeastward-sloping alluvial fans. This upward-coarsening succession accumulated in the Victorville basin between 1.7 and 0.78 Ma as the San Gabriel Mountains were translated northwestward past the Cajon Pass region by the San Andreas fault (ages revised from Weldon et al., 1993, Figure 4). Clast assemblages vary upward through the succession, but each of the three constituent units contains conspicuous detritus of Pelona Schist (Foster, 1980).

Relations between ancestral Mojave River and the Victorville Fan

The Ord River deposits near Deep Creek evidently overlap in age with the adjacent Victorville Fan deposits, as judged by magnetostratigraphy and intertonguing relations (Meisling and Weldon, 1989). However, the Ord River deposits apparently contain little or no detritus of Pelona Schist (Meisling, 1984). With rare exceptions, the upper fluvial unit at GAFB also lacks conspicuous detritus of Pelona Schist, at least within gravelly facies where such material would be easily spotted in the field. This contrast in detrital composition implies that the Victorville Fan and ancestral Mojave River occupied distinct depositional tracts, and that most of the coarse detritus transported northeastward across the San Andreas fault was trapped within the alluvial fans on the southwest flank of the Victorville basin. We suspect that significant amounts of fine-grained (sand-size and smaller) detritus were transported beyond the toe of the Victorville Fan. Using Pelona Schist as a tracer, petrographic studies of the upper fluvial unit might establish what proportion of the fine-grained detritus deposited by the ancestral Mojave River was contributed by drainages of the Victorville Fan.

The fan and river tracts are also distinguished by discordant geomorphic surfaces. The Victorville Fan descends northeastward relatively steeply (mean gradient of 23 m/km) between Cajon Pass and Victorville, and the ground steepens appreciably near the head of the fan. By contrast, remnants of the George surface that cap the Ord River deposits and upper fluvial unit descend very gradually and uniformly to the northwest (mean gradient of 3.6 m/km) between the mouth of Deep Creek and Victorville. The conspicuous steepening at the head of the Victorville Fan may result in part from monoclinal warping adjacent to the San Andreas fault (Meisling and Weldon, 1989; Kenney and Weldon, 1999). Moreover, erosion and sedimentary recycling have extensively modified the surface of the Victorville Fan since it was isolated from its mountain headwaters sometime between 0.78 and 0.5 Ma. Nevertheless, the substantially greater inclination of the overall fan surface reflects a fundamental hydrologic difference between the ancestral Mojave River and the smaller, more ephemeral, or more sediment-laden streams that deposited the Victorville Fan.

Therefore, although deposits of the Victorville Fan and ancestral Mojave River form a composite clastic wedge around the southern margin of the Victorville basin, the fan deposits and river deposits evidently were products of distinct hydrologic systems and may never have combined to form a continuous bajada. Downstream from its confluence with Deep Creek, the Mojave River probably always occupied a distinct, relatively low-gradient channel that skirted the northeast toe of the Victorville Fan (Meisling, 1984).

Geologic maps and cross sections that show deposits of the Victorville Fan extending northward beneath Victorville to the Mojave River (e.g., Bowen, 1954; Dibblee, 1960b, 1967; Bortugno and Spittler, 1986, section A–A′) are inaccurate. Our field studies determined that deposits of the ancestral Mojave River underlie Victorville, and the northeastern boundary of the alluvial-fan deposits lies southwest of the city (Fig. 2). Furthermore, the depositional histories of the river and fan imply that this interface probably lies farther to the southwest in the subsurface. The river advanced northward to GAFB by 1.95 Ma, whereas the fan did not begin growing until ca. 1.7 Ma. When the Victorville Fan finally expanded to the northeast, it probably overran the southwest margin of the river floodplain. After the modern Mojave River began incising its canyon at 60–70 ka, distal fan lobes produced by recycling of the Victorville Fan freely encroached onto the abandoned George surface, shifting the boundary between river and fan deposits yet farther to the northeast (Cox and Tinsley, 1999; Reynolds and Cox, 1999).

Southward-draining braided-stream/alluvial-fan system

The heterogeneous detritus in the lower alluvial unit at GAFB evidently was eroded from a large area of the southern and central Mojave Desert north of the Victorville region. Potential sources for the abundant granitic detritus in the lower alluvial unit are widespread west of the Mojave River (Bortugno and Spittler, 1986). Silicic metavolcanic rocks were derived from Mesozoic-age flows and welded tuffs that crop out extensively east of the Mojave River between Victorville and Barstow. Metasedimentary rocks, including quartzite, marble, and calc-silicate rocks, probably were derived in part from the Shadow Mountains, Iron Mountain, and Quartzite Mountain, which lie northwest, northeast, and east of GAFB (Fig. 2). Hornblende dacite and other nonmetamorphosed volcanic rocks evidently were derived from early Miocene-age extrusive rocks and hypabyssal intrusions that crop out farther to the northwest in the Kramer Hills, and more abundantly to the northeast around Barstow.

Alluvial-fan streams transferred rock detritus from the various upland sources to a large southward-flowing braided stream. The location of the trunk stream channel is locally revealed by deposits of well-stratified sand and pebble-cobble gravel with southward-flowing paleocurrents, which crop out east of the Mojave River between Helendale and Oro Grande (Fig. 12). The latter deposits coincide with one in a series of north-south–trending linear gravity depressions (Mabey, 1960; Biehler et al., 1988; Subsurface Surveys, 1990). If the gravity depressions are a reliable guide to the location of the old stream channel, then its headward reaches may be concealed beneath deposits of the ancestral Mojave River west of Iron Mountain (Figs. 2 and 12). The braided stream terminated down-gradient in a structural depression south of GAFB; this depression originated as the Phelan Peak basin and subsequently evolved into the Victorville basin. A thick bed of well-oxidized clayey silt and fine sand that is intercalated amid coarser fluvial and alluvial-fan deposits near the base of borehole RZ-01 (Fig. 5; 733.3–736.3 m elevation) may represent a tongue of playa sediments that accumulated near the north edge of the Phelan Peak basin.

Magnetostratigraphic data from the lower alluvial unit at GAFB (Figs. 5 and 7) suggest that southward-directed drainage was established prior to 4.2 Ma. Vigorous southward flow probably ended sometime between 1.77 and 1.18 Ma, when northward tilting evidently flattened or reversed the regional paleoslope north of GAFB. With the waning of the trunk stream, local alluvial fans deposited poorly sorted silty sand near the north end of the air base until ca. 1.18 Ma, when the shallow lake or wetland at the terminus of the ancestral Mojave River finally inundated the area.

Escape of ancestral Mojave River from the Victorville basin

The ancestral Mojave River evidently advanced beyond the Victorville basin by overflowing onto level or gently northward-sloping ground at the north end of the basin. By establishing the depth of the buried outlet or sill in relation to the George surface, we can date the river's escape from the basin through reference to the magnetostratigraphic sections in boreholes RZ-02 and RZ-03 (Figs. 5 and 8). The depth of the outlet can be estimated in two ways. First, wetland, lacustrine, and fluviodeltaic sedimentation likely waned as the river progressed beyond the basin. Thus, the buried outlet probably is approximately level with the top of the uppermost thick interval of fine-grained sediments deposited near the north end of the basin. Second, the depth of the outlet may be comparable to the maximum thickness of fluvial sediments exposed directly north of the basin.

Thick fluviodeltaic or floodplain deposits of thinly bedded fine sand, silt, and clay were encountered at depths exceeding 20 m and 22 m in boreholes RZ-02 and RZ-03, respectively (Fig. 5). The sill at the northern basin margin probably extended high enough to confine these deposits. As the top of the fine-grained interval lies above the level of the Brunhes/Matuyama boundary, the river evidently escaped from the Victorville basin sometime after 780 ka. By interpolating between the Brunhes/Matuyama boundary and the George surface (65 ka), the time of the river's departure is estimated to be 475–525 ka.

The northern margin of the basin evidently coincides with the crest of the east-west–trending anticline north of GAFB, which lies 4.3 km north of borehole RZ-02 (Figs. 2 and 6). The thickness of the ancestral Mojave River deposits exposed along an 8-km stretch of the canyon wall directly north of the anticlinal crest mostly oscillates between 15 and 24 m but locally is as little as 10 m near the crest. The deposits locally are capped by remnants of a mature soil profile that is characteristic of the George surface, so they apparently have not been significantly degraded by recent erosion. As the river presumably first overflowed at the lowest point along the basin margin, the thickest fluvial sections (24 m) probably most accurately record the time of its escape. By choosing the thickest deposits, sections attenuated by local uplift and erosion during or following the deposition of the upper fluvial unit can be avoided. Sediments lying 24 m below the George surface in borehole RZ-02 are above the level of the Brunhes/Matuyama boundary, which implies the river escaped after 780 ka. Interpolation refines the estimated time of departure to 575 ka.

Combining the results of the foregoing two approaches, we conclude that the ancestral Mojave River progressed beyond the Victorville basin ca. 475–575 ka. This replaces an earlier estimate of 1 Ma (Cox et al., 1998; Cox and Tinsley, 1999).

Two noteworthy events may have hastened the filling of the Victorville basin, thus promoting the northward advancement of the Mojave River. The first event was the incision of a deep channel at Summit Valley between 0.78 and 0.5 Ma (Meisling, 1984; McFadden and Weldon, 1987); stream erosion may have been induced by arching adjacent to the San Andreas fault (Meisling and Weldon, 1989; Kenney and Weldon, 1999). The cutting of ancestral Summit Valley must have augmented the volume of sediments transported downstream to Victorville and GAFB. The second event, which is deduced from the stratigraphic record at Owens Lake, California (Smith et al., 1997, Figure 6), was an increase in regional precipitation that began ca. 505 ka, after 140,000 yr of relative drought. The change to a wetter climate possibly stimulated erosion in the mountain headwaters of the ancestral Mojave River. This in turn would have accelerated sedimentation in the Victorville basin and might also account for the partial filling of ancestral Summit Valley with fluvial sand and gravel ca. 0.5 Ma (unit Qoa-e of McFadden and Weldon, 1987).

Advancement of ancestral Mojave River to Harper Lake and Lake Manix

Whereas the ancestral Mojave River advanced slowly against an opposing paleoslope on the north side of the Victorville basin, it must have made rapid headway once it encountered a long stretch of level or northward-sloping ground. Therefore, if we have correctly inferred that a broad crustal panel extending from Quartzite Mountain to Harper Lake was tilted to the north in the early Pleistocene, then the Mojave River probably advanced quickly to Harper Lake basin after escaping from the Victorville basin. With the arrival of the river, the center of Harper Lake basin was transformed from a playa to a pluvial lake bordered on the south by a fluvial delta.

Old alluvial-fan deposits exposed near Barstow (Fig. 12) may have formed a sill that temporarily confined the ancestral Mojave River to the Harper Lake basin. If so, then the river likely advanced east of Barstow when a deep pluvial lake in Harper Lake basin overflowed the sill. Ensuing erosion probably lowered the sill, thereby encouraging more frequent eastward excursions of the river. Stratigraphic evidence downstream from Barstow suggests that the river's initial confinement at Harper Lake was fairly brief. Arkosic sediments transported by the Mojave River apparently began accumulating in the Lake Manix basin ca. 500 ka (Jefferson, 1985, 1994, 1999, this volume). Thus, if the river escaped from the Victorville basin ca. 475–575 ka, then its terminus evidently was fixed at Harper Lake no longer than ~75,000 yr. This suggests that the hypothetical sill could have been breached during a single pluvial cycle.

The ancestral Mojave River did not permanently abandon Harper Lake after entering the Lake Manix basin. Relict shorelines and fossil mollusks and ostracodes indicate that the river produced at least two deep lakes in the Harper Lake basin during the late Pleistocene, most recently ca. 25 ka (Meek, 1990, 1999). Vertical movements along the Lenwood and Mount General faults (Fig. 2) may have acted as a switch that diverted the flow of the Pleistocene Mojave River either to Harper Lake or to Lake Manix (Reynolds and Reynolds, 1994b). The Mojave River presently is barred from flowing northward to Harper Lake via Hinkley Valley (Fig. 2) by an artificially reinforced natural levee 5–6 m high. Without human intervention, a major flood or aggradation by the river could easily divert the flow back to Harper Lake (Meek, 1999). More detailed evidence dating the initial arrival and subsequent excursions of the Mojave River at Harper Lake basin probably could be obtained through borehole stratigraphy studies at Harper Lake playa.

Cutting and filling of the Mojave River canyon

The optically stimulated luminescence ages determined on fluvial deposits beneath the George surface suggest that the Mojave River began eroding its present canyon near GAFB ca. 60–70 ka, or shortly thereafter. This event marks the origin of the modern Mojave River in the Victorville region. Sediments beneath the active river channel and inset stream terraces on the canyon walls seem to record three episodes of downcutting alternating with two episodes of aggradation. The depth of the Mojave River canyon now averages ~60 m between Victorville and Iron Mountain. However, the original incision apparently was much deeper. Wells drilled along this stretch of the river penetrate a channel fill of unconsolidated sand and gravel as much as 60 m thick (Stamos et al., 2001, Figure 9, section D–D′). Furthermore, inset late Pleistocene stream terraces and alluvial fans stand as much as 15 m above the level of the active channel. These relations suggest that the river initially excavated a canyon as much as 120 m deep and then refilled it more than halfway to the top. The highest stream terrace is presently undated, but a lower terrace 8 m above the active river channel near GAFB yielded a ^{14}C age of ca. 6 ka (Rector et al., 1983). The latter terrace apparently formed during the second episode of downcutting. Late Holocene-age terraces standing a few meters above the level of the active river channel seem to represent the youngest deposits of the second aggradational episode. The river is now engaged in the third episode of downcutting, which began sometime in the late Holocene.

A well-dated succession of channel cuts and fills along Cajon Creek (McFadden and Weldon, 1987, Figure 2) is intriguingly similar to the arrangement just described for the Mojave River. Two major episodes of downcutting by Cajon Creek occurred 55–14.4 ka and 12.4–1.7 ka. Each of these lengthy incisional episodes was followed by a much shorter episode of aggradation. The initial cycle produced >100 m of erosion followed by ~40 m of fill, while the second cycle produced ~40 m of erosion followed by 10 m of fill. A third, ongoing episode of downcutting that began ~275 yr ago has incised a channel nearly 10 m deep.

The apparent similarities in geomorphic history between Cajon Creek and the Mojave River may indicate that streams on the north and south flanks of the central Transverse Ranges responded synchronously to regional tectonic and climatic events. For example, aggradation of Cajon Creek between 14.4 and 12.4 ka correlates with the last major pluvial episode in the Mojave Desert region and surrounding areas. Wet conditions during this period are recorded by deep-water clay and offshore bar deposits of Koehn Lake, which yielded ^{14}C ages ranging between 14.7 and 12.7 ka (Clark and Lajoie, 1974; Burke, 1979). The initial deep refilling of the Mojave River canyon probably also dates to this period. This in turn suggests that the preceding episode of deep incision may have ended ca. 15 ka. We are presently testing these hypotheses by applying optically stimulated luminescence methods to date terrace deposits of the Mojave River in the Victorville region.

Consequences of Mojave River incision for Harper Lake and Lake Manix basins

We previously suggested that the erosion of Summit Valley probably assisted the northward advancement of the ancestral

Mojave River by accelerating the filling of the Victorville basin (see "Escape of ancestral Mojave River from the Victorville basin"). Similarly, the more recent cutting of the Mojave River canyon between Deep Creek and Hinkley Valley since 60–70 ka probably strongly affected drainage and sedimentation downstream at Harper Lake and Lake Manix basin. One important consequence was the reconfiguration of the drainage network near Harper Lake. Prior to incision, the Mojave River flowed to Harper Lake through two alternate routes, one lying west of Iron Mountain and the other passing south of Iron Mountain and northward along Hinkley Valley. The former route was bypassed when the river cut its modern channel south of Iron Mountain. Thus, a broad fluviodeltaic plain northwest of Iron Mountain (Fig. 2) was permanently abandoned, and flooding of Lake Manix basin probably increased in frequency at the expense of Harper Lake basin.

A second consequence was the redistribution of a large volume of sediment released during the initial erosion of the Mojave River canyon. Much of this material probably was deposited in the Lake Manix basin. During the late Pleistocene, sand and gravel transported by the Mojave River accumulated to form a large delta and adjoining alluvial plain on the southwest margin of Lake Manix. The delta filled in a large area of the lake basin, thereby eventually isolating the modern playa basins of Coyote Lake and Troy Lake (Fig. 1) (Hagar, 1966; Groat, 1967; Meek, 1999).

Near Manix Wash, the upper two units of the Manix Formation comprise a thick unit of lacustrine silt, clay, and sand (member C), and an overlying sheet of fluviodeltaic sand and gravel (member D) (Jefferson, 1985, this volume). Based on isotopic ages from deposits near the base of member D, the delta prograded rapidly eastward to the future site of Manix Wash ca. 60 ka (Jefferson, this volume). Given the timing of this event, it appears likely that the delta expanded in response to the cutting of the Mojave River canyon, which began ca. 60–70 ka based on our optically stimulated luminescence dating near GAFB. Besides constricting the lake basin, the massive influx of sediments may have indirectly hastened the demise of Lake Manix by altering its hydrologic budget. The reduced storage capacity and evaporative surface area of the shrinking lake must have been compensated by increased flow at its outlet. The greater outflow presumably accelerated the erosion that breached the eastern sill of the lake ca. 18 ka (Buwalda, 1914; Meek, 1989, 1999; Jefferson, 1999, this volume).

PALEOTECTONIC IMPLICATIONS

Uplift of the San Bernardino Mountains

We have proposed that the northward-thinning wedge of granitic sand, silt, and gravel that underlies Victorville and George Air Force Base (GAFB) is an orogenic prism that accumulated in response to uplift of the San Bernardino Mountains (Fig. 13, upper fluvial unit). This clastic wedge was deposited by a northwestward-advancing fluvial system that we equate with the ancestral Mojave River, and the underlying diachronous lacustrine stratum apparently accumulated at the northwest tip of the expanding fluvial wedge. This depositional model implies that the inception of uplift in the San Bernardino Mountains should be at least as old as the earliest lacustrine sediments deposited at the tip of the wedge. Therefore, the initial uplift apparently predates 2.55 Ma, which is the age we assign to the deepest lacustrine deposits in borehole RZ-01 (Figs. 5 and 7). Our analysis of the southeastern extension of the clastic wedge suggests that uplift began at least a million years earlier, however. We have inferred that the fluvial system arrived at southernmost Victorville ca. 3.3 Ma, and that it may have first emerged at the mountain front as early as 3.8 Ma.

Previous studies concluded that the modern San Bernardino Mountains began rising sometime after 2–3 Ma in response to transpressional stresses along the San Andreas fault (May and Repenning, 1982; Sadler and Reeder, 1983; Meisling and Weldon, 1989; Matti and Morton, 1993). Key biochronologic evidence was presented by May and Repenning (1982), who found late Pliocene-age (2–3 Ma) vertebrate fossils in the Old Woman Sandstone of Shreve (1968) along the northern front of the range. The fossils were recovered from beds of sand and silt that contain clasts derived from the Mojave Desert. Thus, the deposits evidently predate the development of the steep topography and vigorous northward drainage associated with the modern range front (Sadler, 1982b; Sadler and Reeder, 1983).

However, significant topographic relief may have developed in the San Bernardino Mountains region well before the end of the Pliocene. Sedimentary structures and clast assemblages in the Old Woman Sandstone indicate that some of its fluvial sediments were derived from the south (Shreve, 1968; Powell and Matti, 1998). The source conceivably lay in the ancestral San Gabriel Mountains, which were uplifted south of the San Bernardino Mountains region during the late Miocene and early Pliocene (Matti and Morton, 1975, 1993; Sadler, 1993; Albright, 1999). Such provenance is doubtful, however, because the Old Woman Sandstone apparently lacks detritus of Pelona Schist, Lowe Granodiorite, and other rocks characteristic of the modern San Gabriel Mountains region (Sadler, 1993). The sediments more likely were derived from the vicinity of the north-central San Bernardino Mountains.

Previous paleotectonic models imply that the northern San Bernardino Mountains were uplifted at least a slight amount during the early Pliocene. Sadler and Reeder (1983) inferred that the sedimentary basin of the Old Woman Sandstone originated sometime between 7 and 3 Ma as an initial product of the same transpressional regime that eventually uplifted the northern San Bernardino Mountains. Similarly, Meisling and Weldon (1989) proposed that early episodes of northward-directed thrusting might have elevated the south side of the basin.

The early Mojave River may have been especially responsive to these initial tectonic movements, because its principal tributary, ancestral Deep Creek, drained a very large area of the San Bernardino Mountains. Based on the configuration of the modern drainage network, Sadler and Reeder (1983) inferred that the headwaters of ancestral Deep Creek extended eastward to the core of the range, encompassing the modern watersheds of Big Bear Valley and possibly the uppermost Santa Ana River. The latter watersheds lie directly north of a strongly uplifted tectonic block that encompasses San Gorgonio Mountain (Fig. 1). The mechanism and timing of uplift at San Gorgonio Mountain as compared to the adjacent northern plateau of the San Bernardino Mountains is uncertain (Sadler, 1993; Spotila et al., 1998; Spotila and Sieh, 2000). However, a local restraining bend along the San Andreas fault near San Gorgonio Pass (Fig. 1) apparently controlled the uplift of the entire range (Matti and Morton, 1993; Spotila and Sieh, 2000). Uplift may have propagated northward from this local tectonic knot (Sadler, 1993). If so, then the oldest deposits in the clastic wedge of the ancestral Mojave River may record early uplift of San Gorgonio Mountain. Once significant topographic relief had developed, deeply weathered plutonic rocks beneath the pre-late Miocene regional erosion surface (Oberlander, 1972; Spotila and Sieh, 2000) must have provided the rejuvenated tributaries of ancestral Deep Creek with a copious supply of easily eroded rock detritus.

Deformation of the southern Mojave Desert

During the late Cenozoic, the crust of the Mojave Desert and adjacent Transverse Ranges was shortened by north-south compression (Bartley et al., 1990) and simultaneously was sheared along northwest-trending strike-slip faults (Dibblee, 1961; Dokka, 1983; Dokka and Travis, 1990). Prime examples of the strike-slip faults include the Helendale and Lenwood faults between Victorville and Barstow (Figs. 2 and 12). The contractional structures include thrust and reverse faults and monoclinal flexures along the north flank of the San Bernardino Mountains (Fig. 2) (Meisling, 1984; Miller, 1987; Meisling and Weldon, 1989; Matti et al., 1998a, 1998b; Kenney and Weldon, 1999; Spotila and Sieh, 2000). In addition, numerous east-west–trending contractional structures have been identified in the Mojave Desert (Bartley et al., 1990); these include anticlines along the Helendale and Lenwood faults (Figs. 2 and 12).

Regional compression probably also uplifted low mountains of pre-Cenozoic rocks between Victorville and Barstow, including the Shadow Mountains, Quartzite Mountain, Stoddard Ridge, and the Granite and Ord Mountains (Fig. 2). These ranges form the western part of an elongate belt of mountains that trends west-northwest ~250 km across the southern Mojave Desert. This belt was termed the "Bullion Mountains highlands" by Howard and Miller (1992), who considered it to be a neotectonic basement anticline. Based on the transverse dimensions of the uplands and bordering troughs, Howard and Stuart (1992) proposed that the southern Mojave Desert is underlain by a folded layer of continental crust 7–10 km thick.

Our field and borehole investigations along the Mojave River revealed abundant evidence of north-south contraction during the early and late Pleistocene. Two faults along the southwest margin of GAFB strike northwestward, parallel to strands of the regional dextral fault system (Fig. 12). They are not merely strike-slip faults, however, as they jointly produced at least 50 m of southwest-side-up vertical displacement. They may be southwest-dipping oblique-slip faults that accommodated both contractional strain and dextral shear. Similarly, the southwest side of the Helendale fault locally was uplifted at least 36 m during the Pleistocene, possibly in conjunction with right-lateral slip.

The low, southward-facing anticline north of GAFB (Figs. 2 and 6) apparently marks the crest of the regional fold proposed by Howard and Miller (1992). Associated structures exposed in the Mojave River bluffs record the latest stages of regional contraction that produced the Bullion Mountains highlands (Howard and Cox, 2001). On the southern limb of the anticline, fluvial and alluvial-fan deposits of late Pliocene or early Pleistocene age are tilted gently to the south and are overlain unconformably by early Pleistocene-age deposits of the ancestral Mojave River. Thus, the anticline was arched upward relative to the Victorville basin during the early Pleistocene. On its northern limb, a broad crustal panel extending between GAFB and Harper Lake apparently was tilted very slightly to the north, which reversed the former southward-inclined regional paleoslope. Although the land surface was tilted <1°, this slight amount of rotation was enough to elevate areas near the fold hinge as much as 600 m relative to the distant Harper Lake basin. The anticline was uplifted contemporaneously with the northern front of the San Bernardino Mountains. As the westernmost San Bernardino Mountains near Cajon Pass were uplifted within a *northward*-facing monocline (Meisling and Weldon, 1989; Kenney and Weldon, 1999), the intervening Victorville basin evidently developed as a contractional structure bounded by inward-facing folds.

Stratigraphic and geomorphic relations near GAFB imply that upwarping of the Bullion Mountains highlands began well before the Pleistocene. Early Pleistocene-age stream and alluvial-fan deposits that crop out along the Mojave River near the crest of the anticline (Figs. 6 and 12) are depressed topographically 100–500 m relative to adjacent basement ridges west and east of the river. The old stream and fan deposits apparently accumulated in a Pliocene river valley that was incised into the basement core of the anticline. The paleocurrent trends (Fig. 12) and detrital provenance of the old fluvial deposits suggest that a southward-flowing antecedent stream eroded the valley. The arcuate path of the modern Mojave River between Hesperia and Barstow (Fig. 2) evidently was inherited from this older drainage system, which drained westward along the northern flank and southward around the western nose of the growing anticline.

Sediments deposited in the adjacent synclinal troughs probably record the initial growth of the regional anticline. The trough to the south originated as the greater Phelan Peak basin, which accommodated deposits of the Phelan Peak Formation and Old Woman Sandstone. This basin evidently started subsiding as early as 4.4 Ma, when the basal strata of the Phelan Peak Formation accumulated near Cajon Pass (age revised from Weldon et al., 1993, Figure 8). To the north, a trough near Barstow is filled with

Pliocene and Pleistocene stream and alluvial-fan deposits as much as 600 m thick (Fig. 12) (Zohdy and Bisdorf, 1994; Densmore et al., 1997, Figure 4, unit QTof). Judging from their paleocurrents (Fig. 12) and lateral variations in detrital composition, these deposits accumulated in an east-west–trending basin whose axis lay near the present-day channel of the Mojave River (Cox and Wilshire, 1993). The upper levels of the stratigraphic succession studied in outcrops and boreholes generally coarsen upward, which may reflect concurrent growth of the regional anticline. Tephrochronologic evidence reported herein (see "Regional correlation with deposits north and east of GAFB") suggests that the old stream and fan deposits exposed at the land surface accumulated between 2.5 and 1 Ma. Deeper levels of the basin fill concealed in the subsurface probably accumulated well before 2.5 Ma and possibly are as old as the basal deposits of the Phelan Peak basin.

It is intriguing and possibly significant that the estimated thickness of the old basin-filling sediments near Barstow, approximately 600 m, is identical to the structural relief that we ascribe to broad northward tilting between GAFB and Harper Lake. This similarity may imply that subsidence of the "Barstow basin" was mechanically linked with the regional tilting. Although neither process is precisely dated, the large-scale tilting and local basin development probably were largely coeval, each occurring during the Pliocene and Pleistocene over a span of 3.5–4.5 m.y. The long-term rate of vertical crustal deformation implied by these relations is within the range of 13–17 cm/10^3 yr.

Previous models of the deep structure of the San Bernardino Mountains suggest a likely structural framework and driving mechanism for large-scale folding of the adjacent southern Mojave Desert. From an analysis of regional seismicity, Corbett (1984) determined that the base of seismogenic upper crust dips southward beneath the San Bernardino Mountains. Based in part on this observation, Meisling and Weldon (1989) proposed that the San Bernardino Mountains are underlain by a deep northward-vergent thrust ramp that flattens northward into a décollement ~5 km beneath the floor of the Mojave Desert. They inferred that the range was uplifted partly by northward displacement of the upper plate of this thrust-décollement system. Deep seismic-reflection profiling subsequently confirmed that the base of the brittle upper crust is nearly horizontal beneath the southern Mojave Desert and dips southward beneath the western San Bernardino Mountains (Li et al., 1992). The latter study further concluded that the brittle-ductile transition and hypothetical décollement zone lie 8–13 km beneath the southern Mojave Desert, which is consistent with the 7–10 km thickness of folded crust modeled by Howard and Stuart (1992).

Therefore, besides uplifting the San Bernardino Mountains, northward displacement along the décollement probably buckled the upper crust in the southern Mojave Desert, producing the anticlinal Bullion Mountains highlands and adjacent synclinal troughs (Howard and Cox, 2001). If folding was concentrated north of the San Andreas fault, it may have uplifted the southern Mojave Desert relative to areas south of the Transverse Ranges. This might help explain the origin of the "High-Desert" terrain near Victorville, which lies ~600 m higher than the land surface near San Bernardino. Bowen (1954) inferred that the Mojave River canyon was eroded in response to recent tectonic uplift of the southern Mojave Desert. We endorse this hypothesis and propose that renewed arching of the crust above the deep décollement was responsible for this latest episode of broad uplift.

Alternatively, one could argue that deep incision of the Mojave River between Hesperia and Barstow resulted from climate change rather than tectonic uplift. However, the Mojave Desert and surrounding areas were subjected to extreme climatic fluctuations many times during the middle and late Pleistocene. Smith et al. (1997, Figure 7) showed that five major and numerous minor oscillations of global ice volume during the past 0.5 m.y. correlate with hydrologic cycles at Owens Lake, California. If the cutting of the Mojave River canyon were induced by one of the latest of these regional hydrologic cycles, then several older canyons presumably would have been eroded during the earlier cycles. Yet the modern canyon apparently is unique in the history of the Mojave River. Our studies of boreholes at GAFB and of outcrops in the river bluffs revealed no evidence of deep channeling or major erosional hiatuses within the fluvial succession. We therefore suspect that some mechanism other than climate change—most likely tectonic uplift—provoked incision. However, subsequent aggradational episodes may have been induced by periods of humid climate.

If the late Pleistocene entrenchment of the Mojave River was controlled primarily by tectonic uplift, and climate played only a secondary role, then the duration and depth of incision may approximate the duration and magnitude of uplift. Our previous analysis (see "Cutting and filling of the Mojave River canyon") suggests that the episode of deep incision spanned approximately 50,000 yr (ca. 65–15 ka) and eroded to a maximum depth of 120 m. The corresponding mean rate of incision, and tentative uplift rate, is 2.4 m/10^3 yr. This rate is somewhat remarkable, because it is at least an order of magnitude faster than the long-term rate of Pliocene-Pleistocene vertical deformation inferred from regional stratigraphic and structural relations between Victorville and Barstow.

CONCLUSIONS: SUMMARY OF LATE CENOZOIC GEOLOGIC HISTORY

The stratigraphic succession near Victorville, California records the evolution of a large interior drainage basin in response to transpressional tectonics along the Pacific–North American plate boundary (Fig. 14). Previous studies determined that the Mojave River advanced northward into the Mojave Desert in response to uplift of the central Transverse Ranges, particularly the San Bernardino Mountains. Our findings support this conclusion, but the uplift of the mountain headwaters and northward growth of the river apparently began somewhat earlier than was previously recognized. Moreover, the river's development was strongly influenced by concurrent tectonic movements and preexisting landforms in the Mojave Desert region.

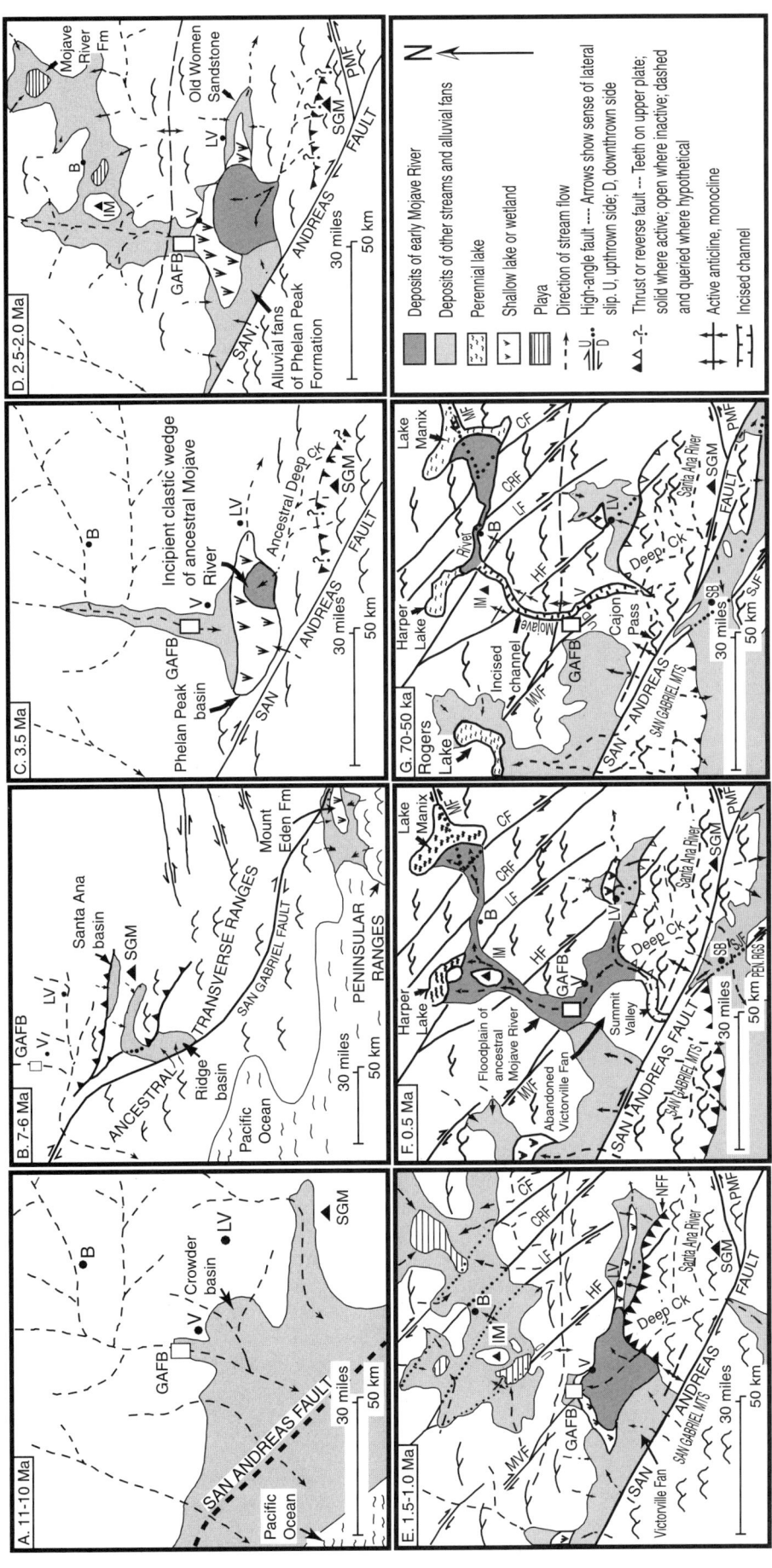

Figure 14. Late Cenozoic paleogeographic evolution of Mojave River drainage basin and surrounding areas. Panels A–G depict depositional and structural relations for seven specific time intervals between 11 Ma and 50 ka (explained in text). Tectonic development of the San Andreas fault system and San Bernardino Mountains is largely based on the reconstructions of Matti and Morton (1993) and Meisling and Weldon (1989). B—Barstow; GAFB—George Air Force Base; LV—Lucerne Valley; SB—San Bernardino; V—Victorville; IM—Iron Mountain; SGM—San Gorgonio Mountain; Pen Rgs—Peninsular Ranges; CF—Calico fault; CRF—Camp Rock fault; HF—Helendale fault; LF—Lenwood fault; MF—Manix fault; MVF—Mirage Valley fault; NFF—north frontal fault system; PMF—Pinto Mountain fault; SJF—San Jacinto fault.

The period of 18–10 Ma in southern California was a time of relative tectonic stability between episodes of early Miocene regional extension and late Miocene contraction. By the end of this period, or ca. 11–10 Ma, prolonged weathering and erosion had molded the landscape of the southern Mojave Desert and surrounding areas into a surface of low relief that drained southwestward across the San Andreas fault to the Pacific Ocean (Fig. 14A). A trunk stream channel crossed the future site of George Air Force Base (GAFB) and deposited sand and gravel in the Crowder basin.

By 7–6 Ma, the southwestward-flowing streams were diverted by uplift and lateral offset of the ancestral Transverse Ranges along the San Gabriel fault and other strands of the San Andreas system (Fig. 14B). This orogenic regime began as early as 10 Ma. During its later stages in the latest Miocene, a nonmarine basin began developing in the San Timoteo Badlands, which then lay far to the southeast of the Victorville region. The fine-grained upper member of the Mount Eden Formation accumulated between 6.0 and 4.5 Ma in the San Timoteo Badlands, probably at the tip of a clastic wedge that prograded southward from the Transverse Ranges. This unit may constitute a paleotectonic marker that is analogous to younger lacustrine and wetland deposits near Victorville, which accumulated at the tip of a *north-ward*-prograding wedge.

A major east-west–trending structural depression, the "Phelan Peak basin," developed at the northern foot of the central Transverse Ranges during the Pliocene, probably in response to north-south compression (Fig. 14C). The basin began subsiding by 4.4 Ma and had accumulated a significant thickness of sediments by 3.5 Ma. It received runoff and sediments from both the Mojave Desert and the Transverse Ranges, and we speculate that it had an eastern outlet. Fine-grained fluviolacustrine and playa sediments accumulated along the basin axis, while coarse sand and gravel derived from the Mojave Desert accumulated along the northern margin of the basin and within the adjacent valley of a southward-flowing trunk stream. The northward-flowing ancestral Mojave River began depositing a wedge of sand and gravel at the southeast margin of the basin between 3.8 and 2.5 Ma, most likely near the beginning of this interval. The river arose during the early phases of a protracted Pliocene-Pleistocene orogenic cycle, possibly during the initial uplift of San Gorgonio Mountain at the head of ancestral Deep Creek.

The San Gorgonio Mountain block or adjacent areas of the ancestral San Bernardino Mountains north of the San Andreas fault apparently continued rising between 2.5 and 2.0 Ma, while mountains south of the fault were sliding along the southwest side of the Phelan Peak basin (Fig. 14D). In response to these movements, the fluvial tract of the ancestral Mojave River and adjoining alluvial fans of the Phelan Peak Formation expanded northward, displacing the axial lacustrine tract northward to the future site of GAFB. Sedimentation also expanded into the Lucerne Valley and Barstow regions during this period. A thick succession of coarse-grained alluvial-fan sediments inundated areas west, south, and northeast of Barstow in response to the growth of a broad, east-west–trending basement anticline in the southern Mojave Desert (Bullion Mountains highlands). Playa basins developed locally amid the alluvial fans. The regional anticline and the alluvial fans on its northern flank may actually have begun developing as early as 4.4 Ma, if they formed contemporaneously with the Phelan Peak basin. The large-scale folding induced sedimentation by increasing topographic relief and by flattening the gradient of the southward-flowing trunk stream on the north flank of the anticline. The ancestral Mojave River and neighboring alluvial fans indirectly influenced sedimentation in the Barstow region by filling the Phelan Peak basin. This raised the base level for the trunk stream draining southward from the Mojave Desert.

The Phelan Peak basin evolved into the modern Victorville basin through three tectonic processes that culminated ca. 2.0–1.5 Ma. First, continued anticlinal arching of the southern Mojave Desert uplifted the northern margin of the basin. The growing arch eventually blocked the southward-flowing regional trunk stream, which severed the connection between depositional tracts in the central and southern Mojave Desert. Second, the San Bernardino Mountains were thrust upward at the southeast margin of the basin. This process reconfigured the basin margin and nourished orogenic clastic wedges deposited by the Mojave River and neighboring streams in Lucerne Valley. Third, lateral movements of the San Andreas fault juxtaposed elevated terrain of the central San Gabriel Mountains against the southwest side of the basin. This induced the accumulation of the Victorville Fan deposits, which contributed significantly to the filling of the Victorville basin.

Uplift of the northern San Bernardino Mountains and the neighboring anticline in the southern Mojave Desert probably waned between 1.5 and 1.0 Ma as a system of northwest-trending right-lateral faults developed across the region (Fig. 14E). However, the large-scale patterns of topography, drainage, and sedimentation induced by previous episodes of compressional deformation persisted, and new fold structures developed locally. Anticlinal warping along the San Andreas fault uplifted the western "wing" of the San Bernardino Mountains, and smaller anticlines developed locally along the Helendale and Lenwood faults. A steep fault scarp developed along the south side of the San Bernardino Mountains as the San Gabriel Mountains receded to the northwest along the San Andreas fault. At ca. 1.5 Ma, a tributary of the early Santa Ana River eroded this escarpment and captured the headwaters of ancestral Deep Creek that lay north of San Gorgonio Mountain. Also ca. 1.5–1.0 Ma, or perhaps earlier, two new playa basins formed west of Barstow in response to the prior large-scale folding, augmented by local extension or constriction associated with strike-slip faults. The ancestral Mojave River filled the Victorville basin at an accelerated pace between 1.2 and 1.0 Ma, which may be a reflection of pluvial conditions that are documented elsewhere in the region.

The northern frontal fault system of the San Bernardino Mountains and the regional anticline in the southern Mojave Desert apparently were virtually inactive during the middle Pleistocene, after 0.78 Ma, while displacements continued on the San

Andreas fault and the northwest-trending strike-slip faults of the Mojave Desert (Fig. 14F). Between 0.78 and 0.5 Ma a tributary of the ancestral Mojave River cut a deep channel into the head of the Victorville Fan near Summit Valley, probably in response to anticlinal warping along the San Andreas fault. The erosion of this fanhead trench, and lateral offset of other trunk streams by the San Andreas fault, terminated the growth of the Victorville Fan. The head of the fan was subsequently destroyed by headward erosion of Cajon Creek after ca. 0.5 Ma.

The sediments eroded from the incision at Summit Valley were redeposited downstream, thus contributing to the filling of the Victorville basin. About 475–575 ka, the ancestral Mojave River escaped from the Victorville basin by aggrading its bed to a saddle at the crest of the regional anticline north of GAFB. This wind gap between Quartzite Mountain and the southeastern Shadow Mountains marked the abandoned, deeply alluviated canyon of the old southward-flowing trunk stream. The route directly north of the saddle had been prepared previously by gentle tilting on the broad northern limb of the anticline, which flattened or reversed the former southward-inclined paleoslope between GAFB and Harper Lake. Therefore, after overtopping the saddle, the river probably advanced quickly to Harper Lake, where it may have lingered briefly before progressing eastward to the Lake Manix basin ca. 0.5 Ma. The terminus of the river subsequently oscillated between the Harper and Manix basins. During pluvial cycles, high fluvial discharge transformed the playas in these basins into perennial lakes. The Mojave River probably also ventured into the Lucerne Valley region occasionally during this and earlier periods.

The upper Mojave River basin acquired its modern form after ca. 60–70 ka, when the river started eroding a canyon between Deep Creek and Barstow (Fig. 14G). The entrenched channel bypassed an old fluvial passage west of Iron Mountain. This reconfiguration of the drainage probably boosted the volume of floodwater delivered to Lake Manix at the expense of Harper Lake. The large volume of sediment eroded from the canyon was mainly deposited in the delta of Lake Manix. The incision of the Mojave River canyon probably resulted from broad arching of the southern Mojave Desert above a midcrustal décollement. Regional north-south contraction during the late Pleistocene is also manifested by local arching and faulting of the George surface at GAFB and along the Helendale fault. To the south near Cajon Pass, a monoclinal fold propagated northwestward along the San Andreas fault. Partly on account of this structure, most streams on the north slope of the eastern San Gabriel Mountains were diverted southeastward to the Santa Ana River, or northwestward toward Rogers Lake, so that the San Gabriel Mountains now contribute little runoff to the Mojave River basin.

The Pliocene and Pleistocene development of the Mojave River basin provides an example of large-scale geomorphic processes associated with a major strike-slip fault system. We hope that this study leads to a better understanding of how landscapes evolve in relation to tectonic movements along continental transform plate boundaries.

ACKNOWLEDGMENTS

Numerous individuals—including S.M. Chrisley, S. Dickey, R.M. Forester, J.H. Foster, K.A. Howard, J.A. Izbicki, G.T. Jefferson, P. Martin, J.C. Matti, N. Meek, D.M. Miller, D.M. Morton, R.E. Powell, M.C. Reheis, R.E. Reynolds, P.M. Sadler, A. Sarna-Wojcicki, E. Scott, K. Springer, and J.C. Tinsley—provided helpful information and advice that influenced the course of our study. S.M. Chrisley and S. Dickey provided access to critical borehole data, cores, and cuttings. K. Springer pointed out two beds of volcanic ash near Lenwood. R.M. Forester examined ostracode samples and provided preliminary interpretations of their ages and paleoenvironmental significance. A. Sarna-Wojcicki provided tephrochronologic analyses and preferred correlations for several samples of volcanic ash. P.M. Sadler, R.E. Powell, and J.C. Matti graciously shared unpublished data and interpretations regarding the geology of the Lucerne Valley region. Reviews by L.B. Albright, K.A. Howard, N. Meek, D.M. Miller, M.C. Reheis, J.C. Tinsley, and especially P.M. Sadler, significantly improved the ideas and organization of the paper. This study is a product of the U.S. Geological Survey, Geologic Division, Southern California Areal Mapping Project. Funding was provided by the Mojave Water Agency and by the National Cooperative Geologic Mapping Program of the U.S. Geological Survey.

REFERENCES CITED

Agenbroad, L.D., Mead, J.I., and Reynolds, R.E., 1992, Mammoths in the Colorado River corridor, in Reynolds, R.E., Old routes to the Colorado: Redlands, California, San Bernardino County Museum Association Special Publication 92-2, p. 104–106.

Aitken, M.J., 1998, An introduction to optical dating: Oxford, Oxford University Press, 267 p.

Albright, L.B., III, 1999, Magnetostratigraphy and biochronology of the San Timoteo Badlands, southern California, with implications for local Pliocene-Pleistocene tectonic and depositional patterns: Geological Society of America Bulletin, v. 111, p. 1265–1293.

Bachman, G.O., and Machette, M.N., 1977, Calcic soils and calcretes in the southwestern United States: U.S. Geological Survey Open-File Report 77-794, 163 p.

Baker, C.L., 1911, Notes on the later Cenozoic History of the Mohave Desert region in southeastern California: University of California Publications, Bulletin of the Department of Geology, v. 6, no. 15, p. 333–383.

Bartley, J.M., Glazner, A.F., and Schermer, E.R., 1990, North-south contraction of the Mojave block and strike-slip tectonics in southern California: Science, v. 248, p. 1398–1401.

Berggren, W.A., Kent, D.V., Swisher, C.C., III, and Aubry, M.-P., 1995, A revised Cenozoic geochronology and chronostratigraphy, in Berggren, W.A., Kent, D.V., Aubry, M.-P., and Hardenbol, J., 1995, Geochronology time scales and global stratigraphic correlation: SEPM (Society for Sedimentary Geology) Special Publication no. 54, p. 129–212.

Biehler, S., Tang, R.W., Ponce, D.A., and Oliver, H.W., compilers, 1988, Bouguer gravity map of the San Bernardino quadrangle, California: California Division of Mines and Geology Regional Geologic Map Series, map no. 3B, scale 1:250,000.

Blythe, A.E., Burbank, D.W., Farley, K.A., and Fielding, E.J., 2000, Structural and topographic evolution of the central Transverse Ranges, California, from apatite fission-track, (U-Th)/He, and digital elevation model analysis: Basin Research, v. 12, p. 97–114.

Bortugno, E.J., and Spittler, T.E., compilers, 1986, Geologic map of the San Bernardino quadrangle, California: California Division of Mines and Geology Regional Geologic Map Series, map no. 3A, sheet 1 of 5, scale 1:250,000.

Bowen, O.E., Jr., 1954, Geology and mineral deposits of the Barstow quadrangle, San Bernardino County, California: California Division of Mines Bulletin 165, p. 1–185.

Burke, D.B., 1979, Log of a trench in the Garlock fault zone, Fremont Valley, California: U.S. Geological Survey Miscellaneous Field Studies Map MF-1028, scale 1:20.

Buwalda, J.P., 1914, Pleistocene beds at Manix in the eastern Mojave Desert region: Bulletin of the Department of Geology, University of California, Berkeley, v. 7, no. 24, p. 443–464.

Cande, S.C., and Kent, D.V., 1995, Revised calibration of the geomagnetic polarity timescale for the Late Cretaceous and Cenozoic: Journal of Geophysical Research, v. 100, p. 6093–6095.

Chrisley, S.M., 1997, Geology and hydrogeology of the George Air Force Base vicinity, California [M.S. thesis]: San Jose, California, San Jose State University, 111 p.

Clark, M.M., and Lajoie, K.R., 1974, Holocene behavior of the Garlock fault: Geological Society of America Abstracts with Programs, v. 6, p. 156–157.

Corbett, E.J., 1984, Seismicity and crustal structure of southern California: Tectonic implications from improved earthquake locations [Ph.D. thesis]: Pasadena, California Institute of Technology, 231 p.

Cox, B.F., and Hillhouse, J.W., 2000, Pliocene and Pleistocene evolution of the Mojave River, and associated tectonic development of the Transverse Ranges and Mojave Desert, based on borehole stratigraphy studies near Victorville, California: U.S. Geological Survey Open-File Report 00-147, 66 p.

Cox, B.F., and Tinsley, J.C., III, 1999, Origin of the late Pliocene and Pleistocene Mojave River between Cajon Pass and Barstow, California, in Reynolds, R.E., and Reynolds, J., eds., Tracks along the Mojave: A field guide from Cajon Pass to the Calico Mountains and Coyote Lake: Redlands, California, San Bernardino County Museum Association Quarterly, v. 46, no. 3, p. 49–54.

Cox, B.F., and Wilshire, H.G., 1993, Geologic map of the area around the Nebo Annex, Marine Corps Logistics Base, Barstow, California: U.S. Geological Survey Open-File Report 93-568, 36 p., scale 1:12,000.

Cox, B.F., and Wilshire, H.G., 1994, Geologic map of the Yermo Annex and vicinity, Marine Corps Logistics Base, Barstow, California: U.S. Geological Survey Open-File Report 94-681, scale 1:12,000.

Cox, B.F., Hillhouse, J.W., Sarna-Wojcicki, A.M., and Tinsley, J.C., III, 1998, Pliocene-Pleistocene depositional history along the Mojave River north of Cajon Pass, California: Regional tilting and drainage reversal during uplift of the central Transverse Ranges: Geological Society of America Abstracts with Programs, v. 30, no. 5, p. 11.

Densmore, J.N., Cox, B.F., and Crawford, S.M., 1997, Geohydrology and water quality of Marine Corps Logistics Base, Nebo and Yermo Annexes, near Barstow, California: U.S. Geological Survey Water-Resources Investigations Report 96-4301, 116 p.

Dibblee, T.W., Jr., 1960a, Geologic map of the Hawes quadrangle, San Bernardino County, California: U.S. Geological Survey Mineral Investigations Field Studies Map MF-226, scale 1:62,500.

Dibblee, T.W., Jr., 1960b, Preliminary geologic map of the Victorville quadrangle: U.S. Geological Survey Mineral Investigations Field Studies Map MF-229, scale 1:62,500.

Dibblee, T.W., Jr., 1960c, Geologic map of the Barstow quadrangle, San Bernardino County, California: U.S. Geological Survey Mineral Investigations Field Studies Map MF-233, scale 1:62,500.

Dibblee, T.W., Jr., 1961, Evidence of strike-slip faulting along northwest-trending faults in the Mojave Desert, U.S. Geological Survey Professional Paper 424-B, p. B197-B199.

Dibblee, T.W., Jr., 1967, Areal geology of the western Mojave Desert, California: U.S. Geological Survey Professional Paper 522, 153 p.

Dibblee, T.W., Jr., 1970, Geologic map of the Daggett quadrangle, San Bernardino County, California: U.S. Geological Survey Miscellaneous Geologic Investigations Map I-592, scale 1:62,500.

Dibblee, T.W., Jr., 1975a, Late Quaternary uplift of the San Bernardino Mountains on the San Andreas and related faults, in Crowell, J.C., ed., San Andreas fault in southern California: California Division of Mines and Geology Special Report 118, p. 127–135.

Dibblee, T.W., Jr., 1975b, Tectonics of the western Mojave Desert near the San Andreas fault, in Crowell, J.C., ed., San Andreas fault in southern California: California Division of Mines and Geology Special Report 118, p. 155–161.

Dohrenwend, J.C., Bull, W.B., McFadden, L.D., Smith, G.I., Smith, R.S.U., and Wells, S.G., 1991, Quaternary geology of the Basin and Range province in California, in Morrison, R.B., ed., Quaternary nonglacial geology: Conterminous U.S.: Boulder, Colorado, Geological Society of America, The Geology of North America, v. K-2, p. 321–352.

Dokka, R.K., 1983, Displacements on late Cenozoic strike-slip faults of the central Mojave Desert, California: Geology, v. 11, p. 305–308.

Dokka, R.K., and Travis, C.J., 1990, Late Cenozoic strike-slip faulting in the Mojave Desert, California: Tectonics, v. 9, no. 2, p. 311–340.

Dutcher, L.C., and Worts, G.F., Jr., 1963, Geology, hydrology, and water supply of Edwards Air Force Base, Kern County, California: U.S. Geological Survey Open-File Report, 225 p.

Enzel, Y., and Wells, S.G., 1997, Extracting Holocene paleohydrology and paleoclimatology information from modern extreme flood events: An example from southern California: Geomorphology, v. 19, p. 203–226.

Foster, J.H., 1980, Late Cenozoic tectonic evolution of Cajon Valley, southern California [Ph.D. thesis]: Riverside, University of California, 243 p.

Fraser, D.M., 1931, Geology of the San Jacinto quadrangle south of San Gorgonio Pass, California: California Division of Mines, Mining in California, v. 27, no. 4, p. 494–540.

Gile, L.H., Peterson, F.F., and Grossman, R.B., 1966, Morphological and genetic sequences of carbonate accumulation in desert soils: Soil Science, v. 101, p. 347–360.

Groat, C.G., 1967, Geology and hydrology of the Troy Playa area, San Bernardino County, California [M.S. thesis]: New York, University of Rochester, 133 p.

Hagar, D.J., 1966, Geomorphology of Coyote Valley, San Bernardino County, California [Ph.D. thesis]: Amherst, University of Massachusetts, 210 p.

Howard, K.A., and Cox, B.F., 2001, Geologic strain partitioning in the eastern California shear zone: Riding crustal waves in the Mojave Desert: Geological Society of America Abstracts with Programs, v. 33, no. 3, p. 73–74.

Howard, K.A., and Miller, D.M., 1992, Late Cenozoic faulting at the boundary between the Mojave and Sonoran blocks: Bristol Lake area, California, in Richard, S.M., ed., Deformation associated with the Neogene Eastern California Shear Zone, southwestern Arizona and southeastern California: Redlands, California, San Bernardino County Museum Special Publication, p. 37–47.

Howard, K.A., and Stuart, W.D., 1992, Compressive origin of topographic waves in the Mojave Desert, California? [abs.]: American Geophysical Union, Chapman Conference on Tectonics and Topography, Snowbird, Utah, p. 28.

Huntley, D.J., Godfrey-Smith, D.I., and Thewalt, M.L.W., 1985, Optical dating of sediments: Nature, v. 313, p. 105–107.

IT Corporation, 1992, Remedial investigation, Operable Unit 2, Jp-4 spill, George Air Force Base, California, v. 1., San Bernardino, California (IT Corporation, 1425 South Victoria Court, Suite A, San Bernardino, CA 92408, USA).

Jefferson, G.T., 1985, Stratigraphy and geologic history of the Pleistocene Manix Formation, central Mojave Desert, California, in Reynolds, R.E., ed., Geologic investigations along Interstate 15: Cajon Pass to Manix Lake, California: Redlands, California, San Bernardino County Museum, p. 157–169.

Jefferson, G.T., 1986, Fossil vertebrates from late Pleistocene sedimentary deposits in the San Bernardino and Little San Bernardino Mountains region, in Kooser, M.A., and Reynolds, R.E., eds., Geology around the margins of the eastern San Bernardino Mountains: Publications of the Inland Geological Society, v. 1, p. 77–80.

Jefferson, G.T., 1994, Stratigraphy and Plio-Pleistocene history of the Lake Manix basin, in McGill, S.F., and Ross, T.M., eds., Geological investigations of an active margin: Geological Society of America, Cordilleran Section Guidebook, 27th Annual Meeting, San Bernardino, California, p. 175–177.

Jefferson, G.T., 1999, Age and stratigraphy of Lake Manix basin, *in* Reynolds, R.E., and Reynolds, J., eds., Tracks along the Mojave: A field guide from Cajon Pass to the Calico Mountains and Coyote Lake: Redlands, California, San Bernardino County Museum Association Quarterly, v. 46, no. 3, p. 109–111.

Kenney, M.D., and Weldon, R.J., 1999, Timing and magnitude of mid to late Quaternary uplift of the western San Bernardino and northeastern San Gabriel Mountains, southern California, *in* Reynolds, R.E., and Reynolds, J., eds., Tracks along the Mojave: A field guide from Cajon Pass to the Calico Mountains and Coyote Lake: Redlands, California, San Bernardino County Museum Association Quarterly, v. 46, no. 3, p. 33–46.

Kirschvink, J.L., 1980, The least-squares line and plane and the analysis of paleomagnetic data: Geophysical Journal of the Royal Astronomical Society, v. 62, p. 699–718.

Li, Y.-G., Henyey, T.L., and Leary, P.C., 1992, Seismic reflection constraints on the structure of the crust beneath the San Bernardino Mountains, Transverse Ranges, southern California: Journal of Geophysical Research, v. 97, no. B6, p. 8817–8830.

Londquist, C.J., Rewis, D.L., Galloway, D.L., and McCaffrey, W.F., 1993, Hydrogeology and land subsidence, Edwards Air Force Base, Antelope Valley, California, January 1989–December 1991: U.S. Geological Survey Water-Resources Investigations Report 93-4114, 74 p.

Mabey, D.R., 1960, Gravity survey of the western Mojave Desert, California: U.S. Geological Survey Professional Paper 316-D, p. 51–73.

Martin, P., 1994, Southern California basins regional aquifer, *in* McGill, S.F., and Ross, T.M., eds., Geological investigations of an active margin: Geological Society of America, Cordilleran Section Guidebook, 27th Annual Meeting, San Bernardino, California, p. 166–169.

Matti, J.C., and Morton, D.M., 1975, Geologic history of the San Timoteo Badlands, southern California: Geological Society of America Abstracts with Programs, v. 7, no. 3, p. 344.

Matti, J.C., and Morton, D.M., 1993, Paleogeographic evolution of the San Andreas fault in southern California: A reconstruction based on a new cross-fault correlation, *in* Powell, R.E., et al., eds., The San Andreas fault system: Displacement, palinspastic reconstruction, and geologic evolution: Boulder, Colorado, Geological Society of America Memoir 178, p. 107–159.

Matti, J.C., Powell, R.E., and Miller, F.K., 1998a, The Blackhawk Mountain massif, southern California: A Quaternary folded-thrust uplift in the left-stepping Helendale fault zone?: Geological Society of America Abstracts with Programs, v. 30, no. 5, p. 53.

Matti, J.C., Powell, R.E., and Miller, F.K., 1998b, The San Bernardino Mountains of southern California: A Quaternary fold-, thrust-, and tear-fault belt: Geological Society of America Abstracts with Programs, v. 30, no. 5, p. 53.

May, S.R., and Repenning, C.A., 1982, New evidence for the age of the Old Woman Sandstone, Mojave Desert, California, *in* Cooper, J.D., compiler, Geologic excursions in the Transverse Ranges, southern California: Geological Society of America, Cordilleran Section 78th Annual Meeting, Anaheim, California, 1982, Volume and Guidebook, Field Trip 6, p. 93–96.

McDaniel, G.E., and Jefferson, G.T., 2002, A late Pleistocene proboscidean site in the Death Valley Lake Tecopa beds near Shoshone, California, *in* Reynolds, R.E., ed., The 2001 Desert Symposium, Zzyzx, California: Fullerton, California, California State University Desert Studies Consortium, Abstracts of Proceedings, p. 72.

McFadden, L.D., and Weldon, R.J., II, 1987, Rates and processes of soil development on Quaternary terraces in Cajon Pass, California: Geological Society of America Bulletin, v. 98, p. 280–293.

Meek, N., 1989, Geomorphic and hydrologic implications of the rapid incision of Afton Canyon, Mojave Desert, California: Geology, v. 17, p. 7–10.

Meek, N., 1990, Late Quaternary geochronology and geomorphology of the Manix basin, San Bernardino County, California [Ph.D. thesis]: Los Angeles, University of California, 212 p.

Meek, N., 1999, New discoveries about the late Wisconsin history of the Mojave River system, *in* Reynolds, R.E., and Reynolds, J., eds., Tracks along the Mojave: A field guide from Cajon Pass to the Calico Mountains and Coyote Lake: Redlands, California, San Bernardino County Museum Association Quarterly, v. 46, no. 3, p. 113–117.

Meisling, K.E., 1984, Neotectonics of the north frontal fault system of the San Bernardino Mountains, southern California; Cajon Pass to Lucerne Valley [Ph.D. thesis]: Pasadena, California Institute of Technology, 394 p.

Meisling, K.E., and Weldon, R.J., 1989, Late Cenozoic tectonics of the northwestern San Bernardino Mountains, southern California: Geological Society of America Bulletin, v. 101, p. 106–128.

Miller, E.L., 1981, Geology of the Victorville region, California: Geological Society of America Bulletin, Part II, v. 92, no. 4, p. 554–608.

Miller, F.K., 1987, Reverse-fault system bounding the north side of the San Bernardino Mountains, *in* Holocene reverse faulting in the Transverse Ranges, California: U.S. Geological Survey Professional Paper 1339, p. 83–95.

Montgomery Watson (geotechnical consultants), 1995, George Air Force Base Installation Restoration Program OU1 predesign study (draft report, July 1995), Walnut Creek, California (Montgomery Watson, 1340 Treat Boulevard, Suite 300, Walnut Creek, CA 94596, USA).

Morton, D.M., and Matti, J.C., 1993a, Extension and contraction within an evolving divergent strike-slip complex: The San Andreas and San Jacinto fault zones at their convergence in southern California, *in* Powell, R.E., et al., eds., The San Andreas fault system: Displacement, palinspastic reconstruction, and geologic evolution: Boulder, Colorado, Geological Society of America Memoir 178, p. 217–230.

Morton, D.M., and Matti, J.C., 1993b, Tectonic synopsis of the San Gorgonio Pass and San Timoteo Badlands areas, southern California: Redlands, California, San Bernardino County Museum Association Quarterly, v. 40, no. 2, p. 3–14.

Nagy, E.A., and Murray, B., 1991, Stratigraphy and intra-basin correlation of the Mojave River Formation, central Mojave Desert, California: Redlands, California, San Bernardino County Museum Association Quarterly, v. 38, no. 2, p. 5–30.

Nagy, E.A., and Murray, B., 1996, Plio-Pleistocene deposits adjacent to the Manix fault: Implications for the history of the Mojave River and Transverse Ranges uplift: Sedimentary Geology, v. 103, nos. 1–2, p. 9–21.

Noble, L.F., 1954, Geology of the Valyermo quadrangle and vicinity, California: U.S. Geological Survey Geologic Quadrangle Map GQ-50, scale 1–24,000.

Oberlander, T.M., 1972, Morphogenesis of granitic boulder slopes in the Mojave Desert, California: Journal of Geology, v. 80, no. 1, p. 1–20.

Pluhar, C.J., Kirschvink, J.L., and Adams, R.W., 1991, Magnetostratigraphy and clockwise rotation of the Plio-Pleistocene Mojave River Formation, central Mojave Desert, California: Redlands, California, San Bernardino County Museum Association Quarterly, v. 38, no. 2, p. 31–42.

Powell, R.E., and Matti, J.C., 1998, Stratigraphic and geomorphic relations in the Mojave Desert and San Bernardino Mountains piedmont, Lucerne Valley, CA, part I: Geological Society of America Abstracts with Programs, v. 30, no. 5, p. 59.

Rector, C.H., Swenson, J.D., and Wilke, P.J., 1983, Archaeological studies at Oro Grande, Mojave Desert, California: Redlands, California, San Bernardino County Museum Association, 181 p.

Reynolds, R.E., 1989, Mid-Pleistocene faunas of the west-central Mojave Desert, *in* Reynolds, R.E., ed., The west-central Mojave Desert: Quaternary studies between Kramer and Afton Canyon: Redlands, California, San Bernardino County Museum Association Special Publication, p. 45–48.

Reynolds, R.E., 1991, Biostratigraphic relationships of Tertiary small vertebrates from Cajon Valley, San Bernardino County, California, *in* Woodburne, M.O., et al., eds., Inland southern California: The last 70 million years: Redlands, California, San Bernardino County Museum Association Quarterly, v. 38, nos. 3 and 4, p. 54–59.

Reynolds, R.E., and Cox, B.F., 1999, Tracks along the Mojave [field trip road log], *in* Reynolds, R.E., and Reynolds, J., eds., Tracks along the Mojave: A field guide from Cajon Pass to the Manix basin and Coyote Lake: Redlands, California, San Bernardino County Museum Association Quarterly, v. 46, no. 3, p. 1–23.

Reynolds, R.E., and Reynolds, R.L., 1994a, The Victorville Fan and an occurrence of Sigmodon, *in* Reynolds, R.E., ed., Off limits in the Mojave Desert: Redlands, California, San Bernardino County Museum Association Special Publication 94-1, p. 31–33.

Reynolds, R.E., and Reynolds, R.L., 1994b, The isolation of Harper Lake basin, *in* Reynolds, R.E., ed., Off limits in the Mojave Desert: Redlands, California, San Bernardino County Museum Association Special Publication 94–1, p. 34–37.

Sadler, P.M., 1982a, An introduction to the San Bernardino Mountains as the product of young orogenesis, *in* Cooper, J.D., compiler, Geologic excursions in the Transverse Ranges, southern California: Geological Society of America, Cordilleran Section 78th Annual Meeting, Anaheim, California, 1982, Volume and Guidebook, Field Trip 6, p. 57–65.

Sadler, P.M., 1982b, Provenance and structure of late Cenozoic sediments in the northeast San Bernardino Mountains, *in* Cooper, J.D., compiler, Geologic excursions in the Transverse Ranges, southern California: Geological Society of America, Cordilleran Section 78th Annual Meeting, Anaheim, California, 1982, Volume and Guidebook, Field Trip 6, p. 83–91.

Sadler, P.M., 1993, The Santa Ana basin of the central San Bernardino Mountains: Evidence of the timing of uplift and strike slip relative to the San Gabriel Mountains, *in* Powell, R.E., et al., eds., The San Andreas fault system: Displacement, palinspastic reconstruction, and geologic evolution: Boulder, Colorado, Geological Society of America Memoir 178, p. 307–321.

Sadler, P.M., and Reeder, W.A., 1983, Upper Cenozoic, quartzite-bearing gravels of the San Bernardino Mountains, southern California: Recycling and mixing as a result of transpressional uplift, *in* Andersen, D.W., and Rymer, M.J., eds., Tectonics and sedimentation along faults of the San Andreas system: Pacific Section, Society of Economic Paleontologists and Mineralogists, p. 45–57.

Scott, E., Springer, K., and Murray, L.K., 1997, New records of early Pleistocene vertebrates from the west-central Mojave Desert, San Bernardino County, California: Journal of Vertebrate Paleontology, v. 17, supplement to no. 3, p. 75A.

Shreve, R.L., 1968, The Blackhawk landslide: Boulder, Colorado, Geological Society of America Special Paper 108, 47 p.

Sibbett, B.S., 1996, Paleo fluvial channels control migration of a hydrocarbon spill on George Air Force Base in the Mojave Desert: Geological Society of America Abstracts with Programs, v. 28, no. 7, p. 393.

Sibbett, B.S., 1999, Pleistocene channels of the Mojave River near Victorville, California, *in* Reynolds, R.E., and Reynolds, J., eds., Tracks along the Mojave: A field guide from Cajon Pass to the Calico Mountains and Coyote Lake: Redlands, California, San Bernardino County Museum Association Quarterly, v. 46, no. 3, p. 65–68.

Singer, B.S., Hoffman, K.A., Chauvin, A., Coe, R.S., and Pringle, M.S., 1999, Dating transitionally magnetized lavas of the late Matuyama Chron: Toward a new $^{40}Ar/^{39}Ar$ timescale of reversals and events: Journal of Geophysical Research, v. 104, p. 679–693.

Smith, G.I., 1984, Paleohydrologic regimes in the southwestern Great Basin, 0–3.2 m.y. ago, compared with other long records of "global" climate: Quaternary Research, v. 22, p. 1–17.

Smith, G.I., Bischoff, J.L., and Bradbury, J.P., 1997, Synthesis of the paleoclimatic record from Owens Lake core OL-92, *in* Smith, G.I., and Bischoff, J.L., eds., An 800,000-year paleoclimatic record from Core OL-92, Owens Lake, southeast California: Boulder, Colorado, Geological Society of America Special Paper 317, p. 143–160.

Spotila, J.A., and Sieh, K., 2000, Architecture of transpressional thrust-faulting in the San Bernardino Mountains, southern California, from deformation of a deeply weathered surface: Tectonics, v. 19, no. 4, p. 589–615.

Spotila, J.A., Farley, K.A., and Sieh, K., 1998, Uplift and erosion of the San Bernardino Mountains associated with transpression along the San Andreas fault, California, as constrained by radiogenic helium thermochronometry: Tectonics, v. 17, no. 3, p. 360–378.

Stamos, C.L., Martin, P., Nishikawa, T., and Cox, B.F., 2001, Simulation of ground-water flow in the Mojave River basin, California: U.S. Geological Survey Water-Resources Investigations Report 01-4002, 129 p.

Subsurface Surveys, Inc., 1990, Inventory of groundwater stored in the Mojave River basins: Unpublished report, prepared for Mojave Water Agency, Apple Valley, California, 47 p. (Subsurface Surveys, Inc., 215 South Highway 101, Suite 203, Solana Beach, CA 92075, USA).

Thompson, D.G., 1929, The Mojave Desert region, California: A geographic, geologic, and hydrologic reconnaissance: U.S. Geological Survey Water-Supply Paper 578, 759 p.

Weldon, R., 1985, Implications of the age and distribution of the late Cenozoic stratigraphy in Cajon Pass, southern California, *in* Reynolds, R.E., ed., Geological investigations along Interstate 15: Cajon Pass to Manix Lake, California: Redlands, California, San Bernardino County Museum, p. 59–68.

Weldon, R.J., II, 1986, The late Cenozoic geology of Cajon Pass: Implications for tectonics and sedimentation along the San Andreas fault [Ph.D. thesis]: Pasadena, California Institute of Technology, 400 p.

Weldon, R.J., II, and Sieh, K.E., 1985, Holocene rate of slip and tentative recurrence interval for large earthquakes on the San Andreas fault, Cajon Pass, southern California: Geological Society of America Bulletin, v. 96, p. 793–812.

Weldon, R.J., II, Meisling, K.E., and Alexander, J., 1993, A speculative history of the San Andreas fault in the central Transverse Ranges, California, *in* Powell, R.E., et al., eds., The San Andreas fault system: Displacement, palinspastic reconstruction, and geologic evolution: Boulder, Colorado, Geological Society of America Memoir 178, p. 161–198.

Williamson, T.N., and Wells, S.G., 1994, Understanding the geomorphic evolution of the Mojave River headwaters: Redlands, California, San Bernardino County Museum Association Quarterly, v. 41, no. 3, p. 32.

Woodburne, M.O., 1975, Cenozoic stratigraphy of the Transverse Ranges and adjacent areas, southern California: Boulder, Colorado, Geological Society of America Special Paper 162, 91 p.

Woodburne, M.O., and Golz, D.J., 1972, Stratigraphy of the Punchbowl Formation, Cajon Valley, southern California: University of California Publications in Geological Sciences, v. 92, 73 p.

Zohdy, A.A.R., and Bisdorf, R.J., 1994, A direct-current resistivity survey near the Marine Corps Logistics Bases at Nebo and Yermo, Barstow, California: U.S. Geological Survey Open-File Report 94-202, 155 p.

MANUSCRIPT ACCEPTED BY THE SOCIETY AUGUST 1, 2002

Stratigraphy and paleontology of the middle to late Pleistocene Manix Formation, and paleoenvironments of the central Mojave River, southern California

George T. Jefferson
Colorado Desert District Stout Research Center, Anza-Borrego Desert State Park, 200 Palm Canyon Drive, Borrego Springs, California 92004, USA

ABSTRACT

The Manix Formation consists of lacustrine, fluvial, and alluvial sediments that were deposited within and adjacent to Lake Manix. This middle to late Pleistocene pluvial lake was probably the largest along the Mojave River drainage system, which heads in the Transverse Ranges of southern California and runs northeast to Death Valley. During high lake stands, the lake covered ~236 km² of the central Mojave Desert. Lake Manix deposits record much of the early hydrological history of the upper Mojave River.

Attaining an exposed thickness of ~40 m, the Manix Formation overlies playa lake deposits of the Pliocene-Pleistocene Mojave River Formation, and is locally overlain by latest Pleistocene and Holocene fluvial deposits. The Manix Formation is mappable as four laterally equivalent members. Generally from oldest to youngest these were deposited as: basin margin fanglomerates, alluvial and fluvial deposits, lacustrine and paralimnic deposits, and fluvial/deltaic deposits. The interfingering relationships of these sediments, especially between the lacustrine and fluvial systems, documents at least four major transgressive/regressive lacustrine events within the last 0.5 m.y. Transgressive lacustrine events are in phase with relatively cool/moist regional climatic conditions inferred to be present during even-numbered marine oxygen isotope stages. The chronology of these deposits is relatively well constrained by paleomagnetic data, U/Th series ages, tephra chronology, and ^{14}C ages.

Molluscan, ostracode, and aquatic and terrestrial vertebrate assemblages have been recovered primarily from the fluvial, paralimnic and lacustrine deposits within the formation. Included are 55 taxa of fossil vertebrates that range between 350+ ka and ca. 20 ka, spanning possibly late Irvingtonian through late Rancholabrean North American Land Mammal Ages. The assemblage comprises extinct and extralocal extant taxa that reflect climatic and biogeographic conditions dramatically different from the present xeric environment.

INTRODUCTION

Lake Manix was probably the largest in a chain of Pleistocene lakes that extended across the central Mojave Desert region of California along the Mojave River drainage (Fig. 1). The lake was located about midway between the headwaters of the Mojave River in the San Bernardino Mountains to the southwest, and Death Valley to the northeast, where the ancestral Mojave River episodically joined the Amargosa River and ended in Pleistocene Lake Manly. Lake Manly also was the terminus of the Owens River system, which drained the east side of the southern Sierra Nevada Mountains and basins within the northern Mojave Desert region (see Blackwelder, 1954; Blanc and Cleveland, 1961a, 1961b; Snyder et al., 1964).

With a highstand of 543 m above mean sea level, Lake Manix occupied a cloverleaf-shaped area of ~236 km² (Meek, 1990) including the present Coyote Lake and Troy Lake playa basins and the Afton basin (Fig. 2). Well-developed wave-cut

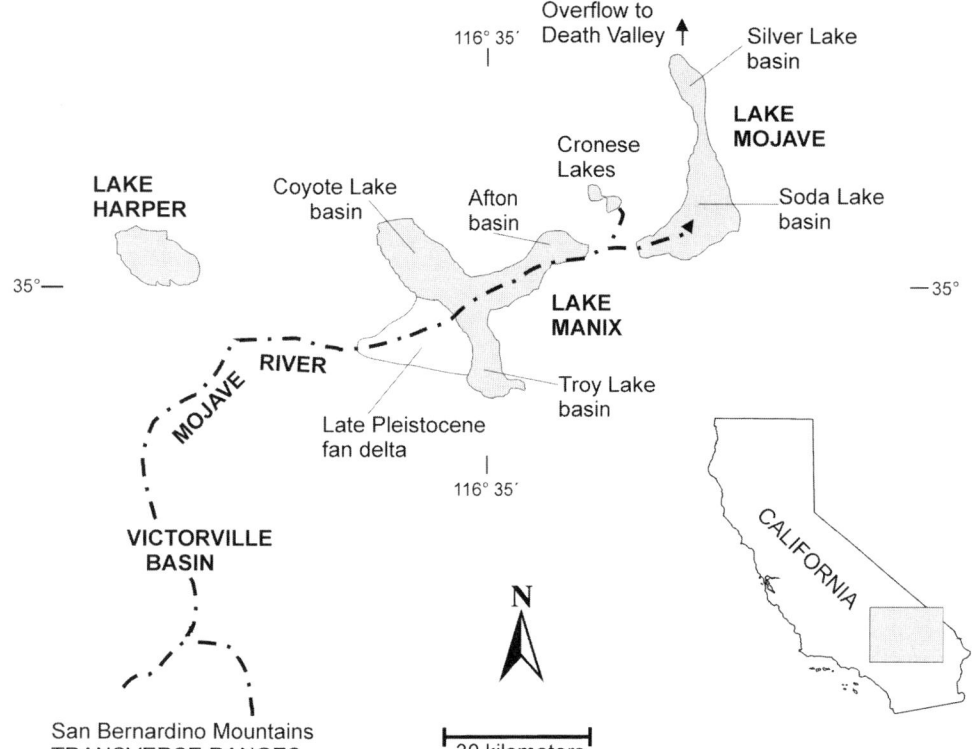

Figure 1. Regional map of Pleistocene lakes along the Mojave River drainage. Pleistocene lakes (shaded) are shown at highest stands. The modern Mojave River course is indicated by a long-short dashed line.

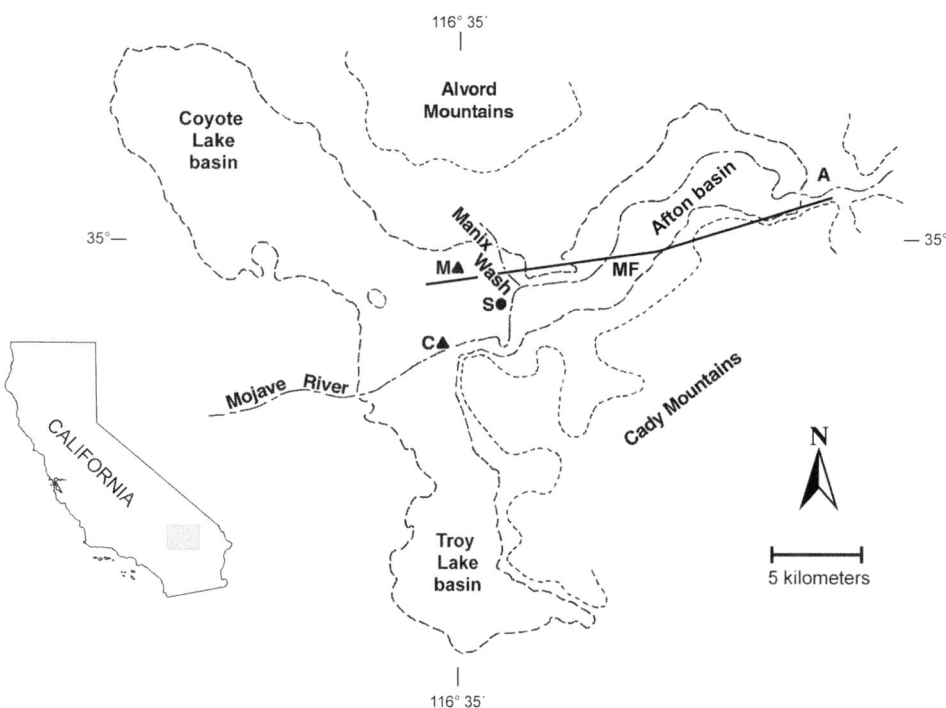

Figure 2. Generalized map of Lake Manix basin. The approximate maximum extent of the lake (shaded) is represented by a long dashed line at the 543 m topographic contour. Short dashed lines encompass mountains. The present Mojave River and Manix Wash drainages are represented by long-short dashed lines. Approximate trace of the Manix fault is shown as a solid line. A—Afton Canyon; C—Camp Cady; M—Manix railroad siding; MF—Manix fault; S—top of the reference stratigraphic section (see Figs. 6, 7).

terraces and beach bars typify its eastern and southern shorelines (Meek, 2000). The ancestral Mojave River began to fill the Lake Manix basin ca. 0.5 Ma, and Lake Manix persisted with several, apparently short-lived low level stands until the late Wisconsin.

Geologic and paleontologic investigations in Lake Manix basin began with Buwalda's (1914) description of the Pleistocene "Manix Beds of the eastern Mojave Desert." Buwalda described the stratigraphy, depositional history, and recognized the climatic implications of such deposits. A year later, in his summary of the extinct faunas of the Mojave Desert, Merriam (1915) confirmed Buwalda's age assessment and compared the terrestrial vertebrate fossils from Manix to those from Rancho La Brea in Los Angeles, California. Since these early works, the geomorphology, stratigraphy, depositional history, and paleontology of the "Manix Beds" have been the focus of numerous studies. Manix basin and the long-lived lake it held, derives its name from the Manix Union Pacific railroad siding (Fig. 2), 32 km east of the city of Barstow, San Bernardino County, California.

Latest Pleistocene and Holocene incision by the Mojave River and local tributaries such as Manix Wash have exposed a 120+ m-thick section of alluvial fan, playa, lacustrine, and fluvial sediments that records much of the Pleistocene history of the central Mojave River. Exposures in this area serve as the type localities for the Pliocene-Pleistocene Mojave River Formation (Nagy and Murray, 1991), the middle to late Pleistocene Manix Formation (Jefferson, 1985a, 1994, 1999), and the Camp Cady faunal assemblage (Winters, 1954; Howard, 1955; Jefferson, 1985b, 1987, 1991a; Seiple, 1994).

The upward-coarsening Pliocene-Pleistocene Mojave River Formation consists primarily of fine-grained sediments deposited in an internally drained playa basin (Nagy and Murray, 1991). These deposits rest on top of the late Miocene and Pliocene fanglomerates that initially filled Manix basin. Within the Manix Formation, which overlies the Mojave River Formation, locally derived alluvial and fluvial sediments, and lacustrine deposits largely transported by the ancestral Mojave River, appear to transgress/regress in consort with climate changes over the past 0.5 m.y. (Jefferson, 1985a, 1994, 1999).

The assemblage of invertebrate and vertebrate fossils recovered from eroded river bluffs and badlands of the Manix Formation dates from ca. 350 ka to 20 ka, and includes extinct and extralocal extant taxa that reflect a more equable climate and diverse biogeographic setting than at present (Buwalda, 1914; Merriam, 1915; Winters, 1954; Howard, 1995; Jefferson, 1985a, 1985b, 1987, 1991a; Jefferson and Steinmetz, 1986; Steinmetz, 1987, 1988; Seiple, 1994).

The following discussion is an amended synthesis of previous studies. It provides a summary of the stratigraphy and depositional history of the Manix Formation, and a brief discussion of deposits and events relevant to the history of Lake Manix and the paleohydrology of the Mojave River. Also included is an analysis of the fossil assemblage recovered from Lake Manix and its paleoenvironmental significance (Jefferson, 1987).

PLIOCENE-PLEISTOCENE STRATIGRAPHY

Basement rocks exposed around the margins of Manix basin are primarily pre-Cenozoic metamorphic rocks and Tertiary volcanic and volcaniclastic deposits (Beyers, 1960; Bassett and Kupfer, 1964). The oldest fill within the basin consists of the late Miocene and Pliocene fanglomerates (older fanglomerate unit of Nagy and Murray, 1991; granitic fanglomerate of Meek and Battles, 1991). These fanglomerates are overlain by Pliocene-Pleistocene playa and fluvial sediments of the Mojave River Formation (Nagy and Murray, 1990, 1991). These units, in turn are usually overlain unconformably by fanglomerates, alluvial, fluvial, and lacustrine sediments of the middle to late Pleistocene Manix Formation (Buwalda, 1914; Ellsworth, 1933; Blackwelder and Ellsworth, 1936; Winters, 1954; Jefferson, 1985a, 1994, 1999; Budinger, 1992). Late Wisconsin and Holocene fluvial sediments (Hagar, 1966; Groat, 1967; Meek, 1990) both cap and are inset into the older sedimentary sequence.

Mojave River Formation, ca. 2.5–1 Ma

The Mojave River Formation consists primarily of gypsiferous, reddish-tan and light gray-green claystones and siltstones that total >80 m in thickness (Nagy and Murray, 1991) (Fig. 3). The base of this upward-coarsening sequence is not exposed in the type area, near the confluence of Manix Wash and the Mojave River (Fig. 2). However, 4 km to the east, these deposits overlie a 30+ m-thick section of fanglomerates (older fanglomerate unit of Nagy and Murray, 1991) that are exposed on the upthrown, north side of the Manix fault (Keaton and Keaton, 1977; McGill et al., 1988) (Fig. 4). To the south, toward the Cady Mountains, the deposits grade up section into fine to coarse-grained lithic arenites and granule to cobble conglomerates.

In the most northern exposures, these deposits are severely deformed and upturned next to the Manix fault, and to the south

Figure 3. Mojave River Formation. The view is to the northwest from approximately 1 km east of the confluence of Manix Wash and the Mojave River. Gypsum-cemented pale gray-green siltstones and claystones form the more resistant horizons (also see exposures in mid-image of Fig. 5). Note geologist for scale.

Figure 4. Fanglomerates and the Mojave River Formation. The view is to the northeast from the Mojave River bed about 4 km east of the mouth of Manix Wash. These units, medium gray fanglomerate (left) and pale reddish-tan siltstones and claystones of the Mojave River Formation (right), are in contact along one of several branches of the Manix fault.

they are folded into a broad east-west–trending syncline (Keaton and Keaton, 1977; McGill et al., 1988; Murray and Nagy, 1990; Nagy and Murray, 1991). Nagy and Murray (1991) maintain that the top of the unit grades up section through a transitional zone into the lowermost member of the Manix Formation (Member A of Jefferson, 1985a). Member A is lithologically similar to the upper Mojave River Formation, and some exposures exhibit this transitional relationship between the two formations (Jefferson, 1994, 1999). However, all other members of the mostly horizontal Manix Formation usually unconformably overlie the Mojave River Formation and all older units (Fig. 5).

Magnetostratigraphic correlations place the base of the Mojave River Formation at 2.48 Ma, and its top at either the Jaramillo (0.92–0.97 Ma) or the Cobb Mountain paleomagnetic events (1.1 Ma) (Pluhar et al., 1991). Sarna-Wojcicki et al. (1980) (Sarna-Wojcicki et al., 1984) correlate the lowermost fine-grained vitric gray ash that occurs in the top half of the Mojave River Formation with the Huckleberry Ridge ash bed (ca. 2.01 Ma, Izett, 1981), and the middle fine-grained vitric white ash with the slightly older Waucoba Beds sequence (ca. 2.3 Ma). The upper fine-grained vitric white ash of Sarna-Wojcicki et al. (1980) (Sarna-Wojcicki et al., 1984) remains to be positively correlated with dated tephra.

Nagy and Murray (1991) consider the presence of two sets of southwest-dipping cross strata, present in the top of the Mojave River Formation (Unit C2 at 19.1 m and 23 m, Nagy and Murray, 1991), as the earliest evidence of fluvial drainage in Manix basin, suggesting an open rather than a closed, internally drained system. However, these strata indicate a southwestward paleocurrent direction (Nagy and Murray, 1991). A change to the eastward flow direction of the present Mojave River apparently postdates deposition of the Mojave River Formation.

Manix Formation, ca. 1–0.2 Ma

The middle to late Pleistocene Manix Formation consists of lacustrine, fluvial, and alluvial sediments that were deposited in and next to Lake Manix (Buwalda, 1914; Ellsworth, 1933; Blackwelder and Ellsworth, 1936; Winters, 1954; Jefferson, 1985a, 1994, 1999; Meek, 1990; Budinger, 1992). Within the type area, near the confluence of Manix Wash and the Mojave River, these deposits unconformably overlie pre-Cenozoic basement rocks and Tertiary volcanic rocks, Miocene-Pliocene fanglomerates, and the Mojave River Formation. The Manix Formation is locally unconformably overlain by latest Pleistocene and Holocene lacustrine and fluvial sediments and Holocene eolian and slope deposits. The Manix Formation has a maximum exposed thickness of 41.5 m and is geologically mappable as four distinct Members A–D (Jefferson, 1985a, 1994, 1999; Budinger, 1992). Generally, from oldest to youngest these are: A, fanglomerates; B, alluvial/fluvial and lacustrine deposits; C, lacustrine and paralimnic deposits; and D, fluvial/deltaic deposits (Figs. 6, 7).

Member A. Member A is exposed south and southeast of Manix siding along both sides of the Mojave River (Fig. 2). The unit lies unconformably on metamorphic basement and volcanic rocks, and overlies the gravely, upper sediments of the Mojave River Formation. It is wedge-shaped, flanks the Cady Mountains, and dips gently to the north and northwest. To the southeast the deposit is >27 m-thick. The uppermost conglomerate beds of Member A interfinger with Member C to the west and north, and with Member B to the north and northeast (Fig. 7). The exposed base of the fanglomerate is estimated to be younger than ca. 0.9 Ma (Nagy and Murray, 1991).

Member A is composed primarily of dark to moderate brown, poorly sorted, subangular, cobble and boulder, clast-supported conglomerate interbedded with silty, fine to coarse-grained lithic arenites. Although locally cross stratified, the arenites are usually massive. Conglomerate beds are generally lenticular and exhibit lateral textural changes over very short distances. Clast imbrications generally indicate northwest paleocurrent directions, and clast lithologies are derived entirely from metamorphic and Tertiary volcanic and volcaniclastic rocks exposed to the south in the western Cady Mountains.

Figure 5. Mojave River Formation and Manix Formation. The view is to the east from Manix Wash (foreground). The gently folded Mojave River Formation appears mid-image and is unconformably overlain by the flat lying Manix Formation. Both units are in contact, along the Manix fault, with older fanglomerates (high left horizon).

Member B. Member B crops out in Manix Wash, and south of Manix Wash along the west side of the Mojave River (Fig. 2). The unit conformably overlies Member A south of Manix Wash, and east of the southern part of Manix Wash, it unconformably overlies the Mojave River Formation. In Manix Wash, the unit is ~30 m thick. To the south it bifurcates into upper and lower wedges (Fig. 7). The lower wedge averages 11 m in thickness and pinches out to the south between Members A and C. The upper wedge averages 2.5 m in thickness and interfingers with Member C and also pinches out to the south.

The base of the lower wedge consists of thinly bedded, pale olive claystones, gray-green siltstones, and fine-grained silty and platy micaceous arenites. These are overlain by coarse-grained arkose with lenses of granule to cobble conglomerate. Clast lithologies are predominately granitic. The uppermost clasts in this upward-coarsening sequence are coated with gray calcareous oncoid stromatolites, indicative of near-shore lacustrine environments typically formed during transgressive lacustrine events (Awramik et al., 2000) (Fig. 8). These sediments are overlain by light gray-green, thinly bedded silty arenite. The remaining upper part of the lower wedge is composed of grayish-orange to pale yellow-brown, silty coarse to medium-grained arkose and interbedded granule conglomerate.

Although both the lower and upper wedges of Member B thicken northward, to the south, the upper wedge generally occurs 13 m above the base of Member C (Fig. 7). It is composed of grayish-orange to pale yellow-brown, silty coarse and medium sand-grained arkose and interbedded granule conglomerate. The conglomerates are cross stratified, indicating a southwest paleocurrent direction, and are scoured into the underlying siltstone of Member C. Clast lithologies are predominately granitic, and are probably derived from the east flank of the Alvord Mountains to the north. An erosional unconformity at the base of the upper part of Member C truncates the top of the wedge.

The lowest exposed beds of Member B are magnetically normal (Kirschvink, 1984, personal commun.), and, given their stratigraphic position, presumably fall within the mid-early Brunhes paleomagnetic event. An infinite U/Th series age of 350+ ka (USGS 80-51) was obtained from an *Equus* ulna fragment (J.L. Bischoff, 1982, 1983, personal commun.) (Table 1) recovered from the top of the lower wedge of Member B, 5 m below the base of Member C (Fig. 6).

Member C. Member C crops out in bluffs along both sides of the Mojave River and Manix Wash (Fig. 2). Here it consists of >32 m of light gray, greenish-yellow and gray-green siltstone and claystone (Fig. 6). The unit is thickest in the middle of the basin and thins at the basin margin. The lowest exposures interfinger with both Members A (Fig. 9) and B. The upper part of Member C locally overlies and pinches out against Member A south of the Mojave River, and with Member B to the north along Manix Wash (Fig. 7).

To the north, the lower and upper parts of Member C are separated by the upper wedge of Member B. These two lacustrine layers of Member C were attributed to two distinct lake phases by Buwalda (1914) and Winters (1954). The two layers are well exposed to the east throughout the Afton basin (Ellsworth, 1933; Blackwelder and Ellsworth, 1936). In eastern Afton basin, Member C interfingers with, and is locally overlain by alluvial fan deposits shed to the north from the eastern Cady Mountains (Blackwelder and Ellsworth, 1936; Meek, 2000).

A 0.5–1 m-thick pale yellowish-orange to grayish-orange, poorly bedded, moderately to well sorted, subrounded, medium-grained, quartz-rich arkose is commonly present where the base of lower Member C laps onto Member A and overlies the lower

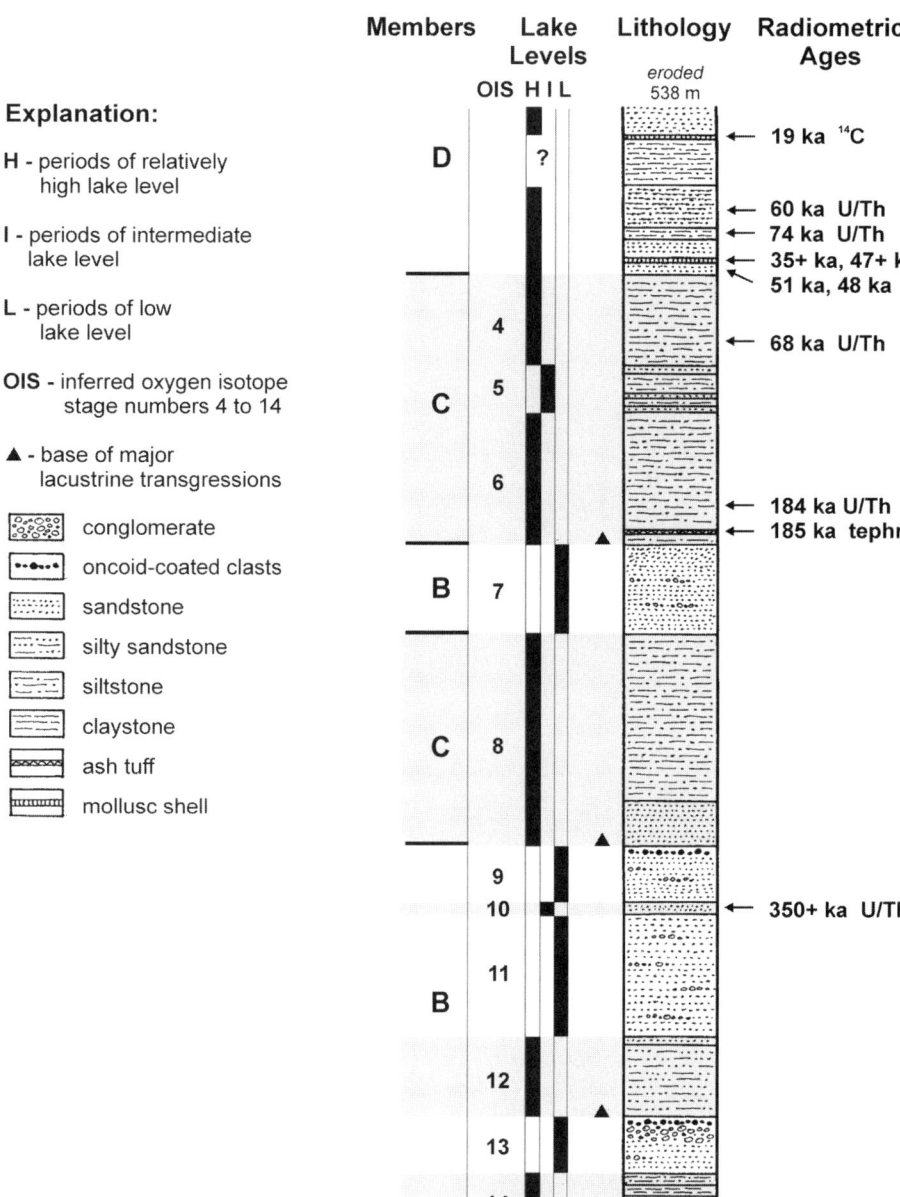

Figure 6. Reference stratigraphic section of the Manix Formation. Shown are Members B through D (see "S" Figs. 2, 7). The top of Member A is not exposed at this location, and only the last three of the four major lacustrine transgressions are indicated (▲). The section totals 40 m, from 498 to 538 m above mean sea level. Lacustrine intervals within Members B and C are shaded. Periods of relatively high (H), intermediate (I), and low (L) lake levels are indicated by the vertical black bars. Radiometric determinations (Table 1) are discussed in the text.

wedge of Member B. The arkose rests on light gray, rugose oncoids that have grown on clasts at the top of Members A and B (Fig. 8). Much of the avian and terrestrial vertebrate assemblage has been recovered from this time-transgressive littoral horizon.

Three thin (0.1–0.15 m-thick), laterally persistent, grayish-orange, moderately well bedded, moderately well sorted, subrounded, medium-grained, quartz-rich arkose beds are present in the middle-upper part of Member C. These sandy horizons, like those at the base of lower Member C, probably reflect littoral, near-shore depositional conditions (Fig. 6).

A 0.3–0.4 m-thick tephra of the Long Valley–Mono Glass Mountain family is present at the base of the upper part of Member C, 0.3 m above the top of the upper wedge of Member B (Fig. 6). It has been chemically correlated with tephra in the Long Canyon source area that date 185.0 ± 15.0 ka (Bacon and Duffield, 1981; Izett, 1981; Sarna-Wojcicki et al., 1984). This white, friable, well-sorted, fine-grained rhyolitic ash is composed of glass shards with minor amounts of biotite and quartz. A *Camelops* scapula fragment recovered 1.5 m above the ash provided a U/Th series age of 183.8 ± 12.0 ka (USGS 81–30) (J.L. Bischoff, 1982, 1983, personal commun.) (Table 1) (Fig. 6).

A U/Th series age of 68.0 ± 4.0 ka (USGS 81–51) (J.L. Bischoff, 1982, 1983, personal commun.) (Table 1), was obtained from a *Camelops* humerus fragment from the mid-upper part of Member C, which was recovered 1 m above the highest of the three near-shore arkosic sands (Fig. 6).

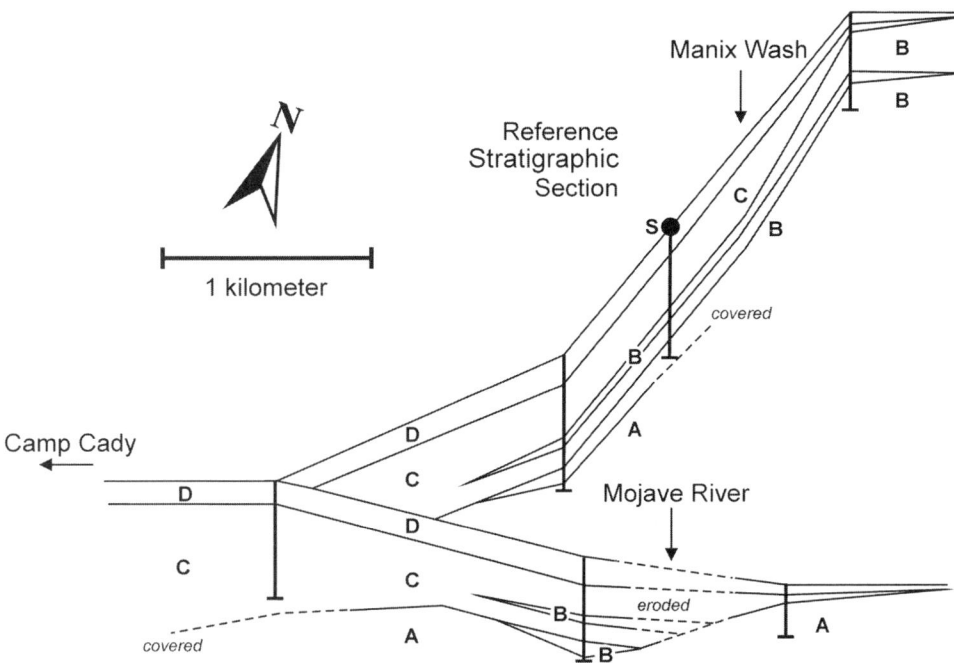

Figure 7. Panel diagram of the surficial exposures of the Manix Formation. Illustrated are the interdigitation and superposition of Members A through D. For location of the reference stratigraphic section (Fig. 6) see "S" in Figure 2. Other measured sections appear as vertical lines. Member C is shaded. The vertical scale is 10 × the horizontal scale. The area within the diagram extends from near Camp Cady, east then north 5 km along the Mojave River (see view in Fig. 9), and 3.5 km from the Mojave River north up Manix Wash drainage.

Member D. Member D conformably overlies Member C, and is overlain by late Pleistocene and Holocene deposits. The member pinches out to the east and north against Members A, B and C, and thickens from ~4.6 m in Manix Wash, to >7.7 m to the west near Camp Cady (Fig. 7). Along the Mojave River, east of Camp Cady, exposures of Member D are composed entirely of light brown, poorly to moderately sorted, silty medium to coarse-grained arkose, and granule to cobble conglomerate lenses. Clasts include a wide variety of igneous and metamorphic lithologies transported by the ancestral Mojave River.

South of Manix, Member D ranges in age from ca. 60 ka to 19 ka. U/Th series ages of 51.2 ± 2.5 ka (USGS 81-49) and 47.7 ± 2.0 ka (USGS 81-48) (J.L. Bischoff, 1982, 1983, personal commun.) were obtained from a *Hemiauchenia* radius/ulna and *Mammuthus* femur fragment recovered from within 0.5 m above the base of Member D (Fig. 6). *Anodonta* shells from 0.7 m above the base of Member D have produced infinite ^{14}C ages of 47+ ka (Y-1993) (Stuiver, 1969; Bassett and Jefferson, 1971) and 35+ ka (R. Berger, 1982, personal commun.). A U/Th age of 74.0 ka on mammalian bone and a U/Th age of 60.3 ka on *Anodonta* shell are reported by Budinger (1992) 1.5 and 2.0 m above the base of Member D respectively (Table 1) (Fig. 6). Given its stratigraphic position, the former age appears to be too old. An *Anodonta* horizon, in the uppermost exposures of Member D south of Manix, has yielded a ^{14}C age of 19.1 ± 0.25 ka (QC-1467) (R. Pardi, 1983, personal commun.). Bivalve remains from stratigraphically correlative strata in eastern Afton basin have produced ^{14}C dates that range in age from ca. 31 ka to 28 ka and from ca. 21 ka to 18 ka (Table 1) (Bassett and Jefferson, 1971; Hastorf and Tinsley, 1981; Meek, 1990, 1999).

PALEONTOLOGY OF THE MANIX FORMATION

The aquatic invertebrate, and aquatic and terrestrial vertebrate taxa that compose the fossil assemblage from Lake Manix (Table 2) have been recovered from the fluvial sediments of both the lower and upper wedges of Member B, the littoral deposits at the base and middle of Member C, and the fluvial sediments of

Figure 8. Oncoid-coated cobbles. The largest cobble is approximately 0.15 m in diameter. This exposed horizon occurs near the base of Member B, Manix Formation, at the inferred 13/12 marine oxygen isotope stage boundary, and represents the second major transgressive lacustrine event (Fig. 6). A similar horizon also is present at the base of Member C, 9/8 oxygen isotope stage boundary (Fig. 6). Highly rugose surfaces develop on the tops of the cobbles.

TABLE 1. RADIOMETRIC AND ABSOLUTE DATES FROM LAKE MANIX BASIN

Age (yr B.P.)	Location	Material	Method	Laboratory	Source(s)
11,810 ± 100	Coyote Lake basin	*Anodonta* shell	^{14}C	UCLA 2609C	Meek, 1990
12,900 ± 120	Coyote Lake basin	*Anodonta* shell	^{14}C	UCLA 2606	Meek, 1990
13,560 ± 145	Coyote Lake basin	*Anodonta* shell	^{14}C	UCLA 2609B	Meek, 1990
13,800 ± 600	Coyote Lake basin	*Anodonta* shell	^{14}C	La Jolla 958	Hubbs et al., 1965
14,230 ± 1,325*	Afton basin	*Anodonta* shell	^{14}C	UCLA 2601	Meek, 1989, 1990
15,025 ± 230	Troy Lake basin	*Anodonta* shell	^{14}C	UCLA 2605	Meek, 1990
15,125 ± 270	Coyote Lake basin	*Anodonta* shell	^{14}C	UCLA 2608	Meek, 1990
16,750 ± 1,000	Central Manix basin	oncoid stromatolite	^{14}C	UCLA 1079	Berger and Libby, 1967
17,950 ± 1,500	Coyote Lake basin	*Anodonta* shell	^{14}C	UCLA 6203	Meek, 1990
18,150 ± 400	Afton basin	*Anodonta* shell	^{14}C	UCLA 2607	Meek, 1990
19,100 ± 250	Manix basin D	*Anodonta* shell	^{14}C	QC 1467	R. Pardi, 1983, personal commun.; Jefferson, 1985a
19,300 ± 400	Afton basin	oncoid stromatolite	^{14}C	UCLA 121	Fergusson and Libby, 1962
19,500 ± 500	Afton basin	oncoid stromatolite	^{14}C	La Jolla 269	Hubbs et al., 1962
19,700 ± 260	Central Manix basin	oncoid stromatolite	^{14}C	UCLA 2600B	Meek, 1990
20,500 ± ?	Afton basin	*Anodonta* shell	^{14}C	Yale	Stuiver, 1969; Bassett and Jefferson, 1971
20,980 ± 345	Central Manix basin	oncoid stromatolite	^{14}C	UCLA 2602	Meek, 1990
21,300 ± 1,710	Afton basin	*Anodonta* shell	^{14}C	-	Meek, 1999
23,090 ± 445	Central Manix basin	*Anodonta* shell	^{14}C	UCLA 2600A	Meek, 1990
28,960 ± 2,490	Afton basin	*Anodonta* shell	^{14}C	UCLA 2601C	Meek, 1999
29,310 ± 310	Afton basin	*Anodonta* shell	^{14}C	CAMS 1856	Meek, 1999
30,650 ± 890	Afton basin	*Anodonta* shell	^{14}C	UCLA 2604	Meek, 1990
30,950 ± 1,000	Afton basin	oncoid stromatolite	^{14}C	La Jolla 895	Hubbs et al., 1965
35 + (infinite)	Manix basin D	*Anodonta* shell	^{14}C	UCLA	R. Berger, 1982, personal commun.; Jefferson, 1985a
47 + (infinite)**	Manix basin D	*Anodonta* shell	^{14}C	Yale	Stuiver, 1969; Bassett and Jefferson, 1971
47,700 ± 2.0	Manix basin D	*Mammuthus* bone	U/Th	USGS 81-48	J. Bischoff, 1982, 1983, personal commun.; Jefferson, 1985a
51,200 ± 2.5	Manix basin D	*Hemiauchenia* bone	U/Th	USGS 81-49	J. Bischoff, 1982, 1983, personal commun.; Jefferson, 1985a
60,300 ± ?	Manix basin D	Mammalia bone	U/Th	USGS	Budinger, 1992
68,000 ± 4.0	Manix basin C	*Camelops* bone	U/Th	USGS 81-51	J. Bischoff, 1982, 1983, personal commun.; Jefferson, 1985a
74,000 ± ?	Manix basin D	*Anodonta* shell	U/Th	USGS	Budinger, 1992
80,000 ± ?	Afton basin	oncoid stromatolite	U/Th	-	Meek, 2000
183,800 ± 12.0	Manix basin C	*Camelops* bone	U/Th	USGS 81-30	J. Bischoff, 1982, 1983, personal commun.; Jefferson, 1985a
185,000 ± 15.0	Manix basin C	Tephra	Chemical	-	Bacon and Duffield, 1981; Izett, 1981
350 + (infinite)	Manix basin B	*Equus* bone	U/Th	USGS 80-51	J. Bischoff, 1982, 1983, personal commun.; Jefferson, 1985a

Notes: Ages are uncorrected. Manix Formation Members (B through D) are designated where known. Meek (1990) provides corrected ^{14}C ages for analyses from Manix basin and a discussion of the reliability of both *Anodonta* and oncoid stromatolite tufa determinations. UCLA—University of California at Los Angeles; USGS—U.S. Geological Survey; QC—Queens College, New York.

* This age has been revised, UCLA 2601C (Meek 1999).

** The previously reported age of 49 + ka for this sample (Jefferson, 1985a) is an error.

Member D of the Manix Formation. Invertebrate fossils are restricted to Members C and D, and not unexpectedly, the remains of fish and aquatic birds are restricted to Member C.

The fossiliferous part of the section ranges in age from >350 ka to ca. 20 ka, and spans late Irvingtonian through late Rancholabrean North American Land Mammal Ages (LMA). Most of the mammalian genera represented in the assemblage range through this entire period, and are found in Members B through D. However, the scimitar-tooth cat *Homotherium* and Antilocapridae occur only at the base of Member C, and *Bison* is restricted to the youngest deposits, Member D. Most vertebrate remains have been recovered from the base of Member C.

Jefferson (1968) named the mammalian fossil assemblage from the Manix Formation the "Camp Cady local fauna" after the historic Union Army post that was located on the Mojave River 3.7 km southwest of Manix siding (Fig. 2). Camp Cady was occupied during and shortly after the United States civil war (Chidester, 1965). In 1968, Lake Manix deposits were though to encompass only the Wisconsin, and the entire assemblage was assigned to the Rancolabrean LMA (Winters, 1954; Howard, 1955; Jefferson, 1968). If retained, this name (see Walsh, 2000, p. 268–270) should apply to only those fossils recovered from the upper wedge of Member C and Member D.

The Lake Manix assemblage is significantly biased taphonomically toward large-sized mammals, although, excluding lower vertebrates, water fowl compose ~20% of the assemblage. Amphibians, lizards, snakes, insectivores, bats, small carnivores, and small ungulates are absent. However, these animals occur

Figure 9. Eroded exposures of the Manix Formation. The view is to the east, from the north side of the Mojave River about 3 km east of Camp Cady. Light gray-green lacustrine silts and clays, Member C, are present in the foreground and to the left of the image. Fanglomerates of Member A form a 7 to 8 m-high bluff in the right mid-image. The Mojave River is to the right, and the Cady Mountains appear on the horizon.

locally in other late Pleistocene assemblages like that from Schuiling Cave (Downs et al., 1959; Jefferson, 1983) ~10 km west of the southern edge of Troy Lake playa basin, or from the ancestral Mojave River deposits at Daggett and Yermo (Reynolds and Reynolds, 1985) ~20 km southwest of Camp Cady.

Avian remains from the lacustrine sediments are often partially articulated (Howard, 1955; Jefferson, 1985b) (Fig. 10). Those from the littoral deposits are not. Most mammalian fossils occur as single, isolated skeletal elements that often exhibit some abrasion due to fluvial transport and/or wave action. Few mammalian osteologic elements are articulated. Under fluvial or littoral high energy depositional conditions, estimates of relative taxonomic abundance within the assemblage (Table 3 and 4) are better represented as a percentage of the total number of identified specimens (NISP) for each taxon, rather than on a calculated minimum number of individuals (Horton, 1984; Badgley, 1986).

Large mammalian herbivores far outnumber carnivores in modern faunas (23:1, Mech, 1966; >250:1, Schaller, 1972). The herbivore/carnivore ratio for the larger mammals (larger than *Canis latrans*) in the Lake Manix assemblage is relatively well balanced with 672 NISP herbivores and 11 NISP carnivores (Table 4), or approximately a 61:1 ratio. All Carnivora in the assemblage are rare, totaling only 1.6% NISP of the mammalian assemblage. These large carnivores, like their modern counterparts, occupied a broad variety of habitats and geographic ranges of often continental scale.

Freshwater invertebrates and aquatic lower vertebrates

Fossil clams and snails have been recovered from several distinctive horizons within fluvial deposits near the base, middle and in the top of Member D (Winters, 1954; Jefferson, 1987, 1991a;

Budinger, 1992), and in deposits of similar age in eastern Afton basin (Meek, 1999). Modern representatives of these fossils forms presently live in a variety of perennial freshwater habitats including streams, rivers, ponds, lakes and bogs or swamps. Some of the poorly sorted, coquina-like sediments in Member D that yield this ecologically diverse assemblage apparently were deposited during flood conditions on the Mojave River fan delta (Fig. 1).

Many of these molluscs are also known from other late Pleistocene localities in the western United States (Winters, 1954; Taylor, 1967), such as China Lake in Inyo County and Lake Cahuilla in Imperial County, California. Interestingly, the snail *Valvata humeralis* presently is endemic to lakes in the San Bernardino Mountains near the headwaters of the Mojave River.

Limnic ostracodes are common in the lacustrine silts and clays of Member C (Jefferson and Steinmetz, 1986; Steinmetz, 1987, 1988). Modern representatives of the five taxa identified (Table 2) indicate the following lake water conditions: mesotrophic, cool temperatures, well oxygenated, clear to muddy/turbid bottom, shallow to deep, and relatively fresh. In Lake Manix, ostracodes were apparently most abundant during mesotrophic conditions at high lake stands. Furthermore, the abundance of these ecologically sensitive organisms appears to peak with a periodicity of ca. 19 ka, suggesting a correlation of favorable habitat conditions with Milankovich-timed, even-numbered marine oxygen isotope stages (Steinmetz, 1987, 1988). The shaded bands on Figure 6 represent such lacustrine conditions.

The tui Mojave chub, *Gila bicolor mojavensis*, presently inhabits lakes and rivers along the Owens and Mojave River systems (Miller, 1973). This taxon is known from late Pleistocene deposits of the Death Valley river system, including the present Owens and Mojave River drainages (Miller, 1948, 1973). The Lake Manix fossils most closely resemble an endemic population living today at Zzyzx Spring (R.R. Miller, 1967, personal commun.) on the southwestern side Soda Lake basin (Fig. 1).

The bones and scales of this fish are the most abundant vertebrate fossil remains from Lake Manix and probably represent many thousands of individuals. Their remains occur throughout the lacustrine clays and silts and are locally concentrated in littoral sands at the base of Member C.

The stickleback, *Gasterosteus aculeatus*, is known from a single specimen recovered from a drill core in Coyote Lake basin (Roeder, 1985). This small fish presently occupies the Mojave River and rivers and streams in Los Angeles and Orange Counties, California.

Clemmys marmorata, the western pond turtle, presently inhabits fresh waters that occur in a narrow band along the West Coast of North America extending from British Columbia, Canada, to Baja California del Sur, Mexico. The eastern most margin of this range includes the Truckee and Carson River drainages of western Nevada and perennial portions of the Mojave River drainage including Afton Canyon (Lovich and Meyer, 2000). This turtle, which feeds mainly on aquatic plants and insects, typically is found in streams, ponds, and lakes and prefers muddy bottoms and marshes (Stebbins, 1966). Although not abundant,

TABLE 2. CAMP CADY PALEOFAUNA TAXONOMIC LIST

Order	Taxonomic identification	Common name
Crustacea		
	Limnocythere bradburyi	ostracode, water flea
	L. ceriotuberosa	ostracode, water flea
	L. platyforma	ostracode, water flea
	L. robusta	ostracode, water flea
	Heterocypris sp.	ostracode, water flea
Pelecypoda		
	Anodonta californiensis	freshwater clam
	Pisidium compressum	freshwater clam
Gastropoda		
	Valvata humeralis	freshwater snail
	Fossaria modicella	freshwater snail
	Planorbella ammon	freshwater snail
	P. subcrenata	freshwater snail
	Planorbella sp.? *P. tenuis*	freshwater snail
	Carinifex newberryi	freshwater snail
	Gyraulus vermicularis	freshwater snail
	Gyraulus sp.	freshwater snail
	Vorticifex effusa	freshwater snail
Osteichthyes		
	Gila bicolor mojavensis	tui (Mojave) chub
	Gasterosteus aculeatus	threespine stickleback
Reptilia		
	Clemmys marmorata	western pond turtle
Aves		
	Gavia sp. cf. *G. arctica*	Arctic loon
	Podiceps sp. cf. *P. nigricollis*	eared grebe
	Aechmophorus occidentalis	western grebe
	Pelecanus sp. aff. *P. erythrorhynchos*	American white pelican
	Phalacrocorax auritus	double-crested cormorant
	Phalacrocorax macropus	large-footed cormorant [†]
	Ciconia maltha	stork [†]
	Phoenicopterus minutus	small flamingo [†]
	Phoenicopterus copei	Cope's flamingo [†]
	Cygnus sp. cf. *C. columbianus*	tundra swan
	Branta Canadensis	Canada goose
	Anas sp. cf. *A. crecca*	green-winged teal
	Anas sp. cf. *A. platyrhynchos*	mallard
	Aythya sp.	greater scaup or canvasback
	Mergus sp. cf. *M. merganser*	common merganser
	Oxyura jamaicensis	ruddy duck
	Haliaeetus leucophalus	bald eagle
	Aquila chrysaetos	golden eagle
	Fulica americana cf. *F. a. minor*	small American coot [†]
	Grus sp.	crane
	cf. *Actitis* sp.	sandpiper
	Phalaropodinae	phalarope subfamily
	Larus sp. cf. *L. oregonus*	Oregon gull [†]
	Larus sp.	gull (large-size)
	Bubo virginianus	great horned owl
Mammalia		
	Megalonyx sp.	ground sloth (medium-size) [†]
	Nothrotheriops sp. cf. *N. shastensis*	Shasta ground sloth (small-size) [†]
	Paramylodon sp.	ground sloth (large-size) [†]
	Mammuthus sp.	mammoth [†]
	Lepus sp.	jack rabbit
	Cricetidae	mice
	Canis sp. cf. *C. dirus*	dire wolf [†]
	C. latrans	coyote
	Arctodus sp.	short-faced bear [†]
	cf. *Ursus* sp.	black bear
	Felis (*Puma*) sp.	mountain lion
	Homotherium sp. cf. *H. crenatidens*	scimitar-tooth cat, robust [†]
	Homotherium sp. cf. *H. serum*	scimitar-tooth cat, gracile [†]
	Equus conversidens	horse (small-size) [†]
	Equus sp.	horse (large-size) [†]
	Camelops sp. cf. *C. hesternus*	yesterday's camel [†]
	Camelops sp. aff. *C. minidokae*	Minidoka camel [†]
	Hemiauchenia macrocephala	llama [†]
	Antilocapridae	prong bucks
	Ovis canadensis	mountain sheep
	Bison sp. cf. *B. antiquus*	antique bison [†]

Note: Data are in part from Jefferson (1985a, 1987).
[†] extinct taxon

plastron and carapace fragments of *C. marmorata* have been recovered from paralimnic deposits of Member C and fluvial deposits of Member D.

Aves

Most of the living species of birds in the assemblage (Howard, 1955; Jefferson, 1985b) presently range throughout southern California and are found seasonally on inland lakes, such as the Salton Sea, Imperial County. The remaining species frequent coastal marine waters or inland areas from the San Joaquin Valley, central California northward (Pyle, 1961; Cogswell and Christman, 1977; Garrett and Dunn, 1981) (Table 2).

Extant species of migratory birds represented in the assemblage are presently absent from southern California during the summer (Pyle, 1961). They migrate northward in the spring following inland portions of the north-south Pacific Coast flyway. During Pleistocene pluvial periods, this flyway would have included the lakes of the Colorado and Mojave Deserts, those east of the Sierra Nevada Mountains, and the western part of Lake Lahontan (Snyder et al., 1964). Many of these species are also known from the late Pleistocene deposits of China Lake, Inyo County, California, and Fossil Lake, Lake County, southeastern Oregon (Jefferson, 1985b).

Two-thirds of the extant species (*Gavia arctica*, *Podiceps nigricollis*, *Aechmophorus occidentalis*, *Pelecanus erythrorhynchos*, *Phalacrocorax auritus*, *Mergus merganser*, *Haliaeetus leucophalus*, *Aquila chrysaetos*, and *Larus* spp.), represented by 80% NISP of the fossil avian specimens (Table 3), presently prefer or feed exclusively on small fish (Cogswell and Christman, 1977; Garrett and Dunn, 1981). *Gila bicolor mojavensis* represents an abundant food source for these predators. Modern representatives of the remaining taxa (*Cygnus columbianus*, *Branta canadensis*, *Anas crecca*, *Anas platyrhynchos*, *Aythya* sp., *Oxyura jamaicensis*, *Fulica americana*, *Grus* sp., *Actitis* sp.), feed on a variety of water plants and freshwater invertebrates (Cogswell and Christman, 1977; Garrett and Dunn, 1981). The owl, *Bubo virginianus*, is the only bird in the assemblage that feeds primarily on small mammals.

Mammalian herbivores

Among the larger herbivores, three taxa of ground sloth, *Megalonyx*, *Paramylodon* and *Nothrotheriops*, are present. All are poorly represented in the assemblage (Table 4). *Megalonyx* is known from the West Coast and eastern two-thirds of the United States, and ranged from South America into Canada and Alaska. Stock (1925) suggested that this animal was adapted to forest or woodland habitats. The habitat and dietary preferences of *Paramylodon*, which also ranged through North and South America, are not well known. Given its association with other presumed grassland animals, Stock (1925, 1930) maintained that *Paramylodon* was a grazer. Based on analysis of the feeding

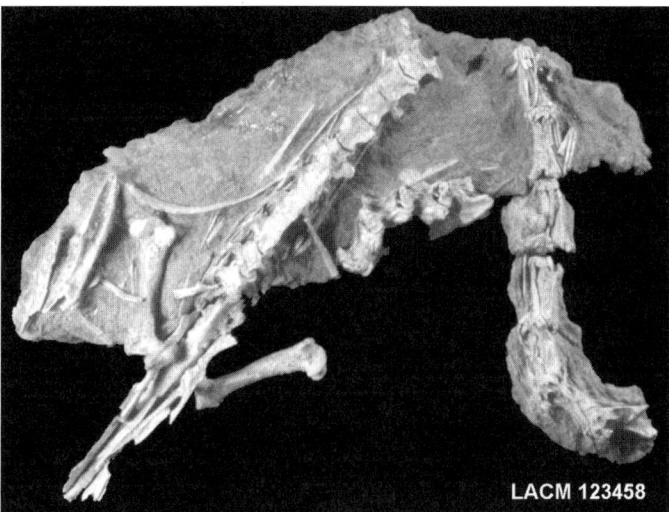

Figure 10. Partial skeleton of *Aechmophorus occidentalis*, western grebe (Natural History Museum of Los Angeles County specimen LACM 123458). The synsacrum and articulated left and right femora (46.1 mm in length) are to the lower left. Body and neck vertebrae occur to the right. A nearly complete skeleton of this animal was recovered from the base of Member C, Manix Formation (Jefferson, 1985b). Note the excellent preservation of delicate osteological structures that is typical of the avian remains from Member C.

mechanism in *Paramylodon*, Naples (1989) argued that it was a grazer-browser. Considering the varied environments within an extensive geographic range, both *Megalonyx* and *Paramylodon* were probably mixed feeders.

Nothrotheriops shastensis ranged throughout the southwestern United States and northern Mexico. It is well represented in cave assemblages from northeastern California through southern Nevada and Arizona where it browsed on desert shrubs and plants typical of a juniper-sage brush savannah habitat (Martin et al., 1961; Hansen, 1978). This selective folivore/browser (Naples, 1987) appears to have inhabited a broad spectrum of floral assemblages, but was limited latitudinally and altitudinally by minimum winter temperatures (McDonald et al., 1996).

Lepus sp. is the only small mammalian herbivore well represented in the assemblage. Jack rabbits typically inhabit grassy and brush covered areas. The order Rodentia is represented by a single cricetid humerus.

The relatively abundant but fragmentary remains of mammoth can not be identified to species. However, they probably represent the well-known middle and late Pleistocene form, *Mammuthus columbi* (= *M. imperator*). *M. columbi* ranged throughout the United States and likely browsed and/or grazed (Davis et al., 1984) in small herds similar to extant elephants. Evans (1961) suggested that juvenile mammoths were the favored prey of the scimitar-tooth cat, *Homotherium serum*. The American mastodon, *Mammut americanum*, is absent in the assemblage, and has not been identified from any Rancholabrean LMA sites within the Mojave Desert or Colorado Desert regions (Jefferson, 1991c).

TABLE 3. RELATIVE ABUNDANCE OF AVIAN TAXA

Taxon	NISP	% NISP
Gavia sp. cf. G. arctica	1	0.7
Podiceps sp. cf. P. nigricollis	4	2.8
Aechmophorus occidentalis	41	29.5
Pelecanus sp. aff. P. erythrorhynchos	12	8.6
Phalacrocorax auritus	15	10.8
Phalacrocorax macropus [†]	2	1.4
Ciconia maltha [†]	6	4.3
Phoenicopterus minutus [†]	14	10.1
Phoenicopterus copei [†]	4	2.8
Cygnus sp. cf. C. columbianus	3	2.1
Branta Canadensis	11	7.9
Anas sp. cf. A. crecca	1	0.7
Anas sp. cf. A. platyrhynchos	2	1.4
Aythya sp.	4	2.8
Mergus sp. cf. M. merganser	2	1.4
Oxyura jamaicensis	3	2.1
Haliaeetus leucocephalus	2	1.4
Aquila chrysaetos	2	1.4
Fulica americana cf. F. a. minor [†]	1	0.7
Grus sp.	1	0.7
cf. Actitis sp.	1	0.7
Phalaropodinae	1	0.7
Larus sp. cf. L. oregonus [†]	3	2.1
Larus sp.	2	1.4
Bubo virginianus	1	0.7

Note: Total number of identified specimens is 139, and total % NISP is 99.5. Data are in part from Jefferson (1985b, 1987). NISP—number of identified specimens; % NISP—number of identified specimens for each taxon divided by the total number of specimens.

[†] extinct taxon.

TABLE 4. RELATIVE ABUNDANCE OF MAMMALIAN TAXA, MANIX FORMATION

Taxon	NISP	% NISP
Megalonyx sp. [†]	1	0.1
Nothrotheriops sp. cf. N. shastensis [†]	2	0.3
Paramylodon sp. [†]	1	0.1
Mammuthus sp. [†]	37	5.2
Lepus sp.	11	1.6
Cricetidae	1	0.1
Canis sp. cf. C. dirus [†]	5	0.7
Canis latrans	5	0.7
Arctodus sp. [†]	1	0.1
cf. Ursus sp.	1	0.1
Felis (Puma) sp.	3	0.4
Homotherium sp. cf. H. crenatidens [†]	1	0.1
Homotherium sp. cf. H. serum [†]	2	0.3
Equus conversidens [†]	52*	7.4
Equus sp. (large-size) [†]	67*	9.5
Camelops sp. cf. C. hesternus [†]	371	54.7
Camelops sp. aff. C. minidokae [†]	5	0.7
Hemiauchenia macrocephala [†]	117	16.6
Antilocapara sp.	2	0.3
Ovis Canadensis	17	2.5
Bison sp. cf. B. antiquus [†]	1	0.1

Note: Total number of identified specimens is 703, and total % NISP is 101.5. Data are in part from Jefferson (1985a, 1987). NISP—number of identified specimens; % NISP—number of specimens identified for each taxon divided by the total number of identified specimens.

* may represent more than one species.

[†] extinct taxon.

Fossil horses are well represented in the assemblage (Table 4), and at least two forms, *Equus conversidens* and *Equus* sp. (large) have been identified. The occurrence of extinct species of *Equus* has been used to infer the presence of grasslands (e.g., Jefferson, 1968). However, feral horses and burros are opportunistic feeders that prefer to graze, although their diet may include up to 80% browse (Hansen, 1976; Ginnett, 1982; Ginnett and Douglas, 1982). Also, young horses from Rancho La Brea apparently were grazer-browsers (Akersten et al., 1988). These data suggest that extinct *Equus* may have periodically browsed as well as grazed, and was not restricted to grassland habitats.

Remains of the relatively small, stout-limbed extinct *Equus conversidens* have been recovered from Members C and D. The species ranged from northern Mexico through the central United States and the northern Great Plains, and has been identified in numerous sites in the Mojave Desert region (Jefferson, 1986, 1989, 1990, 1991b; Scott, 1997, 2000). Some small, specifically nondiagnostic specimens may represent an additional stilt-legged species of *Equus* in the assemblage (Scott, 1996, 1997).

Large fossil horse remains have been recovered from lower Member B through Member D. However, a lack of well-preserved cranial/dental specimens precludes assignment of this material to either the large Irvingtonian LMA form *Equus scotti* (Scott, 1998, 1999), or the large Rancholabrean LMA form, *Equus* sp. cf. *E. occidentalis*. These horses are commonly recovered from middle and late Pleistocene sites respectively throughout the southwestern United States (Jefferson, 1986, 1989, 1990, 1991b; Scott, 1997, 1998, 1999). Both species are probably present in the assemblage; *E. scotti* from the older deposits and *Equus* sp. cf. *E. occidentalis* from the younger.

The extinct camel, *Camelops* sp. cf. *C. hesternus*, is by far the most abundant large herbivore in the assemblage (Table 4), ~55% NISP, and has been found in lower Member B through Member D. It is the most common large mammal in late Pleistocene assemblages of the Colorado Desert and Mojave Desert regions (Jefferson, 1986, 1989, 1990, 1991b). This is in marked contrast to its low abundance in late Pleistocene assemblages from intermontane or coastal southern California (Jefferson, 1988).

Camelops hesternus from Rancho La Brea, the only site where direct dietary data are available, seems to have fed on ~10% monocot (graze) and 90% dicot (browse) plants (Akersten et al., 1988). Although previously considered to be a grazer (e.g., Webb, 1965; Jefferson, 1968), *C. hesternus* was probably a browser or mixed feeder similar to the extant camel, *Camelus* (Gauthier-Pilters and Dagg, 1981). *Camelops minidokae* is represented by only a few specimens from upper Member B through Member D. Otherwise, this more northern form is unknown in southern California Pleistocene assemblages (Jefferson, 1991c).

The llama, *Hemiauchenia macrocephala*, is moderately well represented in the assemblage (Table 4), and has been recovered from Members B through D. The taxon is a common member of late Pleistocene Mojave Desert (Jefferson, 1986, 1989, 1990, 1991b) and intermontane California assemblages like that from McKittrick, Kern County, California (Jefferson, 1988). However, it is very rare in coastal assemblages (Jefferson, 1988). This long-necked, cursorial llama was probably a grazer-browser adapted to open terrain.

Antilocaprids, either extinct *Tetrameryx* or extant *Antilocapra*, are represented by a single specimen recovered from the base of Member C. The other medium-sized browser, *Ovis canadensis*, is moderately well represented in the assemblage. It has been reported historically from the eastern Cady Mountains. Both antilocaprids and ovids are found in other late Pleistocene assemblages from the Mojave Desert (Jefferson, 1986, 1989, 1990, 1991b).

Bison antiquus is rare in the assemblage, and is positively known only from Member D. Pleistocene *Bison* remains are rare throughout the Mojave Desert region (Jefferson, 1986, 1989, 1990, 1991b), which is in contrast to coastal southern California where they are relatively common (Jefferson, 1988). Older bovid remains from the base of Member C, previously referred to this type Rancholabrean LMA taxon (Jefferson, 1987, 1991a), probably represent a large *Camelops*.

Bison antiquus is closely related to the extant *B. bison*. *B. antiquus* is assumed to have had habits similar to the living form that is almost exclusively a grazer but will periodically browse (Meagher, 1973; Peden, 1976). However, analyses of chewed plant remains impacted into the fossettes and fossettids of dentitions from Rancho La Brea indicate that juvenile *B. antiquus* was a mixed feeder (Akersten et al., 1988).

Paleoenvironments

The Lake Manix assemblage includes extinct and extralocal extant forms that reflect ecological conditions dramatically different from the present xeric environment of the central Mojave Desert. Inferences based on the ecology of extant taxa that are closely related to the extinct forms allow paleoenvironmental reconstructions of the local lacustrine and terrestrial paleohabitats. Most of the fossil molluscs represent extant animals that live in a assortment of fluvial, lacustrine, or paralimnic habitats. Some are extralocal, preferring cooler waters. However, others presently inhabit perennial waters in the Mojave Desert region. Although now rare, all lower vertebrates in the assemblage are known from the Mojave River drainage.

The fossil avians clearly suggest the presence of a variety of mildly saline or freshwater lake and lake margin environments. Judging from the food preferences, food procurement methods and nesting habits of extant bird species (Cogswell and Christman, 1977; Garrett and Dunn, 1981), open water, sandy beach flats, and extensive reedy marshlands were the dominant lacustrine habitats in Lake Manix (Jefferson, 1985b). The seasonally extralocal pattern of the migratory forms in the assemblage suggests an overall cooler or more equable climate.

Regional terrestrial vegetation patterns, reconstructed in part from packrat midden data (Spaulding et al., 1984; Spaulding, 1990), permit inferences about the local paleoflora. The alluvial slopes and low hills surrounding Lake Manix probably supported a juniper-sage brushland, and the nearby mountains most likely were covered with a pinyon-juniper woodland (Spaulding, 1980, 1990; Jefferson, 1987, 1991a). Local valley bottoms probably supported patchy semidesert grasslands and desert scrub. These floristic associations are consistent with the inferred browsing habits of the majority of the fossil mammals (Table 4) (Jefferson, 1987, 1991a).

DEPOSITIONAL HISTORY

Upper Mojave River drainage

Before the uplift of the Transverse Ranges, internal drainage typified most basins on the southern and central part of the Mojave Desert block. The southwestern margin of the block apparently drained across the San Andreas fault zone, west to the Pacific Ocean (Meisling and Weldon, 1989). About 3–2 m.y. ago, as the Transverse Ranges adjacent to this margin of the block were elevated along the San Andreas fault, drainage direction shifted to the northeast.

In the headwaters area of the ancestral Mojave River, ~100 km to the southwest of Manix, the Victorville Fan complex was shed off the rising Transverse Ranges northeast into the Victorville basin (Weldon, 1985; Meisling and Weldon, 1989; Kenny and Weldon, 1999) (Fig. 1). Here, magnetostratigraphic data provide a date of 1.95 Ma for the base of the ancestral Mojave River deposits (Cox and Tinsley, 1999; Cox et al., this volume). Cox et al. (this volume) suggest that the river advanced from the Victorville basin northward 50 km to Lake Harper basin sometime after 0.78 Ma and probably between 0.57 and 0.47 Ma. They (Cox and Tinsley, 1999; Cox et al., this volume) argue that the ancestral Mojave River continued to advance eastward, overflowing Lake Harper basin, and (based on Jefferson, 1985a) reached the Manix basin, ~50 km to the east of the Harper basin, no earlier than ca. 0.5 Ma ago (Fig. 1). This places the appearance of the ancestral Mojave River in Manix basin during the deposition of the lower wedge of Member B.

Lake Manix basin

A thick section of late Miocene and Pliocene fanglomerates that presumably reflects regional extensional tectonics, documents the initial formation of Manix basin. The top of the fanglomerates may be laterally equivalent with the base of the Mojave River Formation, which has an age of ca. 2.5 Ma. From prior to 2.5 Ma until ca. 1 Ma, Manix basin was internally drained (Nagy and Murray, 1991). Subaerial oxidization of silts and clays, bedded gypsum, and limestones in the Mojave River Formation reflect the presence of ephemeral saline lakes and/or playas. Fluvial deposits in the top of the upward-coarsening Mojave River Formation, estimated to be ca. 1 Ma (Nagy and Murray, 1991), record a westward-flowing drainage system. Given the drainage history of Victorville and Lake Harper basins (Cox and Tinsley, 1999; Cox et al., this volume), it is unlikely that these early fluvial deposits represent flow from an ancestral Mojave River as Nagy and Murray (1991) have suggested.

No lacustrine deposits ranging in age between ca. 1 and 0.5 Ma have been identified in Manix basin. During this period, Hale (1985) suggested that a large lake, Lake Blackwelder, filled Death Valley, Soda Lake basin (Hooke, 1999) and Manix basin (Hale, 1985). It is then argued that this lake overflowed the southern end of the Troy Lake basin arm of Manix basin southeast into the Colorado River drainage via Bristol and Danby Valleys (Hale, 1985). The existence of this lake has not been confirmed stratigraphically or by the presence of high elevation shorelines in the Manix basin region (Rosen, 1989; Brown and Rosen, 1995; Enzel et al., this volume). However, westward paleocurrents in the fluvial deposits of the upper Mojave River Formation (Nagy and Murray, 1991) are consistent with the flow direction of Hale's proposed drainage system.

A second pulse of alluvial fan development, Member A of the Manix Formation, which was probably tectonically induced (Nagy and Murray, 1991), separates the upper fluvial deposits of the Mojave River Formation from the base of Member B. Fluvial and lacustrine sediments that comprise the lowermost deposits in the lower wedge of Member B, estimated to be ca. 0.5 Ma (Jefferson, 1985a, 1994, 1999), document the appearance of the Mojave River system in Manix basin.

Fluvial/lacustrine deposition in Lake Manix records at least four major transgressive/regressive events over the past 500 ka (Jefferson, 1985a, 1994, 1991b) (the last three events appear on the reference stratigraphic section, Figure 6). Major lacustrine phases (shaded bands Figure 6) presumably were associated with high lake levels. Based on stratigraphic sequence and placement relative to dated horizons, some of these transgressions can be correlated with specific marine oxygen isotope stages (Morley and Hays, 1981; Martinson et al., 1987) (Fig. 6). Oncoid stromatolites also typically occur at the base of the transgressive lacustrine deposits (Awramik et al., 2000).

Although not constrained by radiometric dates, the two transgressive lacustrine events present in the lower wedge of Member B may correlate with the beginning of marine oxygen isotope stage 14 (505 ka, Morley and Hays, 1981) and stage 12 (421 ka, Morley and Hays, 1981). The lacustrine transgression at the base of the lower part of Member C, ~2 m above the U/Th series age of 350+ ka, is tentatively correlated with the beginning of marine oxygen isotope stage 8 (279 ka, Morley and Hays, 1981) (Fig. 6). Member C encompasses marine oxygen isotope stages 8 through 4.

The major lacustrine transgression at the base of the upper part of Member C, 0.3 m below the 185 ± 15 ka tephra (Fig. 6), is essentially coincident with the beginning of marine oxygen isotope stage 6 (189.6 ka, Martinson et al., 1987). Accordingly, the fluvial upper wedge of Member B that separates the two lacustrine phases of Member C (Figs. 6, 7), was deposited during marine oxygen isotope stage 7 (244.2–189.6 ka, Martinson et al., 1987).

Marine oxygen isotope stage 5 (129.8–73.9 ka, Martinson et al., 1987) occurred during the deposition of upper Member C. It may be represented by the three, littoral arkosic beds that appear in the section 1–3 m below a U/Th series age of 68 ± 4 ka (Fig. 6). If so, fluvial deposition within the central part of the lake basin was far less extensive during marine oxygen isotope stage 5 than during marine oxygen isotope stage 7, upper Member B (Figs. 6, 7).

Most of the lacustrine sediments preserved in Afton basin were deposited during marine oxygen isotope stage 6 (Meek, 2000). These are overlain by an extensive wedge of fanglomerate, presumably deposited during marine oxygen isotope stage 5. A U/Th series age of 80 ka, from an oncoid carbonate preserved atop the fanglomerate, indicates that Afton basin was filled by Lake Manix during marine oxygen isotope stage 4 (Meek, 2000).

Late in the deposition of upper Member C, ancestral Mojave River deltaic and fluvial sediments, in part represented by Member D, prograded eastward across the middle of the basin. This fan delta eventually divided the two western arms of the lake into separate Coyote and Troy Lake basins (Hagar, 1966; Groat, 1967; Meek, 1994, 1999; Cox et al., this volume; Enzel et al., this volume) (Fig. 1). Although dates from the base of Member D near Manix, suggest that the Member C/D contact may approximate the age of marine oxygen isotope stage 4/3 boundary (58.9 ka, Martinson et al., 1987) (Jefferson, 1985a), this contact is time-transgressive and older to the west. Cox et al. (this volume) suggest that deposition of the fan delta decreased the evaporative surface and the volume of Lake Manix, increasing overflow to the east into Soda Lake basin.

During the deposition of Member D, between ca. 60 and 19 ka, lake levels fluctuated (Meek, 1990). Detailed studies by Budinger (1992) at the reference section (Fig. 6), record three 0.5 m-thick lacustrine deposits in the lower, upper middle, and upper parts of Member D. These represent interfingering of lacustrine and fluvial/deltaic deposits along the western margin of the lake.

Based on dates obtained from Afton basin (Table 1), Lake Manix was at a highstand between ca. 31 ka and 28 ka (Meek, 1990). Between 28 ka and 21 ka, Meek (1990, 1999) has argued that Lake Manix was at a lowstand, and he further suggests that the ancestral Mojave River may have terminated in Lake Harper basin at this time. This assertion is based on a lack of dates within the 28–21 ka age range from Lake Manix, and the presence of lacustrine deposits in Harper basin of this age. However, evidence of a significant stratigraphic hiatus has not been identified within the fluvial/deltaic and lacustrine deposits in the upper half of Member D. Also, Enzel et al. (this volume) suggest that there may have been sufficient Mojave River flow to sustain lakes in both Harper and Manix basins concurrently.

The youngest accepted age for Lake Manix deposits, 18.1 ka (Meek, 1999), was obtained in Afton basin. Ages of 17.9 ka and younger from Coyote Lake basin and 15.0 ka from Troy Lake basin (Table 1), post date deposition of the top of Member D south of Manix (Figs. 2, 6). By 19 ka, the Mojave River fan delta (Fig. 1) had prograded eastward dividing Coyote and Troy Lakes into separate basins, effectively restricting Lake Manix to Afton basin.

At this time, lake level was high. After 18 ka, probably during the late Wisconsin glacial maximum (17.8 ka, marine oxygen isotope event 2.2, Martinson et al., 1987), Lake Manix breached sill level at the east end of Afton basin (Ellsworth, 1933; Weldon, 1982; Meek, 2000; Cox et al., this volume). The formation of Afton Canyon, presently a >150 m-deep gorge leading east into the Pleistocene Lake Mojave basin (Figs. 1, 2), was initiated.

Meek (1989, 1990, 1999, 2000) has argued that this overflow resulted in catastrophic erosion of the upper part of Afton Canyon (Fig. 2) to the depth of the lake floor in Afton basin, >120 m. This assertion is based on analyses of erosion volumes in Afton basin and deeply buried Wisconsin surfaces in western Lake Mojave basin, a lack of recessional shorelines in Afton basin, a lack of fluvial terraces within the upper walls of Afton Canyon, narrow deeply incised tributaries in Afton basin, and subsurface boulders overlying lacustrine clays downstream from the Canyon. Further incision of Afton Canyon apparently occurred much more slowly over the past <18 ka (Meek, 2000).

However, Wells and Enzel (1994) and Enzel et al. (this volume) point out that an initial rapid incision is unlikely and suggest that the formation of Afton Canyon was time transgressive. This argument is based on the slow westward, upstream migration of the Afton Canyon nick point as evidenced by the existence of lacustrine conditions on the Mojave River fan delta as late as 12–9 ka (Reynolds and Reynolds, 1985), extensive Holocene fluvial features within Afton Canyon (Wells and Enzel, 1994), and recessional shorelines in the Afton basin Enzel et al. (this volume).

Clearly, the erosion of Afton Canyon and incision of the Mojave River through Manix basin has not been a simple process. Although the arguments of Meek (1989, 1999, 2000), Wells and Enzel (1994), and Enzel et al. (this volume) appear well founded, an integrative hypothesis and resolution to this issue must await further field investigations.

Large meander channels, on the top of Member D and incised to varying depths through the Manix Formation east of Camp Cady, suggest that, at least in the central Manix basin, the Mojave River has gradually adjusted to Soda Lake basin base level over the past <18 ka. During latest Pleistocene through early Holocene time, after the disappearance of Lake Manix, the Mojave River continued to intermittently fill Coyote Lake basin (Hagar, 1966; Meek, 1999), Troy Lake basin (Groat, 1967; Meek, 1999) and possibly Afton basin (Ellsworth, 1933; Blackwelder and Ellsworth, 1936; Meek, 1999, 2000). Within historic time, the Mojave River has flowed periodically through Afton Canyon into Cronese Lakes and Soda Lake basin (Wells et al., 1989; Brown et al., 1990; Meek, 1999, 2000; Enzel et al., this volume; Wells et al., this volume) (Fig. 1).

SUMMARY AND CONCLUSIONS

Middle to late Pleistocene deposits within the Manix basin provide a relatively complete record of depositional environments for an important portion of the ancestral Mojave River. Fluvial and lacustrine deposits, which represent the first appearance of the Mojave River in the central Mojave Desert, are recognized near the base of the Manix Formation (Member B). These deposits confirm that the river terminated in the Victorville and/or Harper basins prior to 0.5 Ma (Fig. 1).

The Manix Formation is geologically mappable as four distinct, and largely laterally equivalent lithologic facies. Sequential changes from primarily fluvial to primarily lacustrine deposition (like those represented by the interfingering of Member B and C, Figure 7), are largely climatically driven and not the consequence of motion on the Manix fault (McGill et al., 1988). Major transgressive lacustrine events and high lake levels were the result of increased flow along the ancestral Mojave River. These episodes occurred after marine oxygen isotope stage boundaries and are positively correlated with relatively cool/moist climatic conditions inferred to be present during even-numbered stages. Given the available dates and stratigraphic placement, such events occurred during marine oxygen isotope stages 14, 12, 8, 6, and 4 (Fig. 6).

Late in lake history, a fan delta (in part Member D), deposited by the ancestral Mojave River, spread across the west-central margin of the basin, separating Afton, Coyote Lake and Troy Lake basins (Fig. 1). Pulses of deposition along the prograding delta front were not necessarily climate driven or the result of changes in the upper drainage system of the ancestral Mojave River. About 18 ka, Lake Manix breached the east end of Afton basin, forming Afton Canyon. Erosion of the upper ~120 m of the Canyon may have been rapid, however, the lower part of Afton Canyon was cut more slowly. This has resulted in incision of the Mojave River along its present course, and exposure of the Manix Formation in eroded badlands that extend from Camp Cady to Afton Canyon (Fig. 2).

Fossils recovered from exposures of the fluvial and lacustrine sediments of the Manix Formation (Table 2) are now known to encompass late Irvingtonian through late Rancholabrean LMA time. The relative abundance and paleoecological character of taxa permit a limited but significant reconstruction of paleohabitats. A substantial portion of the extinct and extralocal fossil vertebrates were ecologically tied directly to the lacustrine and paralimnic environments of Lake Manix. These include essentially all lower vertebrate and avian species. Browsers or browser-grazers (75% NISP, Table 4) that take advantage of seasonally available forage were the dominate large mammalian herbivores at Lake Manix and in the central Mojave Desert region during middle to late Pleistocene time.

ACKNOWLEDGMENTS

B.W. Cahill of California State Parks is kindly thanked for his assistance in preparation of the illustrations. The helpful reviews, comments and suggestions of Y. Enzel, D.P. Whistler, and M.O. Woodburne are greatly appreciated. The assistance and cooperation of the many individuals who have supported and participated in research in Manix basin over the past several decades is sincerely acknowledged.

REFERENCES CITED

Akersten, W.A., Foppe, T.M., and Jefferson, G.T., 1988, A new source of dietary data for extinct herbivores: Quaternary Research, v. 30, p. 92–97.

Awramik, S.M., Bucheim, H.P., Leggitt, L., and Woo, K.S., 2000, Oncoids of the late Pleistocene Manix Formation, Mojave Desert region, California, in Reynolds, R.E., and Reynolds, J., eds., Empty basins, vanished lakes: San Bernardino County Museum Association Quarterly, v. 47(2), p. 25–31.

Bacon, C.R., and Duffield, W.A., 1981, Late Cenozoic rhyolites from the Kern Plateau, southern Sierra Nevada, California: American Journal of Science, v. 281, p. 1–34.

Badgley, C., 1986, Counting individuals in mammalian fossil assemblages from fluvial environments: Palaios, v. 1(3), p. 328–338.

Bassett, A.M., and Jefferson, G.T., 1971, Radiocarbon dates of Manix Lake, central Mojave Desert, California: Geological Society of America Abstracts to Meetings, v. 3, no. 2, p. 79.

Bassett, A.M., and Kupfer, D.H., 1964, A geologic reconnaissance in the southeastern Mojave Desert: California Division of Mines and Geology Special Report 83, 43 p.

Berger, R., and Libby, W.F., 1967, UCLA radiocarbon dates VI: Radiocarbon, v. 9, p. 477–504.

Beyers, F.M., Jr., 1960, Geology of the Alvord Mountain quadrangle, San Bernardino County, California: U.S. Geological Survey Bulletin 1089-A, 71 p.

Blackwelder, E., 1954, Pleistocene lakes and drainage in the Mojave region, southern California, in Jahns, R.H., ed., Geology of southern California: California Division of Mines and Geology, Bulletin 170, p. 35–40.

Blackwelder, E., and Ellsworth, E.W., 1936, Pleistocene lakes of the Afton basin, California: American Journal of Science 5th Series, v. 31, p. 453–463.

Blanc, R.P., and Cleveland, G.B., 1961a, Pleistocene lakes of southeastern California, Part I: California Division of Mines and Geology, Mineral Information Service, v. 14(4), p. 1–8.

Blanc, R.P., and Cleveland, G.B., 1961b, Pleistocene lakes of southeastern California, Part II: California Division of Mines and Geology, Mineral Information Service, v. 14(5), p. 72–82.

Brown, W.J., and Rosen, R.M., 1995, Was there a Pliocene-Pleistocene fluvial lacustrine connection between Death Valley and the Colorado River?: Quaternary Research, v. 43, p. 286–296.

Brown, W.J., Wells, S.G., Enzel, Y., Anderson, R.Y., and McFadden, L.D., 1990, The late Quaternary history of pluvial Lake Mojave: Silver Lake and Soda Lake basins, California, in Reynolds, R.E., et al., eds., At the end of the Mojave: Quaternary studies in the eastern Mojave Desert: Redlands, California, Special Publication of the San Bernardino County Museum Association, 1990 Mojave Desert Quaternary Research Center Symposium, May 18–21, 1990, p. 55–72.

Budinger, F.E., Jr., 1992, Targeting early man sites in the western United States: An assessment of the Manix type section, central Mojave Desert, California [M.A. thesis]: Special Major, San Bernardino, California State University, 229 p.

Buwalda, J.P., 1914, Pleistocene beds at Manix in the eastern Mojave Desert region: Bulletin Department of Geology, Berkeley, University of California, v. 7(24), p. 443–464.

Chidester, D.H., 1965, A short history of Camp Cady [Senior thesis]: Department of History, La Verne College, 28 p.

Cogswell, H.L., and Christman, 1977, Water birds of California: Berkeley, University of California Press, Natural History Guides, no. 40, 399 p.

Cox, B.F., and Tinsley, J.C., III, 1999, Origin of the late Pliocene and Pleistocene Mojave River between Cajon Pass and Barstow, California, in Reynolds, R.E., and Reynolds, J., eds., Tracks along the Mojave: San Bernardino County Museum Association Quarterly, v. 46(3), p. 49–54.

Davis, O.K., Agenbroad, L., Martin, P.S., and Mead, J.I., 1984, The Pleistocene dung blanket of Bechan Cave, Utah, in Genoways, H.H., and Dawson, M.R., eds., Contributions in Quaternary vertebrate paleontology: A volume in memorial to John E. Guilday: Carnegie Museum of Natural History Special Publication 8, p. 267–282.

Downs, T., Howard, H., Clements, T., and Smith, G.I., 1959, Quaternary animals from Schuiling Cave in the Mojave Desert, California: Los Angeles County Museum of Natural History, Contributions in Science, no. 29, 21 p.

Ellsworth, E.W., 1933, Physiographic history of the Afton Basin of the Mojave Desert: Des Moines, Iowa, Pan-American Geologist, May 1933, v. 59(4), p. 308–309.

Evans, G.L., 1961, The Friesenhan Cave: Texas Memorial Museum Bulletin, v. 2, p. 3–22.

Fergusson, G.J., and Libby, W.F., 1962, UCLA radiocarbon dates, I: Radiocarbon, v. 4, p. 109–114.

Garrett, K., and Dunn, J., 1981, The birds of southern California status and distribution: Los Angeles Audubon Society, Los Angeles, California, 408 p.

Gauthier-Pilters, H., and Dagg, A.I., 1981, The camel, its behavior, evolution, and relationship to man: Chicago, Illinois, University of Chicago Press, 208 p.

Ginnett, T.F., 1982, Comparative feeding behavior of burros and desert bighorn sheep in Death Valley National Monument: Desert Bighorn Council 1982 Transactions, p. 81–86.

Ginnett, T.F., and Douglas, C.L., 1982, Food habits of feral burros and desert bighorn sheep in Death Valley National Monument: Desert Bighorn Council 1982 Transactions, p. 81–86.

Groat, C.G., 1967, Geology and hydrology of the Troy Playa area, San Bernardino County, California [M.S. thesis]: New York, University of Rochester, 133 p.

Hagar, D.J., 1966, Geomorphology of Coyote Valley, San Bernardino County, California [Ph.D. thesis]: Department Geology, Amherst, University of Massachusetts, 210 p.

Hale, G.R., 1985, Mid-Pleistocene overflow of Death Valley toward the Colorado River, in Hale, G.R., ed., Quaternary lakes of the eastern Mojave Desert, California: Friends of the Pleistocene Pacific Cell Annual Field Trip Guidebook, p. 36–81.

Hansen, R.M., 1976, Foods of free roaming horses in southern New Mexico: Journal of Range Management, v. 29(4), p. 347.

Hansen, R.M., 1978, Shasta ground sloth food habits, Rampart Cave, Arizona: Paleobiology, v. 4(3), p. 302–319.

Hastorf, C.A., and Tinsley, J.C., 1981, Maps and index of radiocarbon-dated samples from southern California: U.S. Geological Survey Miscellaneous Field Studies Map MF-1294, scale 1:500,000.

Hooke, R.L., 1999, Lake Manly(?) shorelines in the eastern Mojave Desert, California: Quaternary Research, v. 52, p. 328–336.

Horton, D.R., 1984, Minimum numbers; a consideration: Journal of Archaeological Science, v. 11, p. 255–271.

Howard, H.H., 1955, Fossil birds from Manix Lake, California: U.S. Geological Survey Professional Paper 264-J, p. 199–205.

Hubbs, C.L., Bein, G.S., and Suess, H.E., 1962, La Jolla natural radiocarbon measurements, II: Radiocarbon, v. 4, p. 204–238.

Hubbs, C.L., Bein, G.S., and Suess, H.E., 1965, La Jolla natural radiocarbon measurements, IV: Radiocarbon, v. 7, p. 66–117.

Izett, G.A., 1981, Volcanic ash beds: Recorders of upper Cenozoic silic pyroclastic volcanism in the western United States: Journal of Geophysical Research, v. 68(B11), p. 10,200–10,222.

Jefferson, G.T., 1968, The Camp Cady local fauna from Pleistocene Lake Manix, California [M.A. thesis]: Department of Geology, Riverside, University of California, 106 p.

Jefferson, G.T., 1983, A fragment of human skull from Schuiling Cave, Mojave Desert, California: Southern California Academy of Sciences Bulletin, v. 82(2), p. 98–102.

Jefferson, G.T., 1985a, Stratigraphy and geologic history of the Pleistocene Lake Manix Formation, central Mojave Desert, California, in Reynolds, R.E., ed., Cajon Pass to Manix Lake: Geological investigations along Interstate 15: Redlands, California, San Bernardino County Museum Association Special Publication, p. 157–169.

Jefferson, G.T., 1985b, Review of the late Pleistocene avifauna from Lake Manix, central Mojave Desert, California: Natural History Museum Los Angeles County Contribution in Science, no. 362, 13 p.

Jefferson, G.T., 1986, Fossil vertebrates from late Pleistocene sedimentary deposits in the San Bernardino and Little San Bernardino Mountains region, *in* Kooser, M., and Reynolds, R.E., eds., Geology around the margins of the eastern San Bernardino Mountains: Redlands, California, Publications of the Inland Geological Society, no. 1, p. 77–80.

Jefferson, G.T., 1987, The Camp Cady local fauna: Paleoenvironment of the Lake Manix Basin: San Bernardino County Museum Association Quarterly, v. 34(3 and 4), p. 3–35.

Jefferson, G.T., 1988, Late Pleistocene large mammalian herbivores: Implications for big game hunters in southern California: Southern California Academy of Sciences Bulletin, v. 87(3), p. 89–103.

Jefferson, G.T., 1989, Late Pleistocene and earliest Holocene fossil localities and vertebrate taxa from the western Mojave Desert, *in* Reynolds, R.E., ed., The west-central Mojave Desert: Quaternary studies between Kramer and Afton Canyon: Redlands, California, Mojave Desert Quaternary Research Center, San Bernardino County Museum Association Special Publication, p. 27–40.

Jefferson, G.T., 1990, Rancholabrean Age vertebrates from the eastern Mojave Desert, California, *in* Reynolds, R.E., et al., eds., At the end of the Mojave: Quaternary studies in the eastern Mojave Desert: Redlands, California, Special Publication of the San Bernardino County Museum Association, 1990 Mojave Desert Quaternary Research Center Symposium, May 18–21, 1990, p. 109–115.

Jefferson, G.T., 1991a, The Camp Cady local fauna: Stratigraphy and paleontology of the Lake Manix basin, *in* Woodburne, M.O., et al., eds., Inland southern California: The last 70 million years: San Bernardino County Museum Association Quarterly, v. 38(3,4), p. 93–99.

Jefferson, G.T., 1991b, Rancholabrean Age vertebrates from the southeastern Mojave Desert, California, *in* Reynolds, R.E., ed., Crossing the borders: Quaternary studies in eastern California and southwestern Nevada: Redlands, California, Mojave Desert Quaternary Research Center, San Bernardino County Museum Association Special Publication, p. 27–40.

Jefferson, G.T., 1991c, A catalog of late Quaternary vertebrates from California: Part two, mammals: Natural History Museum of Los Angeles County Technical Reports, no. 7, 129 p.

Jefferson, G.T., 1994, Stratigraphy and Pliocene-Pleistocene history of the Lake Manix basin, *in* McGill, S.F., and Ross, T.M., eds., Geological investigations of an active margin: Geological Society of America Guidebook, 27th Annual Meeting, San Bernardino, California, p. 175–177.

Jefferson, G.T., 1999, Age and stratigraphy of Lake Manix basin, *in* Reynolds, R.E., and Reynolds, J., eds., Tracks along the Mojave: San Bernardino County Museum Association Quarterly, v. 46(3), p. 109–112.

Jefferson, G.T., and Steinmetz, J.J., 1986, Ostracode biostratigraphy and paleoecology of the late Pleistocene Manix Formation: Current Research in the Pleistocene, v. 3, p. 55–56.

Keaton, J.R., and Keaton, R.T., 1977, Manix fault zone, San Bernardino County, California: California Division Mines and Geology, California Geology, v. 30(8), p. 177–186.

Kenny, M.D., and Weldon, R.J., 1999, Timing and magnitude of mid to late Quaternary uplift of the western San Bernardino and northeastern San Gabriel Mountains, southern California, *in* Reynolds, R.E., and Reynolds, J., eds., Tracks along the Mojave: San Bernardino County Museum Association Quarterly, v. 46(3), p. 33–46.

Lovich, J., and Meyer, K., 2000, Aspects of the ecology of the Western Pond Turtle in the Mojave River, *in* Reynolds, R.E., and Reynolds, J., eds., Empty basins, vanished lakes: San Bernardino County Museum Association Quarterly, Abstracts from the Year 2000 Desert Symposium, v. 47(2), p. 79.

Martin, P.S., Sables, E.B., and Shutler, D., Jr., 1961, Rampart Cave coprolite and ecology of the Shasta ground sloth: Science, v. 259, p. 102–127.

Martinson, D.G., Pisias, N.P., Hays, J.D., Imbrie, J., Moore, T.C., Jr., and Shackelton, J.N., 1987, Age dating and the orbital theory of the ice ages: Development of a high-resolution 0–300,000-year chronology: Quaternary Research, v. 27, p. 1–29.

McDonald, H.G., Jefferson, G.T., and Force, C., 1996, Pleistocene distribution of the ground sloth *Nothrotheriops shastensis* (Xenarthra, Megalonychidae), *in* Reynolds, R.E., and Reynolds, J., eds, The 1996 Desert Symposium, Abstracts of Papers Submitted to the Meetings, San Bernardino County Museum Association Quarterly, v. 43(2), p. 151–152.

McGill, S.F., Murray, B.C., Maher, K.A., Lieske, J.H., Jr., and Rowan, L.R., 1988, Quaternary history of the Manix fault, Lake Manix basin, Mojave Desert, California: San Bernardino County Museum Association Quarterly, v. 35(3 and 4), p. 3–20.

Meagher, M.M., 1973, The bison of Yellowstone National Park: National Park Survey Scientific Monograph Series 1, 171 p.

Mech, L., 1966, The wolves of Isle Royale: Fauna of the National Parks of the United States: Washington, Fauna Series 7, 210 p.

Meek, N., 1989, Geomorphic and hydrologic implications of the rapid incision of Afton Canyon, Mojave Desert, California: Geology, v. 17, p. 7–10.

Meek, N., 1990, Late Quaternary Geochronology and geomorphology of the Manix basin, San Bernardino County, California [Ph.D. thesis]: Department of Geography, Los Angeles, University of California, 212 p.

Meek, N., 1994, The stratigraphy and geomorphology of Coyote basin, central Mojave Desert, California: San Bernardino County Museum Quarterly, v. 41(3), p. 5–13.

Meek, N., 1999, New discoveries about the late Wisconsin history of the Mojave River system, *in* Reynolds, R.E., and Reynolds, J., eds., Tracks along the Mojave: San Bernardino County Museum Association Quarterly, v. 46(3), p. 113–118.

Meek, N., 2000, The late Wisconsin history of the Afton Canyon area, *in* Reynolds, R.E., and Reynolds, J., eds., Empty basins, vanished lakes: San Bernardino County Museum Association Quarterly, v. 47(2), p. 33–34.

Meek, N., and Battles, D.A., 1991, Displacement along the Manix fault: California Division Mines and Geology, California Geology, February, v. 44(2), p. 33–38.

Meisling, K.E., and Weldon, R.J., 1989, Late Cenozoic tectonics of the northwestern San Bernardino Mountains, southern California: Geological Society of America Bulletin, v. 101, p. 106–128.

Merriam, J.C., 1915, Extinct faunas of the Mojave Desert, their significance in a study of the origin and evolution of life in America: Popular Science Monthly, March 1915, p. 245–264.

Miller, R.R., 1948, The cyprinodont fishes of the Death Valley system of eastern California and southwestern Nevada: Miscellaneous Publications of the Museum of Zoology, University of Michigan, no. 68, 155 p.

Miller, R.R., 1973, Two new fishes, *Gila bicolor snyderi* and *Catostomus fumeiventris*, from the Owens River basin, California: Occasional Papers of the Museum of Zoology, University of Michigan, no. 667, 19 p.

Morley, J.J., and Hayes, J.D., 1981, Toward a high-resolution, global deep-sea chronology for the last 750,000 yr: Earth and Planetary Science Letters, v. 53, p. 279–295.

Murray, B.C., and Nagy, E.A., 1990, The relationship of the Manix Formation to the "Mojave River Formation": Mojave Desert Quaternary Research Center Symposium, San Bernardino County Museum Association Quarterly, v. 37(2), p. 32.

Nagy, E.A., and Murray, B.C., 1991, Stratigraphy and intrabasin correlation of the Mojave River Formation, central Mojave Desert, California: San Bernardino County Museum Association Quarterly, v. 38(2), p. 5–30.

Naples, V.L., 1987, Reconstruction of cranial morphology and analysis of function in the Pleistocene ground sloth, *Nothrotheriops shastense* (Mammalia, Megatheriidae): Natural History Museum of Los Angeles County Contributions in Science, no. 398, 21 p.

Naples, V.L., 1989, Feeding mechanism in the Pleistocene ground sloth, *Glossotherium*: Natural History Museum of Los Angeles County Contributions in Science, no. 415, 23 p.

Peden, D.S., 1976, Botanical composition of bison diets on shortgrass plains: American Midland Naturalist, v. 96(1), p. 225–229.

Pluhar, C.J., Kirschvink, J.L., and Adams, R.W., 1991, Magnetostratigraphy and clockwise rotation of the Pliocene-Pleistocene Mojave River Formation, central Mojave Desert, California: San Bernardino County Museum Association Quarterly, v. 38(2), p. 31–42.

Pyle, R.L., 1961, Annotated field list birds of southern California: Los Angeles Audubon Society, revised by A. Small, O. Wade Publisher, Los Angeles, California, 61 p.

Reynolds, R.E., and Reynolds, R.L., 1985, Late Pleistocene faunas from Daggett and Yermo, San Bernardino County, California, *in* Reynolds, R.E., ed.,

Cajon Pass to Manix Lake: Geological investigations along Interstate 15: Redlands, California, San Bernardino County Museum Special Publication, p. 175–191.

Roeder, M.A., 1985, Late Wisconsin records of *Gasterosteus aculeatus* (Three-spine Stickle-back) and *Gila bicolor mojavensis* (Mojave Tui Chub) from unnamed Mojave River sediments near Daggett, San Bernardino County, California, *in* Reynolds, R.E., ed., Cajon Pass to Manix Lake: Geological investigations along Interstate 15: Redlands, California, San Bernardino County Museum Special Publication, p. 171–174.

Rosen, M.R., 1989, Sedimentological, geochemical and hydrological evolution of an intercontinental closed-basin playa (Bristol Dry Lake): A model for playa development and its implications for paleoclimate [Ph.D. thesis]: University of Texas.

Scott, E., 1996, The small horse from Valley Wells, San Bernardino County, California, *in* Reynolds, R.E., and Reynolds, J., eds., Punctuated chaos in the northeastern Mojave Desert: San Bernardino County Museum Association Quarterly, v. 43(1, 2), p. 85–89.

Scott, E., 1997, A review of Equus conversidens in southern California, with a report on a second, previously unrecognized species of Pleistocene small horse from the Mojave Desert: Journal of Vertebrate Paleontology, v. 17(sup. 3), p. 75A.

Scott, E., 1998, *Equus scotti* from southern California: Journal of Vertebrate Paleontology, Abstracts of Papers, v. 18(sup. 3), p. 76A.

Scott, E., 1999, The *Equus (Plesippus)–Equus scotti* transition in western North America: Journal of Vertebrate Paleontology, Abstracts of Papers, v. 19(sup. 3), p. 74A.

Scott, E., 2000, Fossil horses at Fort Irwin, the paleontology of Bitter Springs Playa: Environmental Division of the Directorate of Public Works, Fort Irwin, California, Natural and Cultural Resources Series No. 2, 15 p.

Sarna-Wojcicki, A.M., Bowman, H.R., Meyer, C.E., Russell, P.C., Asaro, F., Michael, H., Rowe, J.J., Jr., Baedecker, P.A., and McCoy, G., 1980, Chemical analyses, correlations, and ages of late Cenozoic tephra units of east-central and southern California: U.S. Geological Survey Open-File Report 80-231, 52 p.

Sarna-Wojcicki, A.M., Bowman, H.R., Meyer, C.E., Russell, P.C., Woodard, M.J., McCoy, G., Rowe, J.J., Jr., Baedecker, P.A., Asaro, F., and Michael, H., 1984, Chemical analyses, correlations, and ages of upper Pliocene and Pleistocene ash layers of east-central and southern California: U.S. Geological Survey Professional Paper 1293, 40 p.

Schaller, G.B., 1972, The Serengeti lion, a study of predator-prey relations: Wildlife Behavior Series, Chicago, Illinois, University of Chicago Press, 480 p.

Seiple, E., 1994, Lake Manix: California Division Mines and Geology, California Geology, March/April, v. 47(2), p. 50–57.

Snyder, C.T., Hardman, G., and Zdenek, F.Z., 1964, Pleistocene lakes in the Great Basin: U.S. Geological Survey Miscellaneous Geological Investigations Map I-416, scale 1:1,000,000.

Spaulding, W.G., 1980, The presettlement vegetation of the California desert: Manuscript on file, U.S. Bureau of Land Management, Riverside, California, 97 p.

Spaulding, W.G., 1990, Vegetational and climate development of the Mojave Desert: The last glacial maximum to the present, *in* Betancourt, J.L., et al., eds., Packrat middens: The last 40,000 years of biotic change: Tucson, Arizona, University of Arizona Press, p. 166–199.

Spaulding, W.G., Leopold, E.B., and Van Devender, T.R., 1984, Late Wisconsin paleoecology of the American southwest, *in* Spaulding, C.P., ed., Late-Quaternary environments of the United States, Volume 1: The late Pleistocene: Minneapolis, University of Minnesota Press, p. 259–293.

Stebbins, R.C., 1966, A field guide to western reptiles and amphibians: The Peterson Field Guide Series: Boston, Houghton Mifflin Company, 279 p.

Steinmetz, J.J., 1987, Ostracodes from the late Pleistocene Manix Formation, San Bernardino County, California, *in* Reynolds, J., ed., Quaternary history of the Mojave Desert, San Bernardino County Museum Association Quarterly, v. 34(3 and 4), p. 46–47.

Steinmetz, J.J., 1988, Biostratigraphy and paleoecology of limnic ostracodes from the late Pleistocene Manix Formation [master's thesis]: Department of Biological Sciences, Pomona, California State Polytechnic University, 64 p.

Stock, C., 1925, Cenozoic gravigrade edentates of western North America with special reference to the Pleistocene Megalonychinae and Mylodontidae of Rancho La Brea: Washington, D.C., Carnegie Institute of Washington, Publication no. 331, p. 206 p.

Stock, C., 1930, Rancho La Brea, a record of Pleistocene life in California: Los Angeles Museum Publication, no. 1, 82 p.

Stuiver, M., 1969, Yale natural radiocarbon measurements, IX: Radiocarbon, v. 11(2), p. 545–658.

Taylor, D.W., 1967, Late Pleistocene molluscan shells from the Tule Springs area, *in* Wormington, H.M., and Ellis, D., Pleistocene studies in southern Nevada: Nevada State Museum Anthropological Papers, no. 13, p. 395–399.

Walsh, S.L., 2000, Eubiobstatigraphic units, quasibiostratigraphic units, and "assemblage zones": Journal of Vertebrate Paleontology, v. 20(4), p. 761–775.

Webb, S.D., 1965, The osteology of Camelops: Natural History Museum of Los Angeles County, Science Series, no. 1, 54 p.

Weldon, R.J., II, 1982, Pleistocene drainage and displaced shorelines around Manix Lake, *in* Cooper, J.D., ed., Geologic excursions in the California desert: Geological Society of America, Cordilleran Section Guidebook, p. 77–81.

Weldon, R.J., II, 1985, Implications of the age and distribution of late Cenozoic stratigraphy in Cajon Pass, southern California, *in* Reynolds, R.E., ed., Cajon Pass to Manix Lake: Geological investigations along Interstate 15: Redlands, California, San Bernardino County Museum Association Special Publication, p. 59–68.

Wells, S.G., and Enzel, Y., 1994, Fluvial geomorphology of the Mojave River in the Afton Canyon area, eastern California—implications for the geomorphic evolution of Afton Canyon, *in* McGill, S.F., and Ross, T.M., eds., Geological investigations of an active margin: Geological Society of America, Cordilleran Section Guidebook, p. 177–182.

Wells, S.G., Anderson, R.Y., McFadden, L.D., Brown, W.J., Enzel, Y., and Miossec, J., 1989, Late Quaternary paleohydrology of the eastern Mojave River drainage, southern California: Quantitative assessment of the late Quaternary hydrologic cycle in large arid watersheds: New Mexico Water Resources Research Institute, Technical Completion Report, Project 14–08–0001-G1312, no. 242, 253 p.

Winters, H.H., 1954, The Pleistocene fauna of the Manix Beds in the Mojave Desert, California [master's thesis]: Department of Geology, California Institute of Technology, Pasadena, California, 58 p.

MANUSCRIPT ACCEPTED BY THE SOCIETY AUGUST 1, 2002

Late Pleistocene lakes along the Mojave River, southeast California

Yehouda Enzel*
Institute of Earth Sciences and Department of Geography, The Hebrew University of Jerusalem, Jerusalem 91904, Israel

Stephen G. Wells
Nicholas Lancaster
Desert Research Institute, 2215 Raggio Parkway, Reno, Nevada 89512-1095, USA

ABSTRACT

Closed and semiclosed basins along the Mojave River in southern California were occupied by pluvial lakes during the latest Pleistocene. The chronologies of Harper Lake, Lake Manix (Coyote Lake and Troy Lake playas and the Afton basin), and Lake Mojave (Soda Lake and Silver Lake playas) are summarized here from available data. We evaluate the chronologies, compare them with each other, and then use them to determine coexistence of lakes within the Mojave River hydrological system. The average annual flow in the lower reaches of the Mojave River that is needed to form and maintain a lake in one of these basins is at least an order of magnitude larger than the present-day average discharge of 9.5 x10^6 m^3. The discharge could have increased by (a) more frequent storms and floods, and/or (b) reduced loss by transmission of flood water along the river length. This reduction in transmission losses could have been caused by longer river reaches either covered by lakes or characterized by base flow that, in turn, was formed by water table near or at the surface. The increase in flood discharge is caused by an increased storm frequency in the headwater of the river. The discharge increase needed to support individual lakes is multiplied when the total lake area of coexisting lakes fed by the Mojave River is considered. It demands an even larger increase of the number of storms in the headwaters than the number needed to support an individual lake. This indicates a large increase in atmospheric moisture transported to this relatively low latitude along the coast of western North America. The coexistence of lakes during the last glacial maximum and the highstands of other lakes in similar latitudes in the southwestern United States indicate that the storm tracks were frequently directed at 32°–34° N latitude at that time.

The chain of lake basins along the Mojave River is an example of a hydrological system that is integrated through time into one large arid river basin. This was possible mainly because of the elevated headwaters located at the San Bernardino Mountains where orographic effects cause heavy storms and large floods. These large floods fill the depositional basins along the river with water and sediments and allow them to overflow downstream. Afton Canyon was formed by such an overflow from the Manix basin to the Lake Mojave basin. The incision of the 150-meter-deep canyon was previously proposed to be rapid and geomorphically catastrophic. Here we propose that a time-transgressive incision lasting over a few thousand of years is more plausible explanation for the formation of this canyon; geologically it is still a rapid event.

---

*E-mail: yenzel@vms.huji.ac.il

Enzel, Y., Wells, S.G., and Lancaster, N., 2003, Late Pleistocene lakes along the Mojave River, southeast California, *in* Enzel, Y., Wells, S.G., and Lancaster, N., eds., Paleoenvironments and paleohydrology of the Mojave and southern Great Basin Deserts: Boulder, Colorado, Geological Society of America Special Paper 368, p. 61–77. © 2003 Geological Society of America.

In addition, we discuss how the findings along the Mojave River reflect upon two long-term hypotheses of (a) mega–Lake Manly that supposedly filled a large area in the Mojave Desert sometimes in the middle Pleistocene, and (b) Lake Manix and/or Lake Mojave overflow to Bristol Lake and to the Colorado River.

INTRODUCTION

The present-day Mojave River (Fig. 1) drainage system (9500 km^2) is the largest hydrological system in the Mojave Desert. It heads at high elevations in the San Bernardino Mountains of the Transverse Ranges of southern California and terminates in the Soda Lake and Silver Lake playas in the heart of the Mojave Desert (Williamson et al., 1856; Wells et al., this volume). The headwaters of the Mojave River in the San Bernardino Mountains located at 34° latitude (Fig. 1) are far south of the average latitude of the North Pacific winter storm track that currently affects western North America (Pyke, 1972). This relatively low latitude makes the Mojave River an excellent candidate to detect past shifts to the south of the storm tracks during glacial times (e.g., Antevs, 1938). The closed basins of the Mojave River provide the key environment for storing lacustrine and playa deposits and therefore the evidence of the past shifts in storm tracks and wetter conditions in the Mojave Desert. Because of the size of the Mojave River drainage basin, the paleohydrology and/or paleoclimate deduced from these chronologies are applicable to the entire Mojave Desert, the southern Great Basin, and perhaps the southwestern United States in general.

The Mojave River crosses several tectonic basins bounded by northwest-southeast–trending right-lateral strike-slip faults (e.g., Garfunkel, 1974), which control both the configuration of the Mojave River groundwater aquifer (Mojave River Agency, 1982, 1985) and the irregular shape of the Mojave River surface-water drainage basin (Fig. 1). These basins and their groundwater are recharged by floodwaters from the Mojave River, which is therefore a central factor in the water resources of the western and central Mojave Desert. This was recognized very early in this century (e.g., McClure et al., 1918) and is the reason for many studies of the subsurface geology (e.g., Martin, 1994).

During the Pleistocene the Mojave River filled these basins with fluvial, deltaic, playa, and lacustrine deposits (see Cox et al., this volume; Jefferson, this volume). These lacustrine deposits record the paleohydrology of the Mojave River and the regional paleoclimatology. The late Pleistocene lakes of Harper Lake, Lake Manix (now the Afton basin, the Coyote and Troy Lake playas), Mojave Lake (now the Soda Lake and Silver Lake playas) (Figs. 1 and 2) that formed in the Mojave River drainage basin raise questions about the degree to which they coexisted, and the pattern, hydrography, and evolution of the overall drainage basin. Earlier studies of the chronologies of these lakes tried to identify coexistence and overflowing from one lake basin to another, and inferred major geomorphic events and changes along the river (e.g., Meek, 1989). From earlier studies we summarize chronologies of various latest Pleistocene and early Holocene paleolakes

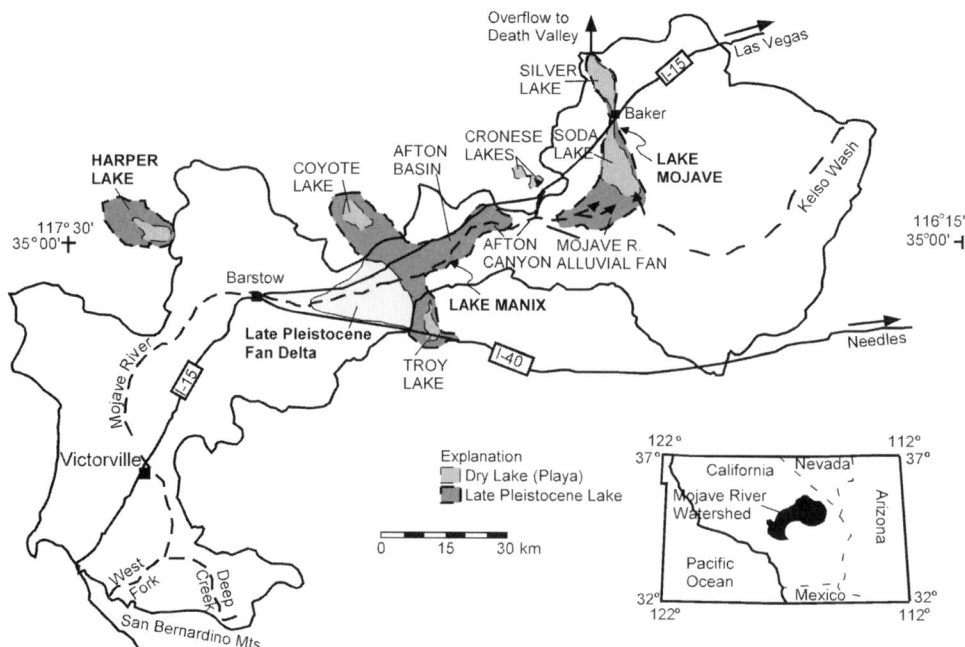

Figure 1. The Mojave River drainage basin and the area draining to Silver Lake playa in southeastern California. Note the late Pleistocene Mojave River fan delta west of the dissected Lake Manix basin. Currently the Mojave River has a large alluvial fan at the southern end of Soda Lake playa. Its floodwaters fill both East and West Cronese Lakes and Silver Lake playas.

in the depositional basins affected by the Mojave River. We mainly concentrate on available information that points to the major paleohydrologic and paleoclimatic conditions. Later we compare the chronologies of these Mojave River lakes with the Great Basin pluvial lakes to the north and other lake chronologies from latitudes similar to Lake Mojave. We finally discuss the paleoclimatic implications of these chronologies.

The summary of chronologies of the individual basins also describes the more general process of integration of several basins into the hydrological system of the present-day Mojave River. Understanding this long-term evolution of the river, in turn, is a necessary stage in deciphering the interplay between hydrology and basin physiography. Only then can the paleohydrology deduced from sediments in the individual basins, be transformed qualitatively or quantitatively into regional paleoclimatology (Wohl and Enzel, 1995). For example, it was pointed out (e.g., Benson and Paillet, 1989) that the surface area of a lake is the most sensitive parameter to variations in climate. Therefore, the areas of one, two, or more coexisting lakes reflect very different hydrologic and climatic conditions and Mojave River inflow for them to be maintained.

Current climatic conditions

Southern California experiences a Mediterranean climate with a relatively hot dry summer and a winter precipitation maximum, which are related to the cyclonic activity over the North Pacific Ocean. In the San Bernardino Mountains, the mean

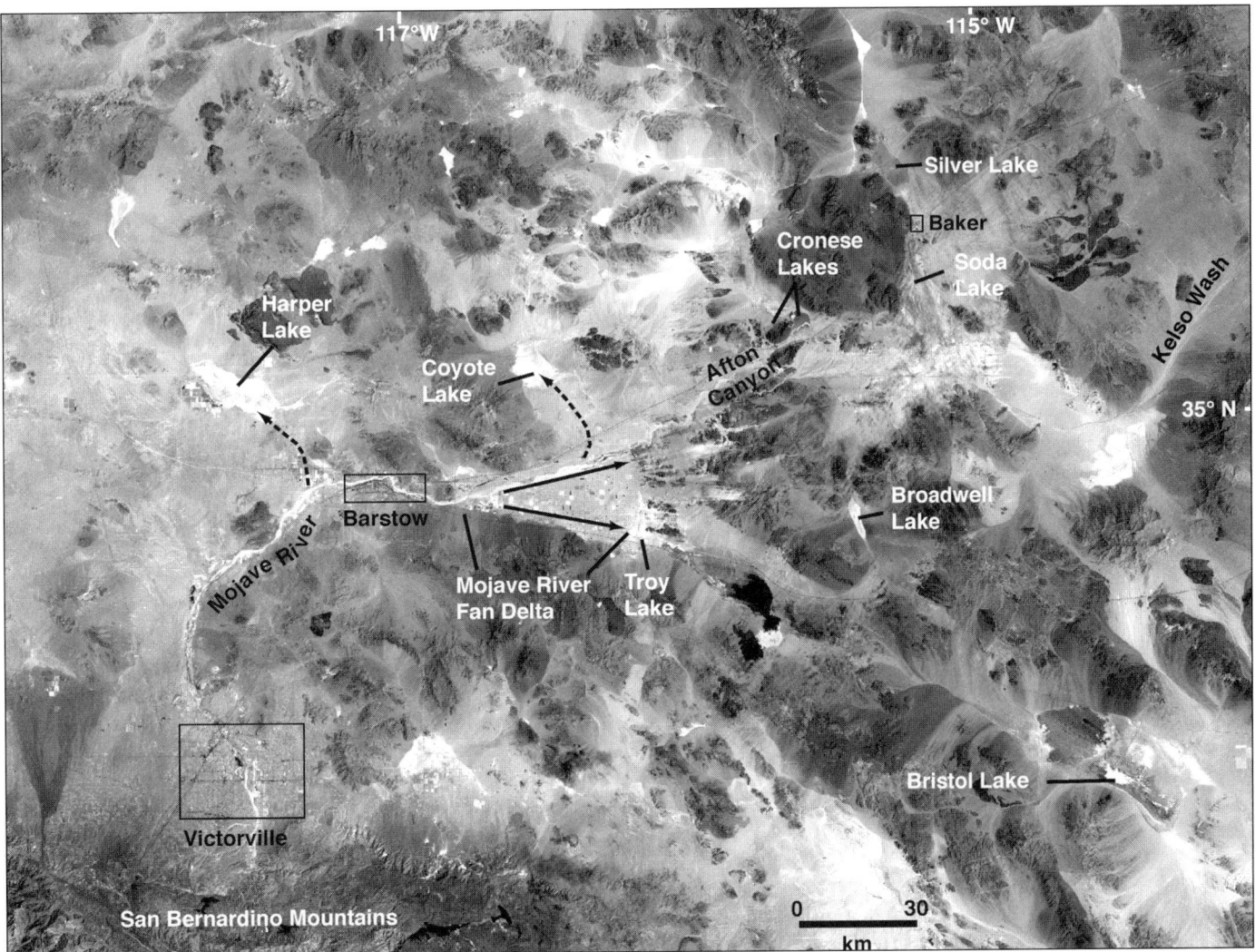

Figure 2. Satellite images (LANDSAT™) of the Mojave River and vicinity. The images were produced from data by Mojave Desert Ecosystem Program of the U.S. Department of Defense (Legacy Program) and the Department of the Interior. Arrows with solid and dashed lines represent proposed Mojave River flow directions during the progradation of the Mojave River fan delta into the Lake Manix subbasins of Coyote Lake, Afton basin, Troy Lake, and during earlier times into the upstream Harper Lake.

annual precipitation exceeds 1000 mm, in the Victorville area it is 125–150 mm, and in the Silver Lake area it is 75 mm. Rainfall seasonality changes across the drainage basin. For example, summer rainfall constitutes 40% and <10% of the annual rainfall in Silver Lake playa and in the headwaters, respectively (Enzel, 1990). Potential lake evaporation in the headwaters area at elevations above 1500 m is 1000 mm/year or less (Crippen, 1965; Mojave River Agency, 1985). The extremely high temperatures at elevations lower than 500 m in Mojave Desert result in ~2000 mm/year equivalent lake evaporation in the terminal basins (Blaney, 1957; Mojave River Agency, 1985; Enzel, 1990). These extremely arid conditions demand a major change in the Pleistocene that introduced into the area the moisture that was necessary to form the paleolakes that occupied several subbasins along the Mojave River.

Anecdotal observations and early research in the lower Mojave River

All early documents that mention the Mojave River seem to assume that it flowed to the Colorado River. For example, in 1776 Fray Francisco Garcés, who crossed Soda Lake playa and continued through Afton Canyon to San Gabriel Mission, thought that the river flowed to the Colorado River (Coues, 1900). He named it "*Arroyo De Los Martires*," which is the name on a map from 1777 that shows that the Mojave River flowing to the Colorado River (Font's map; "*R. de Los Martires*"; Preston, 1988). Jedediah Smith mentioned the playa in 1825 and 1826. However, he (Morgan, 1953) and later Captain Ewing Young and Kit Carson in 1829 and even later the Frémont Expedition in 1844 (e.g., Dale, 1918; Sabin, 1914; Jackson and Spence, 1970), did not notice that the Mojave River terminates in dry playas. Frémont claimed that he continued all the way to "the end of the river." He did not mention, however, the playas at that end; Thompson (1929) claimed that they did not reach the playas.

Williamson et al. (1856), who explored the Mojave River during the Pacific Railroad Expeditions in 1853, first documented that (a) the Mojave River does not flow to the Colorado River, and (b) the playas presently named Soda Lake and Silver Lake are the actual termini of the river. In 1853, even after the Williamson discovery, the Whipple party of the Pacific Railroad Expeditions was still looking for the confluence of the Mojave River and the Colorado River.

Hydrological and paleohydrological observations began much later. Gale (1914), Buwalda (1914), and Free (1914, 1916) identified the presence of an ancient lake in the playas, and Free (1916) concluded that the outflow from this ancient lake was "both small and transient." Huntington (1915) proposed that the Mojave River was more vigorous in the past, overtopping its terminal basin and flowing into the Amargosa River in Death Valley. Thompson (1921, 1929), in his seminal study of the area, termed the ancient lakes that occupied Soda/Silver Lake playas and the East and West Cronese playas "Lake Mojave" and "Little Mojave Lake," respectively.

Buwalda (1914) suggested that ancient Lake Manix, once the terminus of the Mojave River, overtopped its hydrologic barrier, cut a channel, and drained rather rapidly into the Lake Mojave basin. Thompson (1921, 1929) agreed with Buwalda that Lake Manix was once the terminus of the Mojave River. He hypothesized that Lake Manix and Lake Mojave did not exist at the same time and that the formation of Afton Canyon and filling of Lake Mojave occurred relatively recently as a consequence of the draining of Lake Manix. Antevs (1937) estimated that Lake Mojave existed between 25 and 10 ka, and Crozer-Campbell et al. (1937) and Hubbs and Miller (1948) postulated that Lake Mojave and Lake Manix might have been contemporaneous. These opposing assertions have major implications for the estimation of past regional climates: There is a need for lower Mojave River discharge to support only one lake at a time if lakes are not contemporaneous. Thompson's (1921, 1929) hypothesis may have been the basis for the conclusion of a recent study on the history of Lake Manix and the formation of Afton Canyon (Meek, 1989, 1990). These studies conclude that late Wisconsin Lake Manix did not overflow into Soda Lake and Silver Lake basins and that Lake Manix was rapidly drained by the rapid incision of Afton Canyon between 13.8 and 13.3 ka (Meek, 1989, 1990). To address this issue, we will present the chronologies of Lake Mojave and Lake Manix and then discuss the incision of Afton Canyon. The Mojave River is a good example of a river that through time integrates a few subbasins into one hydrological system that reaches further downstream into a closed basin. We stress, however, that not all the data exist to resolve this issue completely. Therefore, the main goal is to summarize available information and to discuss the possible conclusions based on them.

OBSERVATIONS FROM THE MOJAVE RIVER BASINS

Harper Lake playa

Thompson (1929) was the first to raise the possible existence of a paleolake within the Harper basin. He reported shells and blue clays in boreholes from the basin. Dibblee (1960, 1968) identified the highest shoreline at the 2160 foot (~658 m) elevation contour and a few lower recessional shorelines in this basin. The preservation of these features indicates that the lake that filled the basin occurred recently. Dibblee considered unlikely the possibility that Harper Lake received overflow from Lake Thompson (which previously occupied the present-day Rogers and Rosamond Dry Lakes, 70 km southwest of Harper Lake playa). According to Meek (1999), who surveyed the highest wave-cut cliff of Harper Lake, the elevation of the highest lake level is 656.9 m (2155 feet). Reynolds and Reynolds (1994) raised a question regarding the source of water for the formation of such a lake. They observed that the highest shoreline is only 6 m below the wide alluvial saddle that separates the Mojave River drainage basin from the Harper Lake basin. This low water divide and the lacustrine fauna led Reynolds and Reynolds

(1994) to suggest: (a) a past connection between the Mojave River and the Harper Lake basin, (b) that minor vertical tectonic activity was responsible for connecting the basins and later the isolation of the Harper Lake, and (c) that the isolation of the Harper Lake from the Mojave River occurred during the late Pleistocene. Meek (1999) proposed a possible route for the Mojave River to flood Harper Lake (Fig. 2) under the reasonable assumption that groundwater could not have caused the late Wisconsin Harper Lake stand.

Recently, Meek (1999) dated two *Anodonta* samples from the same horizon, ~2 m below the highest shoreline, that yielded ages of 24,440 ± 2190 (UCLA-2627a) and 25,000 ± 310 (UCR 2867) ^{14}C yr B.P. (all ages in this manuscript are not calibrated). A third age of an *Anodonta* from ~5 m below the shoreline resulted in >30,000 ^{14}C yr B.P. (UCLA-2627b). Beach deposits that contain lacustrine clays yielded abundant ostracodes with *Limnocythere bradburyi* dominating (Meek, 1999). This taxon has been found in other late Wisconsin lakes along the Mojave River (Wells et al., this volume), but it was not identified in the 300–40 ka lacustrine units of Lake Manix (Steinmetz, 1988, 1989; Meek, 1999). In Lake Mojave, this taxon was identified only in deposits associated with the last glacial maximum stand of Lake Mojave I phase (Wells et al., 1989, this volume).

Lake Manix

Late Pliocene to early Pleistocene sediments from the Manix basin indicate stream flow to the southwest (Nagy and Murray, 1991; Jefferson, this volume). This observation, in turn, indicates that the eastward-flowing Mojave River did not reach that area and probably terminated in one of the upstream basins (Cox and Tinsley, 1999; Jefferson, this volume; Cox et al., this volume). During the period 2.5–1 Ma, the Manix basin was probably internally drained and was occupied occasionally by ephemeral saline lakes or playas (Jefferson, 1985, this volume). Although older sediments contain gypsiferous deposits, the earliest known lakes in the Manix basin are ~500,000 yr old (Jefferson, this volume). The first appearance of these full-lake conditions with their diverse environmental conditions (Jefferson, 1985, this volume; Steinmetz, 1988), indicates the first large discharge of the Mojave River. Steinmetz (1988) suggested that the appearance and abundance of sensitive ostracodes in the various beds of Lake Manix peak with expanded favorable habitats during glacial conditions. Shlemon and Budinger (1990) emphasized the potential importance of the Lake Manix paleoenvironments to early humans in the region.

The following paragraphs summarize the latest Pleistocene chronology of Lake Manix. Earlier lake phases in the Manix basin, from ca. 500 ka to the latest Pleistocene, are summarized by Jefferson (this volume). Although a large number of radiocarbon ages exist for Lake Manix subbasins and vicinity (Fergusson and Libby, 1962; Hubbs et al., 1962, 1965; Bassett and Jefferson, 1971; Jefferson, 1985; Reynolds and Reynolds, 1985; Meek, 1990, 1999), the latest Pleistocene chronology of Lake Manix is yet not well established, even though it is crucial to understanding the paleohydrology of the Mojave River.

In the Lake Manix basin, ages have been derived from radiocarbon analyses on *Anodonta*, tufa, bone and rock-varnish samples, U/Th analyses on identified faunal fragments, and "cation ratios" on rock varnish (Jefferson, 1985; Meek, 1990, 1999; Berger and Meek, 1992). In this summary we do not use the calibrated ages based on the rock-varnish cation ratios or the radiocarbon ages from rock-varnish samples. These ages were found to be problematic (Beck et al., 1998; Watchman, 2000; Gillespie, 2003) and even R. Dorn, who conducted the rock-varnish cation ratios analyses on samples from Lake Manix (Meek, 1990), de-emphasized their importance (Whitley and Dorn, 1993). In his recent discussion of Lake Manix, Meek (1999) also avoids using the various rock-varnish ages. We choose also not to use the 480-years correction for *Anodonta* suggested by Meek (1990) and Berger and Meek (1992), not because we do not agree with them on the need for correction, but because the value proposed for the reservoir correction is based on limited data.

Meek (1990, Plate 4) used the elevations of dated samples rather than dated morphostratigraphy of shore features from Lake Manix to produce a lake-level curve. Because the sequence of events represented by the various shore features is based on the ages of the shore features and not on stratigraphic relations, this curve is subject to modifications with additional ages, different ages, and whether or not the varnish ages are included. Meek (1990) defined four highstands of Lake Manix that reached 543 m in the Afton basin: ca. 31–30, 21–20, 18, and 15–14 ka. Based on two additional age determinations, the 15–14 ka stand was recently discarded (Meek, 1999) and the same beach ridge from which the earlier age came is now dated at ca. 29 ka (Meek, 1999). This change emphasizes the problem of using ages without a clear stratigraphic context; a beach ridge once considered the youngest in the sequence is now considered older than other late Pleistocene lacustrine features.

The highstands of Lake Manix all reached repeatedly the same elevation of 543 m (Meek, 1990, Plate 4), probably indicating that the lake reached a topographic threshold of either its spillway elevation downstream or a sudden increase of area within Lake Manix at that elevation. We suggest that Lake Manix probably reached its spillway elevation a few times during the late Pleistocene and overflowed downstream to Lake Mojave basin. Low lake stands are not defined for Lake Manix, although Meek (1999) argues that a gap in ages means either a drop in lake level and/or Mojave River discharge was diverted to another basin. We suggest that both are possible but too simplistic assertions.

Coyote Lake and Troy Lake playa basins

The Coyote and Troy Lake playas were part of the area inundated when Lake Manix was at its highstand but, because of limited exposures of lacustrine deposits (Meek, 1990), they have attracted much less attention than the Afton basin. Some lacustrine features in Lake Coyote and Troy Lake basins were discussed by Thompson (1929) and Blackwelder and Ellsworth (1936), and mapped by Byers (1960) and Dibblee and Bassett (1966a, 1966b). Hagar (1966) mapped the surface characteristics of the Coyote Lake playa and described shore features and the stratigraphy and sedimentology of the Mojave River delta front in Coyote Lake basin; Groat (1967) mapped in detail the Troy Lake basin. Subsurface information is available for Coyote Lake in Thompson (1929), Dyer et al. (1963), and Meek (1990). Meek (1990) indicates the existence of ~100 boreholes from the basin that have not been published or interpreted. Several ages of shore features are available from these basins, and one age has been determined from the base of a delta that prograded into the lake (Meek, 1990). The ages on samples collected from shore features of Coyote Lake basin are 13,800 ± 600 (LJ-958); 13,560 ± 145 (UCLA-2609b); 12,900 ± 120 (UCLA-2606); and 11,810 ± 100 (UCLA-2609c) ^{14}C yr B.P.; from the base of the exposed portion of the delta that prograded into the Coyote Lake basin and on the lakebed, the radiocarbon ages on Anodonta are 17,590 ± 1500 (UCLA-2603) and 15,125 ± 270 (UCLA-2608) ^{14}C yr B.P. (Hubbs et al., 1965; Meek, 1990). From these ages and their elevations, but without any stratigraphic relationships, Meek (1990) interpreted a fluctuating lake in the Coyote Lake basin between ca. 15 and 11 ka. The age from the base of the delta indicates that a lake existed in the Coyote Lake basin also ca. 17 ka. The two ages from radiocarbon (Dorn et al., 1986) and a cation-ratio analysis of rock varnish (Meek, 1990) from the Coyote Lake shore features are problematic, as discussed above, and therefore are not used here. It should be noted that all the ages acquired from Coyote Lake are ca. 17 ka or younger (Fig. 3). However, the thickness of lake deposits identified in a few of the 100 borehole cores indicates that earlier lakes occupied the basin (Meek, 1990). Only one age is available from Troy Lake basin, 15,025 ± 230 ^{14}C yr B.P. (UCLA-2605) (Meek, 1990). This age is directly related to the Lake Manix levels in the Afton basin because both basins are fed directly by the Mojave River (Meek, 1990). The 15 ka age from the Troy Lake basin becomes the youngest age associated with the 543-meter shoreline that is common around the Lake Manix basin.

The Mojave River fan delta

When the Mojave River began flowing into Lake Manix basin sometime in the early to middle Pleistocene it probably formed a delta along the western margins of that lake. The large delta that is observed at the surface and termed "Mojave River braid delta" (e.g., Meek, 1990) is of late Quaternary age (Figs. 1 and 2). The oldest age (U-series age) from any fan-delta deposits in the Afton basin is >68 ka (Jefferson, 1985; Meek, 1990). Other prograding delta sediments were deposited in that basin since 55–60 ka (G. Jefferson, 2000, written commun.), between ca. 55 and 19 ka, and immediately after 19 ka (Jefferson, 1985; Meek, 1990). Part of this late Quaternary delta prograded north into Coyote Lake (Hagar, 1966, p. 128) soon after 17 ka (see above; Meek, 1990). Hagar (1966) analyzed the stratigraphy and sedi-

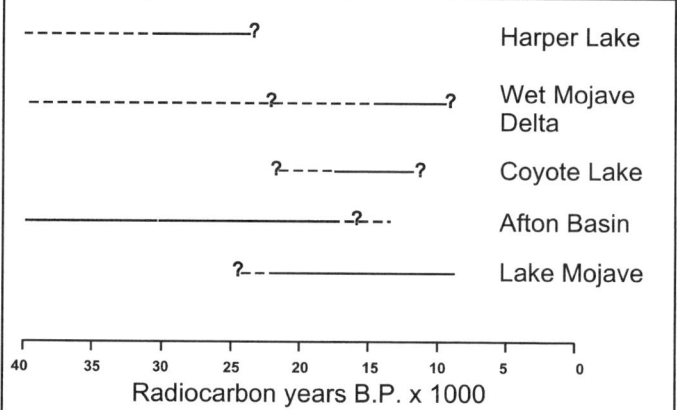

Figure 3. A chart showing the range of ages for the various lakes that were fed by the Mojave River during the late Pleistocene: Lake Mojave, Harper Lake, and Lake Manix (Afton, Coyote Lake, and Troy Lake subbasins). The Mojave River fan delta upstream of Lake Manix indicates repeated wet depositional environment. Proposed lowstands and gaps in radiocarbon ages are not marked. See Figures 1 and 2 for locations of the basins.

mentology of the delta front of the Mojave River in the Coyote Lake playa and suggests that the deltaic area in the Coyote Lake basin is not typical of deltas because of the presence of an arcuate bar. He reported that the gravel content decreases, the sands thin, and the silt layers increase in clay content from the delta-front toward the present-day margins of the Coyote Lake playa. The delta deposits interfinger with the lacustrine deposits (e.g., Hagar, 1966) indicating progradation into an existing lake throughout the time of deposition. The surface of the delta that prograded into the Coyote Lake exhibits remnants of meandering and braided channels (Hagar, 1966, Plate 1) that are associated with the final stages of the Mojave River fluvial activity prior to a later incision. From the available data, it is not possible to determine whether the delta propagated into a lowstand or highstand of Lake Manix. Sediments deposited during this late Quaternary deltaic progradation are probably responsible for the topographic separation of the Afton, Coyote Lake, and Troy Lake basins of Lake Manix. This separation allows for independent lake histories in the Coyote Lake basin and perhaps in the Troy Lake basin after the incision of the Afton Canyon.

The Mojave River today flows in a channel incised into its late Pleistocene delta. Reynolds and Reynolds (1985) suggest that the incision in the western part of the delta occurred around or soon after 7.5 ka. They provide ages and sedimentological and paleontological data that indicate that the Mojave River was flowing on the wide plain of the delta before ca. 13 ka and 9 ka (Fig. 3). The eastward increase in abundance of mesic and aquatic fauna, relative to xeric fauna, during that time interval indicates that wetter environments in the delta increase to the east, with abundant freshwater riparian, oxbow, fluvial, and lacustrine environments (Reynolds and Reynolds, 1985). These environments may represent the persistence of groundwater either near or at the surface at the eastern part of the delta and at somewhat greater depth at the western part of the delta where xeric indictors are more abundant during the same period.

Lake Mojave

The Lake Mojave chronology (Ore and Warren, 1971; Wells et al., 1989; Wells et al., this volume) is determined by dating of two depositional environments. Most of the dates are from the shore and beach environments of the lake and are based on samples of lustrous whole pelecypod shells (*Anodonta*) and lithoid tufa coating on gravel clasts (Ore and Warren, 1971; Wells et al., 1989, this volume; Brown et al., 1990). A few dates are from disseminated organic matter, extracted from fine-grained lacustrine deposits obtained in core samples from Silver Lake playa (Wells et al., 1989, this volume; Enzel et al., 1989; Brown et al., 1990).

The range of all radiocarbon ages acquired from Lake Mojave is >20 ka to 8.5 ka (Fig. 3), and ages from shore features range between 16.2 ^{14}C ka and 8.7 ^{14}C ka (Wells et al., this volume). The oldest date from the beach environment is 16,270 ± 310 ^{14}C yr B.P. (Beta-29553), from the lower of two beach-ridge complexes at the northeast of Silver Lake playa (Wells et al., this volume). No age is available for the older and higher beach ridge. However, the existence of a higher beach ridge points to two conclusions (Wells et al., this volume) crucial to the regional paleohydrology: (a) prior to 16.2 ^{14}C ka, and probably during the last glacial maximum, Lake Mojave existed as a higher and deeper lake than later in its history, and (b) during that earlier period, the lake reached the elevation of the highest shoreline (287–288 m), which is controlled by the bedrock-spillway elevation, and the lake probably overflowed toward the Amargosa River and Death Valley.

The higher and lower beach ridges are related respectively to an older and a younger lake stand (Wells et al., this volume). Wells et al. (1989) tested the existence of the earlier lake stand by drilling core SIL-G (Fig. 23 of Wells et al., 1989) into the center of the small basin between the two beach ridges. About 3 m of lacustrine sediments were recovered at this site. Sedimentologic and paleontologic evidence from these deposits indicate that they are related to Lake Mojave I lithozone (Wells et al., this volume). Their thickness indicates at least a few thousands of years of lacustrine environment prior to 16.2 ka (Wells et al., this volume). Therefore, we conclude that Lake Mojave I lithozone was deposited by a deep, overflowing lake with only minor fluctuations during the last glacial maximum (we use this term as the uncalibrated radiocarbon age of 18,000 yr B.P.).

Ages from the Silver Lake playa core SIL-I indicate that a lacustrine environment dominated the basin during the last glacial maximum (Wells et al., this volume). From the two ages of 20,320 ± 740 (Beta-21801) and 14,660 ± 260 (Beta-21800) ^{14}C yr B.P., Wells et al. (this volume) extrapolated the sedimentation rate down the core and concluded that a lake existed in this basin by 22 ka or earlier. To test the use of these rates, they also extrapolated the rate up the core to the well-recognized and dated transition from lacustrine to playa deposits. This transition marks the desiccation of Lake Mojave, which was dated to ca. 8500 ^{14}C yr B.P. by Ore and Warren (1971) who used shore features. An additional age from immediately below the sediments associated with the lake desiccation in core SIL-M (Enzel et al., 1989; Wells et al., 1989; Enzel and Wells, 1997) indicates that the desiccation occurred soon after 9330 ± 95 ^{14}C yr B.P. (Beta-24342; Wells et al., this volume). The extrapolation of the sedimentation rates resulted in an estimated age of 8700 ^{14}C yr for the desiccation. This similarity among the ages for desiccation indicates that the 22 ka estimated age for the lake's initiation in the Silver Lake basin is reasonable. In turn, it also indicates that a lake already existed in the Soda Lake subbasin of Lake Mojave by 22,000 ^{14}C yr B.P. The internally consistent ages from both the shore and the lacustrine environments of Silver Lake playa indicate that these ages are acceptable.

Subsurface information indicates that water in Soda Lake subbasin of Lake Mojave was almost twice as deep as the Silver Lake subbasin (Wells et al., this volume). A bedrock sill just south of Baker, California separates the two subbasins (Williamson et al., 1856; Thompson, 1929; Wells et al., 1989, this volume). Therefore, Soda Lake was filled before any water could flow to Silver Lake. This indicates that the lower part of the late Pleistocene lacustrine sequence from Soda Lake was probably deposited even earlier than 22 ka. However, no direct age control is available.

In boreholes, from the area now termed the Mojave Sink, between Afton Canyon and Soda Lake playa, Dickey et al. (1979) identified relatively thick lacustrine gypsiferous silt-clay below the latest Wisconsin Lake Mojave deposits. This small moist playa and/or shallow lake was temporally more persistent, although evidently they were smaller than was Lake Mojave (Brown and Rosen, 1995). There is no information on the age of these deposits or whether this lake was fed by Mojave River water.

DISCUSSION

Age of the Afton Canyon incision

We agree with earlier studies (Buwalda, 1914; Thompson, 1929; Muessig et al., 1957; Jefferson, 1985; Meek, 1989, 1990) that Afton Canyon is geologically young. However, Wells and Enzel (1994) concluded that there is little conclusive evidence in Afton Canyon and downstream basins to indicate that the canyon formed shortly after 14 ka during a single relatively rapid incision, as was advocated by Meek (1989). The rapid incision was estimated to take as little as "10 days" or "three weeks" (N. Meek, 1988, 1990, written commun.), or as much as 500 yr (Meek, 1989, 1990). The rapid-incision idea was based mainly on the lack of recessional shorelines in the Afton basin, the lack of terraces within the canyon, and the presence of deeply incised tributary channels into the canyon (Meek, 1989). Wells and Enzel (1994) analyzed the geomorphology of Afton Canyon and these specific observations, and concluded that an alternative hypothesis such as time-transgressive incision during at least a few thousands of years is a more realistic explanation. The supportive evidence to this explanation is: (a) the slow upstream migration of the nickpoint formed by the incision (Meek, 1999), (b) the existence of marsh and lacustrine conditions in the Mojave River delta in western Lake Manix as late as 12–9 ka (Reynolds and Reynolds, 1985), indicating that the river was not yet incised back that far by that period (note that at its eastern end Lake Manix could have overflowed at the same time and even earlier), and (c) the existence of pronounced Holocene terraces within the canyon (Wells and Enzel, 1994), indicating that the incision of Afton Canyon to its present form was not a continuous but episodic and a relatively long process. In addition, we observe recessional shorelines in the Afton basin at elevations below the highest level of Lake Manix, suggesting a longer period of downcutting.

The timing of the incision (Meek, 1989) was based on the comparison of the youngest age (ca. 14 ka) from the shore features of the Afton basin of Lake Manix and the ages of Lake Mojave, with the assumption that these two lakes did not coexist (Meek, 1989). The 14 ka age is now considered problematic (Meek, 1999). The chronology of Lake Mojave discussed above and in Wells et al. (this volume) indicates that Lake Mojave was overflowing prior to 16.2 ka and that a lacustrine environment existed in Silver Lake playa as early as 22 ka, and somewhat earlier in the Soda Lake basin. Lake Dumont north of Silver Lake playa (Fig. 4), which was supported by Lake Mojave overflow existed during the last glacial maximum (Anderson, K., and Wells, this volume) and probably overflowed to Death Valley (Anderson, D., and Wells, this volume). These overlapping ages indicate that both Lake Manix and Lake Mojave were full during the last glacial maximum, indicating a much larger water supply by the Mojave River.

Because of the importance of the single 14 ka age for the youngest shore feature in the Afton basin of Lake Manix, recently Meek (1999) reexamined the evidence for it, discarding the 14 ka

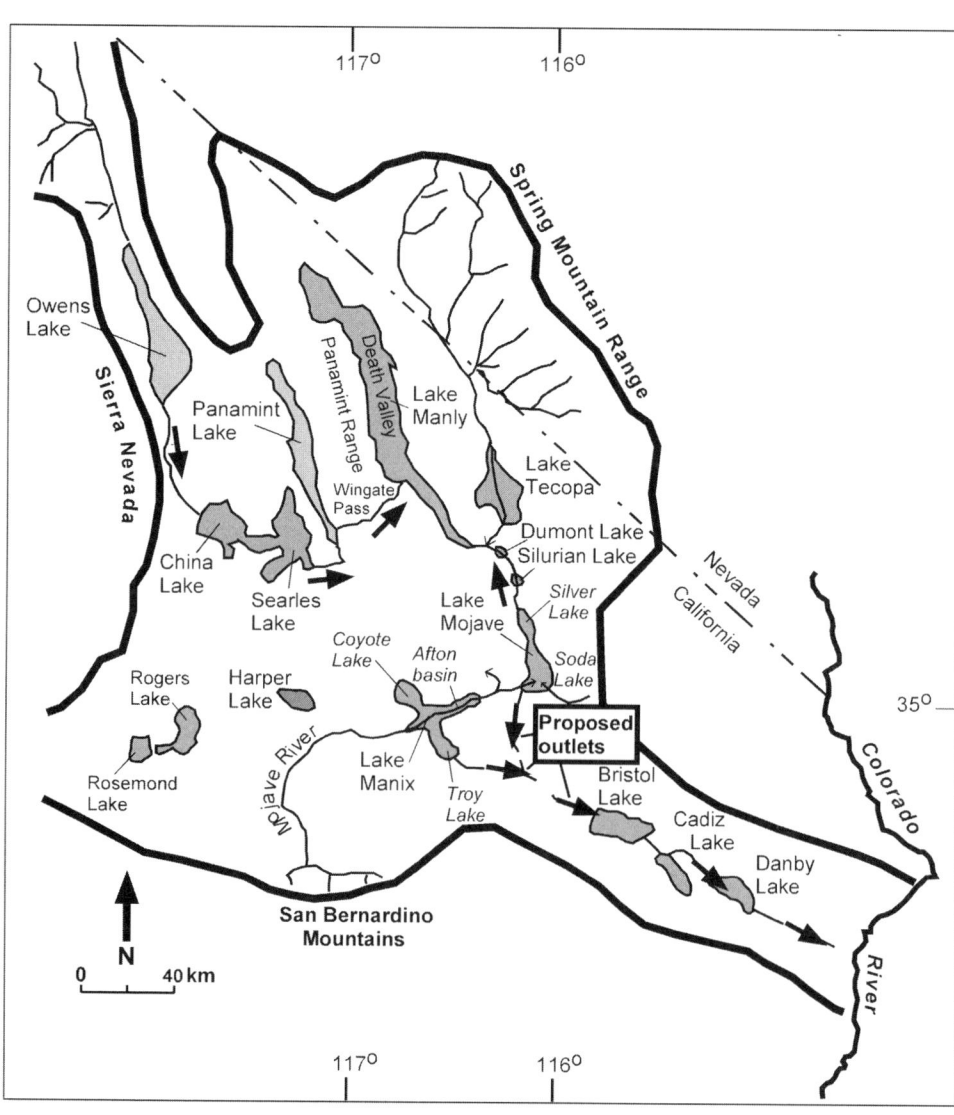

Figure 4. Pleistocene drainage paths in the Mojave Desert as proposed by Blackwelder (1954). It presents the hypotheses: (a) Mojave River flow through Troy Lake into Bristol, Cadiz, and Danby Lake to the Colorado River, and (b) overflow through the southern end of Soda Lake playa of a mega–Lake Manly into Bristol Lake.

age, and providing two new ages from the same shell deposit [28,960 ± 2490 (UCLA-2601c) and 29,310 ± 310 ^{14}C yr B.P. (CAMS-1856); Meek, 1999]. These dates fit the timing of an earlier lake stand in the Afton basin dated by Meek (1990) to 31–28 ka. Discarding the 14 ka age leaves an 18.1 ka age as the youngest age in the Afton subbasin, and a 15 ka age as the youngest age from Troy Lake basin based on *Anodonta* (Meek, 1999). We ask again the key question: When did the Afton basin of Lake Manix drain (i.e., the beginning of the lowering of its highest stand from 543 m by spillway incision)? Catastrophically ca. 13 ka (Meek, 1989)? Or after 18 ka (Meek, 1999)? Was it in fact a catastrophic event, or was it a time-transgressive process that continued into the early Holocene (Wells and Enzel, 1994)? Specific answers to these questions will require additional research. However, if the Afton Canyon incision occurred sometimes after 20 ka, Lakes Manix and Mojave and probably Lake Dumont coexisted and the Mojave River thus supported a very large joint lake area.

We agree with Meek (1999) that understanding the history of the different lakes in all the basins affected by the Mojave River discharge is crucial to understanding the paleohydrology of this system. In fact, two major conclusions can already be drawn from the current knowledge on Lake Manix, Lake Mojave and the other basins: (1) Lake Mojave is a good recorder of the paleohydrology of the basin during the latest Pleistocene, and (2) Lakes Manix and Mojave coexisted during the last glacial maximum, at least for a short period, to form a joint lake area that could have reached up to ~500 km^2 that was supported by the Mojave River. Observation (2) requires a very large atmospheric moisture transport to this part of the world at that time, nearly double the earlier discharge estimates (Wells et al., this volume) for the late Pleistocene Mojave River that were calculated based on the area and volume associated with Lake Mojave alone. If Lake Dumont, downstream of Lake Mojave also received Mojave River discharge during that time (K. Anderson and Wells, this volume) the total lake area thus supported is even greater.

The Mojave River fan delta and Coyote Lake

The aquatic environments at the broad Mojave River fan delta as late as 13–9 ka indicate that the water table was still high and not yet experiencing the drop that should be associated with incision of the Mojave River channel or the draining of the downstream Afton basin. The delta was also synchronous with (a) the fully or partially incised Afton Canyon, (b) an already drained Afton basin, and (c) an upstream propagation of the nickpoint formed by the incision, exposing erodible sediments of Lake Manix. These observations may indicate either only partial incision at Afton Canyon at 13–9 ka or a very slow knickpoint propagation upstream through unconsolidated deposits.

An overlap of ages exists among the various shore features of Coyote Lake and the Mojave River prograding delta (Meek, 1990; Reynolds and Reynolds, 1985). The delta that prograded into the Coyote Lake basin was probably part of the Mojave River fan delta (Fig. 2). If so, this coexistence of the Mojave River delta and Coyote Lake indicates that the Mojave River was still feeding and maintaining lacustrine conditions in that basin as recently as 11 ka (the youngest age from Coyote Lake) and perhaps even more recently. According to N. Meek (1990, written commun.): (1) Coyote Lake basin could have received Mojave River water only as an overflow from the Afton and Troy basins after these basins reached a sill elevation of 543 m, and (2) only after Coyote Lake also filled to that elevation could the joint Lake Manix have risen to higher levels. Therefore, Meek (1990, written commun.) suggested that the lakes in the Coyote Lake basin after 13 ka (i.e., after his proposed draining of Afton basin ca. 13.5 ka) are residual water bodies that were trapped as remnants of the larger Lake Manix after the formation of Afton Canyon. However, the retraction of the 14.5 ka age means that all the ages from Coyote Lake are <18.1 ka, which is the youngest age from Afton basin (Meek, 1999). We postulate that simple calculations will show that even for lowest estimations of latest Pleistocene evaporation rates, Coyote Lake would dry up within a few to tens of years were it not being fed continuously by discharge of the Mojave River. The overlapping ages of the unincised Mojave River delta, the lakes in Coyote Lake basin, and the prograding delta into that basin all indicate that a direct link between the Mojave River and the Coyote Lake persisted until ca. 13–11 ka.

When lakes occupied the Coyote Lake basin, full-lake phases Lake Mojave I (ca. 18–16 ka), and Lake Mojave II (ca. 14–12 ka), and Intermittent Lake II (ca. 16–14 ka) occupied Soda Lake and Silver Lake playas (Wells et al., this volume). After 13 ka the Afton basin of Lake Manix is supposedly already drained (Meek, 1989, 1990, 1999), but the Mojave River delta still included active fluvial, lacustrine, and marshy environments (Reynolds and Reynolds, 1985) and was capable of delivering water simultaneously to Coyote Lake and, through an incising Afton Canyon, to Lake Mojave. The persistence of wet environments indicates that the upstream migration of incision of the Mojave River after the formation of Afton Canyon reached the delta only after ca. 9 ka. After the Mojave River incised a channel in its delta ca. 7.5 ka (or soon thereafter), its floods could no longer reach either Coyote Lake or Troy Lake, and river discharge reached only the playas of Soda Lake, Silver Lake, and Cronese Lakes.

Overlapping ages of lakes

Meek (1999) estimated that Lake Harper had a maximum surface area of 255 km^2 and storage capacity of water of ~6 km^3. This large volume of water and the nonoverlapping ages discussed above led him to conclude that Harper Lake was not contemporaneous with late Wisconsin highstands of Lake Manix. He interpreted that the gaps in the Manix dates could be explained by dates from Lake Harper suggesting to him that Lake Harper was filled when Lake Manix was low. This is a very plausible hypothesis, but we think its implications are so important for paleohydrological interpretations of the Mojave River and for the regional paleoclimatological interpretations that intensive stratigraphic and dating efforts are needed to confirm or deny it. In this regard, we note that during at least one episode, the Mojave River was able to support Lake Manix and Lake Mojave at the same time. Lake Mojave's area (~290 km^2) is ~1–1.5 times the estimated area of Harper Lake. The large delta of the Mojave River, the low water divide (if it existed) between the Harper Lake basin and the Mojave River floodplain, and the magnitudes of the Mojave River floods could all have contributed one or more of the following scenarios: (1) shifting of the river between basins on millennial time scales (Meek, 1999), (2) the river feeding both basins all the time, and/or (3) frequent shifting of the river on its delta resulting in a permanent but shallow body of water in each basin. In the deltaic sediments of the Manix Formation's type section, which span the time interval of the proposed shift from Lake Manix to Harper Lake, G. Jefferson (2000, written commun.) did not identify an obvious break in the stratigraphy. Therefore, the possibility of coexisting Manix and Harper lakes should be examined thoroughly, including dating of the continuous lacustrine record so that its paleohydrological interpretation is clear. Chronologies based on shoreline features are by definition episodic and therefore cannot be used for detailed correlation between basins. The migration of the Mojave River between various subbasins well may have occurred; we consider it to be an untested hypothesis.

The mega-lake and alternative Mojave River routes

Two longstanding hypotheses in the Quaternary geology of the Mojave Desert are associated with the Mojave River and its current terminal basins. The first hypothesis is that a mega–Lake Manly covered very large areas from Death Valley to south of Soda Lake playa. This large lake was proposed in various forms, most recently by Hooke (1999). The second hypothesis indicates that the Mojave River was connected through a topographic trough from Troy Lake through Bristol Lake (Figs. 2 and 4) to the lower Colorado River (Thompson, 1929, p. 112; Miller, 1946; Hubbs and Miller, 1948; Hewett, 1954, p. 19). These two hypotheses are sometimes mixed in the literature as the mega-lake was suggested (Blackwelder, 1933; Hale, 1985) to overflow from the southern Soda Lake area through Broadwell Lake to Bristol Lake playas, thence to the Colorado River (Figs. 2 and 4). In the paragraphs below we suggest that available information does not support either of these hypotheses.

The mega–Lake Manly hypothesis

In deep boreholes along the Mojave River Wash in the southern Soda Lake playa, Dickey et al. (1979) encountered fine-grained, brownish to greenish clastic sediments with minor amounts of evaporite minerals beneath the latest Wisconsin Lake Mojave deposits. They suggested that a small, moist playa or a shallow lake existed during the early-middle Pleistocene. This "Ancestral Soda Lake" was temporally more persistent, although areally more restricted, than latest Quaternary Lake Mojave (Brown and Rosen, 1995). However, sedimentologic evidence from continuous, 326-meter-deep cores in the center of the Soda Lake playa basin indicates only one lacustrine sequence in this largest subbasin of pluvial Lake Mojave (Muessig et al., 1957; Smith, 1991a; Brown and Rosen, 1995). This sequence at a depth of 14–36 m from the playa surface is related to the latest Pleistocene Lake Mojave sediments (Wells et al., 1989, this volume; Brown and Rosen, 1995). The single lacustrine sequence supports Thompson's (1921, 1929) hypothesis that Lake Mojave formed only later in the geologic history of the Mojave River (e.g., Smith, 1991a).

Below this late Pleistocene sequence, the Soda Lake cores contain playa, alluvial fan, and eolian sediments (Muessig et al., 1957; Smith, 1991a; Brown and Rosen, 1995). Brown and Rosen (1995) used various sedimentation rates to estimate that the age of the bottom of the deep core range from 0.82 Ma to 3.2 Ma. They conclude that at least since the early Pleistocene, there is no evidence of the supposedly large lake that occupied vast areas in the Mojave Desert as suggested by Blackwelder (1933, 1954), Hubbs and Miller (1948); Miller (1981), Hale (1985), and more recently by Hooke (1999). Soda Lake, which (a) is located in the center of the proposed lake and (b) constitutes the essential link between the basins that such a lake supposedly occupied, does not contain any known sedimentary sequence to support the mega–Lake Manly hypothesis (Muessig et al., 1957; Smith, 1991a; Brown and Rosen, 1995).

The recent attempt by Hooke (1999) to revive a version of the "mega–Lake Manly" hypothesis by adding regional large-scale tilting to fit the elevations of the various shore features of the hypothetical mega-lake is not supported by the available data from Soda Lake and Silver Lake playas (Enzel et al., 2002). He proposed that the tilting was at a rate of 2 mm/year. Over the last 18–12 ka, which is the age of the shorelines associated with Lake Mojave, such a rate should have lowered northern Silver Lake playa by 24–36 m relative to southern Soda Lake playa (a distance of 35–40 km). All the highest shoreline features around Silver Lake and Soda Lake playas were surveyed in detail relative to existing benchmarks (Wells et al., 1989, this volume). No tilting or any other deformation was observed from this survey. Actually, this survey indicates that the most pronounced shorelines around Silver Lake and Soda Lake playas are within 10–20 cm of the 287 m and 285.5 m shorelines A and B, respectively of Wells et al. (1987). Furthermore, there are no shorelines at elevations higher than the 287 m, which is the spillway elevation at the northern end of Silver Lake playa and therefore also the expected shoreline elevation if tilting did not occur. Therefore, the significant tilting proposed by Hooke (1999) seems not to include Silver Lake and Soda Lake, so it apparently did not occur. In addition, all ages from the highest shoreline features around the Soda Lake and Silver Lake playas are associated with the latest Pleistocene lake (Wells et al., this volume). Not one age supports the idea that an earlier lake occupied this basin as proposed by Hale (1985) and Hooke (1999). We also visited all the Mesquite Hills deposits mentioned by Hooke (1999) and concluded that (a) none has the characteristics of shore deposits, (b) they are probably spring deposits along an active fault, and (c) their absolute and/or relative ages are not yet determined. Therefore, we suggest that the mega-lake hypotheses do not stand the test of field evidence.

The Mojave River–Bristol Lake–Colorado River route

Thompson (1929), Miller (1946), Hubbs and Miller (1948), and Hewett (1954) suggested that after the filling of Lake Manix, the Mojave River overflowed into Bristol Lake basin through a spillway at southeast Troy Lake (Fig. 4). It then overflowed from Bristol Lake to the basins of Cadiz and Danby Lakes and reached the Colorado River.

The summaries of Jefferson (this volume), Cox et al. (this volume), and Smith (1991a) indicate that Lake Manix could have overflowed only after 500 ka, when the Mojave River water first filled that lake. Therefore, Bristol Lake basin could have received Mojave River water only after that period.

Bristol Lake cores show that throughout the Pleistocene, the basin was occupied by only shallow lakes fluctuating between a saline lake to a playa (Rosen, 1989; Brown and Rosen, 1995). A deep overflowing lake that could have overflowed sometime during the last 0.5 ka to downstream basins and to the Colorado River is not documented in the Bristol Lake cores (Rosen, 1989; Brown and Rosen, 1995). Therefore, we note that currently, evidence does not support this hypothesis.

Estimating Mojave River discharge needed for coexisting lakes

During the latest Pleistocene it appears that the Mojave River fed more than one lake at the same time. The clearest example is a Lake Mojave-Lake Manix coexistence during the last glacial maximum. To get an estimate of the Mojave River discharge needed to support these two lakes we use the estimates of discharge calculated for Lake Mojave alone (Wells et al., this volume).

Average present-day annual flow, recorded at the lower reaches of the Mojave River, is ~9.5×10^6 m^3 (Enzel, 1990; Wells et al., this volume). Rare floods (>10 yr recurrence time) that reach the lower Mojave River deliver ~75×10^6 m^3 per flood on average. To form Lake Mojave during the last glacial maximum demands an order of magnitude greater river discharge than today's average annual flow, even with 50% reduction in evaporation (Wells et al., this volume). Fifty percent reduction in evaporation is probably an overestimation of the evaporation during the late Pleistocene. Annual flow into the lake basins in the magnitude of twice the average flow of the eight largest modern extreme floods is necessary to form a lake. Reducing modern rates of evaporation by 50%, and doubling modern rainfall would result in a full lake almost at the elevation of the Lake Mojave shoreline. To feed and form simultaneously an additional lake with Mojave River discharge, and therefore doubling the total lake area supported by the river, would demand a further increase of the annual discharge to ~300×10^6 m^3. This annual discharge is approximately a factor of three larger than the discharge of the 1938 flood, the largest flood that reached Afton Canyon in the last 110 yr.

Presently, the Mojave River experiences large transmission losses into its alluvial aquifer (e.g., Buono and Lang, 1980; Enzel et al., 1989; Enzel, 1990, 1992; Enzel and Wells, 1997) and only the largest floods reach the terminal playas. When they reach the playas they have greatly reduced peaks and total discharges. We assume that during the late Pleistocene, with the increased flood frequency and discharge, higher water tables, and longer reaches with base flow or even occupied by lakes, the transmission losses of today were dramatically reduced. Large transmission losses are observed today in the Barstow-Afton reach of the Mojave River (Enzel, 1990). During the late Pleistocene most of this reach was occupied either by the fan delta, with aquatic environments at the surface (Reynolds and Reynolds, 1985), or by Lake Manix, indicating a limited potential transmission losses. In contrast with the Holocene conditions (Enzel, 1992), under which only the largest floods reach the lower Mojave River, under late Pleistocene conditions most floods that originated at the headwaters could have reached and contributed to the two coexisting lakes. Observations at the late Pleistocene delta of the Mojave River west of Lake Manix support the existence of very high groundwater levels at this time, probably at or near the surface (Reynolds and Reynolds, 1985), elevations that are much higher than under present-day conditions. Therefore, the floods routed along the Mojave River channel probably were attenuated but without the high transmission losses that characterize Holocene and present conditions.

We note that flood discharges of the magnitude demanded by these calculations have occurred during the twentieth century (Enzel, 1990), in the upper reaches of the Mojave River. For example, upstream of Victorville, California (Fig. 1) ~30% of the modern floods have discharges >100×10^6 m^3. Ten percent of the floods have discharges >300×10^6 m^3. If these floods were more frequent, did not experience the present-day transmission losses in downstream reaches, and if the evaporative losses were at the proposed late-Pleistocene levels, a large lake could form.

Implication for paleoclimatology

Increase in Mojave River discharge probably requires more frequent heavy storms over its upper reaches and its headwaters in the San Bernardino Mountains, located at a latitude of 34°N (Enzel et al., 1990). A persistent winter North Pacific storm track delivering moisture to relatively low latitude along the west cost of North America is also indicated. As coexistence of lakes is most pronounced during the last glacial maximum, the moisture transport to this latitude was probably most effective during that time. During later stages of the deglaciation (14–9 ka) Lake Mojave was fluctuating more and probably overflowed less permanently to downstream reaches. This indicates less-frequent water supply by storms at its headwaters than during the glacial maximum. We suggest that this indicates that the average storm track was located farther north.

This implication to the changes in the moisture transported to the area during glacial maximum and deglaciation, is supported by the deeper water in lakes and "wetter" conditions interpreted from other basins at similar latitudes. Quade et al. (this volume) indicate that the wettest conditions associated with discharge to the surface in valleys in southern Nevada (Fig. 5) were recorded during the last glacial maximum. Although discharge to the surface in this area continued also during 14–12 ka, conditions were then drier than during the last glacial maximum (Quade et al., this volume). The chronology of Lake Estancia in central New Mexico (Fig. 5; latitude 34°–35°N) (Allen, 1991; Allen and Anderson, 2002) shows that the highest and most persistent lake stand occurred during the glacial maximum, with mostly lower and less-persistent lake stands occurring later, in the latest Wisconsin. Other paleolakes also show maximum highstands during the last glacial maximum: (a) Lake Cloverdale (Fig. 5) occupied the southern Animas Valley at 32° N latitude (Krider, 1998) between ca. 20 and 18 ka; (b) a paleolake occupied the San Agustin Plains (Fig. 5; 34°N) in west-central New Mexico between 22 and 17 ka (Markgraf et al., 1984; Phillips et al., 1992); (c) Lake Cochise (Fig. 5) in southeastern Arizona (32°30′ N) experienced a highstand during full glacial times (Long, 1966), but latest Wisconsin lake stands probably reached the same elevation (Waters, 1989) in this relatively shallow lake basin; and (d) Laguna Diablo located in northern Baja California, Mexico (Fig. 5; 31°N) was high during the glacial maximum (Y. Enzel, L. Ely, B. Allen, and M. Palacios, 1993, written commun.).

All these high lake stands at latitudes of 32°–35°N indicate intensive moisture transport to that region during the glacial maximum, diminishing during the later stages of the deglaciation. The source of the moisture that supported these lakes was probably winter precipitation from a North Pacific source (e.g., Van Devender et al., 1987; Enzel, 1990; Smith, 1991b; Allen and Anderson, 1993; Stute et al., 1995; Anderson et al., 2002). So far there is no evidence for the later deglaciation lakes in the San Agustin Plains (Phillips et al., 1992) and Lake Cloverdale (Krider, 1998) that were observed in Lake Estancia (Allen and Anderson, 1993), Lake Cochise (Waters, 1989), Lake Mojave (Wells et al., this volume), and in southern Nevada as discharge deposits (Quade et al., 1998, this volume).

In contrast to the highstand of lakes in the 32°–35° latitude range in the southwestern United States, the northern Great Basin Lake Lahontan and Lake Bonneville of Utah and Nevada at latitude of 40°N, experienced their highest stands at 14–12 ka (Benson and Thompson, 1987; Benson et al., 1990, 1995; Oviatt et al., 1992; Adams and Wesnousky, 1998). The last glacial maximum lake stands in these basins were somewhat smaller (Benson and Thompson, 1987; Benson et al., 1990, 1995; Oviatt et al., 1992).

Antevs (1938, 1952) hypothesized that the growth and retreat of the Laurentide Ice Sheet caused shifts in the average position of the polar jet and storm track along the western coast of North America. As maximum precipitation, cloud coverage, lower maximum temperatures, and reduced evaporation are associated with the core of the jet stream and the area just north of it

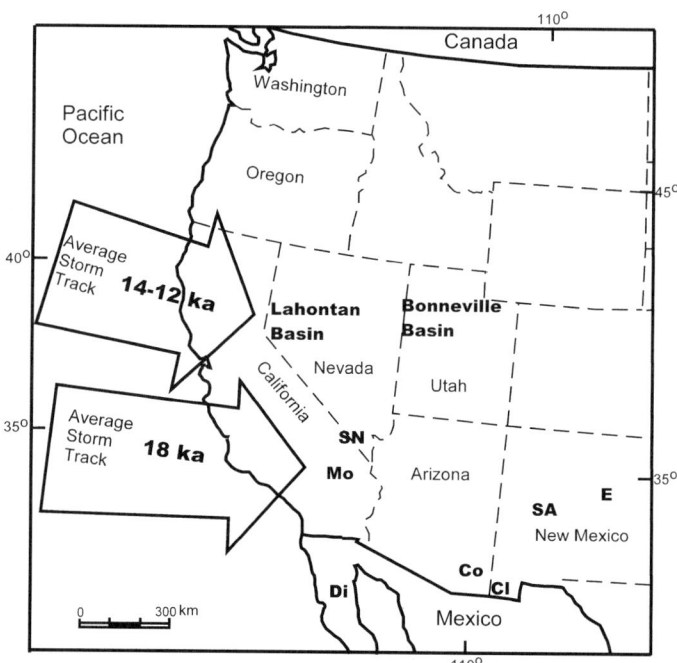

Figure 5. Proposed average storm track over the West Coast of North America during the last glacial maximum (18 ka) and late deglaciation (14–12 ka). SN—southern Nevada stream deposits; Mo—Lake Mojave; SA—San Agustin Plains; E—Lake Estancia; Co—Lake Cochise; Cl—Lake Cloverdale; Di—Laguna Diablo.

(Starrett, 1949), a northward shift of the jet stream following the glacial maximum is a reasonable cause for the rise of Great Basin pluvial lakes at this time (e.g., Hostetler and Benson, 1990; Benson et al., 1995). The results of atmospheric general circulation models (Manabe and Broccoli, 1985; Kutzbach and Wright, 1985; Kutzbach and Guetter, 1986; Kutzbach, 1987; Kutzbach et al., 1998) indicated that glacial-age boundary conditions such as the size and shape of Laurentide Ice Sheet, were sufficient to produce the effect that Antevs had earlier hypothesized (Benson et al., 1995). As a result, shifts in the polar jet positions were advocated as the cause for the Great Basin and Mojave Desert lakes by other researchers (e.g., Harrison and Metcalf, 1985; Benson and Thompson, 1987; Enzel et al., 1989; Hostetler and Benson, 1990; Benson et al., 1995). Note, however, that the shifts indicated by the global climate models were to 30°N latitudes (e.g., Kutzbach et al., 1998); i.e., to even lower latitudes than indicated by these available lake chronologies.

At present, the average annual position and the episodic shifts to the south of the jet stream are related to persistent anomalies in the sea surface temperature of the North Pacific Ocean (e.g., Namias et al., 1988). Smith (1991b) used sea-air interaction and suggested that the compressed isotherms of the sea-surface temperatures of the North Pacific at 30°–40° latitudes during the winter for 18 ka (CLIMAP, 1981) might be the explanation for the increased moisture transported into the southwestern United States. Furthermore, he argued that this configuration allows for storms from warmer, lower latitudes of the North Pacific to reach the area carrying much larger quantities of moisture. The interaction of similar subtropical moisture sources with the polar jet and associated mid-latitude low pressure systems was proposed by Enzel et al. (1989) as a major mechanism to resolve the moisture quantities needed during the late Pleistocene. In light of this hypothesis and model results, we postulate that the systematic changes in timing of highest stands of late Pleistocene lakes in western United States reflect systematic latitudinal differences in the average storm track and polar jet position along the western coast of North America. During the glacial maximum (ca. 18 ka) the storm track was frequently farther to the south, probably at 32°–35° latitudes (Fig. 5). During the latest Wisconsin (14–12 ka) the average storm track was at ~40° latitude, feeding the Lake Lahontan and lake Bonneville basins.

The polar jet position varies along the western coast of North America from intraseasonal (Pyke, 1972) to probably millennial scales. The passage (and average position) of the jet stream over a specific drainage basin in the Great Basin and Mojave Deserts for hundreds to thousands of years will sustain a lake or a pronounced spring discharge during the late Pleistocene (Benson and Thompson, 1987; Enzel et al., 1989; Benson et al., 1995).

Relatively high-resolution lake records from 34° to 35° N latitude (Lake Estancia in New Mexico [Allen and Anderson, 1993] and Lake Mojave [Wells et al., this volume]) indicate decadal to centennial lake-level changes even during the last glacial maximum. Therefore, Allen and Anderson (1993) argue that the shifts in the average storm track cannot be attributed solely to the millennial-scale forcing boundary conditions used in the models (i.e., growth and decay of the Laurentide Ice Sheet and astronomical parameters) because of the different temporal scales of variations in the ice sheets and lakes. We suggest that the shift to the south of the polar jet and storm tracks is caused partially by the ice sheet, but that the high-frequency lake fluctuations are caused by seasonal to centennial variations in storm-track position that are superimposed on the lower-frequency, larger-scale shifts. Lakes located farther to the south will record more variations because the average storm track is located north of them, and only episodically would they have experienced persistent storms over seasons, years, or decades. The causes for these shorter episodes are unknown but probably are related to oceanic and atmospheric conditions over the North Pacific Ocean or upwind in East Asia.

The summary above indicates that millennial-scale shifts in the position of the storm track during the latest Pleistocene affected very large regions; consequently, high-resolution records are needed along the coast of North America to identify the temporal and spatial structure of the shifts.

Integration of the Mojave River hydrological basin

The Mojave River is a hydrological system that integrates closed basins to one relatively long arid river basin. In this volume Cox et al. present data on the late Pliocene and early Pleistocene evolution of the upstream reaches of the Mojave River. They associate this evolution with tectonic activity at the Transverse Ranges of southern California (e.g., Meisling and Weldon, 1989). They summarize earlier work (see also Jefferson, this volume) on the first appearance of Mojave River water and sediments at ca. 500 ka in the Manix basin in the middle reaches of the river. In this paper we have discussed our observations of the Mojave River integration to the downstream basins of the Soda Lake and Silver Lake playas during the latest Pleistocene. Although they are rare, present-day continuous flows of the Mojave River terminate at the Silver Lake basin. Anderson, D., and Wells (this volume) and Anderson, K., and Wells (this volume) indicate that, during the latest Pleistocene, the Mojave River overflowed and contributed to the Lake Dumont and the Death Valley lakes (Fig. 4) as proposed by Free (1914) and Huntington (1915). The breaching of the sill of Lake Dumont after 18 ka allowed more Mojave River discharge to reach Death Valley (Anderson, D., and Wells, this volume; Anderson, K., and Wells, this volume). Whereas Soda Lake basin contains a very thick sequence of deposits that were generated from local sources prior to the impact of the Mojave River (Brown and Rosen, 1995; Wells et al., this volume), it contains only relatively thin layer of sediments from the Mojave River, the oldest dating from the last glacial maximum.

Since its inception in result of the uplift of the San Bernardino Mountains, the Mojave River required >1 million years to fill its upstream basins with sediments and an additional 0.5 million years to reach the Manix basin. It then took 0.5 million years to fill the Manix basin and to overflow and integrate this basin with the Soda Lake and Silver Lake basins downstream. This last major integration was accomplished by incising the sediments at the Lake Manix spillway in the Afton Canyon area. This incision caused the loss of the storage capacity of the Manix basin (Meek, 1989) that was already reduced owing to the 0.5 million years of discharge of Mojave River sediments. The storage capacities of the Soda Lake and Silver Lake basins were reduced by sedimentation at much faster rates than were the basins upstream. From 22 ka to 8.5 ka storage was reduced from ~7 km^3 to ~1 km^3 (Wells et al., this volume). Since 8.5 ka the storage capacity of these basins was further reduced to ~0.15 km^3 (Enzel, 1990). This rapid loss of storage was probably related to the incision of the Manix basin (Meek, 1989) and transport of the eroded sediment to the Soda Lake and Silver Lake basins. The spillway of Lake Mojave did not and probably will not erode in the same way as the spillway of Lake Manix, because it is composed of metamorphic rocks, and not by weaker sediments. The emptying of sediments from one basin into a downstream basin can accelerate integration and probably is associated with active faulting (i.e., Manix fault) that uplifted unconsolidated basin-fill sediments to the spillway elevation during prelake times.

The integration of the Mojave River into Death Valley, its ultimate terminal basin, is in its final stages. Under Holocene conditions large floods generated in the Mojave River headwaters deposit most of their load of sand and gravel in southern

Soda Lake and their finer fraction in northern Soda Lake and Silver Lake playas. Although some of the sand and finer-grained deposits are blown out of the basins by wind, net Holocene deposition (e.g., Enzel and Wells, 1997) is observed and the deposition of the coarse alluvium is rapidly migrating north on top of the Soda Lake playa sediments. When the terminal basins of the Mojave River ultimately fill to the spillway elevation, Mojave River floodwaters will flow downstream even under Holocene conditions. According to Enzel (1992), the possibility that modern floods will overcome the transmission losses along the alluvial reaches between Silver Lake and Death Valley is not realistic under present-day conditions.

CONCLUSIONS

The available data on chronologies of the late Pleistocene lakes along the Mojave River (Harper Lake, Lake Manix and its subbasins, and Lake Mojave) and its late Pleistocene fan delta, indicate that high lake stands in these individual lake basins could have coexisted. If so, the Mojave River, which at present rarely delivers water to its terminal basin in Silver Lake playa, was able to support a very large lake area during the late Pleistocene. In turn, this relation indicates that southern California and the Mojave Desert experienced frequent, intensive North Pacific storms that draw moisture from the tropical and subtropical Pacific Ocean, the most probable moisture source for these large lakes. We think that this warm, moisture laden subtropical source is the only one that can explain the very large water volume needed to support lakes along the Mojave River. The similarity in the chronology of late Pleistocene highstands of the Mojave River lakes, and lakes in closed basins in southern Arizona and New Mexico, indicates a regional hydrological cause of this shift. The observed chronologies require that the storm tracks along the West Coast of North America were at a very low latitude of ~34°–35° N during the last glacial maximum. Later in the deglaciation the storm track shifted to the north, consistent with the observed highstands of the northern Great Basin lakes and the fluctuating lake levels in the southwestern United States.

The vigorous floods of the Mojave River originating in the San Bernardino Mountains delivered large quantities of sediments that progressively filled a series of basins. The reduction in storage capacity of these basins caused the lakes that these floods maintained to overflow downstream and to integrate the hydrological system during the Pleistocene. The overflow of Lake Manix to Lake Mojave caused the incision of Afton Canyon during a geologically short episode during the latest Pleistocene, but this incision lasted longer and was less catastrophic than has been proposed in earlier studies.

The Mojave River as we know it today is a young and evolving hydrological system. It owes its evolution to late Pliocene to early Pleistocene tectonic uplift of the Transverse Ranges, heavy storms at these elevated areas, transport of sediments by the resulting flows, and the filling of the downstream basins by those sediments. The integration during the Pleistocene of a few closed basins across the Mojave Desert formed the modern river. The last integration occurred during the late Pleistocene, when Lake Manix overflowed to the Soda Lake basin of Lake Mojave, and ultimately to Death Valley.

ACKNOWLEDGMENTS

The manuscript was written when YE was visiting the Division of Earth and Ecosystem Sciences, Desert Research Institute, Reno, Nevada; the support of this institute is greatly appreciated. Tim Minor and Alan Gillespie helped produce Figure 2. Several of the ideas presented here were reshaped by discussions over the years with Bruce Allen, Roger Anderson, Ray Weldon, Bruce Harrison, Les McFadden, George Jefferson, and Bob Reynolds. Tom Bullard, Alan Gillespie, George Smith, and John Tinsley provided excellent and constructive reviews.

REFERENCES CITED

Adams, K.D., and Wesnousky, S.G., 1998, Shoreline processes and the age of the Lake Lahontan highstand in the Jessup embayment, Nevada: Geological Society of America Bulletin, v. 110, p. 1318–1332.

Allen, B.D., 1991, Effect of climatic change on Estancia Valley, New Mexico: Sedimentation and landscape evolution in closed-drainage basin: New Mexico Bureau of Mines and Mineral Resources Bulletin 137, p. 166–171.

Allen, B.D., and Anderson, R.Y., 1993, Evidence from western North America for rapid shifts in climate during the last glacial maximum: Science, v. 260, p. 1920–1923.

Allen, B.D., and Anderson, R.Y., 2000, A continuous, high-resolution record of late Pleistocene climate variability from the Estancia Basin, New Mexico: Geological Society of America Bulletin, v. 112, p. 1444–1458.

Anderson, R.Y., Allen, D.B., and Menking, K.M., 2002, Geomorphic expression of abrupt climate change in southwestern North America at the Glacial termination: Quaternary Research, v. 57, p. 371–381.

Antevs, E., 1937, Age of Lake Mojave culture, in Crozer-Campbell, W.E., et al., eds., The archaeology of Pleistocene Lake Mohave: A symposium: Los Angeles, California, Southwest Museum Publishing, Southwest Museum Papers, v. 11, 118 p.

Antevs, E., 1938, Postpluvial climatic variations in the southwest: Bulletin of the American Meteorological Society, v. 19, p. 190–193.

Antevs, E., 1952, Cenozoic climates of the Great Basin: Geologische Rundschau, v. 40, p. 94–108.

Bassett, A.M., and Jefferson, D.T., 1971, Radiocarbon dates of Manix Lakes, central Mojave Desert, California: Geological Society of America Abstracts to Meeting, v. 3, no. 2, p. 79.

Beck, W., Donohue, D., Jull, A.J., T., Burr, G., Broecker, W.S., Berani, G., Hajdas, I., and Malotki, E., 1998, Ambiguities in direct dating of rock surfaces using radiocarbon measurements: Science, v. 280, p. 2132–2135.

Benson, L.V., and Paillet, F.L., 1989, The use of total lake-surface area as an indicator of climatic change: Examples from the Lahontan basin: Quaternary Research, v. 32, p. 262–275.

Benson, L.V., and Thompson, R.S., 1987, The physical record of lakes in the Great Basin, in Ruddiman, W.F., and Wright, H.E., Jr., eds., North America and adjacent oceans during the last deglaciation: Boulder, Colorado, Geological Society of America, The Geology of North America, v. K-3, p. 241–260.

Benson, L.V., Currey, D.R., Dorn, R.I., Lajoie, K.R., Oviatt, C.G., Robinson, S.W., Smith, G.I., and Stine, S., 1990, Chronology of expansion and contraction of four Great Basin lake systems during the past 35,000 yr: Palaeogeography, Palaeoclimatology, Palaeoecology, v. 78, p. 241–286.

Benson, L., Kashgarian, M., and Rubin, M., 1995, Carbonate deposition, Pyramid Lake subasin, Nevada, 2: Lake levels and polar jet stream positions reconstructed from radiocarbon ages and elevations of carbonates (tufas) deposited in the Lahontan basin: Palaeogeography, Palaeoclimatology, Palaeoecology, v. 117, p. 1–30.

Berger, R., and Meek, N., 1992, Radiocarbon dating of *Anodonta* in the Mojave River basin: Radiocarbon, v. 34(3), p. 578–584.

Blackwelder, E., 1933, Lake Manly, an extinct lake in Death Valley: Geographical Review, v. 23, p. 464–471.

Blackwelder, E., 1954, Pleistocene lakes and drainage in the Mojave region, southern California, in Jahns, R.H., ed., The geology of southern California: California Division of Mines and Geology Bulletin, v. 170, p. 35–40.

Blackwelder, E., and Ellsworth, E.W., 1936, Pleistocene lakes of the Afton basin, California: American Journal of Science, v. 231, p. 453–463.

Blaney, F.H., 1957, Evaporation study at Silver Lake in the Mojave Desert, California: Eos (Transactions, American Geophysical Union), v. 38, p. 209–215.

Brown, W.J., and Rosen, R.M., 1995, Was there a Pliocene-Pleistocene fluvial-lacustrine connection between Death Valley and the Colorado River?: Quaternary Research, v. 43, p. 286–296.

Brown, W.J., Wells, S.G., Enzel, Y., Anderson, R.Y., and McFadden, L.D., 1990, The late Quaternary history of pluvial Lake Mojave: Silver Lake and Soda Lake basins, California, in Reynolds, R.E., et al., eds., At the end of the Mojave: Quaternary studies in the eastern Mojave Desert: Redlands, California, Special Publication of the San Bernardino County Museum Association, 1990 Mojave Desert Quaternary Research Center Symposium, May 18–21, 1990, p. 55–72.

Buono, A., and Lang, D.J., 1980, Aquifer discharge from the 1969 and 1978 floods in the Mojave River basin, California: U.S. Geological Survey Water-Resources Investigations Report 80-207, 25 p.

Buwalda, J.P., 1914, Pleistocene beds at Manix in the eastern Mojave Desert region: University of California Publication, Bulletin of the Department of Geology, v. 7, no. 24, p. 443–464.

Byers, F.M., Jr., 1960, Geology of the Alvord Mountain quadrangle, San Bernardino County, California: U.S. Geological Survey Bulletin 1089-A, 71 p.

CLIMAP Project Members, 1981, Seasonal reconstruction of the Earth's surface at the last glacial maximum: Geological Society of America Map and Chart Series 36.

Coues, E., 1900, On the Trail of a Spanish Pioneer—The diary and itinerary of Francisco Garcés in his travels through Sonora, Arizona, and California 1775–1776: New York, Francis P. Harper, 608 p.

Cox, B.F., and Tinsley, J.C., III., 1999, Origin of the late Pliocene and Pleistocene Mojave River between Cajon Pass and Barstow, California, in Reynolds, R.E., and Reynolds, J., eds., Tracks along the Mojave: Quarterly of the San Bernardino County Museum Association, v. 46, no. 3, p. 49–54.

Crippen, J.R., 1965, Natural water loss and recoverable water in mountain basins of southern California: U.S. Geological Survey Professional Paper 417-E, 24 p.

Crozer-Campbell, W.E., Cambell, H.W., Antevs, E., Amsden, A.C., Barbierry, A.G., and Bode, D.F., editors, 1937, The archaeology of Pleistocene Lake Mohave: A symposium: Los Angeles, California, Southwest Museum Publishing, Southwest Museum Papers, v. 11, 118 p.

Dale, H.C., 1918, The Ashley-Smith exploration and the discovery of a central route to the Pacific, 1822–1829: Cleveland, Ohio, Arthur H. Clark Company, 352 p.

Dibblee, T.W., Jr., 1960, Geology of the Rogers Lake and Kramer quadrangles, California: U.S. Geological Survey Bulletin 1089-B, 139 p.

Dibblee, T.W., Jr., 1968, Geology of the Fremont Peak and Opal Mountain quadrangles, California: California Division of Mines and Geology Bulletin 188, 64 p.

Dibblee, T.W., Jr., and Bassett, A.M., 1966a, Geologic map of the Cady Mountains quadrangle, San Bernardino County, California: U.S. Geological Survey Miscellaneous Geologic Investigations Map I-467.

Dibblee, T.W., Jr., and Bassett, A.M., 1966b, Geologic map of the Newberry quadrangle, San Bernardino County, California: U.S. Geological Survey Miscellaneous Geologic Investigations Map I-461.

Dickey, S.K., Neimeyer, R.A., and Sholes, R.C., 1979, Soda Lake groundwater investigations—phase II: Los Angeles, California, Southern California Edison Company, 72 p.

Dorn, R.I., Bamforth, D.B., Cahill, T.A., Dohrenwend, J.C., Turrin, B.D., Donahue, D.J., Jull, A.J.T., Long, A., Macko, M.E., Weil, E.B., Whitley, D.S., and Zabel, T.H., 1986, Cation-ratio and accelerator radiocarbon dating of rock varnish on Mojave artefacts and landforms: Science, v. 231, p. 830–833.

Dyer, H.B., Bader, J.S., and Giessner, F.W., 1963, Wells and springs in lower Mojave Valley area, San Bernardino County, California: California Department of Water Resources Bulletin 91-10.

Enzel, Y., 1990, Hydrology of a large, closed arid watershed as a basis for paleohydrology and paleoclimatological studies in the Mojave River drainage system, southern California [Ph.D. thesis]: Albuquerque, University of New Mexico, 316 p.

Enzel, Y., 1992, Flood frequency of the Mojave River and the formation of late Holocene playa lakes, southern California: The Holocene, v. 2, p. 11–18.

Enzel, Y., and Wells, S.G., 1997, Extracting Holocene paleohydrology and paleoclimatology information from modern extreme flood events: An example from southern California: Geomorphology, v. 19, p. 203–226.

Enzel, Y., Cayan, R.D., Anderson, R.Y., and Wells, S.G., 1989, Atmospheric circulation during Holocene Lake stands in the Mojave Desert: Evidence of a regional climatic change: Nature, v. 341, p. 44–48.

Enzel, Y., Anderson, R.Y., Brown, W.J., Cayan, D.R., and Wells, S.G., 1990, Tropical and subtropical moisture and southerly displaced North Pacific storm track: Factors in the growth of late Quaternary lakes in the Mojave Desert, in Betancourt, J.L., and MacKay, A.M., eds., Proceedings of the Sixth Annual Pacific Climate (PACLIM) Workshop: Sacramento, California, California Department of Water Resources Technical Report 23, p. 135–139.

Enzel, Y., Knott, J.R., Anderson, K., Anderson, D.E., and Wells, S.G., 2002, Is there any evidence of mega–Lake Manly in the eastern Mojave Desert during Oxygen Isotope Stage 5e/6?: Quaternary Research, v. 52, p. 173–176.

Fergusson, G.J., and Libby, W.F., 1962, UCLA radiocarbon dates I: Radiocarbon, v. 4, p. 113.

Free, E.E., 1914, The topographic features of the desert basin of the United States with reference to the possible occurrence of potash: U.S. Department of Agriculture Bulletin 54, p. 45.

Free, E.E., 1916, An ancient lake basin on the Mohave River: Washington, D.C., Carnegie Institute of Washington, Yearbook No. 15, p. 90–91.

Gale, H.S., 1914, Note on the Quaternary lakes of the Great Basin, with special references to the deposition of potash and other salines, in White, D., ed., Contribution to economic geology, Part I: Metals and nonmetals except fuels: U.S. Geological Survey Bulletin 540, p. 399–406.

Garfunkel, Z., 1974, Model for the late Cenozoic tectonic history of the Mojave Desert, California, and for its relation to adjacent region: Geological Society of America Bulletin, v. 86, p. 1931–1944.

Gillespie, A.R., 2003, Rock-art dating at the end of the millennium: Epigraphic Society Occasional Papers, v. 24 (in press).

Groat, C.G., 1967, Geology and hydrology of the Troy playa area, San Bernardino County, California [M.S. thesis]: Amherst, University of Massachusetts, 133 p.

Hagar, D.J., 1966, Geomorphology of Coyote Valley, San Bernardino County, California [Ph.D. thesis]: Amherst, University of Massachusetts, 210 p.

Hale, G.R., 1985, Mid-Pleistocene overflow of Death Valley toward the Colorado River, in Hale, G.R., ed., Quaternary Lakes of the eastern Mojave Desert, California: Friends of the Pleistocene, Pacific Cell Field Trip Guidebook, p. 36–81.

Harrison, S.P., and Metcalf, S.E., 1985, Spatial variations in lake levels since last glacial maximum in the Americas north of the equator: Zeitschrift fur Gletscherkund und Glazialzeologie, Band 21, p. 1–15.

Hewett, D.F., 1954, General geology of the Mojave Desert region, California, in Jahns, R.H., ed., Geology of southern California: California Division of Mines, Bulletin 170 p. 5–20.

Hooke, R.LeB., 1999, Lake Manly(?) shorelines in the eastern Mojave Desert, California: Quaternary Research, v. 52, p. 328–336.

Hostetler, S., and Benson, L.V., 1990, Paleoclimatic implications of the high-stand of Lake Lahontan derived from model of evaporation and lake level: Climate Dynamics, v. 4, p. 207–217.

Hubbs, C.L., and Miller, R.R., 1948, The Great Basin, II: The zoological evidence: University of Utah Bulletin, v. 38, p. 17–166.

Hubbs, C.L., Bien, G.S., and Suess, H.E., 1962, La Jolla natural radiocarbon measurements II: Radiocarbon, v. 4, p. 204–238.

Hubbs, C.L., Bien, G.S., and Suess, H.E., 1965, La Jolla natural radiocarbon measurements IV: Radiocarbon, v. 7, p. 66–117.

Huntington, E., 1915, The curtailment of rivers by desiccation: Washington, D.C., Carnegie Institution of Washington, Yearbook for 1915, p. 96.

Jackson, D., and Spence, M.L., 1970, The expedition of John Charles Frémont, v. 1: Chicago, Illinois, University of Illinois Press, p. 674–677.

Jefferson, G.T., 1985, Stratigraphy and geologic history of the Pleistocene Lake Manix Formation, central Mojave Desert, California, in Reynolds, R.E., ed., Cajon Pass to Manix Lake: Geological investigations along Interstate 15: Redlands, California, San Bernardino County Museum Association Special Publication, p. 157–169.

Krider, P.R., 1998, Paleoclimatic significance of late Quaternary lacustrine and alluvial stratigraphy, Animas Valley, New Mexico: Quaternary Research, v. 50, p. 283–289.

Kutzbach, J.E., 1987, Model simulation of climatic patterns during the deglaciation of North America, in Ruddiman, W.F., and Wright, H.E., Jr., eds., North America and adjacent oceans during the last deglaciation: Boulder, Colorado, Geological Society of America, The Geology of North America, v. K-3, p. 425–446.

Kutzbach, J.E., and Guetter, P.J., 1986, The influence of changing orbital parameters and surface boundary conditions on climate simulations for the past 18,000 yr: Journal of the Atmospheric Sciences, v. 43, p. 1726–1759.

Kutzbach, J.E., and Wright, H.E., 1985, Simulation of the climate of 18,000 yr B.P.: Results for North America/North Atlantic/European Sector: Quaternary Science Review, v. 4, p. 147–187.

Kutzbach, J., Gallimore, R., Harrison, S., Behling, P., Selin, R., and Laarif, F., 1998, Climate and biome simulations for the past 21,000 yr: Quaternary Science Reviews, v. 17, p. 473–506.

Long, A., 1966, Late Pleistocene and recent chronologies of playa lakes in Arizona and New Mexico [Ph.D. thesis]: Tucson, University of Arizona, 141 p.

Manabe, S., and Broccoli, A.J., 1985, The influence of continental ice sheet on the climate of an ice age: Journal of Geophysical Research, v. 90, p. 2167–2190.

Markgraf, V., Bradbury, J.P., Forester, R.M., Singh, G., and Sternberg, R.S., 1984, San Agustin Plains, New Mexico: Age and paleoenvironmental potential reassessed: Quaternary Research, v. 22, p. 336–343.

Martin, P., 1994, Southern California basins regional aquifer, in McGill, S.F., and Ross, T.M., eds., Geological investigations of an active margin: Geological Society of America, Cordilleran Section Guidebook, p. 166–169.

McClure, W.F., Sourwine, J.A., and Tait, C.E., 1918, Report on the utilization of Mojave River for irrigation in Victor Valley, California: Sacramento, California, State of California, Department of Engineering Bulletin No. 5, 93 p.

Meek, N., 1989, Geomorphologic and hydrologic implications of the rapid incision of Afton Canyon, Mojave Desert, California: Geology, v. 17, p. 7–10.

Meek, N., 1990, Late Quaternary geochronology and geomorphology of the Manix Basin, San Bernardino County, California [Ph.D. thesis]: Los Angeles, University of California, 212 p.

Meek, N., 1999, New discoveries about the late Wisconsin history of the Mojave River system, in Reynolds, R.E., and Reynolds, J., eds., Tracks along the Mojave: Quarterly of the San Bernardino County Museum Association, v. 46, no. 3, p. 113–117.

Meisling, K., E., and Weldon, R.J., 1989, Late Cenozoic tectonics of the northwestern San Bernardino Mountains, southern California: Geological Society of America Bulletin, v. 101, p. 106–128.

Miller, R.R., 1946, Correlation between fish distribution and Pleistocene hydrography in eastern California and southwestern Nevada, with map of the Pleistocene waters: Journal of Geology, v. 54, p. 43–53.

Miller, R.R., 1981, Coevolution of deserts and pupfishes (Genus *Cyprinodon*) in the American Southwest, in Naiman, R.J., and Soltz, D.L., eds., Fishes in North American deserts: New York, John Wiley and Sons, p. 39–94.

Mojave River Agency, 1982, Report on historic and present conditions, Newberry Groundwater Basin: Victorville, California.

Mojave River Agency, 1985, Historic and present conditions, Upper Mojave River Basin: Victorville, California.

Morgan, L.D., 1953, Jedediah Smith and the opening of the west: The Bobbs-Merrill Company, New York, New York, p. 205–207.

Muessig, S., White, N.G., and Byers, M.F., 1957, Core Logs from Soda Lake, San Bernardino County, California: U.S. Geological Survey Bulletin 1045-C, p. 81–96.

Nagy, E.A., and Murray, B.C., 1991, Stratigraphy and intrabasin correlation of the Mojave River Formation, central Mojave Desert, California: San Bernardino County Museum Association Quarterly, v. 38(2), p. 5–30.

Namias, J., Yuan, X., and Cayan, D.R., 1988, Persistence of North Pacific sea surface temperature and atmospheric flow patterns: Journal of Climate, v. 1, p. 682–703.

Ore, H.T., and Warren, C.N., 1971, Late Pleistocene–early Holocene geomorphic history of Lake Mojave, California: Geological Society of America Bulletin, v. 82, p. 2553–2562.

Oviatt, C.G., Currey, D.R., and Sack, D., 1992, Radiocarbon chronology of Lake Bonneville, eastern Great Basin, USA: Palaeogeography, Palaeoclimatology, Palaeoecology, v. 99, p. 225–241.

Phillips, F.M., Campbell, A.R., Kruger, C., Johnson, P., Roberts, R., and Keyes, E., 1992, A reconstruction of the response of the water balance in western United States lake basins to climatic changes: New Mexico Water Resources Research Institute Report No. 269, v. 2, 259 p.

Preston, R.N., 1988, Early California Atlas—Southern edition, 2nd edition: Portland, Oregon, Binford and Mort Publishers, 76 p.

Pyke, C.B., 1972, Some meteorological aspects of the seasonal distribution of precipitation in the western United States and Baja California: Los Angeles, California, University of California Water Resources Center Contribution No. 139.

Quade, J., Forester, R.M., Pratt, W.L., and Carter, C., 1998, Black mats, spring-fed streams, and late-glacial-age recharge in southern Great Basin: Quaternary Research, v. 49, p. 129–148.

Reynolds, R.E., and Reynolds, R.L., 1985, Late Pleistocene faunas from Daggett and Yermo, San Bernardino County, California, in Reynolds, R.E., ed., Cajon Pass to Manix Lake: Geological investigations along Interstate 15: San Bernardino County Museum Association Special Publication, p. 175–191.

Reynolds, R.E., and Reynolds, R.L., 1994, The isolation of Harper Lake Basin, in Reynolds, R.E., ed., Off limits in the Mojave Desert: Field trip guidebook: San Bernardino County Museum Association Special Publication 94-1, p. 34–37.

Rosen, M.R., 1989, Sedimentologic, geochemichal and hydrologic evolution of and intracontinental closed-basin playa (Bristol Dry Lake): A model for playa development and its implications for paleoclimate [Ph.D. thesis]: Austin, Texas, University of Texas, 256 p.

Sabin, E.L., 1914, Kit Carson days: Chicago, Illinois, A.C. McLurg and Company, 686 p.

Shlemon, R.J., and Budinger, F.E., 1990, The archaeological geology of the Calico site, Mojave Desert, California, in Lasca, N.P., and Donahue, J., eds., Archaeological Geology of North America: Boulder, Colorado, Geological Society of America, Centennial Special Volume 4, p. 301–313.

Smith, G.I., 1991a, Stratigraphy and chronology of Quaternary-age lacustrine deposits, in Morrison, R.B., ed., Quaternary nonglacial geology: Conterminous U.S.: Boulder, Colorado, Geological Society of America, The Geology of North America, v. K-2, p. 339–346.

Smith, G.I., 1991b, Continental paleoclimatic records and their significance, in Morrison, R.B., ed., Quaternary nonglacial geology: Conterminous U.S.: Boulder, Colorado, Geological Society of America, The Geology of North America, v. K-2, p. 35–41.

Starrett, L.G., 1949, The relation of precipitation patterns in North America to certain types of jet stream at the 300-millibar level: Journal of Meteorology, v. 6, p. 347–352.

Steinmetz, J.J., 1988, Biostratigraphy and paleoecology of limnic ostracodes from the late Pleistocene Manix Formation [M.Sc. thesis]: Department of Biological Sciences, Pomona, California State Polytechnic University, 64 p.

Steinmetz, J.J., 1989, Ostracoda of the late Pleistocene Manix Formation, central Mojave Desert, California, in Reynolds, R.R., ed., The west-central Mojave

Desert: Quaternary studies between Kramer and Afton Canyon: Redlands, California, Mojave Desert Quaternary Research Center, San Bernardino County Museum Association Special Publication, p. 70–77.

Stute, M., Clark, J.F., Schlosser, P., Broecker, W.S., and Bonani, G., 1995, A 30,000 yr continental paleotemperature record dervied from noble gases dissolved in groundwater from the San Juan Basin, New Mexico: Quaternary Research, v. 43, p. 209–220.

Thompson, G.D., 1921, Pleistocene lakes along the Mojave River, California: Journal of the Washington Academy of Science, v. 11, no. 17, p. 423–424.

Thompson, G.D., 1929, The Mojave Desert region, California: A geographic, geologic, and hydrologic reconnaissance: U.S. Geological Survey Water-Supply Paper 578, p. 1–143 and p. 371–572.

Van Devender, T.R., Thompson, R.S., and Betancourt, J.L., 1987, Vegetation history of the deserts of southwestern North America; the nature and timing of the late Wisconsin–Holocene transition, *in* Ruddiman, W.F., and Wright, H.E., Jr., eds., North America and adjacent oceans during the last deglaciation: Boulder, Colorado, Geological Society of America, The Geology of North America, v. K-3, p. 323–352.

Watchman, A., 2000, A review of the history of dating rock varnishes: Earth-Science Reviews, v. 49, p. 261–277.

Waters, M.R., 1989, Late Quaternary lacustrine history and paleoclomatic significance of pluvial Lake Cochise, southeastern Arizona: Quaternary Research, v. 32, p. 1–11.

Wells, S.G., and Enzel, Y., 1994, Fluvial geomorphology of the Mojave River in the Afton Canyon area, eastern California: Implications for the geomorphic evolution of Afton Canyon, *in* McGill, S.F., and Ross, T.M., eds., Geological investigations of an active margin: Geological Society of America, Cordilleran Section Guidebook, p. 177–182.

Wells, S.G., McFadden, L.D., and Dohrenwend, J.C., 1987, Influence of late Quaternary climate changes on geomorphic and pedogenic processes on desert Piedmont, eastern Mojave Desert, California: Quaternary Research, v. 27, p. 130–146.

Wells, S.G., Anderson, R.Y., McFadden, L.D., Brown, W.J., Enzel, Y., and Miossec, J-L., 1989, Late Quaternary paleohydrology of the eastern Mojave River drainage basin, southern California: Quantitative assessment of the late Quaternary hydrologic cycle in a large arid watershed: New Mexico Water Resources Research Institute, Technical Report 242, 250 p.

Whitley, D.S., and Dorn, R.I., 1993, New perspectives on the Clovis versus pre-Clovis controversy: American Antiquity, v. 58, p. 626–647.

Williamson, R.S., Blake, W.P., Durand, E., and Hilgard, T.C., 1856, Reports of explorations and surveys to ascertain the most practicable and economical route for a railroad from the Mississipi River to the Pacific Ocean in 1853–1854, v. 5: Washington, D.C., House of Representatives, Ex. Doc. No. 91, p. 30–34.

Wohl, E.E., and Enzel, Y., 1995, Data for paleohydrology, *in* Gregory, K.J., et al., eds., Continental Paleohydrology, John Wiley and Sons, New York, p. 23–59.

MANUSCRIPT ACCEPTED BY THE SOCIETY AUGUST 1, 2002

Late Quaternary geology and paleohydrology of pluvial Lake Mojave, southern California

Stephen G. Wells
Desert Research Institute, 2215 Raggio Parkway, Reno, Nevada 89512, USA

William J. Brown
Tecumseh Professional Associates, Inc., One Sycamore Plaza, 5600 Wyoming Blvd. NE, Suite 150, Albuquerque, New Mexico 87109, USA

Yehouda Enzel
Institute of Earth Sciences and Department of Geography, Hebrew University of Jerusalem, Jerusalem 91904, Israel

Roger Y. Anderson
Leslie D. McFadden
Department of Earth and Planetary Sciences, University of New Mexico, Albuquerque, New Mexico 87131, USA

ABSTRACT

A complex history of lake-level fluctuations is recorded in subsurface, cored lake deposits and shoreline features of Silver Lake and Soda Lake depositional basins of southeastern California. These basins are the location of former pluvial Lake Mojave and the present terminus of the Mojave River. The Silver Lake depositional basin is relatively shallow and has minimal relief across the pre-lake basin floor, resulting in a high resolution stratigraphic sequence that can be correlated in the subsurface and to the surface features. Radiocarbon-dated lake sediments from Silver Lake indicate that episodic flooding of the basin began as early as 22 ka with prolonged highstands lasting between 2000 and 3000 yr. Two major high and persistent lake stands occurred in the Silver Lake basin, and presumably in the Soda Lake basin, between ca. 18.4 ka and 16.6 ka (Lake Mojave I) and 13.7 ka and 11.4 ka (Lake Mojave II). These pluvial periods resulted from significantly increased precipitation and annual large-scale floods. The floods originated in the upper Mojave River drainage basin and reached Afton Canyon with discharge values two to three times larger than modern extremes. Periods of intermittent lake conditions during which the Silver Lake basin experienced several desiccation events separated the higher stands and more continuous Lake Mojave phases. The most significant drying event is recorded at 15.5 ka as large desiccation cracks infilled with windblown sand.

During the earlier phases of its existence, Lake Mojave was the second of two large desert lakes sustained by the Mojave River; the other, Lake Manix, occurred upstream from Lake Mojave. Overflow from Lake Manix sustained Lake Mojave I which stabilized at the A-shoreline (elevation 287–288 m). The beginning of Lake Mojave II appears to have coincided with the incision of Afton Canyon and subsequent draining of Lake Manix, an event which significantly increased sediment loading, reducing total

Wells, S.G., Brown, W.J., Enzel, Y., Anderson, R.Y., and McFadden, L.D., 2003, Late Quaternary geology and paleohydrology of pluvial Lake Mojave, southern California, *in* Enzel, Y., Wells, S.G., and Lancaster, N., eds., Paleoenvironments and paleohydrology of the Mojave and southern Great Basin Deserts: Boulder, Colorado, Geological Society of America Special Paper 368, p. 79–114. © 2003 Geological Society of America.

lake volume and increasing evaporative surface area. This condition resulted in significantly greater overflow of Lake Mojave into the Death Valley basin. This overflow produced controlled downcutting of the Lake Mojave outlet spillway between 12 and 11 ka and ultimately stabilized at an elevation of 285.5 m (B-shoreline). The majority of shoreline features currently found around the margins of Silver Lake and Soda Lake date to Lake Mojave II, as the shallow lake conditions resulted in modification and erosion of the older Lake Mojave I landforms. A transition to a drier climatic regime resulted in the total drying of Lake Mojave by ca. 8.7 ka, with playa conditions dominating Silver Lake and Soda Lake basins following this event.

Analysis and correlation of the surface and subsurface environments of pluvial Lake Mojave yield a detailed reconstruction of the lake level elevation history as influenced by the discharge and floods of the Mojave River. Using a simplified, precipitation-discharge–evaporation model, we infer that the late Pleistocene hydrologic conditions resulting in Lake Mojave overflow at Spillway bay in Silver Lake lie between two sets of conditions: (1) a 50% increase in precipitation in the headwater catchment resulting in annual flood events reaching Afton Canyon with discharges three times that of modern extreme floods; or (2) a 100% increase in catchment precipitation with a 50% decrease from modern evaporation combined with annual flood events reaching Afton Canyon with discharges two times that of modern extreme floods.

INTRODUCTION

Insights into the paleoclimatic and paleohydrologic nature of arid and semiarid basins can be inferred from detailed reconstructions of lake fluctuations recorded in subsurface sedimentary sequences and surrounding shoreline environments (Smith and Street-Perrott, 1983; Bradley, 1999). The terminal basins of the Mojave River in southern California contain a well-preserved record of depositional events associated with latest Quaternary pluvial Lake Mojave (Wells et al., 1989; Brown, 1989). Numerous investigators have recognized the abundance of geological and archaeological features encircling the terminal depositional basins associated with Lake Mojave: Soda Lake and Silver Lake playas (Fig. 1; e.g., Ore and Warren, 1971). Because the upper elevations of the Mojave River watershed were not glaciated, the late Pleistocene and early Holocene Mojave River watershed and its associated fluvial-lacustrine system were more sensitive to climatic variations because direct glacial storage effects neither influenced runoff nor served as a buffer to shorter-term climatic fluctuations recorded in lake sediments and shoreline features (Sharp et al., 1959).

In this paper, we elucidate the shoreline geomorphology and stratigraphy, basin-fill geometry, and subsurface stratigraphy of pluvial Lake Mojave's depositional basins, Soda Lake and Silver Lake. By establishing a high resolution subsurface stratigraphy and correlating the subsurface and shoreline stratigraphy, we provide a detailed geologic and hydrologic history of Lake Mojave during the latest Pleistocene and early Holocene. Our paper builds upon the pioneering work of Crozer-Campbell et al. (1937) and Ore and Warren (1971) who first attempted to reconstruct the chronology of Lake Mojave shoreline deposits in relation to early human occupation of this region.

The playas of Soda Lake and Silver Lake occur in the two modern depositional basins at the terminus of the Mojave River (Figs. 1, 2). The two playas are separated by a broad sill that appears to be related to pre-lake topography (i.e., local bedrock high near the town of Baker; Figures 1, 3). Remnants of shorelines of prehistoric and historic lakes occur topographically higher than the playa floors, delineating the depositional boundaries of lacustrine events associated with pluvial Lake Mojave and younger, shorter duration lake events. Remnants of the late Pleistocene to early Holocene pluvial Lake Mojave shoreline features, consisting of erosional and depositional landforms, occur between 287 and 283 m and can be traced around the margins of the modern Silver Lake and Soda Lake playas. These shorelines and their associated subsurface lacustrine deposits are the primary focus of this study. The largest and most extensively preserved shoreline features are typically found along the northern margins of both basins (e.g., Plates 1 and 2; GSA Data Repository files A and B[1]).

In addition to pluvial Lake Mojave, floodwaters of the Mojave River inundated Soda Lake and Silver Lake playas creating ephemeral, mid-to-late Holocene lakes and deep, long-lasting lakes in the latest Pleistocene and early Holocene (Wells et al., 1989; Enzel and Wells, 1997). Historic Mojave River flooding occasionally has created short-term shallow lakes. Shoreline features associated with the mid-Holocene and younger lakes only have been found at the northern end of Silver Lake playa (Fig. 2;

---

[1]GSA Data Repository item 2003069—(A) detailed stratigraphic description of units at the Silver Lake quarry site shown in Figures 7 and 12 (from Brown, 1989; Wells et al., 1989); (B) lithologic description of two outcrops exposed in the Baker Dump quarry, northern Soda Lake; fence diagram of sediments exposed in the Baker Dump quarry, A-shoreline beach ridge, northern Soda Lake playa and outcrop sites of two lithologic descriptions; (C) detailed stratigraphic descriptions of cores Sil-E, Sil-F, Sil-G, Sil-H, Sil-I, Sil-J, Sil-L, and Sil-M taken from the Silver Lake depositional basin (see Figure 7 for locations; from Brown, 1989); and (D) plate summarizing major stratigraphic features and bounding surfaces in selected cores in the northern Silver Lake depositional basin and the correlation of these features (Plate 4 from Brown, 1989)—is available on request from Documents Secretary, GSA, P.O. Box 9140, Boulder, CO 80301-9140, USA, editing@geosociety.org, or at www.geosociety.org/pubs/ft2003.htm.

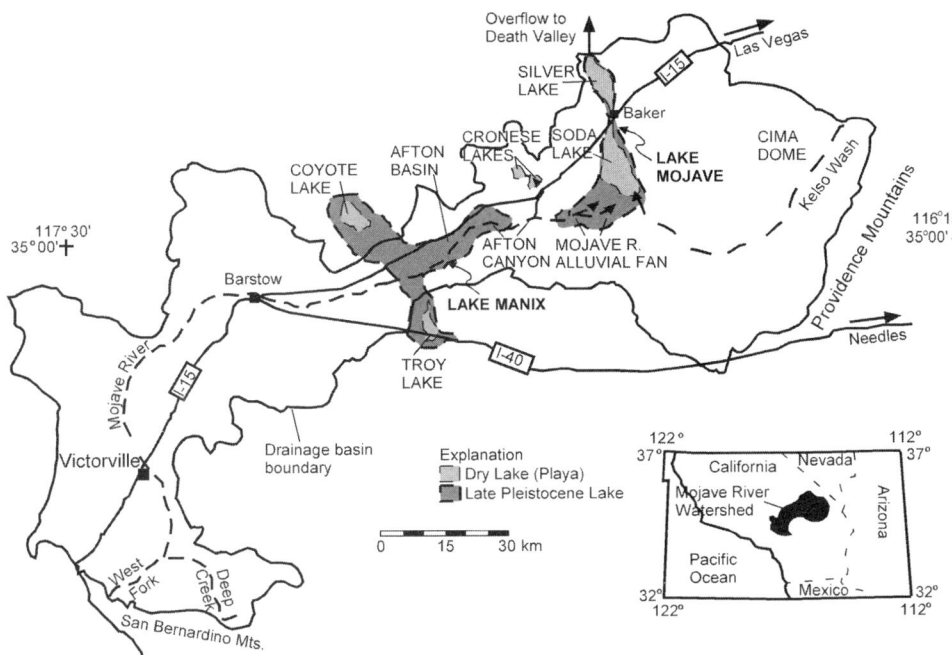

Figure 1. The Mojave River drainage basin showing maximum extent of latest Quaternary Lake Mojave and Lake Manix (now dissected), as well as key geographic features of the study; modified from Wells et al. (1989) and Meek (1989).

Wells et al., 1987, 1989; McFadden et al., 1992). These shorelines are not related temporally with pluvial Lake Mojave and are not the focus of this paper.

Mojave River watershed

The Mojave River drainage basin (Fig. 1) is ~9500 km^2 in size. The Mojave River flows from the San Bernardino Mountains (with elevations above 3000 m) 200 km eastward to a fan-delta region downstream of Afton Canyon where river discharge flows north into the Cronese Lakes basins (326 m) or east into the Silver Lake and Soda Lake basins (276 m). The mountainous headwater region that comprises <5% of the total drainage basin area accounts for >90% of the total basin precipitation (Wells et al., 1989; Enzel, 1990, 1992). During historic times (1894–2001), at least 10 temporary lakes (lasting 2–18 months) formed in the Silver Lake basin as a result of large-magnitude winter storms in the Transverse Ranges and subsequent runoff of the Mojave River (Enzel et al., 1989; Enzel and Wells, 1997). During these large flood events, the Mojave River crossed Soda Lake playa via a series of anastomosing channels. The channels terminated at the northern end of Silver Lake playa, and the floodwaters filled Silver Lake basin to produce these historic ephemeral lakes. The most extreme events resulted in back flooding into the Soda Lake basin. One historical flood and associated ephemeral lakes in the Silver Lake basin and the northern end of Soda Lake playa are shown in Figure 2.

The Soda Lake and Silver Lake basins are surrounded by numerous mountain ranges with range-top elevations >2000 m (e.g., Providence Mountains, Figure 1). The largest ephemeral stream draining these desert mountains is Kelso Wash, originating

Figure 2. Oblique aerial photograph of Silver Lake playa and northern Soda Lake playa near Baker, California, after the 1938 flood. The 1938 ephemeral lake in Silver Lake playa was fed by overflow through a broad, shallow channel connecting the northern end of Soda Lake basin with Silver Lake basin. The level of these historic, ephemeral lakes did not reach the level of the prehistoric shorelines that surround the playa floors (Wells et al., 1987). Photograph by Spence Air Photos, courtesy of the Department of Geography, University of California, Los Angeles.

between Cima Dome and the Providence Mountains (Fig. 1). During the highstands of pluvial Lake Mojave, Soda Lake and Silver Lake basins were not the terminus of the Mojave River because waters overflowed toward Death Valley (Fig. 1). The maximum elevation of ancient Lake Mojave was controlled by an outlet spillway (at a current elevation of 285.4 m) which developed

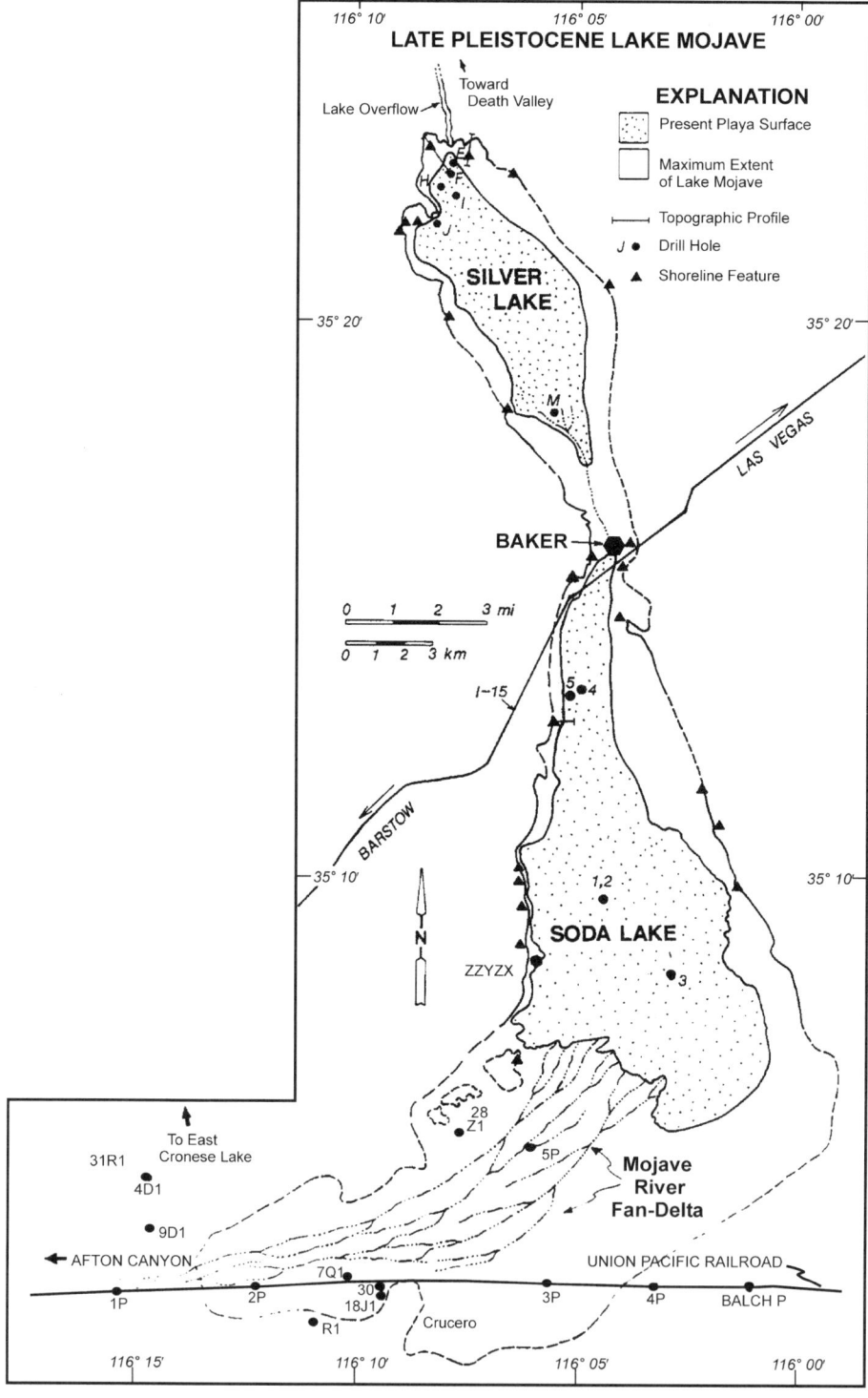

Figure 3. Map showing the maximum extent of latest Pleistocene Lake Mojave in Silver Lake playa and Soda Lake playa depositional basins. The location of the shoreline is dashed where it is buried or eroded. Major shoreline features are indicated by a solid triangle, and drill holes are indicated by a solid circle (note: (1) prefix Sil- has been omitted on figure such that Sil-J, for example, is represented by J only, and (2) numbers 3P or 4P, for example, are drill holes in the Soda Lake basin). The present-day playa surface is shown with a stippled pattern.

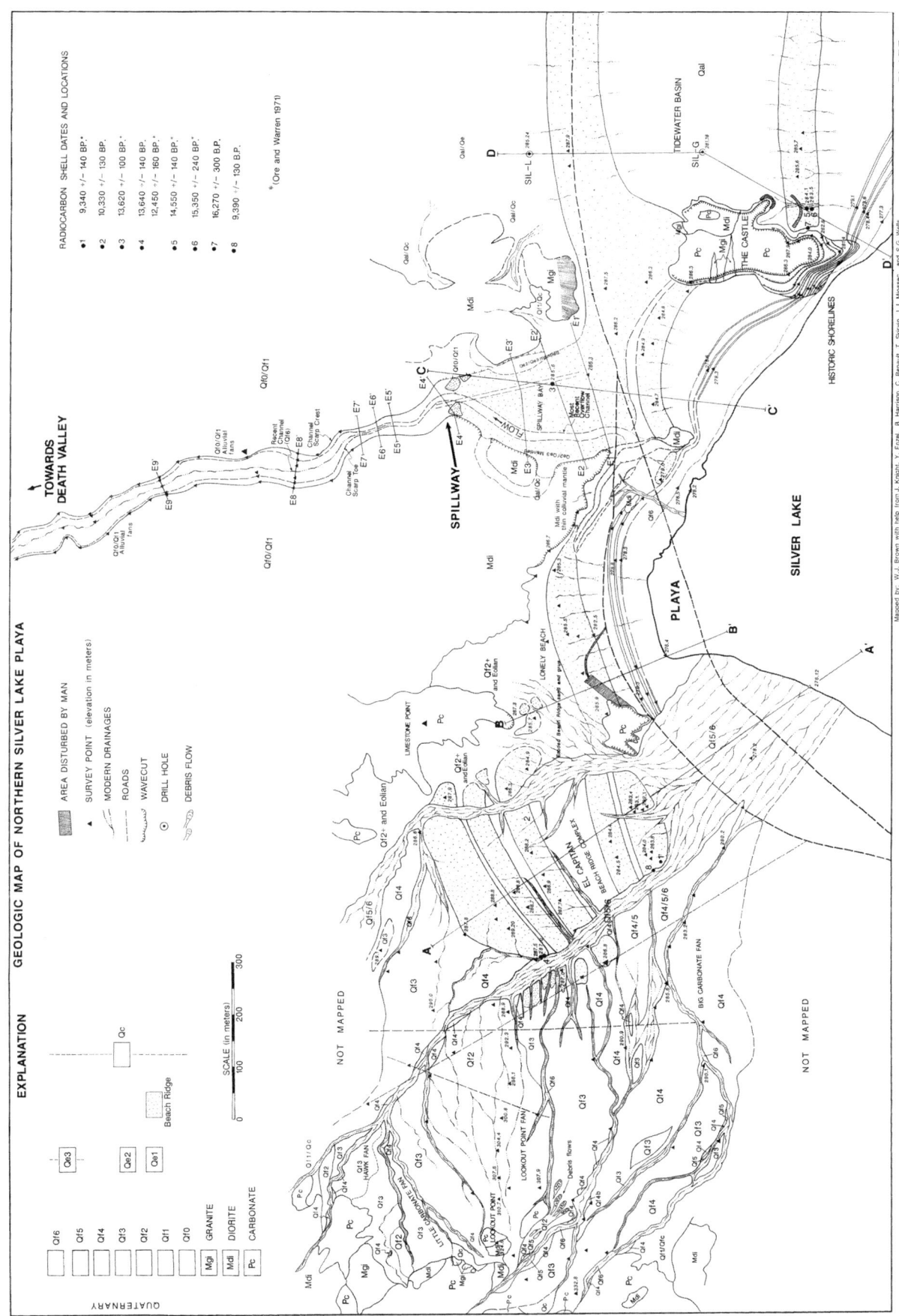

Plate 1. Surficial geologic map of northern Silver Lake playa showing major stratigraphic and geomorphic features and elevations of shoreline features.

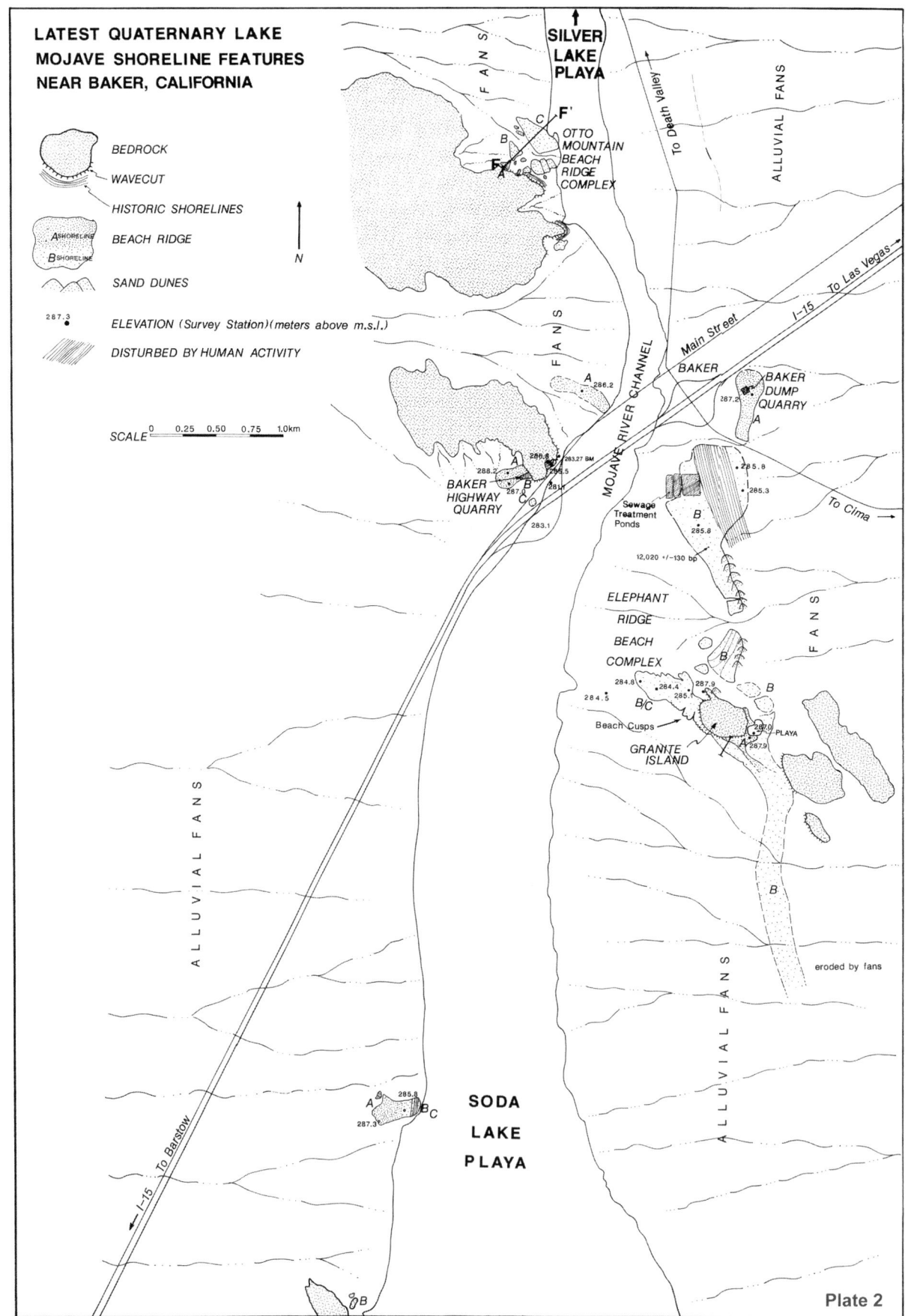

Plate 2. Geomorphic map of northern Soda Lake playa showing elevations of shoreline features.

in bedrock at the extreme north end of Silver Lake (Wells et al., 1989). During maximum lake stages, overflow from this spillway drained northward toward Dry Lake and Silurian Lake and eventually toward Death Valley (e.g., Free, 1914; Huntington, 1915; Blackwelder and Ellsworth, 1936; Blackwelder, 1954). Prominent shoreline features are found around the margins of Silver Lake and Soda Lake playas. These features indicate at least two major high and persistent lake stands at 287–288 m and 285.5 m (Ore and Warren, 1971; Wells et al., 1984, 1987, 1989; Enzel et al., 1988; and Brown et al., 1990).

No deep drilling has occurred in the Silver Lake basin, and little is known about the depth to the bedrock or configuration of the bedrock underlying the basin. This study provides the first detailed results of shallow drilling in the Silver Lake basin. Geophysical studies in the southern Soda Lake basin indicate that the depth to bedrock is >700 m (Dickey et al., 1979). Drilling information from 326 m in central Soda Lake playa and nearly as deep in southern Soda Lake playa supplement this data (Muessig et al., 1957; Dickey et al., 1979; Calzia, 1991). The Soda Lake cores indicate only one major, basin-wide, lacustrine sediment package at depths of 10–36 m (Brown and Rosen, 1995). This sediment package corresponds in age to latest Quaternary Lake Mojave shoreline features (Brown, 1989). Older basin-fill sediments composed of oxidized silts, sands, and minor clays indicate that playa and distal fan depositional environments predominated in the Soda Lake basin since at least the early Pleistocene (Brown and Rosen, 1995). Drill cores within the extreme southern portion of the Soda Lake basin and in the Cronese basins reveal blue and green lacustrine sediments at depth (Dickey et al., 1979; Brown and Rosen, 1995). No large evaporite deposits are found in any of the cores or drill holes from either the Soda Lake or Silver Lake basins suggesting that neither basin is a true hydrologically closed basin.

Climate of the watershed

California experiences a warm, dry summer and a winter precipitation maximum related to Pacific cyclonic activity, referred to as a Mediterranean regime (Barry, 1983). The precipitation distribution within the Mojave River drainage basin is highly variable. Within the headwaters of the Mojave River in the San Bernardino Mountains, mean annual precipitation exceeds 1000 mm/year. Ninety percent of the drainage area, which contributes runoff to Silver Lake playa, typically receives 125 mm or less in annual precipitation. In the terminal playas, mean annual precipitation is <100 mm, an order of magnitude less than in the Mojave River headwaters (Enzel, 1990). Furthermore, the seasonality of precipitation changes from the headwaters to the terminal playas. Within the Victorville area (Fig. 1), for example, precipitation occurs during the winter. In contrast, there are two seasons of precipitation downstream of Barstow. The primary season is during the winter months (principally February), but a secondary season occurs during the summer months (principally August) (Pyke, 1972). Precipitation during the winter season is usually during storms that cover larger portions of the Mojave River watershed, whereas precipitation during the summer season is highly variable across the eastern regions of the watershed. During the summer months, evaporation dominates over precipitation for the majority of the Mojave River drainage basin.

Previous work

Thompson (1929) examined Silver Lake and Soda Lake playas and found remnants of a pluvial lake, which he named Lake Mojave. He suggested that Lake Manix (Fig. 1) had once been the terminus of the Mojave River and that Silver Lake and Soda Lake basins became the terminus after the incision of Afton Canyon. Subsequently, researchers postulated that both lakes may have existed contemporaneously (Crozer-Campbell et al., 1937; Hubbs and Miller, 1948). Recent work by Meek (1989, 1999) and the results of this study support the latter conclusion and indicate that both lakes coexisted for a period of a few thousand years (see also Enzel et al., this volume). Studies of the Soda Lake basin by Muessig et al. (1957), Dickey et al. (1979), and Calzia (1991) revealed only one prolonged basin-wide lacustrine period in sediments from cores drilled up to 326 m. Faunal assemblages in these lacustrine clays (at depths of 10–35 m below the present playa surface of Soda Lake) have been found to correlate with similar sediments drilled in Silver Lake (Brown, 1989). Playa sediments are found above and below these lacustrine deposits.

Ore and Warren (1971) refined the Lake Mojave chronology with extensive radiocarbon dating of pelecypod shells and lithoid tufa found in shoreline features in the Silver Lake area, demonstrating that Lake Mojave existed episodically between 15 and 8 ka. Extensive work on the geomorphic history of the surrounding alluvial fans and their relations to the two prominent Lake Mojave shorelines was undertaken by Wells et al. (1984, 1987) in order to establish a dated chronology of alluvial fan deposition during the Pleistocene-Holocene transition. Soil geomorphic studies of the Silver Lake Holocene and Pleistocene shorelines and beach ridges exposed in northwestern Silver Lake allowed regional correlation of shoreline deposits, revealing a sequence of younger Holocene shoreline deposits in addition to the latest Pleistocene shorelines (McFadden et al., 1992).

RESULTS

Geometry of pluvial Lake Mojave depositional basins

In order to understand the depositional and hydrological history of pluvial Lake Mojave, the three-dimensional geometry of the ancient lake basin was defined using surface and subsurface data. Remnants of pluvial Lake Mojave shorelines surrounding the two depositional basins of Silver Lake and Soda Lake basins were used to delineate the uppermost elevations and surface boundaries of the Lake Mojave depositional system (Figs. 3, 4). The depositional system of pluvial Lake Mojave is related to the following:

- Two major shorelines—A (288–287 m) and B (285–286 m)
- One minor shoreline—C (283 m)

The geomorphology and stratigraphy of these shoreline features are discussed in detail later in this paper.

In order to delineate the subsurface boundaries of the Lake Mojave depositional system, we combined the data obtained from cores obtained in Silver Lake during this study with core and borehole data obtained from the Soda Lake basin (Fig. 3; Muessig et al., 1957; Moyle, 1967; Dickey et al., 1979). Cores from the Silver Lake basin were drilled utilizing a weight-driven coring rig (Cores Sil-A, Sil-B, Sil-C, Sil-D, Sil-E, Sil-G, Sil-L, Sil-M; represented as A-M in Figure 3) and a truck-mounted, continuous auger coring device using a 1.52 m split-barrel sampler (Cores Sil-F, Sil-H, Sil-I, and Sil-J; represented as F-J in Figure 3). Recovery for the above 13 cores was 95% or greater for each hole. In addition to the above coring operations, lacustrine sediments were studied in 1–2 m deep trenches. Detailed logging was undertaken on eight of the 12 cores at a 1:1 scale and later reduced to a manageable size for publication (see GSA Data Repository item C; see footnote 1). Accelerator mass spectrometry (AMS) radiocarbon ages were obtained from bulk samples taken from selected subsurface lacustrine sediments (Table 1).

Core and borehole data support that Lake Mojave was composed of two depositional basins within which lacustrine sedimentation was the thickest (Figs. 3, 4). The Soda Lake depositional basin was deeper than the Silver Lake basin during most of the lake history (Fig. 5). During the time of the deepest lacustrine deposits, ancient floodwaters filled Soda Lake basin first then overflowed the sill near Baker into Silver Lake basin (Wells et al., 1989), forming a hydrologically integrated Lake Mojave as illustrated in Figure 3. During the Holocene, greater sedimentation rates in Soda Lake basin compared to Silver Lake basin apparently resulted in a higher elevation of the Soda Lake playa surface. As a result, present-day floodwaters now pass over Soda Lake playa into the Silver Lake basin before hydrologic ponding backs into Soda Lake (Fig. 5). Figures 4 and 5 also indicate that the fan-delta, lacustrine, and playa sediments filling the two basins during the latest Quaternary reduced the lake water storage capacity dramatically. Storage capacity is the volume between the surface defined by the sloping playa (or lake) floor and the horizontal surface defined by the elevation of the overflow spillway (287 m). Total water storage capacity of pluvial Lake Mojave during the early phase of lacustrine deposition was ~7 km^3 (Fig. 6; Wells et al., 1989; Enzel, 1992). Currently, the lake basins can store only 0.15 km^3 of water and sediments before overtopping the ancient spillway at the northern end of Silver Lake playa (Fig. 6).

Geomorphology and stratigraphy of Lake Mojave shorelines

The primary goals of this part of our study are to (1) determine the timing of highstands that existed during pluvial Lake Mojave and (2) reconstruct the depositional and erosional processes

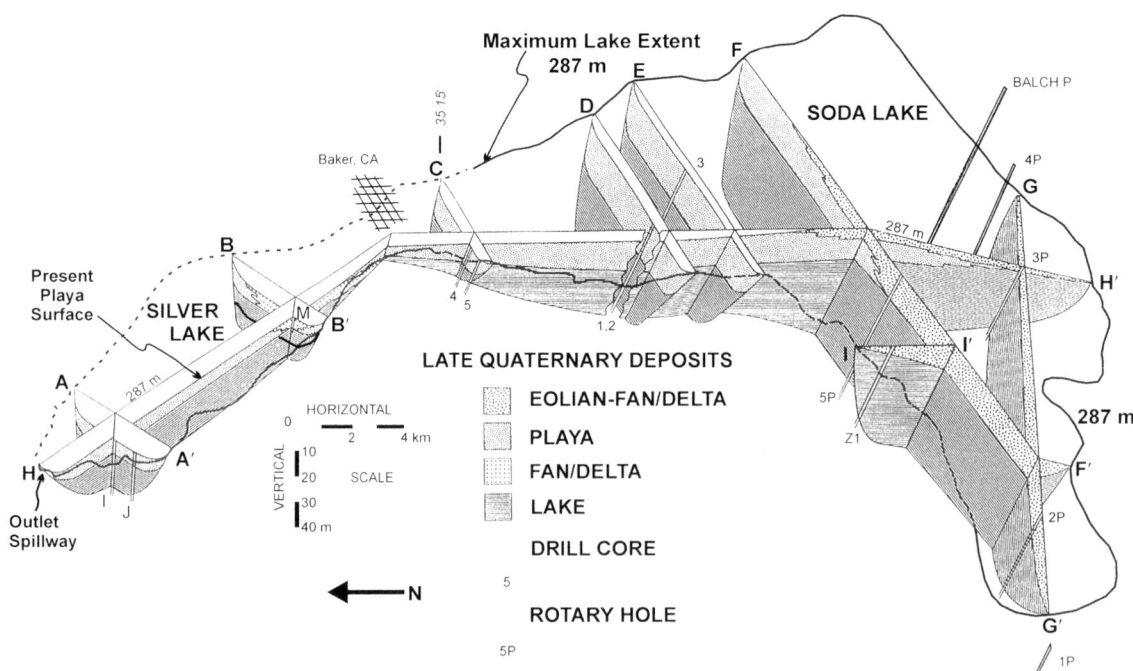

reconstruction is based on subsurface data from 14 rotary drill holes in southern Soda Lake, 19 drill cores from Silver Lake and central Soda Lake, and mapped shoreline features (Fig. 3). Pluvial Lake Mojave sediments form the base of the section, which, in turn, is overlain by playa deposits. The difference between the modern playa surface and the maximum lake extent represents the present-day storage capacity of these basins. Note that the section shown at core Sil-M (B–B′) is expanded vertically (not to scale) to illustrate stratigraphic relations; dark line at base of section at B–B′ is approximate depth of drill hole.

TABLE 1. SUMMARY OF UNCALIBRATED RADIOCARBON DATES OBTAINED FROM SAMPLES
OF LAKE MOJAVE SHORELINE DEPOSITS AND LAKE SEDIMENTS

Location	Dating method	Elevation of sample (masl)	Type of material dated	Radiocarbon age (yr B.P.) and sample number	Study
Silver Lake fan delta Depth of 15–17 cm	C-14	~278	Organic carbon in sediments	390 ± 90 Beta-25634	W
Silver Lake Sil-M core Depth of 55–59 cm	AMS	~277.5	Organic carbon in sediments	3,620 ± 70 Beta-25341	W
Silver Lake Sil-M core Depth of 3.12–3.15 m	AMS	~274.9	Organic carbon in sediments	9,330 ± 95 Beta-24342	W
Silver Lake Sil-I core Depth of 6.0 m	AMS	270.65	Organic carbon in sediments	14,200 ± 145 Beta-25339	W
Silver Lake Sil-I core Depth of 9.8 m	AMS	267.75	Organic carbon in sediments	14,660 ± 260 Beta-21800	W
Silver Lake Sil-I core Depth of 15.9 m	AMS	260.70	Organic carbon in sediments	20,320 ± 740 Beta-21801	W
Beach Ridge V El Capitan complex	C-14	282.5	Whole pelecypod shells	9,390 ± 120 Beta-29552	M
Beach Ridge III El Capitan complex	C-14	286–286.5	Whole pelecypod shells	10,330 ± 120 Beta-21200	W
Beach Ridge–Soda Lake Elephant Ridge complex	C-14	~285	Whole pelecypod shells	12,020 ± 130 Beta-21199	W
Beach Ridge I El Capitan complex	C-14	287	Pelecypod shell fragments	13,640 ± 120 Beta-26456	B
Tidewater Basin Beach Ridge II–Silver Lake	C-14	282	Whole pelecypod shells	16,270 ± 310 Beta-29553	B
Top of gravel pit–Silver Lake	C-14	Unit no. 10	Tufa coats on gravel	9,160 ± 400 LJ-935	H
(See Figure 12 for stratigraphic locations.)	C-14	Unit no. 10	Tufa coats on gravel	8,350 ± 300 LJ-929	H
	C-14	Unit no. 8	Pelecypod shells	10,580 ± 100 Y-1593	O/W
	C-14	Unit no. 7	Tufa coats on gravel	9,900 ± 100 Y-1592	O/W
	C-14	Unit no. 6	Pelecypod shells	10,700 ± 100 Y-1591	O/W
	C-14	Unit nos. 6,7,8 Comb.	Tufa coats on gravel	10,870 ± 450 LJ-930	H
	C-14	Unit nos. 6,7,8 Comb.	Pelecypod shells	10,260 ± 400 LJ-932	H
	C-14	Unit nos. 6,7,8 Comb.	Pelecypod shells	10,000 ± 300 I-444	H
	C-14	Unit nos. 6,7,8 Comb.	Pelecypod shells	9,640 ± 240 LJ-200	H
	C-14	Unit no. 5	Tufa coats on gravel	11,320 ± 120 Y-1590	O/W
	C-14	Unit no. 5	Tufa coats on gravel	11,630 ± 500 LJ-934	H
	C-14	Unit no. 4	Pelecypod shells	13,150 ± 350 I-443	H
	C-14	Unit no. 4	Pelecypod shells	13,290 ± 550 Y-1589	O/W
	C-14	Unit no. 4	Pelecypod shells	13,670 ± 550 LJ-933	H
	C-14	Unit no. 1	Tufa coats on gravel	13,190 ± 500 LH-931	H

Continued on following page.

TABLE 1. SUMMARY OF UNCALIBRATED RADIOCARBON DATES OBTAINED FROM SAMPLES
OF LAKE MOJAVE SHORELINE DEPOSITS AND LAKE SEDIMENTS (continued)

Location	Method	Unit/Elevation	Material	Date	Source
Bottom of gravel pit–Silver Lake	C-14	Unit no. 1	Tufa coats on gravel	13,040 ± 120 Y-1588	O/W
Beach Ridge V El Capitan complex	C-14	283.2 depth = 1	Pelecypod shells	9,340 ± 140 Y-2407	O/W
Beach Ridge I El Capitan complex	C-14	288.2 depth = 0.3	Pelecypod fragments	12,450 ± 160 Y-2408	O/W
North-central Silver Lake–Spillway Bay	C-14	Avg = 285.7 depth = 0-.6	Pelecypod fragments	13,620 ± 100 Y-1585	O/W
Tidewater Basin Beach Ridge II	C-14	284.1	Pelecypod shells	14,550 ± 140 Y-1586	O/W
Tidewater Basin Beach Ridge II	C-14	283.5	Pelecypod shells	15,350 ± 240 Y-1587	O/W
Bench Mark Bay	C-14	286.5	Tufa coats gravel	9,960 ± 200 Y-2410	O/W
Bench Mark Bay—below stone artifacts	C-14	281.9 depth = .3-.5	Pelecypod fragments	10,270 ± 160 Y-2406	O/W
Northwest Silver Lake El Capitan BR II	C-14	~287	Pelecypod shells	11,860 ± 95 DIC-2824	W
Northern Silver Lake High Shoreline El Capitan BR II	C-14	~287	Pelecypod shells	11,970 ± 160 Beta, written commun.	RW
A-Shoreline—Northwest Silver Lake	C-14	~286.5	Tufa	10,850 ± 75 DIC-2823	W

Note: Reference source for these dates (study) are coded: W—Wells et al., 1989; M—McFadden et al., 1992, B—Brown, 1989, H—Hubbs et al., 1965, O/W—Ore and Warren, 1971, and RW—Weldon, 1982.

creating the shoreline features. In order to accomplish this goal, shoreline features in Silver Lake and northern Soda Lake basins were initially mapped using aerial photographs of selected study sites (Fig. 7). Subsequently, many of the shoreline features recorded in these photographs were mapped and surveyed with an EDM-Total Station to determine exact elevations and distances. All survey measurements were tied to local U.S. Geological Survey benchmarks. Individual beach ridge units were distinguished using stratigraphic relations, height above playa surface, correlation with various wave-cut features, and lithology of beach ridge material. Conventional radiocarbon analyses were performed on lustrous pelecypod shells showing little or no recrystalization or replacement of the original aragonite structure (Table 1).

The best preserved and laterally most extensive shoreline features were formed during lake highstands, which were controlled by the elevation of the outlet spillway. Lake overflow during the A-shoreline period resulted in erosion of unconsolidated sediments at the outlet spillway and formation of the younger B-shoreline that was stabilized on bedrock (Bode, 1937; Ore and Warren, 1971; Wells et al., 1987, 1989). Therefore, these shoreline features are found around elevations of 287–288 m (A-shoreline) and 285.5 m (B-shoreline) (Figs. 4, 5). Topographically lower, less well preserved shoreline features are found below the A- and B-shorelines and formed below the elevation of the spillway control. The highest topographically of the younger shoreline features occurs at ~283 m (C-shoreline; Wells et al., 1987, 1989; McFadden et al., 1992; Enzel et al., 1992).

The largest, laterally most extensive, and best preserved shoreline features typically are found along the northern margins of both basins (Fig. 7; Plates 1 and 2) because (1) the maximum fetch of the lake in a north-south direction combined with prevailing southwesterly winds produced the strongest wave activity and therefore the largest features along the northern margins of each lake; and (2) the proximity of bedrock outcrops in these areas protected the shore features from erosion or burial by active alluvial fan streams. Four sites were selected for detailed studies at the northern end of the Silver Lake basin (the Tidewater basin, Spillway bay, El Capitan complex, and Silver Lake quarry), as well as several sites selected in the northern Soda Lake basin (Fig. 7).

Tidewater basin

Two large beach ridges are preserved in the northeastern portion of Silver Lake in the area known as Tidewater basin (Fig. 7 and Plate 1). Both of these ridges are composed predominantly of fine gravel to coarse, sand-sized grus. The highest ridge is 287.9 m amsl at its maximum elevation and appears to be associated with the topographically highest lacustrine depositional features associated with pluvial Lake Mojave. Detailed topographic measurements of beach deposits show that they are topographically

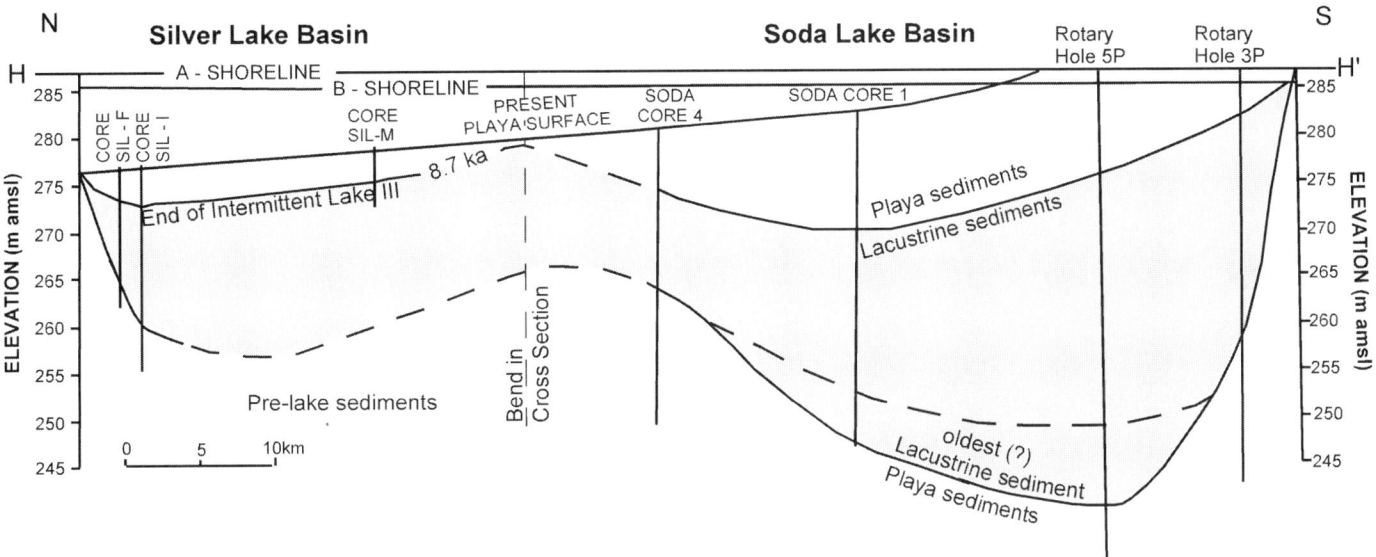

Figure 5. Cross section (H–H' from Fig. 4) showing pluvial Lake Mojave deposits in the Silver Lake and Soda Lake depositional basins using selected core data as well as the slope of the modern playa floor and the elevation of the two highest shorelines (A and B) associated with Lake Mojave.

higher. Detailed stratigraphic studies, however, show that many of the beach deposits are mantled by younger eolian sediments (McFadden et al., 1992), which does not reflect an accurate measurement of shoreline processes.

The highest beach ridge, which merges westward into the A-shoreline features in Spillway bay (Plate 1) and eastward into a bedrock outcrop, has not been dated directly. No shell material was observed in exploratory pits excavated in the highest ridge. To the north of this ridge lies a small topographic depression that slopes gradually uphill to a broad saddle (289.5 m) that marks the drainage divide between Silver Lake basin and Dry Lake basin to the north. Core Sil-L, which was obtained from this basin (Plate 1), yielded no lacustrine sediments and indicates that bedrock is <2 m below the surface (Brown, 1989). We infer that the highest beach ridge served as a barrier to Lake Mojave at its highest stand.

A larger topographic depression (Tidewater basin) lies between the two beach ridges (Fig. 7 and Plate 1). Core Sil-G was drilled to a depth of 6.25 m in the Tidewater basin (Fig. 7 and Plate 1) and reveals several meters of green clays, interpreted as lacustrine in origin (Brown, 1989; Wells et al., 1989). The topographically lower ridge starts in an embayment of the Castle (Plate 1), a large bedrock outcrop, and trends eastward.

A complex depositional history is recorded in a stream-cut exposure at the extreme western edge of the topographically lower beach ridge (285–286 m; Figure 8). Eleven lithofacies, ranging from cobbles and pebbles to clayey silt, are exposed at this locality along the walls of the streams that dissect the beach ridge. These lithofacies indicate changing depositional environments as well as fluctuating water levels (Table 2). Several green, fine-grained clay and silt-rich lithofacies (units IV, VI, VIII, and XI) are interbedded with coarser sand and gravel units (units V,

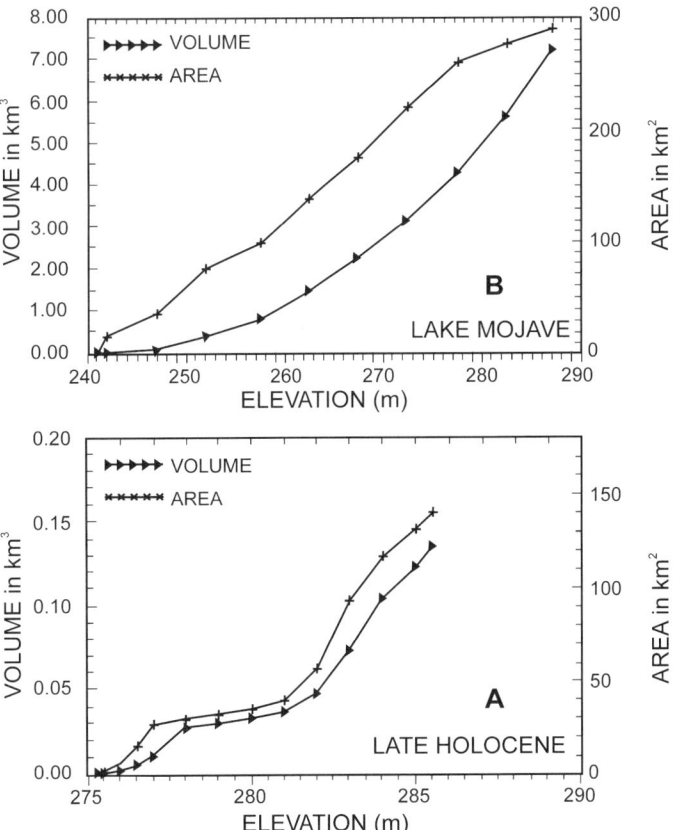

Figure 6. Lake elevation, lake volume, and lake area curves (two sets) for highstand (A-shoreline, see Fig. 5) of late Pleistocene Lake Mojave (B) and shallower and shorter-duration Holocene lakes (A, late Holocene, modified from Enzel, 1990). Note that elevation of the A-shoreline is 287 m and the B-shoreline is 285.5 m.

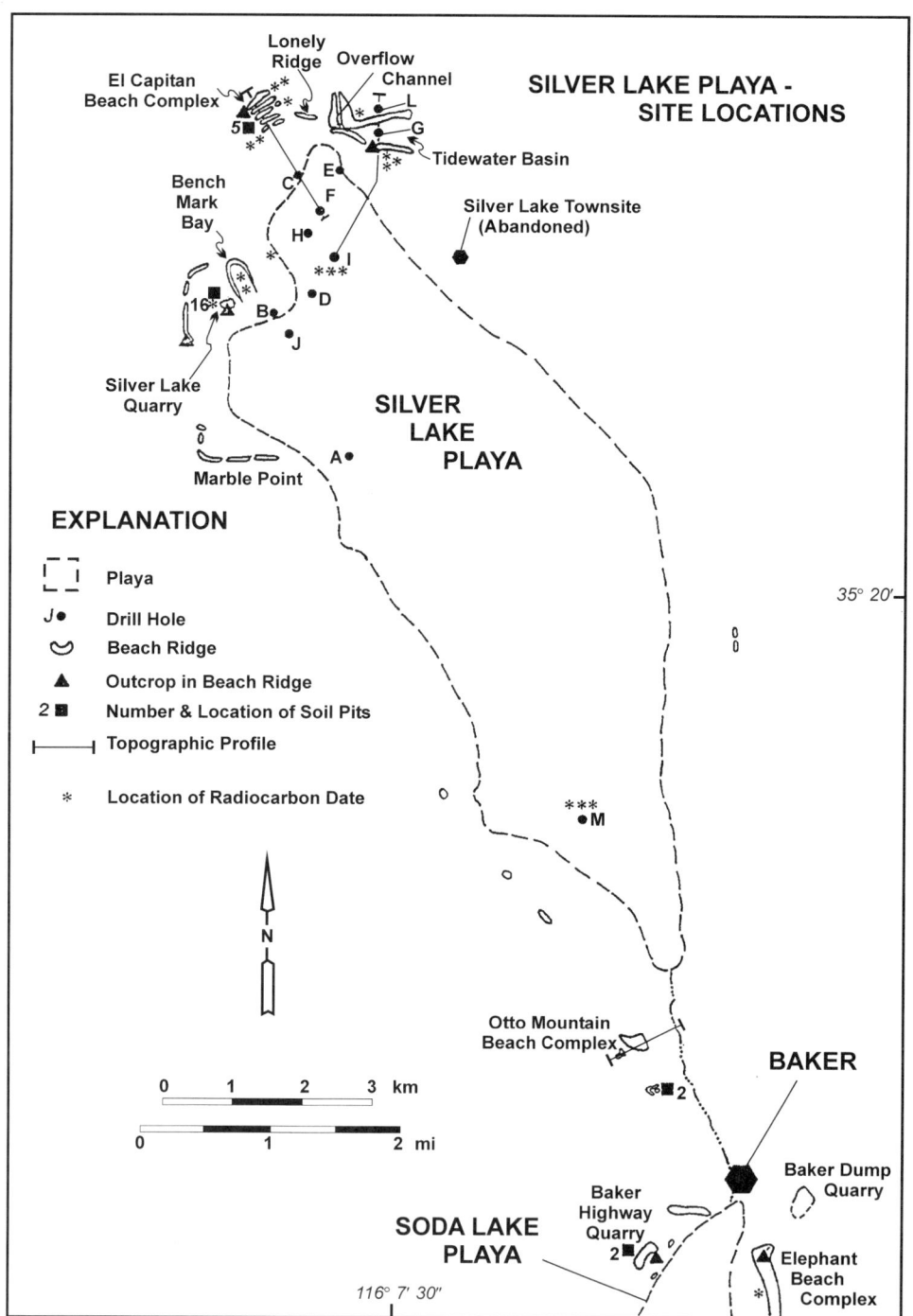

Figure 7. Map of Silver Lake playa and northern Soda Lake playa, showing major study sites (i.e., Tidewater basin, major shoreline features, location of key samples used in radiocarbon dating, and sites of cores investigated during this study. Radiocarbon ages are provided in Table 1.

VII, IX, and X) and indicate a low-energy deposition environment (e.g., offshore, backbasin, or lagoonal). At least two of these fine-grained lithofacies are truncated laterally by lakeward-dipping, one-to-two clast thick cobble lags, interpreted as erosional (ravinement) surfaces and overlying transgressive cobble lags. We believe that these lags formed during rising lake levels as the shoreface moved upward and landward (northward). Similar stratigraphic relations have been described for Quaternary marine coastal sequences (Nummedal and Swift, 1987; Kraft and Chrzastowski, 1987). One of these bounding surfaces truncates a section within unit VII immediately above sediments dated at 15,350 ± 240 yr B.P. and below sediments dated at 14,550 ± 140 yr B.P. From this stratigraphic section (Fig. 8), we infer that lake levels receeded below 283 m circa 15.4 ka before rising circa 14.5 ka.

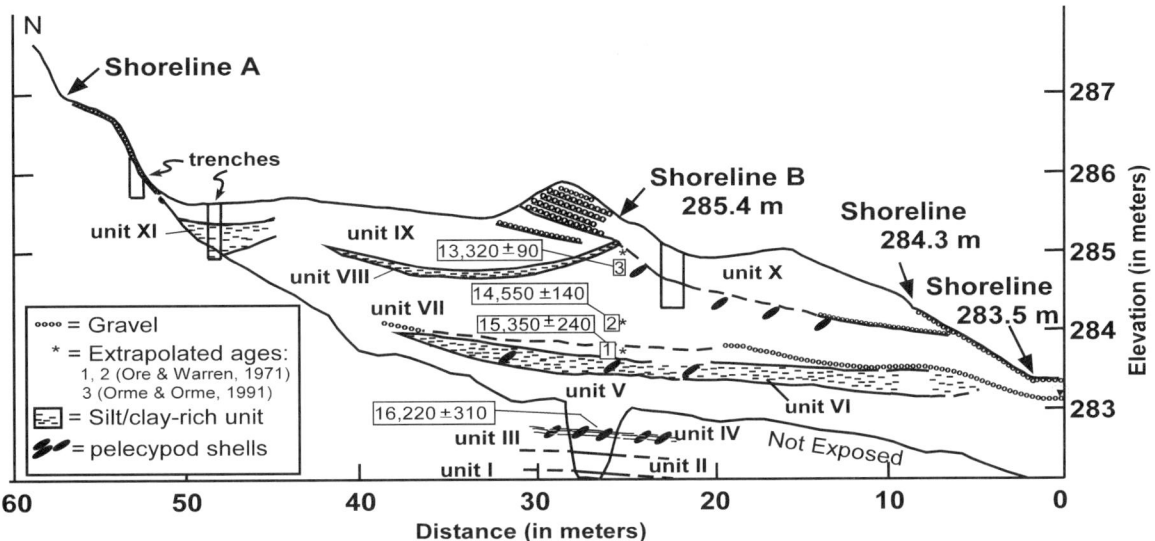

Figure 8. Stratigraphic cross section of Lake Mojave beach ridge deposits in the Tidewater basin area of northern Silver Lake playa (see Fig. 7). Eleven distinct lithofacies are present in the stream-cut exposure at the westernmost edge of the beach ridge. Four fine-grained green deposits are present in the outcrop (i.e., lithofacies IV, VI, VIII, XI), and all other units are composed of gravel and coarse sand. Prominent erosion (ravinement) surfaces preserved as gravel lags indicating at least three major transgressions and regressions are recorded at this locality. Shells collected by Ore and Warren (1971) and Orme and Orme (1991) were obtained several meters to the east of this outcrop from unit VII and were extrapolated stratigraphically into this cross section. Ages are reported in radiocarbon years B.P. (see Table 1 for details).

Abundant ostracodes were present in the fine-grained sediments of these four units (IV, VI, VIII, and XI; Figure 8), supporting our interpretation that these fine-grained lithofacies are lacustrine in origin (R. Forester, 1987, 1988, personal commun.; Steinmetz, 1988, 1989). Unit IV contains abundant *L. ceriotuberosa* and minor amounts of *L. bradburyi*. Unit VI contains abundant *L. ceriotuberosa* and common *L. bradburyi*. Units VIII and XI contain only one *L. ceriotuberosa*. Steinmetz (1989) correlates an abundance of *L. ceriotuberosa* in stratigraphic sections of Lake Manix sediments (Fig. 1) with an increase in the trophic state of the lake and lake filling. He has found peak abundances at times of high lake levels within a stratigraphic section that correlates with the beginning of the Wisconsin glaciation. *L. bradburyi* occurs in lakes that are relatively shallow, turbid, and isothermal with ambient air (Forester, 1987; Steinmetz, 1988). The co-occurrence of *L. bradburyi* and *L. ceriotuberosa* within sediments of the Coyote subbasin of Lake Manix suggests that climatic seasonality was not extreme and that the lake was relatively shallow during that period (Steinmetz, 1989). Based upon a comparison with Steinmetz's study, we infer that unit IV formed under relatively deep water conditions and that unit VI was deposited under relatively shallow water conditions. The later inference is consistent with our interpretation of the ravinement surface in unit VII that truncates unit VI; this relation reflects lake shallowing and shoreline recession after the deposition of unit VI and then shoreline transgression that formed the ravinement surface and lake filling to form unit VII.

Whole pelecypod shells, in growth position and in an excellent state of preservation, were collected from unit IV of this exposure (Fig. 8) and yielded a radiocarbon age of 16,270 ± 310 yr B.P. (Beta-29553, Table 1). This is the oldest known radiocarbon age obtained from deposits associated with any of the shoreline landforms related to pluvial Lake Mojave. The preservation of the pelecypod shells in growth position supports the inference that unit IV formed under deeper water conditions (below wave base/offshore).

Ore and Warren (1971; Table 1) dated shells from two superimposed sets of cross strata located ~20 m west of the above location. These units are composed of coarse-sand-sized to granule-sized clasts that dip ~26° toward N10°W. These two shell-bearing horizons are located at elevations of 283.5 m and 284.1 m and yield age dates of 15,350 ± 240 and 14,550 ± 140 yr B.P., respectively (Fig. 8). At the same location as the Orr and Warren site, Orme and Orme (1991) generated a radiocarbon age of 13,320 ± 90 yr B.P. (Beta-36305) from shells obtained at 285.2 m, clearly below unit X but stratigraphically above the shells found by Ore and Warren (1971). Based upon detailed topographic measurements and numerous stream-cut exposures, we extrapolated the stratigraphic position of these three radiocarbon ages westward to the outcrop illustrated in Figure 8, suggesting the dates are from lithofacies that are no older than unit VI and younger than unit VIII.

The Tidewater basin site is critical to understanding the complex history of Lake Mojave and reconstructing the formation

TABLE 2. LITHOLOGIC DESCRIPTION OF AN EXPOSURE IN THE LOWER BEACH RIDGE,
TIDEWATER BASIN SITE, NORTHERN SILVER LAKE BASIN, FAUNAL INFORMATION

Unit no.	Description of Outcrop-Tidewater Basin Beach Ridge II
XI	**66 cm thick Gravelly, Silty Clay.** (Not the same location as Units I–X.) Unit is irregularly laminated to massive in nature. Bottom not reached. Lower portion of unit is very strongly indurated and contains abundant disseminated $CaCO_3$. Occasional shell fragments found throughout. Top of unit is moderately sorted with subangular to subrounded granitic grus and marble clasts. Bottom of unit is poorly sorted with angular to subrounded granitic grus and marble clasts. Boundary between these subunits is gradational. Color 2.5Y 6/4. Unit contains ostracodes (*L. ceriotuberosa*).
X	**0–45 cm Gravel.** This unit includes crest and youngest beach deposits. Cobbles to pebbles (gravel–90%) with a coarse sand matrix composed almost entirely of grus. 7/5YR 6/4. Alternating laminations of coarse sand averaging about 15–20 cm thick. Each layer coarsens upward. Reddening at base of unit (5YR 7/4) 4–5 cm above unit 4. Coarse sand laminae dip about 5° to the south.
IX	**45–66 cm Gravel Cobbles to Pebbles with Minor Sand Matrix.** Unit is composed almost entirely of grus. Coarse sand-gravel laminations dip gently (3°–5°) lakeward and exhibits moderate sorting within individual laminar. Minor scattered shell fragments. Color 7.5YR 5/4 (dry).
VIII	**66–72 cm Fine Sand and Silt with Evaporite Minerals.** Unit is variable in thickness (4–8 cm). Pinches out about 1 m south of described outcrop locality. Abundant disseminated $CaCO_3$. Irregular patches of clay-silt with abundant carbonate. Locally preserved seeds. Overall moderately to well sorted and moderately stratified. Sharp upper boundary. Color 5Y 6/2 (dry), 5Y 5/2 (moist). Sample collected from the most clayey part of this unit contain abundant ostracodes (*L. ceriotuberosia*) as well as *L. heterocypris* sp.
VII	**72–187 cm Gravel Granules to Cobbles Composed Almost Entirely of Grus.** Occasional bedding or laminations dipping almost horizontal. Overall, unit is about 70% subrounded to subangular granules. Laminar units are well sorted and usually about 2–3 cm thick. Sharp upper boundary. Color 10YR 4/4 (moist).
VI	**187–207 cm Gravelly, Silty Sand.** Moderately well sorted fine-medium sand matrix with varying amounts of granules that increase in percentage near boundaries. Moderate amounts of silt and clay near middle. Upper 4–5 cm contain cobbles to pebbles and pelecypod shells. Possible bioturbation in upper 10 cm of unit. Color 5Y 7/2. Locally abundant evaporate minerals $CaCO_3$ and mirabilite in thin horizontal laminations. Lower boundary gradational, upper boundary sharp. Entire unit effervesces strongly. Abundant ostracodes (*L. bradburyi* and *L. ceriotuberosia*).
V	**207–255 cm Silty, Sandy Gravel.** Pebbles and granules in basal 8–10 cm grading upwards to finer-grained granules and sand with a silt-rich component that increases in abundance near the top of the unit. Overall the unit is composed of grus and displays mottled oxidized (7.5YR 6/6 dry) and reduced (5Y 5.5/3 dry), roughly horizontal zones. Stratification dips gently north (3°) away from the lake. Lower boundary is irregular and slightly wavy. Upper boundary is gradational.
IV	**255–269 cm Sand–Gravel.** Basal coarser zone in lower 5–8 cm with angular clasts up to 2.5 cm in size. Finer-grained alternating thin bands of reddened, oxidized (7.5YR 5/6 dry) granules and reduced gray-green (5Y 6.5/3 dry) sand-sized material overlie the coarser zone. The sandy zones contain abundant whole pelecypod and gastropod shells preserved in growth position. Strong disseminated $CaCO_3$ is present in the lower portion of this unit, extending from the base of the shell horizon to a depth of 33 cm. Sorting is moderate within individual lamination zones. Lower boundary shows evidence of scour. Moderately abundant ostracodes present (*L. cariotuberosa* with minor *L. bradburyi*).
III	**269–282 cm Gravel.** Subrounded to subangular clasts (up to 4 cm in size) of granite and lesser amount of limestone. The unit coarsens upward. Fine-grained matrix composed of sand- to granule-sized grus. Overall gray-green, reduced color (5Y 6/4 dry). Unit is strongly cemented by disseminated $CaCO_3$.
II	**282–312 cm Granule-sized Gravel.** Moderately well sorted with thinly stratified bends dipping gently lakeward. Locally horizontal aligned oxidized zones, although unit is an overall gray, reduced color (5Y 6/4 dry). Lower gradational boundary unit is composed almost entirely of grus.
I	**312–327+ cm Gravel, Pebbles to Cobbles.** Subrounded angular, composed of limestone and granite. Matrix is composed of sand-sized material of grus and limestone. Overall unit is a reduced, green-gray color (5Y 6/6 moist, 5Y 6/4 dry).

Note: from R. Forester (personal commun., 1987, 1988).

and age of the highest shoreline. Although no age is available for the older and higher beach ridge (287.9 m) complex at Tidewater basin, this site appears to provide the oldest radiocarbon date of lacustrine sediments within any of the studied Lake Mojave shoreline environments (Figs. 8 and 9; Table 1). The existence of this older, undated beach ridge (Fig. 9) suggests that (1) prior to 16.2 ka, and probably during the last glacial maximum, Lake Mojave was a higher and deeper lake than later in its history; and (2) during this highstand period, the lake reached the elevation of 287–288 m that is controlled by a spillway and overflow channel (see profile E4–E4′; Plates 1 and 3). During this time period, Lake Mojave overflowed toward Death Valley (Fig. 3).

Based upon these stratigraphic and geomorphic relations, the higher (287.9 m) beach ridge appears to represent a distinctly different (older) lake stand than that represented by the lower (285.6 m) beach ridge. We hypothesized that an older (at least 16.2 ka and older) stand of Lake Mojave produced the higher beach ridge at the level of shoreline A (287–288 m) during this time and under conditions in which the lower beach ridge either did not exist or was not emergent (e.g., a spit). Under this environment, lacustrine clays (such as those in unit IV) would be deposited in relatively deeper waters at the present site of the Tidewater basin topographic depression. We tested this hypothesis by drilling Core Sil-G in the topographic depression between the two ridges (GSA Data Repository item C; Figure 9; Plate 1).

This hypothesis is supported by the recovery of 3 m of green clays from Core Sil-G. Thus, we infer that the clays were deposited during the same time (at least 16.2 ka and older) and under conditions similar to unit IV (GSA Data Repository item C; Plate 3). In order to determine if these clays were lacustrine and not playette, groundwater discharge, or lagoonal in origin, the paleontology of the sediments was examined by J.P. Bradbury (1988, personal commun.). He determined that the basal part of these green clays contained diatoms (*Fraglilaria constuens v. subsaline*), which are found in abundance within the stratigraphic top of an older phase of deep-lake deposits (dated ca. 16 ka and defined as Lake Mojave I later in this paper). These paleontologic data support the hypothesis that a lacustrine environment existed within the Tidewater basin topographic depression prior to ca. 16.0 ka, and that an older and higher lake existed within the Tidewater basin prior to formation of the 285.6 m beach ridge starting at ca. 14.6 ka. During this older Lake Mojave highstand, the green clays were deposited in water up to, but not exceeding, 10–12 m (Plate 3, Figure 9). In that there is no evidence for younger lacustrine sediments in the Tidewater basin, we infer that the highest beach ridge formed at the time of this older highstand and that this site was not modified significantly by younger highstands reaching approximately the same elevation in other sites along the Silver Lake basin. The lack of lacustrine or lagoonal deposits north of the highest beach ridge (288 m; Plate 3) indicates that this ridge was the barrier to Lake Mojave in this area and that water did not overtop the ridge. The only other course for water to overflow during the older highstand is the overflow channel at Spillway bay that drains toward Death Valley (Fig. 7).

We believe that the data above support the hypothesis that the ~288 m beach ridge is older than the ~285 m beach ridge, and this interpretation allows us to reconstruct a simplified geomorphic evolution of the shoreline environment of Tidewater basin (Fig. 10). Based upon radiocarbon data obtained from cores discussed later in this paper, we infer that the geomorphic history starts at ca. 18 ka and, based upon radiocarbon data from the northern Silver Lake sites, continues through 10 ka. Between 18 ka and 16 ka, the following conditions prevailed: Lake Mojave extended into the Tidewater basin; the beach ridge at ~288 m (A-shoreline) formed and stabilized; spits prograded laterally from local bedrock and alluvial fan topographic highs at the current site of the lower beach ridge; and any highstands of Lake Mojave overflowed through Spillway bay and the overflow

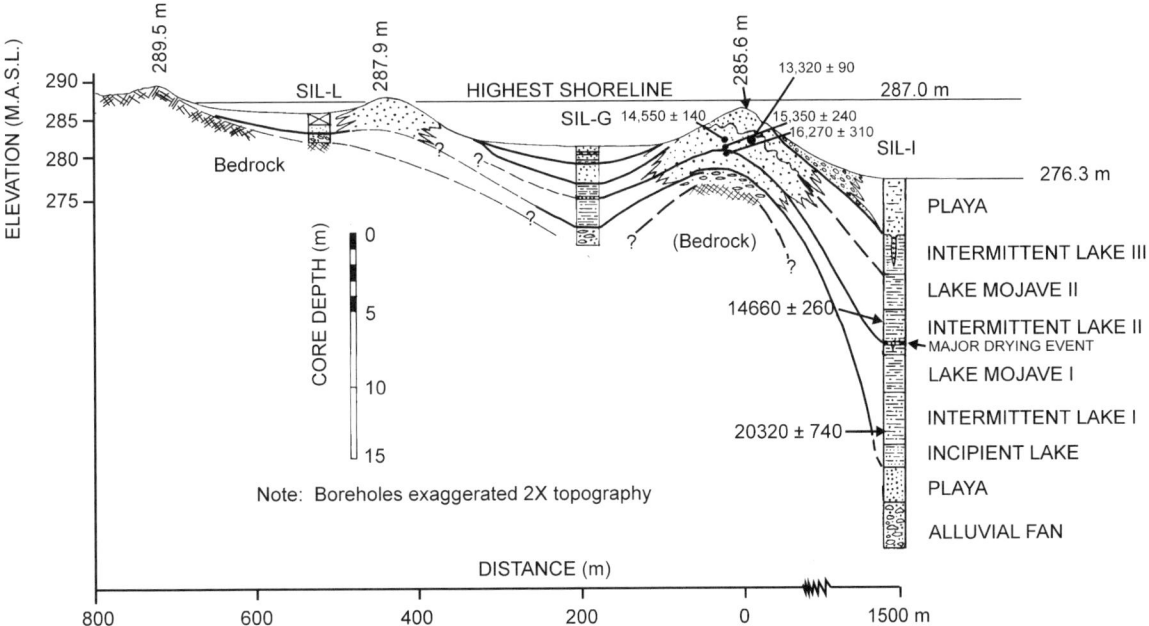

Figure 9. Cross section showing correlation of shoreline features and cored-lake deposits in northern Silver Lake deposition basin. Correlations are based on biostratigraphy, lithostratigraphy, and radiocarbon dating of pelecypod shells in shoreline sediments and lacustrine sediments in cores Sil-I, Sil-G and Sil-L (see Fig. 7 for location). Bold solid and dashed lines represent isochrones in near-shore and subaqueous environments, and dots are locations of radiocarbon-dated samples.

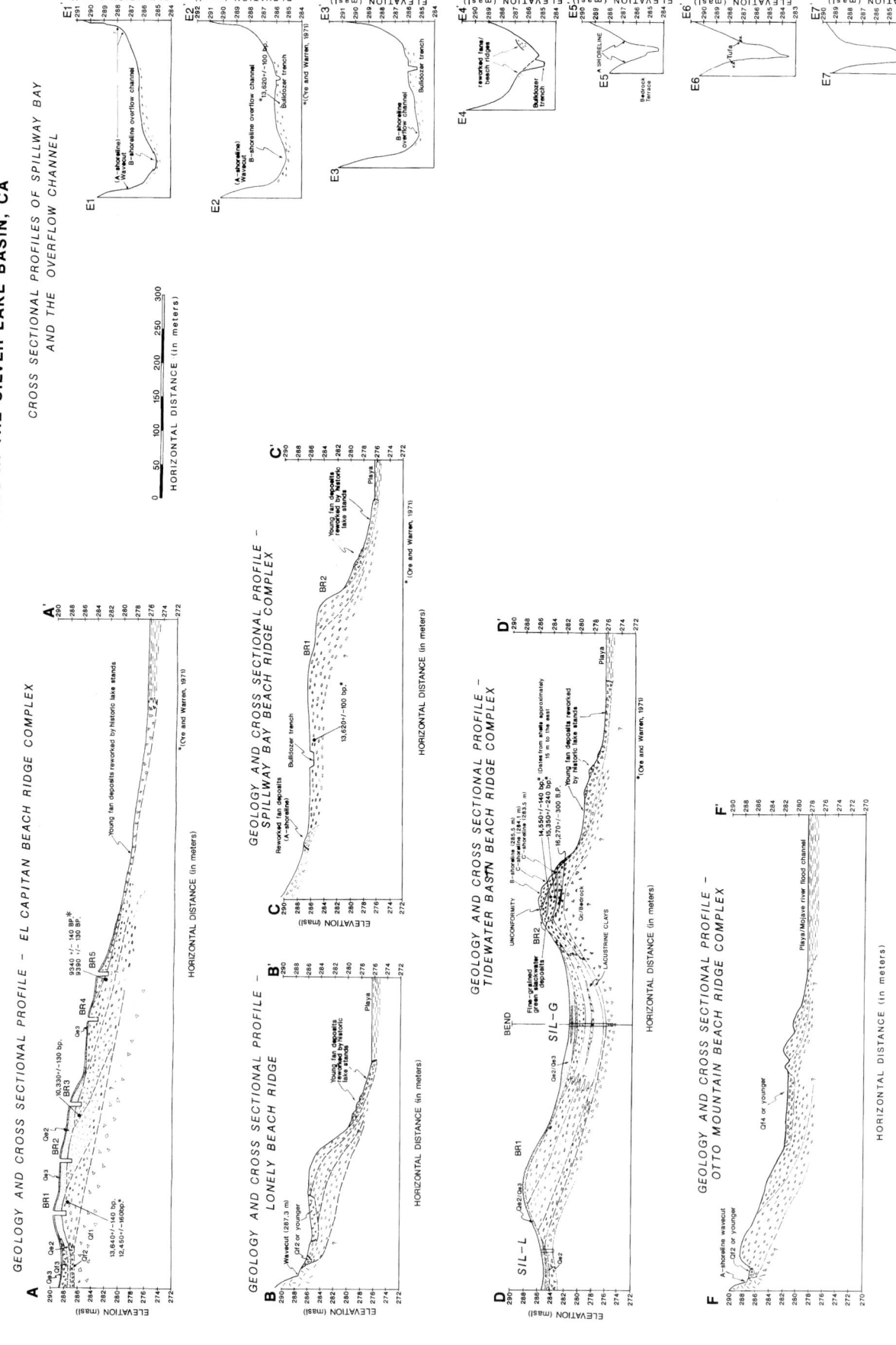

Plate 3. Geologic and topographic profiles of selected sites in Silver Lake depositional basin (see Plate 1 for locations).

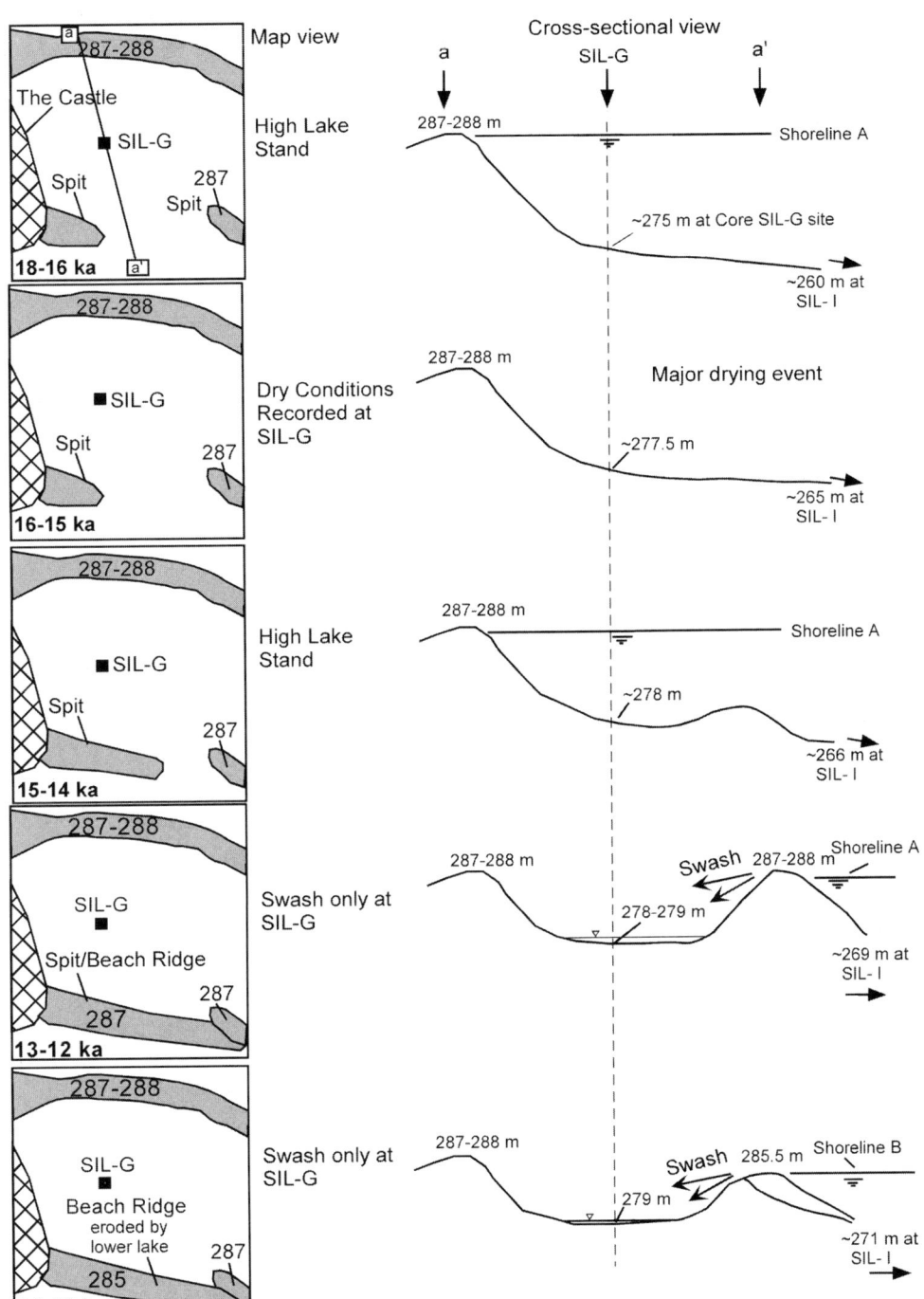

Figure 10. Schematic map and cross sections showing interpretation of spit and beach ridge evolution in the Tidewater basin (see Fig. 7) between 18 ka and 10 ka.

channel toward Death Valley (Fig. 10). Circa 15.4 ka, a major drying event occurred with total lake drying in the Silver Lake basin evidenced by eolian sediments found in cores Sil-G and Sil-I (see GSA Data Repository item C for detailed stratigraphy; Figures 9, 10). After 15 ka and before 14.6 ka, lake levels rose forming a major bar/spit complex that emerged between 14 ka and 13 ka, blocking lake access to the Tidewater basin and ending lacustrine deposition within the basin. Lacustrine processes apparently modified the emergent bar (287 m) until circa 11–10 ka when Lake Mojave overflow resulted in downcutting of the spillway to an elevation of 285.5 m. Episodic transgressions eroded the upper parts of the beach ridge complex attached to the bedrock high west of the Tidewater basin, resulting in an elevation of 285–286 m (Fig. 10).

Spillway bay

Topographic profiles at the northern end of the bay (Spillway site; E-4 in Plate 3) show two distinct spillway levels at 287 and 285.5 m that we correlate with the A- and B-shorelines, respectively. This correlation indicates that overflow occurred during both stands of pluvial Lake Mojave. There are no natural exposures of the shore-margin deposits in the Spillway bay area (outlet channel bay of Ore and Warren, 1971) (Plates 1, 3), but two key observations can be made from the geomorphology of the Spillway bay: (1) weakly developed wavecut cliffs present on both sides of the bay are most pronounced at 287–288 m in elevation; and (2) a broad, shallow paleochannel is cut into the bay's deposits with its thalwag at 285.5 m, exactly at the elevation of the B-shoreline.

Small pits excavated along the shallow walls of a trench (Plate 3) indicate that bay sediments consist of granular grus deposited in both beach and fluvial environments. The fluvial deposits suggest fluvial reworking by channels that formed during overflow discharge and migrated laterally across the entire width of the bay. Ore and Warren (1971) collected reworked pelecypod shells from fluvial sediments 30–70 cm below the surface and at an elevation of ~285 m (i.e., B-shoreline elevation; see Plates 1, 3). The radiocarbon age of these reworked shells is 13,620 ± 160 yr B.P. These shells suggest that the Spillway bay site was last occupied by Lake Mojave stands at the B-shoreline elevation and that fluvial channels during overflow reworked these shoreline sediments after ca. 13.6 ka.

Detailed field studies indicate that overflow during lake high stages only exited the basin through the overflow channel north of the Spillway bay beach-ridge complex. The overflow channel has no visible terraces (Plate 3) and is cut into alluvial fan deposits inferred by Wells et al. (1987) to be correlative with either unit Qf0 or Qf1 (mid- or late-Pleistocene, respectively) based on the degree of soil development. The alluvial fan deposits slope away from the local, low relief hillslopes composed of marble, dioritic, and mafic lithologies. The channel walls are very steep and expose stage III+ to IV pedogenic carbonate horizons in fan units. Field observations suggest that these deposits and petrocalcic soils probably were altered extensively by groundwater movement or streamflow (i.e., dissolution and recementation features) during lake highstands. After the last overflow event, the channel acted as a sediment trap for eolian material that blankets the floor to depths >1 m. Field observations and interpretation of aerial photographs suggest that the channel previously continued several kilometers north toward Lake Dumont (K. Anderson and Wells, this volume). Overflow of the Silver Lake basin has not occurred during historic times (past ~150 yr; McFadden et al., 1992).

El Capitan beach ridge complex

Five separate beach ridges are present in the El Capitan complex (formerly the northern beach ridge sequence of Bode [1937] and northwest beaches of Ore and Warren [1971]), more than in any other beach ridge sequence in the Lake Mojave area (Figs. 7, 11). Topographic profile A–A′ constructed perpendicular to the long axis of the ridges indicates that these ridges formed on a long, gentle slope (Plate 1). The highest ridges—BR I, BR II, and BR III—have well-developed stone pavements overlying an eolian unit ~1 m thick. This unit contains a soil similar to that described on unit Qe2 (Wells et al., 1987; McFadden et al., 1992). Below this eolian unit, imbricated beach gravels dip gently toward the former lake basin floor.

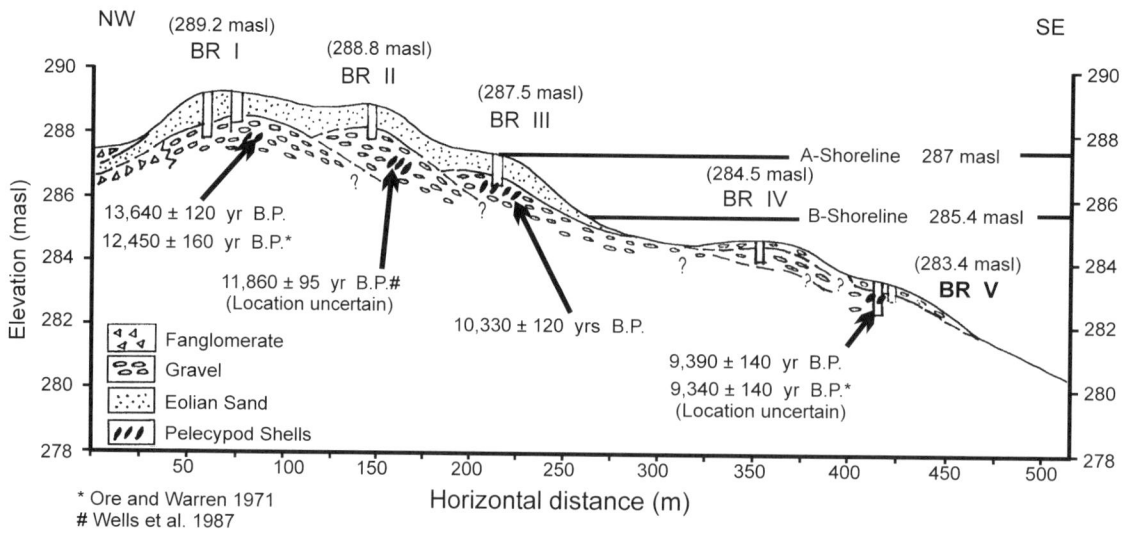

Figure 11. Stratigraphic cross section of the El Capitan beach ridge complex, northern Silver Lake playa (see Fig. 7) emphasizing beach ridge stratigraphy, location of radiocarbon dates on pelecypod shells, and shoreline elevations (modified from McFadden et al., 1992).

The lithology of the beach material is predominately mafic metamorphic rock, diorite, and granodiorite with minor amounts of limestone/marble. A significant increase in the amount of grus is observed in the BR III deposits. Abraded carbonate coatings on many of the gravel clasts in the upper three ridges suggest reworking of the nearby fan units. Tufa is common on many of the larger clasts, and all five ridges contain pelecypod shells. Shells from seven locations in the El Capitan complex have been radiocarbon dated (Table 1; Figure 11) making this one of the most well-dated series of beach ridges in southern California. The El Capitan complex shells become progressively older with increasing elevation and distance from the playa (Fig. 11). Shell ages from the upper three ridges are (1) 13,640 ± 120 yr B.P. (this study) and 12,450 ± 160 yr B.P. (Ore and Warren, 1971) from BR I (Fig. 11); (2) 11,970 ± 160 yr B.P. (Weldon, 1982) and 11,860 ± 95 yr B.P. (Wells et al., 1987; location uncertain) from BR II (Fig. 11); and (3) 10,330 ± 120 yr B.P. (this study) from BR III (Fig. 11). The lowest two beach ridges are Holocene in age (Orr and Warren, 1971; McFadden et al., 1992). Radiocarbon dating of the soil carbonates on the beach ridges supports the above ages and is summarized in McFadden et al. (1994) and Amundson et al. (1994).

It is important to note that surface elevations of the three highest beach ridges at the El Capitan site do not reflect the actual shoreline elevations due to an early Holocene eolian accretionary mantle overlying the beach deposits (Fig. 11; McFadden et al., 1992). We infer that the two highest beach ridges (BR I and BR II, elevation 288 m; dated between ca. 13.6 and 11.8 ka) formed during Lake Mojave stands at the A-shoreline. Based upon data from the El Capitan beach ridges and our observations at Tidewater basin, we believe that the A-shoreline has been occupied episodically from >16 ka through ca. 12 ka. Thus, shoreline deposits at the A-shoreline elevation (287–288 m) may be of differing geologic ages at different geographic locations, but all are older than ca. 12 ka. The El Capitan and Spillway bay data together indicate that the B-shoreline (285.5 m) is no older than ca. 12 ka, and one radiocarbon age in a beach ridge associated with the elevation of the B-shoreline is 10.3 ka. We infer from the study sites along the northern Silver Lake basin that the B-shoreline is no older than 12 ka and was stabilized by ca. 10.3 ka to allow beach ridge formation.

These observations demonstrate that deposits of significantly different ages can occur at the same elevation along a shoreline as well as within a single landform along the shoreline margin. Clearly, the results derived from age dating must be carefully combined with detailed stratigraphic and geomorphic analyses when reconstructing paleolake histories.

Silver Lake quarry

A small gravel quarry was excavated in a spit composed of coarse-to-fine grained deposits at the northwestern corner of the Silver Lake playa (Figs. 7, 12). The quarry exposes a series of interbedded gravel, sand, and silty clay beds deposited by pluvial Lake Mojave (Ore and Warren, 1971). The top of the quarry is ~8 m below the A-shoreline, which is preserved as a wave-cut cliff in dioritic bedrock flanking the quarry. The top of the spit slopes southwest from the bedrock. A detailed chronology of these deposits has been reconstructed from radiocarbon ages obtained from tufa-coated gravels and locally abundant pelecypod shells (Fig. 12) (Ore and Warren, 1971; also discussed by Wells et al., 1987). These ages show a regular progression from the oldest in unit 1 (13,670 ± 550 yr B.P.) to the youngest at the top of the outcrop (8350 ± 300 yr B.P.).

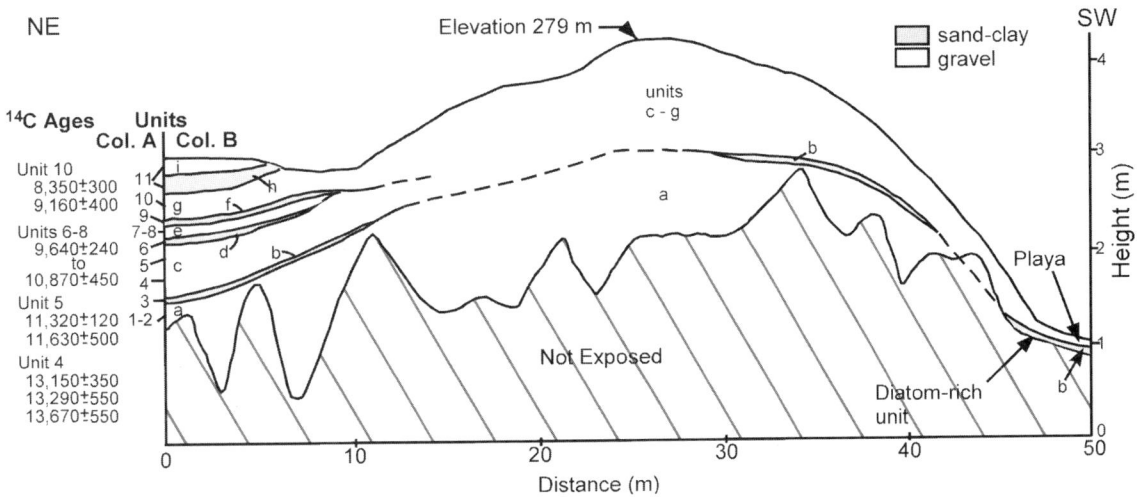

Figure 12. Simplified stratigraphic cross section of the Silver Lake quarry site (see Fig. 7) at the northern end of Silver Lake Playa highlighting stratigraphic positions of radiocarbon dates obtained by Ore and Warren (1971) and this study (Table 1). Green sandy clay rich in pelecypod shells indicated by shading.

We observe two stratigraphic packages (unit a and units b–i; Figure 12) of sediments within the spit separated by a sharp bounding surface (see GSA Data Repository item A for stratigraphic description). Unit a (undated) forms the lower stratigraphic package and is composed predominately of weathered angular to subrounded pebbles and cobbles. The upper package (units b–i) consists of alternating layers of silty clay, sand, and unweathered gravels along the flanks of the spit and ranges in age from ca. 13.7 ka to 8.3 ka (Fig. 12). The difference in weathering of the two packages of sediment indicates a significant hiatus in the deposition of the two units; this contrasts with the conclusion first proposed by Wells et al. (1987) suggesting that no major diastem exists in this sequence. The core of the spit, represented by unit a, clearly predates 13.7 ka; and these deposits were exposed subaerially during lake-lowering events to allow the weathering of the clasts, which was probably enhanced by salts derived from the lake. We infer that these lower lake levels and period of subaerial weathering most likely occurred between the older (>16.2 ka) and younger (<14.6 ka) stands of Lake Mojave observed at Tidewater basin.

Northern Soda Lake basin

Extensive shoreline landforms are preserved along the northern margin of Soda Lake basin near Baker, California (Plate 2). Field investigations focused on three separate beach ridges, corresponding in height to the A-shoreline stands of Lake Mojave (287 m). Sand and gravel quarries at each of these sites provide an excellent three-dimensional view of the shore-margin stratigraphy (Wells et al., 1989). The Baker Highway quarry site (Plate 2) contains sediments that can be divided into four units. The upper three units contain abundant pelecypod shells (or shell fragments, depending on the outcrop) and are of lacustrine origin. These units unconformably overlie sediments composed of angular, poorly sorted, gravel-to-sand sized clasts that are interpreted to be of alluvial fan origin. Ostracodes collected from sediments directly above the lacustrine–alluvial fan unconformity have been identified as *L. ceriotuberosa* and are common to lacustrine deposits of pluvial Lake Mojave (R. Forester, 1988, personal commun.).

The Baker dump quarry, which was excavated in a broad beach ridge deposit southeast of the town of Baker, exposes beach sediments unconformably overlying alluvial fan and eolian deposits (GSA Data Repository item B). This beach ridge extends as high as 287 m, forms part of the A-shoreline, and is undated. A soil developed on topographically lower ridge deposits exhibits stage I+ pedogenic carbonate and displays oxidized horizons (color = 7.5YR 5/8, dry). Ostracodes identified as *L. ceriotuberosa* (R. Forester, 1988, personal commun.) were found throughout the beach sediments in addition to locally abundant pelecypod shells. A very large, almost continuous sand ridge extends from the beach ridge at the Baker dump quarry southward in an arc toward a bedrock outcrop (Elephant Ridge beach complex; Plate 2). This beach ridge has a maximum elevation of ~285.5 m, corresponding to the elevation of the B-shoreline. Locally, the ridge has been eroded by alluvial fan channels, covered by small eolian dunes, and extensively modified by human activity. Pelecypod shells from 40 cm below the land surface of this beach ridge were radiocarbon dated at $12,020 \pm 130$ yr B.P., representing the oldest ages obtained from the B-shoreline complex at this site (Plate 2). Surrounding Granite Island are several higher beach ridges and locally preserved wavecut scarps that correspond in elevation to the A-shoreline (287–288 m).

On the western side of Soda Lake, wavecuts are found on bedrock headlands and older alluvial fan segments that extend several km north of Zzyzx (Fig. 3; Harvey and Wells, this volume). The absence of major bedrock outcrops south of the Zzyzx area combined with the prograding fan delta of the Mojave River from Afton Canyon and extensive eolian deposition have resulted in little preservation of latest Quaternary Lake Mojave shoreline features in the southern Soda Lake basin (Fig. 3). Core data discussed in the section below, however, indicates that Lake Mojave extended south all the way to Crucero and west toward Afton Canyon (Fig. 3).

Geologic and chronologic relationships of shoreline deposits at the north end of Soda Lake basin are similar to those observed in the northern Silver Lake basin (Plate 2; GSA Data Repository item B). The A-shoreline is at the same elevation in the Soda Lake basin as it is in the Silver Lake basin, demonstrating a lack of regional tilting. In addition, the Soda Lake basin A-shoreline is older than 12 ka based on the radiocarbon age from the topographically lower and stratigraphically younger B-shoreline (consistently at ~285 m elevation in the Soda Lake basin). Also, deposits associated with the Soda Lake basin A-shoreline, exposed in the Baker dump quarry, unconformably overlie a buried soil developed in alluvial fan sediments resting upon eolian deposits. This stratigraphic sequence is very similar to that observed throughout the Cronese basins, which were part of pluvial Lake Mojave during the late Quaternary (Fig. 1; Clarke et al., 1996; Wells and Anderson, 1998; Lancaster and Tchakerian, this volume). In the Cronese basins (Fig. 1), these eolian deposits range in age from 30 ka to 15 ka, with three deposits dating from 22 ka to 23 ka (Lancaster and Tchakerian, this volume).

Subsurface sedimentology and stratigraphy within the Silver Lake depositional basin

We located the sites of drill cores within the northern Silver Lake basin to document geographic variations with the sediments and stratigraphy from the lake axis to the lake margin and in order to better understand variations in lake levels during pluvial Lake Mojave. Cores Sil-I and Sil-H were drilled near a north-south axis of the playa and show evidence of deeper water conditions (greatest thickness of lacustrine clays) than any of the other core sites (Fig. 7). Cores Sil-J, Sil-F, and Sil-E were drilled in areas farther from the axis and closer to the margins of pluvial Lake Mojave, where the thickness of lacustrine sediments should be less. Sediments in Core Sil-E, which was drilled near the present playa-fan interface, indicate the shallowest, most lake-margin conditions of any core site drilled in Silver Lake basin. Core Sil-M was drilled in the fan-delta area at the southern end of Silver Lake basin (Fig. 7).

Sediments observed in the Silver Lake and Soda Lake basin cores are grouped into three chronostratigraphic units (see Table 3 for definition; see GSA Data Repository item C for core descriptions) based upon radiocarbon ages from Silver Lake basin (Table 1) and regional correlation of sediments between Silver Lake and Soda Lake basins (Figs. 4 and 5). The three units are as follows: (1) a pre-Lake Mojave chronostratigraphic unit, the oldest characterized by sand and gravel deposited as alluvial fan and eolian sediments; (2) the Lake Mojave chronostratigraphic unit dominated by clastic lacustrine sediments composed mainly of reduced green-to-blue, clay-size particles indicating persistent standing water; and (3) a post-Lake Mojave chronostratigraphic unit characterized by oxidized, massive, silt-to-clay-size sediments indicating playa conditions with episodic wetting and drying events (Enzel et al., 1992). Although the sedimentary character and stratigraphic properties (facies) vary laterally across the Silver Lake basin, the boundaries between these units are defined in terms of the formation and timing of late-Pleistocene, early-Holocene Lake Mojave.

Pre-Lake Mojave paleogeography in northern Silver Lake basin

Seven cores are utilized to define basal contact of the Lake Mojave chronostratigraphic unit (i.e., pre-lake topography) and sedimentary environments underlying the pre-lake surface in the northern Silver Lake basin (Fig. 13). These cores show that the pre-Lake Mojave sediments immediately below the Lake Mojave package are predominantly alluvial fan and playa in origin. Fine-grained, playa-like deposits occur at the base of core Sil-H and Sil-I but are not observed in cores Sil-J, Sil-F, or Sil-E (Fig. 13). Thus, the areal extent of the pre-lake playa was smaller than the current playa in the northern part of the Silver Lake basin. Within the subsurface, pre-lake environment, alluvial fans flank the playa and consist of angular-to-subrounded, gravel-to-sand sized clasts that are moderately to strongly stratified and sorted. Pedogenic carbonate (stage I to II and stage IV) is present on clast surfaces and is disseminated in the matrix, indicating that the pre-lake surface was subaerially exposed with sufficient time to allow soil development. Based upon the two stages of pedogenic carbonate observed in the subsurface fan sediments, we infer that two different ages of alluvial fan units formed in this basin prior to Lake Mojave. The oldest, with stage IV pedogenic carbonate, is interpreted to be an alluvial fan deposit that correlates with unit Qf0 of Wells et al. (1987). The alluvial fan unit with stage I–II pedogenic carbonate may be equivalent to unit Qf1 of Wells et al. (1987) and is inferred to be inset within and/or overlie the older alluvial fan unit (Fig. 13).

The maximum relief of the playa deposits forming the pre-lake topography between cores J and E near the basin axis (Fig. 13) is ~1 m. The relief between the paleo-playa floor and the flanking alluvial fan deposits is >5 m. These cores suggest that the pre-lake landscape of northern Silver Lake basin was similar to the current playa/alluvial fan landscapes of Silver Lake basin: low relief and shallow gradients except near topographic highs created by bedrock or older alluvial fan units (Fig. 13).

Sedimentology and stratigraphy of Lake Mojave chronostratigraphic unit

Detailed stratigraphic and sedimentologic studies of the Lake Mojave chronostratigraphic unit, including the nature of internal bounding surfaces, provide data for a high resolution geologic and hydrologic reconstruction of the Mojave River terminal basin (Silver Lake) during the late Pleistocene. We refer to our reconstructions as high resolution because core measurements were made at a scale of 1:1, allowing us to record variations in strata at the millimeter scale and correlate these strata from one core to another (GSA Data Repository items C and D, see footnote 1). The primary sedimentary features within the strata of the Lake Mojave

TABLE 3. DEFINITION OF SEDIMENTOLOGIC AND HYDROLOGIC TERMINOLOGY USED TO DESCRIBE DIFFERENT SCALES OF SEDIMENTARY FEATURES AND ASSOCIATED HYDROLOGIC EVENTS WITHIN THE SILVER LAKE BASIN, CALIFORNIA

Term	Definition
Bounding surface	A bedding plane or surface, classified into 12 types (see Table 4), separating sediments of differing properties
Sedimentary layer	Fundamental and discrete body of sediment with specific characteristics and separated by bounding surfaces
Banding	Thin zone of chemical precipitates within sediments and/or along the margin of a bounding surface
Sedimentary package	Grouping of lithofacies sedimentary layers in cores or surface exposures according to common, major attributes
Hydrologic condition (basin-wide)	Five relative lake levels reconstructed from the correlation of bounding surfaces and stratigraphic features among cores that vary geographically from the depocenter of pluvial Lake Mojave
Lake phase	Hydrologic condition interpreted from the lithologic and geographic variation of a chronozone identified in cores from the northern Silver Lake depositional basin

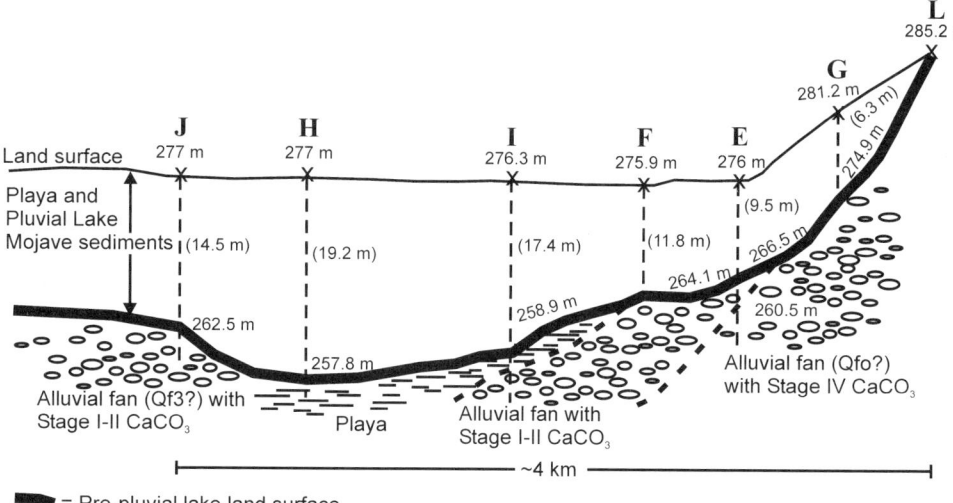

Figure 13. Schematic cross section reconstructed from selected cores (Sil-E through Sil-J, and Sil L) at the northwestern edge of the Silver Lake playa (see Fig. 7), showing the paleogeography of the pre-Lake Mojave depositional surface (bold line). Depth from the playa floor to the base of the pluvial Lake Mojave sediments is given in parentheses, and includes both Holocene playa and Lake Mojave sediments. Note the low relief and shallow gradient in the basin center.

unit include layers and bounding surfaces between these layers (see Table 3 for definitions). We distinguish layers within these strata from bedding and laminae because we focus on changes in key properties from layer to layer and less on the significance of individual layer properties (e.g., thickness).

Sediments recovered from the cores of the Lake Mojave chronostratigraphic unit include both detrital sediment and chemical precipitates. These sediments are dominated by detrital sediments from the Mojave River and are predominantly clay and silt-sized. Within the detrital lacustrine sediments, the following two types of layers (see Table 3 for definitions) dominate: (1) upward-fining sequences of very fine sand and silt grading to fine silt and clay, and (2) laminated-to-sublaminated clay beds varying in thickness from one to several centimeters (GSA Data Repository item C). In addition to these two types of layers, cores contain a third type of thin layer that consists of nonlaminated silt to very fine sand, which appears to be slightly coarser-grained relative to the two types discussed above. This third type of layer occurs at irregular intervals within the laminated-to-sublaminated, clay-sized sediments (Table 4). This third type of nonlaminated layer typically is accompanied by color changes, is almost always rich in silt, and contains a distinct lithologic fingerprint of phlogopite-rich sediments. The source of the phlogopite is within the upper Mojave River drainage basin (Brown, 1989).

We interpret that the phlogopite-rich layers were deposited during large-scale flooding of the Mojave River into pluvial Lake Mojave (see GSA Data Repository item C). This interpretation is supported by observations of exceptionally high proportions of phlogopite in Mojave River flood sediments near the mouth of Afton Canyon and the present-day Silver Lake and East Cronese Lake fan. Several stream sediment samples collected near Victorville, California (Fig. 1) also contained abundant phlogopite particles. In contrast, local floods from precipitation in the mountains surrounding Silver Lake and Soda Lake basins carry sediments that reflect local lithologies with little or no phlogopite.

During logging of cores taken from Silver Lake basin, we recognized that both the layers and nature of the bounding surfaces within the cores were critical in constructing the vertical sequence of geologic events. We observed that layers composed of different sediment types are separated by bounding surfaces (see Table 3 for definition) with well-defined properties. The properties of the bounding surfaces are used to classify them into 12 distinct types (Table 4). These bounding surface types reflect subtle but significant changes in depositional conditions (Brown, 1989, p. 24–25; Wells et al., 1989, p. 76–82; see GSA Data Repository items C and D). We infer that thin horizons of coarser silt-rich material, thin carbonate and sulfate-rich evaporite bands, and desiccation cracks reflect lake fluctuations such as lake lowering or drying events (see below; Table 4; see Smoot and Lowenstein, 1991, for example). These bounding surfaces are, in turn, overlain by lacustrine sediments representing deeper water conditions. The vertical sequence of these layers and boundary types has been correlated from core to core in the Silver Lake basin (Fig. 14; GSA Data Repository item D). Some of the higher-order bounding surfaces maintain their general character among the cores, whereas some of the lower-order bounding surfaces show significant variations. These variations are systematic from the center of the ancient lake to its margin and are critical to interpretation of the latest Quaternary hydrologic responses of Lake Mojave and the ancestral Mojave River. Successful correlation of the lake deposits are due, in part, to low relief of the pre-lake topography (Fig. 13). For example, the elevation difference of the pre-lake surface between cores Sil-H and Sil-I (which show the best correlation as illustrated in Figure 14 and GSA Data Repository item D) appears to be <1 m and >600 m. Correlations become slightly more difficult for those cores near the lake margin that suggest a 9 m elevation change in pre-lake topography over a distance of ~1000 m.

At some of the bounding surfaces, we infer drying, partial drying, or lake lowering from the presence of evaporites and mud cracks. Within a clastic matrix, primary chemical precipitates are

TABLE 4. TYPES AND PROPERTIES OF BOUNDING SURFACES OBSERVED IN PLUVIAL LAKE MOJAVE SEDIMENTS AND USED IN THE STRATIGRAPHIC ANALYSIS OF CORES OBTAINED FROM SILVER LAKE BASIN

Bounding surface type	Description
1	Dark coarser clay with minor silt-sized phlogopite overlying lighter colored finer grained clay
2	Silt containing phlogopite overlying clay
3	Sandy silt overlying silty clay
4	Fine sand overlying silty clay
5	Fine sand overlying clay
6A	Horizontal laminations of $CaCO_3$
6B	Horizontally aligned blebs of $CaCO_3$
7A	Horizontal laminations of thenardite and/or mirabilite
7B	Horizontally aligned blebs of thenardite and/or mirabilite
7C	Disseminated thenardite and/or mirabilite
8	Gravel overlying sand
9	Sediment (interpreted as being locally derived sediment from runoff/turbidite processes) consisting of diorite, carbonate, and/or granitic clasts overlying sediment of any size

Note: The letter D is used in conjunction with the bounding surface types to indicate a mudcrack associated with a drying event related to a boundary surface, and the letter S is used with these types to indicate a mud crack possibly caused by syneresis (i.e., nondrying) processes.

found at or just below bounding surfaces throughout the core as millimeter-to-centimeter thick bands of nonorganic carbonate, Na_2SO_4 (thenardite) and $Na_2SO_4 \cdot 10H_2O$ (mirabilite), including combinations of these precipitates (Tables 3 and 4; GSA Data Repository items C and D). Microscopic analysis of selected carbonate bands indicates that they are composed of microcrystalline calcite (Brown, 1989). In addition, displacive halite (Li et al., 1997) is widely disseminated throughout most of the upper parts of the cores. The frequency of the carbonate, thenardite, and mirabilite bands in individual cores increases with increasing proximity of the core to the paleo-shoreline (shallower conditions). For example, core Sil-H, which represents the deepest water conditions of any core in the lake, has the fewest bands of precipitates. In contrast, core Sil-E near the lake margin (Fig. 13) has individual carbonate, thenardite, and mirabilite bands up to 12 cm thick (GSA Data Repository item C).

Stratigraphically, these bands of chemical precipitates often are found directly below a bounding surface associated with mud cracks, suggesting that they are caused by lake drying episodes. If thenardite and mirabilite appear together, the carbonate is always found stratigraphically below them within the same band. This sequence within a band is typical when evaporation and lowering of lake waters produce supersaturated conditions of these mineral species. During the summer months, for example, evaporation rate, biogenic activity, and water temperature in a lacustrine environment increase greatly, thus decreasing values of carbonate solubility. These conditions are ideal for producing supersaturated solutions of $CaCO_3$ and Na_2SO_4; $CaCO_3$ will precipitate first because it is less soluble than Na_2SO_4. The precipitate bands are correlated from core to core (GSA Data Repository item D) and typically change in composition from carbonate in the cores associated with the deeper lake environment to thenardite in the cores associated with lake margins, thus suggesting the characteristic "bulls eye" pattern of evaporite facies changes found in many dry lake basins (Hunt et al., 1966; Hardie et al., 1978; Eugster and Hardie, 1975).

Mud cracks are observed commonly in subsurface sediments from Silver Lake basin (Fig. 14; GSA Data Repository item C). Mud cracks or shrinkage cracks form by the following two processes in a saline lacustrine environment: desiccation (subaerial conditions) and syneresis (subaqueous conditions) (Plummer and Gostin, 1981). Determining which processes (subaerial or subaqueous) formed a given set of shrinkage cracks can be difficult in cross-sectional profiles such as lake cores (Glaessner, 1969; Plummer and Gostin, 1981). In order to address this difficulty, we carefully examined the sediments infilling the lowest part of the mud cracks. Many of the shrinkage cracks preserved in the cores are filled with coarser, often oxidized (eolian and fluvial) sediment or are associated with evaporite bands. These features indicate a subaerial origin rather than a subaqueous origin. The above observations suggest that the majority of shrinkage cracks found in Silver Lake cores formed during desiccation of the lake.

Chronozones of the Lake Mojave chronostratigraphic unit, Silver Lake basin

Core data for the Lake Mojave chronostratigraphic unit (GSA Data Repository items C and D) yield a high resolution stratigraphy that is primarily based on the identification of the three types of sedimentary layers and the 12 types of boundary surfaces. Our goal is to use this high resolution stratigraphy and its systematic geographic variations to define correlative, basin-wide, relative

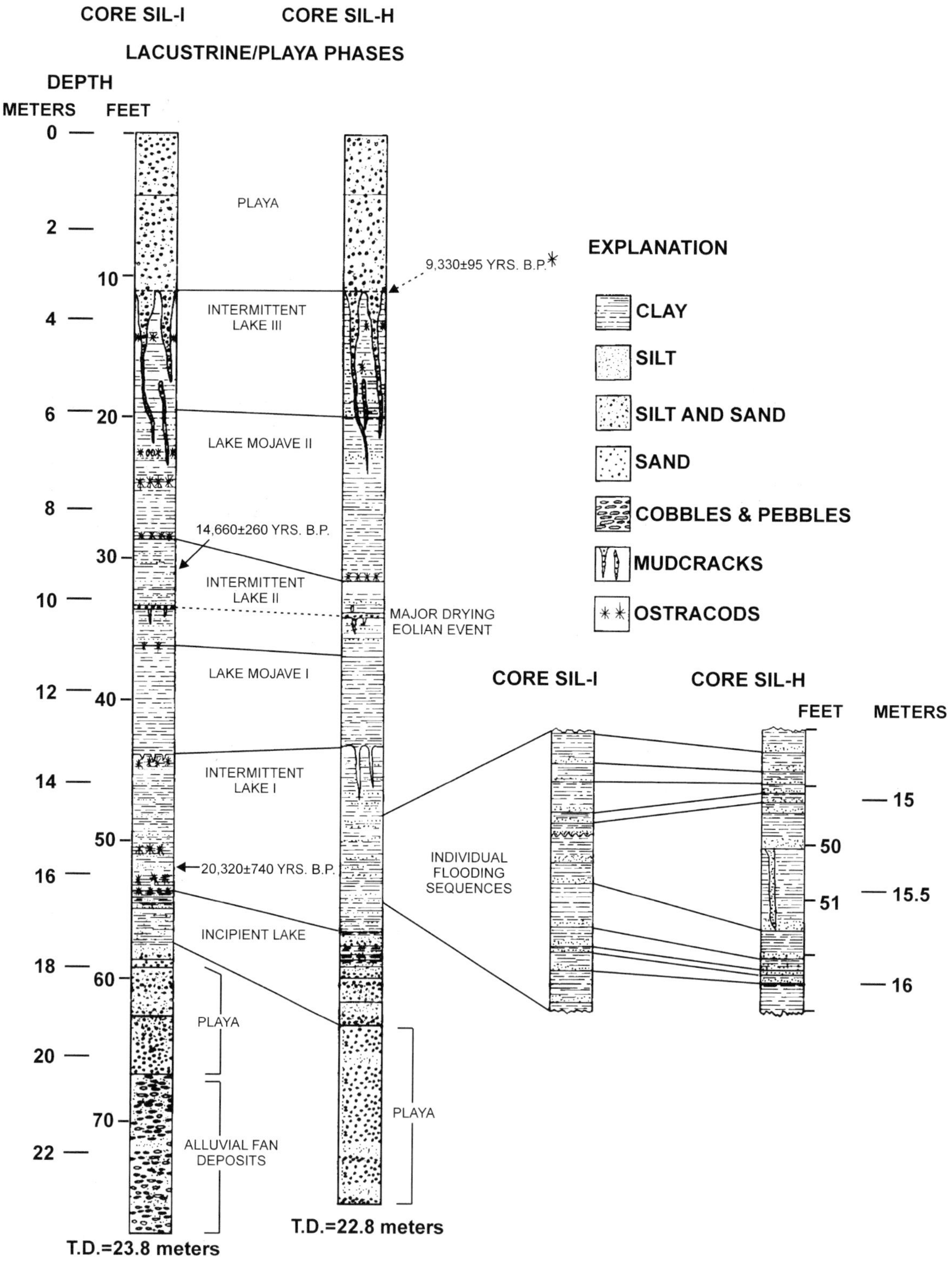

Figure 14. Simplified stratigraphy of cores Sil-I and Sil-H in the northern Silver Lake depositional basin, showing major lake chronozones (phases) and an expanded view of the zone between 14 and 16 m depth to emphasize the detailed nature of event correlations between the two core locations. The location of radiocarbon dates also are shown in core Sil-I (see Table 1 for details).

hydrologic conditions within the Silver Lake depositional basin (Fig. 3). Using the vertical sequences of these layers and bounding surfaces, we define five primary and one secondary informal chronozones on the basis of physical features that we interpret to reflect general synchronicity, including the frequency or absence of bounding surfaces with mud cracks and evaporites (Table 4; GSA Data Repository items C and D). Three chronozones are dominated by features indicating lake-drying events, and two lack such sedimentary features. Cores Sil-I and Sil-H exhibit the five primary chronozones (herein defined as Intermittent Lake I, Lake Mojave I, Intermittent Lake II, Lake Mojave II, and Intermittent Lake III from oldest to youngest) and the secondary chronozone (defined as Incipient Lake) that is stratigraphically below Intermittent Lake I and the oldest chronozone (Fig. 14). Those chronozones that show no signs of drying are Lake Mojave I and Lake Mojave II (Fig. 15); in contrast, Intermittent Lakes I, II, and III show frequent signs of drying and lake lowering conditions (Fig. 15).

The five primary chronozones of pluvial Lake Mojave overlie deposits collectively defined as Incipient Lake (Fig. 14). The Incipient Lake deposits are characterized by many upward-fining sequences (10–20 cm thick), consisting of sand to coarse, silt-sized particles at the base grading to generally nonlaminated, fine, silt- and clay-sized particles at the top. We interpret these upward-fining sequences to represent individual flooding sequences within the Silver Lake depositional basin. These sediments are characteristically gray to pale green in color indicating reducing depositional conditions. The top of the Incipient Lake I chronozone is the least distinguishable boundary; these deposits grade upward into the overlying Intermittent Lake I chronozone. Thus, we classify the Incipient Lake I chronozone as secondary because it may not be distinguishable from the overlying chronozone. The Intermittent Lake I chronozone is characterized by nonlaminated to laminated, clay-sized sediments. Less frequent desiccation cracks suggest relatively longer-term lake stands interrupted by short-term drying events. An absence of drying events distinguishes the Lake Mojave I chronozone from the underlying and overlying Intermittent Lake chronozones (Figs. 14 and 15). Diatoms from the Lake Mojave I chronozone suggest generally deeper, less saline water conditions than lake conditions in the younger chronozones (J.P. Bradbury, 1988, personal commun.). Therefore, we conclude that continuous water existed in Silver Lake basin in the areas of Cores Sil-I and Sil-H during the period of Lake Mojave I.

Sediments of the Intermittent Lake II chronozone rest conformably on the Lake Mojave I sediments and exhibit sediment and faunal characteristics similar to those of the Intermittent Lake I chronozone. More mud cracks are found in the sediments of the Intermittent Lake II chronozone than in any other lake phase. One of most extensive drying periods during the history of pluvial Lake Mojave occurred during the deposition of this chrono-

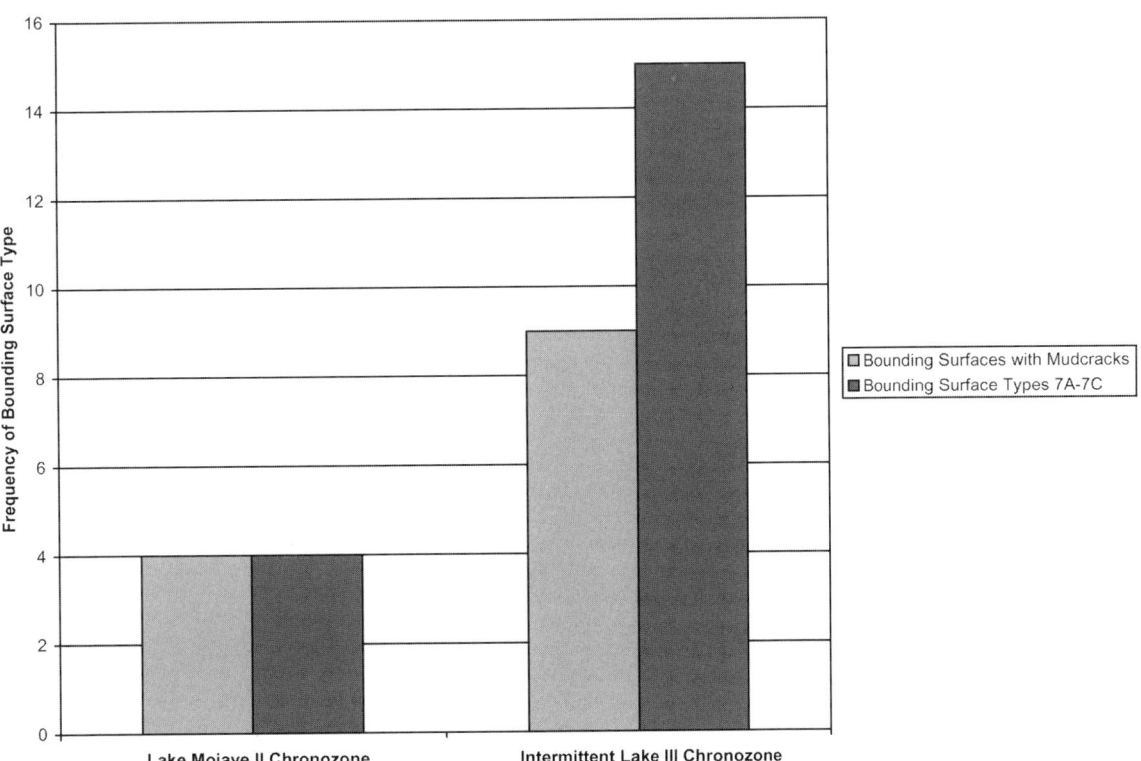

Figure 15. Comparison of bounding surface types in two different phases of pluvial Lake Mojave (Lake Mojave II and Intermittent Lake Phase III). Note significantly lower frequency of bounding surface types which indicates drying or partial drying during Lake Mojave II.

zone. This drying event produced large, distinctive desiccation cracks infilled and overlain with several centimeters of well-sorted, fine-medium eolian sand in most cores from the northern Silver Lake area (e.g., Figure 14).

Overlying the Intermittent Lake II chronozone is the Lake Mojave II chronozone. Major drying events are absent from the Lake Mojave II chronozone, indicating long-term stands of water; however, evidence of numerous partial drying or lake lowering events are found, especially in the lake-marginal cores (Brown, 1989).

Overlying the sediments of Lake Mojave II is the last major lacustrine chronozone, Intermittent Lake III. Sediments of the Intermittent Lake III chronozone are similar in nature to the sediments of Lake Mojave II but contain mud-crack and other desiccation features. Numerous desiccation cracks infilled with eolian and fluvial sands mark the end of pluvial Lake Mojave and the Intermittent Lake III chronozone as well as establish the beginning of the playa conditions, which have persisted to the present.

Thermophilic ostracodes and diatoms in the sediments from these cores and in shoreline features indicate that lake water during the Lake Mojave I and Intermittent Lake I chronozones was generally warmer (especially during the winter months) and less turbid or less saline than lake waters of the three younger lacustrine chronozones (R. Forester and J.P. Bradbury, 1987, personal commun.). The presence of these ostracodes in the Lake Mojave I and Intermittent Lake I phases does not necessarily indicate warmer climatic conditions but may reflect the greater buffering effect on seasonal water temperature fluctuations in a larger, deeper lake system.

Radiocarbon dates were obtained from three of the chronozones (Fig. 14; Table 1). The oldest date (20,320 ± 740 yr B.P.) comes from a depth of ~15.9 m below the playa surface in core Sil-I within the basal part of the Intermittent Lake I chronozone. Also from core Sil-I and at a depth of 9.8 m is a radiocarbon date of 14,660 ± 260 yr B.P. from the upper section of the Intermittent Lake II chronozone. We obtained a radiocarbon date of 9330 ± 95 yr B.P. from core Sil-M at a depth of slightly >3 m below the playa surface (Fig. 7). This age was obtained from sediment in that core that is just below the boundary of the Intermittent Lake III chronozone and the overlying, post-Lake Mojave chronostratigraphic unit; the relative stratigraphic position of this age is illustrated in core Sil-H in Figure 14 although the dated sample was not taken from this core. This date closely limits the end of pluvial Lake Mojave (Fig. 14). Radiocarbon dates from the cores in northern Silver Lake basin indicate that the five primary chronozones forming the Lake Mojave chronostratigraphic unit were deposited during a period lasting at least 11,000–12,000 yr.

DISCUSSION

Age estimates and sedimentation rates of pluvial Lake Mojave

Radiocarbon dating of shell material and tufa from shoreline features and paleoenvironmental reconstructions suggest that pluvial Lake Mojave existed in the Silver Lake basin prior to ca. 16 ka and ended between ca. 8.3 and 9.1 ka. Radiocarbon dates from core samples allow us to refine this timing. Organic materials from Core Sil-I at depths of 9.8 m and 15.9 m below the playa floor were dated at 14,660 ± 260 yr B.P. and 20,320 ± 740 yr B.P., respectively (Fig. 14; GSA Data Repository item C). A radiocarbon age of 9330 ± 95 yr B.P. was obtained from samples at a depth of 3 m in Core Sil-M (its relative stratigraphic position is illustrated in Core Sil-H, Figure 14). These ages are consistent with an independently developed chronology of pluvial Lake Mojave determined from the various beach ridges and wavecut shorelines around the lake (Figs. 7, 8, 11, and 12). The validity of a third radiocarbon date (14,200 ± 145 yr B.P.), taken from core Sil-I at a depth of ~6 m, is considered suspect because its stratigraphic position is very close to the final desiccation of pluvial Lake Mojave, which has been independently dated from samples in the shoreline environments at ca. 8.7 ka and in the subsurface at ca. 9.3 ka. Thus, the radiocarbon age from core Sil-I contradicts a chronology of other dates obtained from the lake margins and subsurface, and we have chosen not to incorporate this radiocarbon date into our studies.

Using the stratigraphic separation between the two radiocarbon ages obtained from a single core (core Sil-I, a vertical column of sediment), we calculate an average, long-term sedimentation rate of 1.08 m/1000 yr. We acknowledge the limitations of our age estimates derived from this method (e.g., uncalibrated versus calibrated radiocarbon dates; boundaries between layers reflecting a hiatus in the sedimentation rate; and variable processes over time such as flooding, lake lowering, and total drying events). The average sedimentation rate, however, is based on data derived from core Sil-I near the deepest part of Silver Lake basin (Figs. 13, 14). During this study, Sil-I has been used as a baseline to relate all other cores drilled in the lake, and core recovery during drilling of Sil-I was nearly 100%.

Application of the 1.08 m/1000 yr rate to the sediment column above these dates yields an estimated age of 8.7 ka for termination of pluvial Lake Mojave. When this age is compared to the Lake Mojave termination based on the radiocarbon age of tufa (a deposit that does not always yield reliable radiocarbon ages) from the Silver Lake quarry (Fig. 12; Table 1; Ore and Warren, 1971; Wells et al., 1987), these two ages differ by only 0.2%.

Allowing a test of the value of the average sedimentation rate derived from core Sil-I and supporting the concordance of these two independently derived age estimates is a radiocarbon age derived from bulk sediments between 3.1 and 3.2 m below the playa surface in Core Sil-M (Figs. 3, 14) (Enzel et al., 1992; Enzel and Wells, 1997). The age of 9330 ± 95 yr B.P. was obtained just below the bounding surface that reflects the end of pluvial Lake Mojave (Table 1; GSA Data Repository item C). Combining this radiocarbon date and its associated depth below the surface with the two radiocarbon dates from core Sil-I (14,660 ± 260 yr B.P. and 20,320 ± 740 yr B.P.) at depths of 9.8 m and 15.9 m, an estimated sedimentation rate of 1.16 m/1000 yr is calculated. This value, calculated from three radiocarbon samples in two different cores, is similar to the rate of 1.08 m/1000 yr calculated using two dates from the same core. In addition, the radiocarbon age from core Sil-M is similar to the ages derived from shells (9390 ± 140 yr B.P.; 9340 ± 140 yr B.P.) in the topographically lower portions of the El Capitan beach ridge and tufa in Silver Lake quarry (9160 ± 400 yr B.P.; Figure 11).

Radiocarbon ages within the youngest shoreline deposits of pluvial Lake Mojave (some ages on tufa are as young as 8350 ± 300 yr B.P. and others on shells are 9340 ± 140 yr B.P.) vary with the type of sample and from depositional site to depositional site in response to natural variations in lacustrine processes and sediment preservation. The radiocarbon date from the youngest deposits in one core is 9330 ± 95 yr B.P., which may differ from another core similar to shoreline environments and thus explain the 600 year difference estimated by the average sedimentation rate. We infer that the subsurface environment provides the most reliable estimate of when pluvial Lake Mojave ended, which we estimate to have occurred between 8.7 ka and 9.3 ka.

The similarity between surface and subsurface ages as well as tufa and shell radiocarbon dates suggests that use of an average sedimentation rate for age estimations of lacustrine phases is reasonable. Using age estimation by sedimentation rates, the earliest lake sediments in the Silver Lake basin are inferred to be ca. 22 ka. This age is consistent with the stratigraphic relations observed in the Baker Dump quarry (Wells et al., 1989; GSA Data Repository item B). The earliest deposits of pluvial Lake Mojave at this site cover a sequence of alluvial fan deposits overlying eolian deposits; the eolian deposits are inferred to be 22–23 ka based upon correlation with results from Lancaster and Tchakerian (this volume).

We use the average sedimentation rate to estimate the age of the Lake Mojave sedimentary package and to provide a more refined estimate of the hydrologic conditions leading to the layering and bounding surfaces in the cores. We recognize, however, that both of these sedimentary properties as well as the sedimentation rates will vary with (1) the pre-Lake Mojave topographic conditions, (2) the core location within the depositional basin that impacts drying events (Fig. 13; GSA Data Repository items C and D), and (3) climatic conditions and erosion rates in the Mojave River headwaters. Because cores Sil-H and Sil-I are near the basin axis and had little topographic relief (Fig. 13), their sedimentation rates were probably very similar (GSA Data Repository items C and D). Thus, we infer that the entire depositional history of Lake Mojave in the Silver Lake basin is recorded within cores Sil-H and Sil-I. As the Silver Lake basin progressively filled with sediment, gradients between high and low topography became subdued and geographic variations in sedimentation rate probably became comparable for the later stages of basin filling (Fig. 13).

Figure 16 is a time-series plot of sedimentation processes including thickness between bounding surfaces, frequency of bounding surfaces, and cumulative sediment thickness during the formation of Lake Mojave's five chronozones. In Figure 16, cumulative sediment thickness is plotted on the y-axis, and time in radiocarbon years is plotted on the x-axis. The time scale is based upon the average sedimentation rate calculated from core

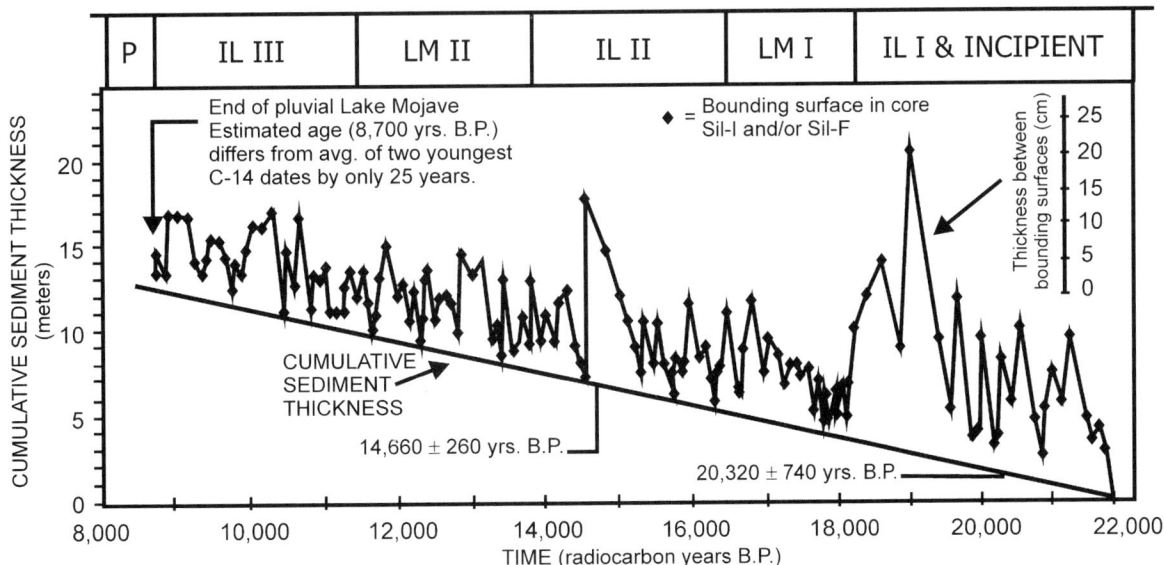

Figure 16. Reconstructed chronology of Lake Mojave sedimentation based on sediment characteristics from Silver Lake cores Sil-I, Sil-F and Sil-H (see Fig. 7 for locations) showing cumulative thickness over time, major bounding surfaces, and thickness of sediments between bounding surfaces. Chronology based on average sedimentation rates calculated from accelerator mass spectrometry core dates and correlations of subsurface stratigraphy with dated shoreline features (see text for further description). Timing of lake phases are shown above the cumulative thickness plot. IL I—Intermittent Lake I; LM I—Lake Mojave I; IL II—Intermittent Lake II; LM II—Lake Mojave II; IL III—Intermittent Lake III; P—Holocene playa.

Sil-I. The average accumulation rate of sediment within Silver Lake basin at the core Sil-I location is shown by the straight line increasing from 0 m at the base of the Intermittent Lake I (and Incipient Lake) chronozone to ~13 m at the top of the Intermittent Lake III chronozone. The frequency of boundary surfaces preserved in the combined records of cores Sil-I and Sil-F are shown by the black diamond symbols (Fig. 16). The frequency of these surfaces illustrates the number of times that significant sedimentologic changes occurred within the record. These changes include (1) change from low or dry conditions to standing water; (2) change in the depositional process during standing water conditions; and (3) flooding events. The thicknesses of sediment accumulation between bounding surfaces was measured vertically from the cumulative sediment line, using the scale in the upper right-hand portion of Figure 16. That is, the time-series plot has been "tilted" to parallel the accumulation of sediment.

The Lake Mojave I and Lake Mojave II chronozones display more frequent, more regularly spaced bounding surfaces whereas the Intermittent Lakes I, II, and III chronozones show fewer surfaces and greater variability in the stratigraphic distance between the surfaces. These patterns suggest that the processes producing the two types of Lake Mojave chronozones were more consistent in frequency and amount of water and sediment input into the Silver Lake basin. Figure 16 also graphically shows that the average sedimentation rate does not fully explain the episodic nature and variability of sedimentation processes in Silver Lake basin during its occupation by pluvial Lake Mojave.

Latest Pleistocene Mojave River floods and sedimentation in the Silver Lake basin

Pluvial Lake Mojave consisted of two major depositional basins the configuration of which influenced sedimentary processes at the terminus of the ancestral Mojave River and consequently the sedimentologic record of these basins during the late Pleistocene (Figs. 3, 4, 5). During the deposition of the Lake Mojave chronostratigraphic unit, the latest Pleistocene Mojave River fan delta prograded eastward from the mouth of Afton Canyon (Fig. 1). Lake Mojave extended ~40 km northward from the terminal reaches of the fan delta to the spillway at the northern end of Silver Lake basin.

The boundary surfaces identified in the cores represent a sedimentologic change. We interpret these changes to indicate either (1) flooding of this basin to produce standing water after low or dry conditions or (2) flooding into a body of standing water that created a silt-rich plume containing phlogopite. In both cases, we infer that large floods of the Mojave River produced these sedimentologic changes. The northernmost location of the cores within Silver Lake basin is the most geographically distant point from the fan-delta complex of the Mojave River where water and sediment first entered Lake Mojave within Soda Lake basin. Consequently, when pluvial Lake Mojave was low (e.g., Intermittent Lake I, II, and III chronozones), thicker accumulation of sediment without any significant sedimentologic change occurred in the most northern, or distal, portions of Silver Lake basin (Fig. 16) because the effects of flood discharge and sedimentation were mostly confined to Soda Lake basin. During higher stands of pluvial Lake Mojave, accumulation of sediment between bounding surfaces in northern Silver Lake basin was thinner, and bounding surfaces reflect frequent flood input (Fig. 16). The sequence of these floods as recorded in the stratigraphic record are illustrated in cores Sil-I and Sil-H of Figure 14.

Variations in hydrologic conditions and lake levels in Silver Lake basin during the latest Pleistocene

Comparison of the sedimentology of individual cores also was used to reconstruct basin-wide depositional environments and, in turn, to infer the major hydrologic conditions during pluvial Lake Mojave. Several characteristic features that record hydrologic conditions, varying from floods to desiccation, are observed in all cores and were used to reconstruct lake levels in Silver Lake basin. Because correlated sedimentary layers from individual core locations in Silver Lake basin represent different environments in the lake at a single point in time, hydrologic conditions (i.e., flooding, partial drying or lake lowering, and total drying) have been recorded differently in the various core locations. For example, a total drying event in core Sil-E may be recorded as a partial drying or lake lowering event in core Sil-F and not at all in core Sil-I, respectively located from the shore margin to the basin axis (see Figures 7 and 13 for geographic locations). Thus, environments proximal to the shore record a greater number of lake fluctuations than environments proximal to the basin center during the same period of time (GSA Data Repository item D).

In order to quantify hydrologic conditions across Silver lake basin during the latest Pleistocene, we defined five relative lake levels based upon a comparison of hydrologic conditions (as interpreted from sediments and bounding surfaces) at core sites Sil-I and Sil-F (Fig. 17). The chronology of these relative lake-level types is based upon age estimates calculated from average sedimentation rates (Fig. 17). Lake-level type 5 represents high lake water conditions throughout Silver Lake basin, and most likely the Soda Lake basin lacks any evidence of drying or partial drying. Such conditions probably resulted in spillway overflow. Quantitative constraints on the water levels, however, are not available on this scale other than maximums represented by high shorelines. Lake-level type 4 indicates partial drying or other lake lowering events near the lake margin within the Silver Lake basin (i.e., reduction of lake volume). Type 3 represents partial drying or other lake lowering conditions near the center of pluvial Lake Mojave in Silver Lake basin. Type 2 involves total drying along the lake margin in Silver Lake basin, whereas Type 1 represents total drying of Silver Lake basin. Because Soda Lake basin is deeper than Silver Lake basin and the two are separated by a sill (Fig. 4), the paleohydrologic conditions associated with Types 1–4 may not necessarily apply to the southern part of pluvial Lake Mojave in Soda Lake basin.

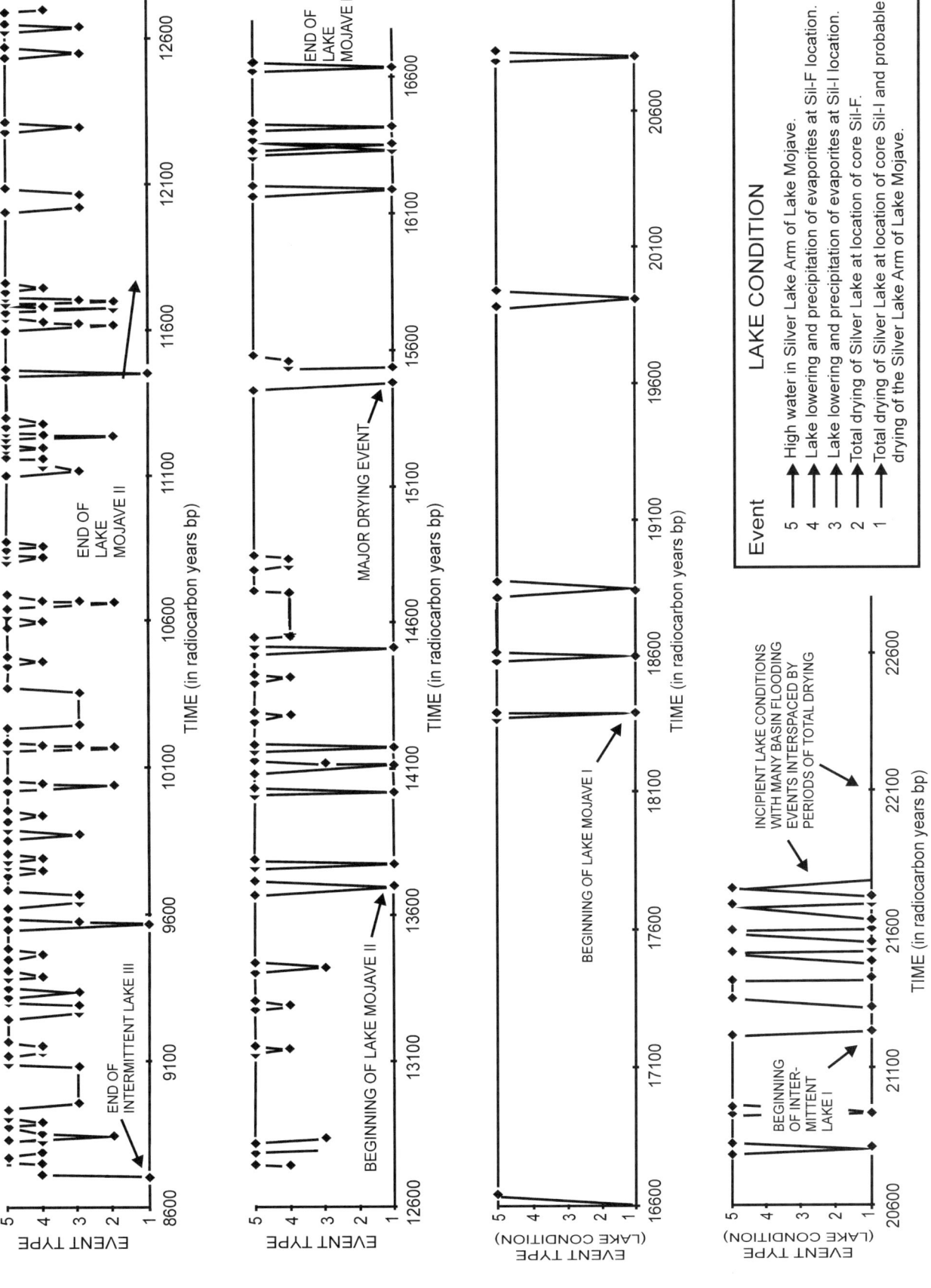

Figure 17. Reconstructed hydrology of Lake Mojave based on five types of lake levels inferred from comparison of stratigraphy of cores Sil-I and Sil-F located at the northern end of Silver Lake depositional basin (modified from Brown, 1989).

Based on the variation and timing of lake levels, pluvial Lake Mojave can be divided into the following six major lacustrine phases: Incipient Lake, Intermittent Lake I, Lake Mojave I, Intermittent Lake II, Lake Mojave II, and Intermittent Lake III (Fig. 14). Five of these lacustrine phases correlate with the chronozones discussed above. The sedimentology of the Incipient Lake phase suggests that waters from the Mojave River inundated Silver Lake basin in an episodic yet progressive style. There is no evidence in the subsurface that catastrophic processes rapidly inundated the basin. Thus, there is no subsurface evidence to support the hypothesis that Lake Manix drained catastrophically into Lake Mojave (Meek, 1989).

Cores Sil-J and Sil-F (Fig. 7) contain sediments similar in nature to those found in Sil-I and Sil-H (GSA Data Repository items C and D). Significantly more drying events (desiccation cracks) and partial-drying lake-lowering events, however, are evident in cores Sil-J and Sil-F, supporting the interpretation that these cores are proximal to the paleo-shoreline and represent shallower conditions than indicated by Sil-I or Sil-H (Fig. 13). Cores near the paleo-shoreline record smaller but more frequent fluctuations in this lake system. Core Sil-F, when compared to core Sil-I, shows the most detailed record of pluvial Lake Mojave fluctuations of any core drilled in Silver Lake. Exceptional preservation of $Na_2SO_4/Na_2SO_4 \cdot 10\ H_2O$ (partial drying or lake lowering) bands and desiccation cracks in Core Sil-F indicate significant and rapid fluctuations in lake levels during the Intermittent Lake III phase.

Using the chronology established by the average sedimentation rate and the variations in sedimentologic and hydrologic process (Figs. 16 and 17), the approximate ages of the lake phases (or chronozones) are estimated as follows:
- Incipient and Intermittent Lake I = ca. 22.6 ka to 18.4 ka
- Mojave I = 18.4 ka to 16.6 ka
- Intermittent Lake II = 16.6 ka to 13.7 ka
- Mojave II = 13.7 ka to 11.4 ka
- Intermittent III = 11.4 ka to 8.7 ka

Changes in lake volume and storage capacity of Lake Mojave

The sediment accumulated in Silver Lake and Soda Lake basins during the history of pluvial Lake Mojave reduced the storage capacity and impacted the hydrology of both basins (Figs. 4, 5, and 6). Estimates of potential lake volume and storage capacity in the basins are based on spatial relationships identified in drill cores and rotary holes. These data indicate that at the beginning of pluvial Lake Mojave, both Silver Lake and Soda Lake basins would have held >7 km³ of water before overflowing the spillway in northern Silver Lake basin. Also, the Soda Lake basin could have contained an incipient Lake Mojave for a substantial period of time before the rising lake level deposited the first lacustrine clays in Silver Lake basin (Figs. 4, 5). By the end of the Intermittent Lake III phase, Silver Lake and Soda Lake basins contained <40% of their original volume. The loss of storage capacity in response to sedimentation, combined with downcutting of the spillway, apparently enhanced the sensitivity of the lake system during the latter phases of its existence. Therefore, the Intermittent Lake III and Lake Mojave II phases were more sensitive to temporarily drier conditions in the upper Mojave basin. This hydrologic phenomenon, and internal response, can explain the increase in partial-drying lake-lowering events observed in the latter phases of Lake Mojave (Fig. 17), as well as the accumulation of salts observed in the younger lake sediments (GSA Data Repository items C and D), without requiring change in an external forcing factor such as regional climate.

The decrease in water storage capacity with time also implies that if climatic (temperature and evaporative) conditions were relatively constant in the terminal basins during the latest Quaternary, then (1) significantly larger volumes of Mojave River discharge would have been required to fill the Lake Mojave I phase to the overflow elevation and (2) larger quantities of lake water would have exited the basin via the spillway during the Lake Mojave II phase.

The location and incised nature of the outflow spillway channel (Plates 1 and 3) imply sustained periods of overflow during which discharge was confined to the overflow channel; field observations support that lake outflow did not occur at any other site forming the northern boundary of Silver Lake basin. It would appear that "typical" lake overflow could easily produce this channel without requiring large-scale, catastrophic flood volumes. We infer that downcutting of the spillway from the A-shoreline (287–288 m) to the B-shoreline (285.5 m) occurred during Lake Mojave II (after 12 ka but prior to 10 ka) when overflow conditions were more easily produced because of decreased lake storage. The timing of the downcutting also is supported by the ages of the different shorelines preserved in the El Capitan complex (Fig. 11).

Correlation of pluvial Lake Mojave shoreline and lacustrine deposits

Within the subsurface environment, the basin axis typically yields a well-preserved and relatively continuous record of depositional history but generates limited data on the absolute magnitude of lake-level fluctuations. Shoreline sediments and landforms, however, yield the best data for reconstructing lake area and lake-level elevations but only produce a discontinuous record. The use of both types of data to correlate between the surface and subsurface environments of paleolakes yields the most complete history (Wohl and Enzel, 1995). In our reconstruction of the detailed geologic history of pluvial Lake Mojave, we correlate the shoreline deposits and core stratigraphy along northern Silver Lake basin. We base our correlations on stratigraphic relations, faunal distribution, radiocarbon dating, and soil geomorphic data. These data indicate that Lake Mojave I sediments were deposited earlier than previously recognized and that during this time, lake levels were at or near the spillway elevation (288–287 m), although little erosion of the spillway occurred. During the subsequent highstand of Lake Mojave II, most of the deposits and faunal remains associated with the older lake phase were eroded. Of the known outcrops of beach deposits around the Lake Mojave basin, only those deposits within Tidewater basin (Figs. 8, 9, 10) appear to record the shoreline levels associated with Lake Mojave I.

Figure 9 shows a correlation between the lacustrine sediments of core Sil-I and Sil-G and the two beach ridges preserved in Tidewater basin (Fig. 7). Samples collected from the clay units and several of the upper sandy units in Core Sil-G have been analyzed by J.P. Bradbury (1988, personal commun.) for diatoms and other microfauna. The lowermost clays in this core contain abundant *Fragilaria construens* v. *subsalina*; and this diatom and others found in Silver Lake cores, as well as within lower portions of lacustrine clays in Soda Lake Core 1, occur only in Lake Mojave I sediments (Wells et al., 1989). Thus, we correlate the basal green clays in Tidewater basin (core Sil-G) with the top of the Lake Mojave I phase in core Sil-I (Fig. 9). In addition, diatom distributions within the fine-grained deposits in the topographically lower beach ridge at Tidewater basin support this correlation (Fig. 8). The upper beach ridge (287.9 m) in the Tidewater basin corresponds to the level of the A-shoreline, and the lower Tidewater basin beach ridge (285.6 m) corresponds to the level of the B-shoreline. No shells have been dated from the upper ridge at this locality. Based on a detailed reconstruction of paleoenvironments in the Tidewater basin (Fig. 10), however, the 287–288 m beach ridge (A-shoreline) predated constuction of a topographically lower spit that formed the core and platform for the construction of the 285–286 m shoreline (B-shoreline) (Fig. 8). Thus, the A-shoreline marks the highest level of both Lake Mojave I and Lake Mojave II. Only during the later phases of Lake Mojave II, or during Intermittent Lake III and after downcutting of the overflow spillway, was the B-shoreline stabilized over a much older spit (Plates 1, 3). These relations support our interpretation that the A-shoreline was occupied during both Lake Mojave I and Lake Mojave II.

Shell dates of 14,550 ± 140 yr B.P. and 15,350 ± 240 yr B.P. (Ore and Warren, 1971) from buried sandy gravels in the 285.6 m beach ridge at Tidewater basin indicate that they are correlative, in part, to Intermittent Lake II. Stratigraphically below the above dated sediments are two green, finer-grained units containing *L. bradburyi* and *L. ceriotuberosa* (units IV and VI, Figure 8; Table 2). The older fine-grained unit contains shells dated at 16,270 ± 310 yr B.P., an age that is close to the age of the boundary between Lake Mojave I and Intermittent Lake II given the errors in dating and estimating based upon sedimentation rates. These data suggest that a highstand in lake level (288 m) and overflow at the spillway occurred during the latest stages of Lake Mojave I or the earliest stage of Intermittent Lake II. Because the majority of these early shoreline features at 288–287 m were eroded or buried by the later highstands of the lake, perhaps this early highstand did not last very long and therefore did not produce significant shoreline deposits. In the Tidewater basin, the youngest shoreline deposits (Intermittent Lake III phase) of the pluvial Lake Mojave complex are represented by the uppermost unit in the B-shoreline ridge (unit X, Figure 8; Table 2), which unconformably overlies older units.

An unconformity is recognized in both cores Sil-I and Sil-G by the presence of eolian sand filling and covering mud cracks at a major bounding surface (Figs. 9, 14). Based upon sedimentation rates (Fig. 17), the approximate age for this eolian deposit is 15.5 ka. These features indicate a major drying event during Intermittent Lake II (Figs. 9, 10, and 17). The presence of eolian sands on the lake basin floor strongly suggests that Lake Mojave in the Silver Lake basin was completely desiccated at this time (15.5 ka). Within the deposits of the 285–286 m beach ridge in the Tidewater basin is an unconformity within unit VII that is dated between 15.4 ka and 14.6 ka (Fig. 8). This unconformity indicates at least a lowering of the lake level below 283 m, followed by a rise in the lake level. We correlate this unconformity with the major drying event recorded in cores Sil-I and Sil-G (Fig. 9).

Water balance of pluvial Lake Mojave

Analysis and correlation of the surface and subsurface environments of pluvial Lake Mojave yield a detailed reconstruction of the lake-level elevation history as influenced by the discharge and floods of the Mojave River. In this section, we provide our interpretations of the hydrologic conditions of the Mojave River watershed that created volumes of water to sustain pluvial Lake Mojave, including overflow at the spillway in Silver Lake basin (Fig. 3).

The water balance calculation presented below is based on a simplified precipitation-discharge/evaporation model (see references in Enzel, 1992). The goal of these calculations is not to detect the climatic conditions of the past but to present several scenarios demonstrating the general changes and order of magnitude of changes that must occur in the hydrology to fill Lake Mojave I to its overflow spillway (i.e., last glacial maximum conditions). The model is as follows:

$$Ve = D + P(R) - E$$

where Ve equals equilibrium volume at a given lake elevation, D equals annual Mojave River discharge at Afton Canyon (Fig. 1), R equals runoff contribution from rainfall (P) over the basin in addition to river discharge, and E equals annual evaporation. For modern data, see Enzel (1990, 1992). Our assumptions include the following: (1) groundwater inflow, leakage out, and overflow were negligible; and (2) the calculated Ve is probably a minimal estimate. Afton Canyon was selected for the site to model the annual discharge input from the Mojave River as the river fed directly into the Soda Lake basin of Lake Mojave at the eastern end of the canyon.

The Mojave fluvial-lacustrine system is characterized by a high mountain catchment within the San Bernardino Mountains that generates high magnitude runoff events in a very different climatic zone than the desert catchments immediately surrounding the pluvial lake (Enzel, 1990). During late Pleistocene climate changes, we assumed differing types of climate and runoff changes for each area (e.g., contributing catchment and the area directly contributing into the lake). For example, in the present climatic regime, a relatively large portion of rainfall is transformed into runoff at the headwaters; whereas, the runoff from the desert catchments surrounding the two playas is very low and can be considered negligible. As a result, present-day potential lake evaporation is much greater than the local runoff entering the lake (Enzel, 1992). In the late Pleistocene, Lake Mojave reached an evaporative surface area of nearly 300 km². We assume that the main sources of water input to pluvial Lake Mojave were Mojave River discharge from its mountainous catchment, relatively minor local runoff from the desert catchments feeding directly into the lake basins, and direct rain over the lake. Our efforts are directed at estimating the main input source and roughly estimating the volume of river discharge into the lake. We assume several evaporation and precipitation scenarios and demonstrate the magnitude of change needed to produce Lake Mojave I at the spillway elevation.

We combined elevation-area-volume curves (Fig. 6) with assumed values for possible precipitation and evaporation to demonstrate the magnitude of change needed to produce Lake Mojave I overflow at the spillway elevation (Fig. 18; Table 5). The assumed values in the lake water budget calculations are based upon three different scenarios (Table 5). Our calculations indicate that whatever scenario is chosen to produce a significant body of water in Soda Lake and Silver Lake basins demands a significant increase in Mojave River discharge at Afton Canyon. A 50% reduction in evaporation over the lake still requires more than an order of magnitude increase in annual river discharge for the lake level to reach the spillway. This 50% evaporation reduction is larger than any values inferred for the American Southwest evaporation reductions during the latest Pleistocene (for summary of sources, see Smith and Street-Perrott, 1983).

We stress that even greater reductions in lake evaporation would not affect our conclusion about the required dramatic increase in Mojave River discharge. We also stress that the assumed values for the climate-related variables are only illustrative; each assumption could be challenged and the value could be refined. We believe that the values used, however, when combined with the elevation of the modern playa and shorelines as well as the bathymetry of Lake Mojave, are adequate for estimating the relative importance of climatic and hydrologic parameters in producing observed responses in the terminal basins.

Given the pre-Lake Mojave basin geometry (Fig. 5) and assuming an increase in annual discharge of the Mojave River at Afton Canyon, modern values of hydroclimatic parameters would produce only a very shallow lake (i.e., playa conditions) on the floor of Soda Lake basin at an elevation between 245 and 250 m (see scenario D, Fig. 18). We believe that a lake cannot be produced in the terminal basins of the Mojave River (with the Lake Mojave configuration [overflow, specific geometry, etc.]) without an order of magnitude increase in river discharge resulting from the mountainous catchment (see highest values for parameters in scenarios A and B, Fig. 18; Table 5). This amount of discharge can be achieved by increasing the number of storms that affect the Mojave River headwaters to produce large flow volumes at the lower Mojave River reaches.

Currently, the Mojave River experiences large transmission losses into its alluvial aquifer (Enzel et al., 1989; Enzel, 1992; Enzel and Wells, 1997), and only a small portion of the largest floods are able to reach the terminal playas. We assume that during the late Pleistocene, with increased Mojave River discharge, the transmission losses were dramatically reduced and smaller floods could have reached the lake. As shown above, the stratigraphy observed in the cores supports a flood-related sedimentology. Currently, discharge of the largest 10% of the floods in the Mojave River headwaters (Fig. 1) exceeds 250 million m³ (Enzel, 1992). Assuming (1) no loss along the length of the Mojave River on its route to Soda Lake basin, (2) an annual occurrence of these flood discharges, and (3) a 40–50% reduction in lake evaporation, a lake will be produced with sufficient volumes to create overflow at the spillway (see scenario A, Figure 18). If we assume a tenfold increased frequency of modern extreme flood events within the San Bernardino Mountains resulting in floodwaters that reach Soda Lake and Silver Lake basins, reduced evaporation alone will result in an overflowing lake at the spillway. We believe that the most likely scenarios have been parameters that lie between A (+50% P, ~ −50% E, and about three times the value of modern extreme floods that make it annually to Afton Canyon, Figure 1) and B (+100% P, ~ −50% E, and about twice the modern values of extreme floods that make it annually to Afton Canyon) (Fig. 18).

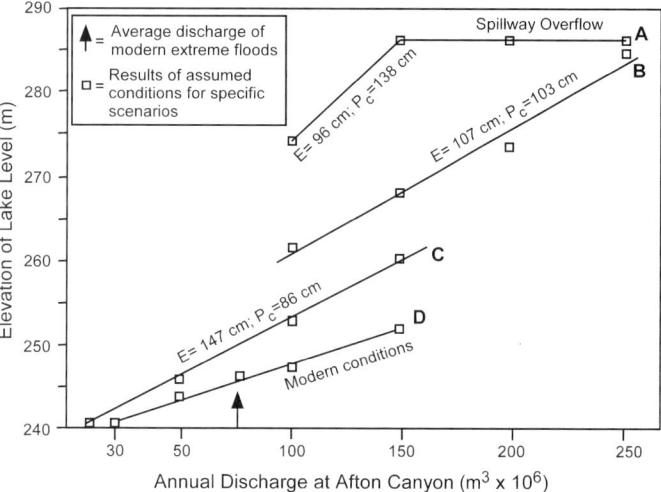

Figure 18. Modeling results of four different hydrologic conditions and associated climatic scenarios (A–D) resulting in the flooding of the terminal basins (D—modern conditions) to filling of pluvial Lake Mojave to the A-shoreline elevation (A—spillway overflow). Evaporation (E) is assumed for the topographically lower elevations of ancient Lake Mojave area, and precipitation (Pc) is assumed for the topographically higher catchment area in the San Bernardino Mountains (see Fig. 1 for locations).

TABLE 5. OBSERVED AND ASSUMED CLIMATIC AND HYDROLOGIC PARAMETERS
FOR TWO REGIONS WITHIN THE MOJAVE RIVER WATERSHED

	Modern conditions		Assumed condition		
	Annual average	Extreme storms	Pc + 25%	Pc + 50%	Pc + 100%
A. Headwater catchment area					
Precipitation (cm)	69	86	103	138	
Runoff (%)	34	38	43	51	63
Discharge at Afton ($m^3 \times 10^6$)	9.4	75	84	150	250
	Modern conditions		Assumed conditions*		
	Annual average				
B. Lake area					
Precipitation (cm)	10		15	20	30
Runoff (%)	0–1		0	12	8
Evaporation (cm)	203		147	107	96

Note: Modern extreme storms and/or floods are summarized in Enzel (1990, 1992) and Enzel and Wells (1997). Pc—assumed precipitation in the headwater catchment of the San Bernardino Mountains, and Pc + 25% = 25% increase in the annual average precipitation in the headwater catchment.
  A. The topographically higher headwater catchment area in the San Bernardino Mountains.
  B. The lake and surrounding topographically lower catchments within the Mojave River watershed.
  *assumed conditions that are independently varied for each parameter (precipitation, runoff, and evaporation) in the lake area.

CONCLUSIONS

Late Quaternary geologic and hydrologic history of pluvial Lake Mojave

Our reconstruction of the late Quaternary history of Lake Mojave is derived from subsurface and surface geology within Silver Lake basin. This history is shown in Figure 19, illustrating the timing of changes in lake levels and storage capacity, formation and modification of highstand shorelines, and the geomorphic processes operating peripherally to the lake basin. Details on the late Quaternary geomorphic history of the piedmonts surrounding Lake Mojave are provided by Wells et al. (1987), Lancaster and Tchakerian (this volume), McDonald et al. (this volume), and Harvey and Wells (this volume).

From its beginning in the latest Pleistocene, Lake Mojave was the second of two large desert lakes fed by Mojave River stream discharge (Enzel et al., this volume). An abandoned spillway south of present-day Afton Canyon probably provided Lake Mojave with Lake Manix overflow during this period (Wells and Enzel, 1994). Lake Mojave, sustained by the ancestral Mojave River, experienced a significant reduction in total lake volume and surface area in response to the draining of Lake Manix. Meek (1989) estimated the surface area of Lake Manix prior to draining at 215 km². Estimates of the surface area and water storage volume of Lake Mojave have been based on subsurface core and drill hole data as well as surficial mapping of shoreline features. Initially, Lake Mojave would have been able to hold ~7 km³ of water with a surface area of ~290 km² at the A-shoreline elevation (Brown, 1989). With the incision of Afton Canyon and draining of Lake Manix, the Mojave River flowed directly into Lake Mojave, and the total lake surface area sustained by the Mojave River may have decreased from ~500 km² to ~290 km². The decrease in storage capacity within the Silver Lake basin during the latest Quaternary is illustrated in Figure 19. The amount of lake water lost through evaporation would have been reduced drastically, and a much greater portion of water would have overflowed pluvial Lake Mojave draining north toward Death Valley (Fig. 1).

Based upon sedimentation rates, we estimate that Lake Mojave waters first reached Silver Lake basin circa 22.6 ka. In that Soda Lake basin is deeper and closer to the input point of Mojave River water, it is probable that Lake Mojave existed in Soda Lake basin long before reaching Silver Lake basin. Unfortunately, the lack of reliable core data limits our ability to estimate when Lake Mojave first occupied Soda Lake basin. Wells and Enzel (1994) infer that Lake Manix waters spilled into Soda Lake basin prior to any significant downcutting of Afton Canyon (Fig. 1). Incipient lake conditions in Silver Lake basin lasted until ca. 21 ka when the sedimentology of the cores support formation of Lake Mojave phases (and chronozones) with more variable hydrologic conditions (Intermittent Lakes I, II, and III) and those with high lake stands and a paucity of drying events (Lake Mojave I and Lake Mojave II).

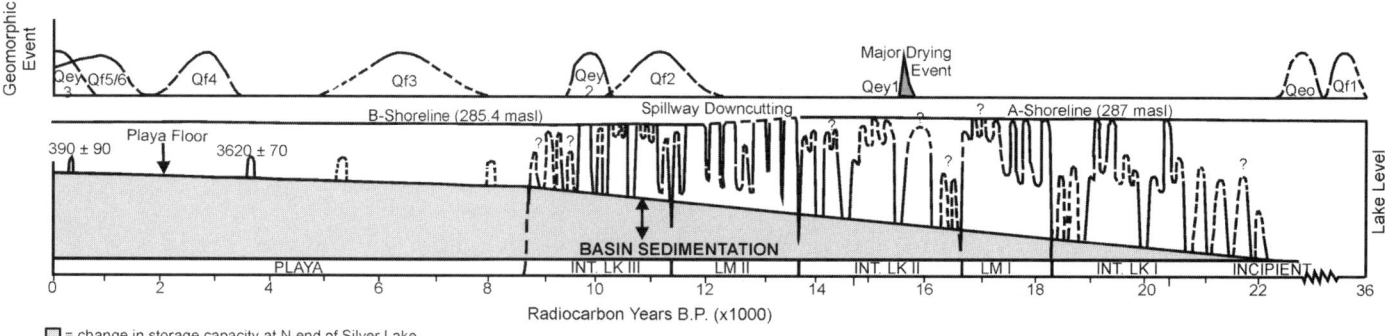

Figure 19. Simplified history of pluvial Lake Mojave fluctuations and piedmont depositional history in the Silver Lake depositional basin during the last 36,000 years showing change in storage capacity in the Silver Lake basin, periods of pluvial lake phases, and major drying events.

With the beginning of Intermittent Lake I, a series of lakes formed and lasted from a few hundred years up to thousands of years in Silver Lake basin between ca. 21.1 ka and 18.4 ka (Figs. 17, 19). During their existence, these lakes experienced no partial drying or total drying events. Circa 18.4 ka, an increased frequency and magnitude of large-scale Mojave River flooding events may have produced more frequent overflow of Lake Manix and the first prolonged lake stands with no drying events (Lake Mojave I) in Silver Lake basin. This stand persisted until ca. 16.6 ka, and only one shoreline feature at 287–288 m (A-shoreline, Tidewater basin; Figures 9, 10) may have formed during this stand.

Transition to a drier climatic regime led to decreased annual discharge of the Mojave River. Reduced frequency of large-magnitude flood events and several drying events in Silver Lake basin occurred between 16.6 ka and 13.7 ka (Intermittent Lake II). During this time, the most extensive, and perhaps prolonged, drought in the history of Lake Mojave occurred (ca. 15.5 ka). Deep subaerial mud cracks formed, and widespread eolian sedimentation occurred on the exposed lake floor and adjoining piedmonts. In addition, ostracodes and diatoms from sediments of Intermittent Lake II and younger phases indicate seasonally colder water conditions coupled with relatively greater turbidity and salinity than existed during Lake Mojave I or older phases (R.M. Forester, 1987, personal commun.; J.P. Bradbury, 1987, personal commun.).

The Lake Mojave II phase began at ca. 13.7 ka. This lake phase lasted ~2300 yr and may have coincided with the final stages of the draining of Lake Manix and increased Mojave River discharge resulting from higher magnitude, more frequent, large-scale flooding events. During this time, Lake Mojave repeatedly reached the elevation of the A-shoreline (~287 m), likely reworking shoreline sediments of Lake Mojave I. Partial drying events were more frequent in Silver Lake basin during Lake Mojave II than Lake Mojave I due to the reduction of storage capacity by sediment infilling and Mojave River delta progradation. Such conditions aided in the modification of older shoreline deposits, increased the frequency of overflow conditions at the spillway, and resulted in downcutting of the spillway to an elevation of ~285.5 m and formation of the B-shoreline. Radiocarbon dated shorelines in the El Capitan beach complex in northern Silver Lake indicate that downcutting of the outlet spillway by 1.5 m probably occurred after 12 ka and before 10 ka. The maximum expansion of shallow lakes in Death Valley occurred during this same period (D. Anderson and Wells, this volume), partially in response to Lake Mojave overflow.

Increasing loss of lake volume due to increased sediment storage combined with decreasing Mojave River discharge resulted in a more sensitive hydrologic system during Intermittent Lake III, which began ~11,400 yr ago. This final phase of Pleistocene Lake Mojave was characterized by steadily deteriorating lake conditions until the eventual transition to the prolonged playa environment of post-Lake Mojave. During the early Holocene, eolian processes reworked near-shore sediment, mantling the beach ridges and other shoreline features (Fig. 19) (McFadden et al., 1992).

SUMMARY POINTS

In summary, we offer the following points as key observations and interpretations on the research approach and results of this study:

1. Late Quaternary sedimentary sequences within the depositional basins associated with Silver Lake and Soda Lake playas can be divided into three chronostratigraphic units: pre-Lake Mojave, Lake Mojave, and post-Lake Mojave.

2. Sediments from the Silver Lake cores yield a complex, but detailed, stratigraphy from which conditions of lake flooding, lake lowering, partial drying, and total drying can be inferred. Based upon the sedimentology of the strata, including layers (three types) and bounding surfaces (12 properties) as well as the vertical and lateral changes in these features, the Lake Mojave chronostratigraphic unit contains two chronozones (Lake Mojave I and Lake Mojave II) that represent high lake stands with few drying or partial drying events and four chronozones (Intermittent Lakes I, II, and III and Incipient Lake) that represent highly variable hydrologic conditions including total lake desiccation.

3. Because the depositional basins of Lake Mojave, especially Silver Lake basin, are shallow and have minimal relief on

the pre-lake surface, sediments in these basins yield a high resolution record.

4. The high resolution record within Silver Lake basin is enhanced by our ability to correlate subsurface stratigraphy with shoreline features and deposits (Fig. 9). Radiocarbon dating is useful for such correlations, but is limited because younger lake events commonly occupied the same shoreline elevations, thus modifying or partially removing older sediments. Event stratigraphic units, such as the 15.5 ka eolian deposit, aid correlation because they were preserved in both surface and subsurface environments.

5. One of the most significant geomorphic events recorded in the cores of the Silver Lake basin indicates a major drought during the Intermittent Lake II phase, during which eolian sediments were deposited directly on the exposed lake floor at 15.5 ka.

6. Using a simplified precipitation-discharge/evaporation model, we infer that the late Pleistocene hydrologic conditions resulting in Lake Mojave overflow at Spillway bay in Silver Lake lie between the following two scenarios (Fig. 18):

• 50% increase in precipitation in the headwater catchment resulting in annual flood events reaching Afton Canyon with discharges three times that of modern extreme floods

• 100% increase in catchment precipitation with a 50% decrease from modern evaporation combined with annual flood events reaching Afton Canyon with discharges two times that of modern extreme floods.

7. The changes that have occurred in Soda Lake and Silver Lake basins due to sedimentation and loss of storage capacity are so profound that if hydrologic conditions change such that another Lake Mojave could form, we believe that the majority of water would pass through the system and flow into Death Valley.

REFERENCES CITED

Amundson, R., Wang, Y., Chadwick, O., Trumbore, S., McFadden, L., McDonald, E., Wells, S., and DeNiro, M., 1994, Factors and processes governing the carbon-14 content of carbonate in desert soils: Earth and Planetary Science Letters, v. 125, p. 385–405.

Barry, R.G., 1983, Late Pleistocene Paleoclimatology, in Wright, H.E., and Porter, S.C., eds., Late Quaternary environments of the United States, Volume 1: The late Pleistocene: Minneapolis, University of Minnesota Press, p. 390–407.

Blackwelder, E., 1954, Pleistocene lakes and drainage in the Mojave region, southern California, in Jahns, R.H., ed., Geology of southern California: Sacramento, California, California Division of Mines, p. 35–40.

Blackwelder, E., and Ellsworth, E.W., 1936, Pleistocene lakes of the Afton basin, California: American Journal of Science, v. 31, p. 453–463.

Bode, F.D., 1937, Geology of Lake Mohave outlet channel, in Crozer-Campbell, E., ed., The archaeology of Pleistocene Lake Mohave: A symposium: Los Angeles, California, Southwest Museum Publishing, Southwest Museum Papers, v. 11, p. 109–118.

Bradley, R.S., 1999, Paleoclimatology—Reconstructing climates of the Quaternary, International Geophysics Series: New York, Academic Press, 613 p.

Brown, W.J., 1989, Late Quaternary stratigraphy, paleohydrology, and geomorphology of pluvial Lake Mojave, Silver Lake and Soda Lake basins, southern California [M.S. thesis]: Albuquerque, University of New Mexico, 266 p.

Brown, W.J., and Rosen, R.M., 1995, Was there a Pliocene-Pleistocene fluvial-lacustrine connection between Death Valley and the Colorado River?: Quaternary Research, v. 43, p. 286–296.

Brown, W.J., Wells, S.G., Enzel, Y., Anderson, R.Y., and McFadden, L.D., 1990, The late Quaternary history of pluvial Lake Mojave: Silver Lake and Soda Lake basins, California, in Reynolds, R.E., et al., eds., At the end of the Mojave: Quaternary studies in the eastern Mojave Desert: Redlands, California, Special Publication of the San Bernardino County Museum Association, 1990 Mojave Desert Quaternary Research Center Symposium, May 18–21, 1990, p. 55–72.

Calzia, J., 1991, Geophysical, lithologic, and water quality data from Soda dry lake, San Bernardino, California: Reston, Virginia, U.S. Geological Survey Open-File Report 91-0266, 1 sheet.

Clarke, M.L., Wintle, A.G., and Lancaster, N., 1996, Infrared stimulated luminescence dating of sands from the Cronese basins, Mojave Desert: Geomorphology, v. 17, no. 1–3, p. 199–206.

Crozer-Campbell, W.E., Campbell, H.W., Antevs, E., Amsden, A.C., Barbierry, A.G., and Bode, D.F., editors, 1937, The archaeology of Pleistocene Lake Mohave: A symposium: Los Angeles, California, Southwest Museum Publishing, Southwest Museum Papers, v. 11, 118 p.

Dickey, S.K., Neimeyer, R.A., and Sholes, R.C., 1979, Soda Lake groundwater investigations—phase II: Southern California Edison Company, 72 p.

Enzel, Y., 1990, Hydrology of a large, closed arid watershed as a basis for paleohydrological and paleoclimatological studies in the Mojave River drainage system, southern California [Ph.D. thesis]: Albuquerque, University of New Mexico, 316 p.

Enzel, Y., 1992, Flood frequency of the Mojave River and formation of late Holocene playa lakes, southern California, USA: The Holocene, v. 2, p. 11–18.

Enzel, Y., and Wells, S.G., 1997, Extracting Holocene paleohydrology and paleoclimatology information from modern extreme flood events: Examples from southern California: Geomorphology, v. 19, no. 3–4, p. 203–226.

Enzel, Y., Brown, W.J., Anderson, R.Y., and Wells, S.G., 1988, Late Pleistocene–early Holocene lake stand events recorded in cored lake deposits and in shore features, eastern Mojave Desert, southern California: Geological Society of America Abstracts with Programs, v. 20, no. 3, p. 158.

Enzel, Y., Cayan, D.R., Anderson, R.Y., and Wells, S.G., 1989, Atmospheric circulation during Holocene lake stands in the Mojave Desert: Evidence of regional climatic change: Nature, v. 341, p. 44–48 and 21.

Enzel, Y., Brown, W.J., Anderson, R.Y., McFadden, L.D., and Wells, S.G., 1992, Short-duration Holocene lakes in the Mojave River drainage basin, southern California: Quaternary Research, v. 38, no. 1, p. 60–73.

Eugster, H.P., and Hardie, L.A., 1975, Sedimentation in an ancient playa-lake complex: The Wilkens Peak Member of the Green River Formation of Wyoming: Geological Society of America Bulletin, v. 86, p. 319–334.

Forester, R.M., 1987, Late Quaternary paleoclimatic records from lacustrine ostracodes, in Ruddimann, W.F., and Wright, H.E., Jr., eds., North America and adjacent oceans during the last deglaciation: Boulder, Colorado, Geological Society of America, The Geology of North America, v. K-3, p. 261–276.

Free, E.E., 1914, The topographic features of the desert basin of the United States with reference to the possible occurrence of potash: U.S. Department of Agriculture Bulletin 54, p. 45.

Glaessner, M.F., 1969, Trace fossils from the Precambrian and basal Cambrian: Lethaia, v. 2., p. 369–393.

Hardie, L.A., Smoot, J.P., and Eugster, H.P., 1978, Saline lakes and their deposits: A sedimentological approach, in Matter, A., and Tucker, M.E., eds., Modern and ancient lake sediments: Oxford, England, Blackwell Scientific Publications, p. 7–41.

Hubbs, C.L., and Miller, R.R., 1948, The Great Basin, II: The zoological evidence: University of Utah Bulletin, v. 38, p. 17–166.

Hubbs, C.L., Bien, G.S., and Suess, H.E., 1965, LaJolla natural radiocarbon measurements: Radiocarbon, v. 7, p. 66–117.

Hunt, C.B., Robinson, T.W., Bowles, W.A., and Washburn, A.I., 1966, Hydrologic basin, Death Valley, California: U.S. Geological Survey Professional Paper 494B.

Huntington, E., 1915, The curtailment of rivers by desiccation: Washington, D.C., Carnegie Institution of Washington, Yearbook for 1915, p. 96.

Kraft, J.C., and Chrzastowski, M.J., 1987, The transgressive barrier-lagoon coast of Delaware: Morphostratigraphy, sedimentary sequences and responses to relative rise in sea level, *in* Nummedal, D., et al., eds., Sea-level fluctuation and coastal evolution: A symposium: Special Publication, Society of Economic Paleontologists and Mineralogists, v. 41, p. 129–145.

Li, J., Lowenstein, T.K., and Blackburn, I.R., 1997, Responses of evaporite mineralogy to inflow water sources and climate during the past 100 k.y. in Death Valley, California: Geological Society of America Bulletin, v. 109, p. 1361–1371.

McFadden, L.D., Wells, S.G., Brown, W.J., and Enzel, Y., 1992, Soil genesis on beach ridges of pluvial Lake Mojave: Implications for Holocene lacustrine and eolian events in the Mojave Desert, southern California: Catena, v. 19, p. 77–97.

McFadden, L.D., Wells, S.G., Brown, W.J., Enzel, Y., Amundson, R., Wang, Y., and Trumbore, S., 1994, Formation and pedogenic isotope studies of soils on beach ridges of Silver Lake playa, Mojave Desert, California, *in* McGill, S.F., and Roos, T.M., eds., Geological investigations of an active margin: Geological Society of America, Cordilleran Section Guidebook, San Bernardino County Museum Association, p. 195–200.

Meek, N., 1989, Geomorphic and hydrologic implications of the rapid incision of Afton Canyon, Mojave Desert, California: Geology, v. 17, p. 7–10.

Meek, N., 1999, New discoveries about the late Wisconsin history of the Mojave River system, *in* Reynolds, R.E., and Reynolds, J., eds., Tracks along the Mojave: Quarterly of the San Bernardino County Museum Association, v. 46, no. 3, p. 113–117.

Moyle, W.R., 1967, Water wells and springs in Soda, Silver and Cronese valleys, San Bernardino County, California: California Department of Water Resources Bulletin 91-13, 16 p.

Muessig, S., White, N.G., and Byers, M.F., 1957, Core logs from Soda Lake, San Bernardino County, California: U.S. Geological Survey Bulletin 1045-C, 96 p.

Nummedal, D., and Swift, D.J.P., 1987, Transgressive stratigraphy at sequence-bounding unconformities: some principles derived from Holocene and Cretaceous examples in sea-level fluctuation and coastal evolution, *in* Nummedal, D., et al., eds., Sea-level fluctuation and coastal evolution: A symposium: Special Publication, Society of Economic Paleontologists and Mineralogists, v. 41, p. 241–260.

Ore, H.T., and Warren, C.N., 1971, Late Pleistocene–early Holocene geomorphic history of Lake Mojave, California: Geological Society of America Bulletin, v. 82, p. 2553–2562.

Orme, A.J., and Orme, A.R., 1991, Relict barrier beaches as paleoenvironmental indicators in the California desert: Physical Geography, v. 12, no. 4, p. 334–346.

Plummer, P.S., and Gostin, V.A., 1981, Shrinkage cracks: Desiccation or synaeresis?: Journal of Sedimentary Petrology, v. 51, no. 4, p. 1147–1156.

Pyke, C.B., 1972, Some meteorological aspects of the seasonal distribution of precipitation in the western United States and Baja California: Los Angeles, California, University of California Water Resources Center Contribution No. 139, 205 p.

Sharp, R.P., Allen, R.C., and Meier, F.M., 1959, Pleistocene glaciers on southern California mountains, American Journal of Science, v. 257, p. 81–94.

Smith, G.I., 1979, Subsurface stratigraphy and geochemistry of late Quaternary evaporates, Searles Lake, California: U.S. Geological Survey Professional Paper 1043, 130 p.

Smith, G.I., and Street-Perrott, F.A., 1983, Pluvial lakes in the western United States, *in* Wright, H.E., and Porter, S.C., eds., Late Quaternary environments of the United States, Volume 1: The late Pleistocene: Minneapolis, University of Minnesota Press, p. 190–214.

Smoot, J.P., and Lowenstein, T.K., 1991, Depositional environments of nonmarine evaporates, *in* Melvin, J.L., ed., Evaporites, petroleum and mineral resources: Developments in sedimentology, v. 50: New York, Elsevier, p. 189–347.

Steinmetz, J.J., 1988, Biostratigraphy and paleoecology of limnic ostracodes from the late Pleistocene Manix Formation [thesis]: Department of Biological Sciences, Pomona, California State Polytechnic University, 64 p.

Steinmetz, J.J., 1989, Ostracoda of the late Pleistocene Manix Formation, central Mojave Desert, California, *in* Reynolds, R.E., ed., The west-central Mojave Desert: Quaternary studies between Kramer and Afton Canyon: Redlands, California, Mojave Desert Quaternary Research Center, San Bernardino County Museum Association Special Publication, p. 70–77.

Thompson, G.D., 1929, The Mojave Desert region, California: A geographic, geologic, and hydrologic reconnaissance: U.S. Geological Survey Water-Supply Paper 578, p. 1–143 and p. 371–572.

Weldon, R.J., 1982, Pleistocene drainage and displaced shorelines around Manix Lake, *in* Cooper, J.D., ed., Geologic excursions in the California desert: Geological Society of America, Cordilleran Section Guidebook, p. 77–81.

Wells, S.G., and Enzel, Y., 1994, Fluvial geomorphology of the Mojave River in the Afton Canyon area, eastern California: Implications for the geomorphic evolution of Afton Canyon, *in* McGill, S.F., and Ross, T.M., eds., Geological investigations of an active margin: Geological Society of America, Cordilleran Section Guidbook, p. 177–182.

Wells, S.G., and Anderson, K.C., 1998, Late Quaternary geology and geomorphology of the lower Mojave River/Silurian Valley System and southern Death Valley area, southeastern California, *in* Byrd, B.F., ed., Springs and lakes in a desert landscape: Archaeological and paleoenvironmental investigations in the Silurian Valley and adjacent areas of southern California: Encinitas, California, ASM Affiliates, p. 137–264.

Wells, S.G., McFadden, L.D., Dohrenwend, J.C., Bullard, T.F., Feilberg, B.F., Ford, R.L., Grimm, J.R., Miller, J.R., Orbock, S.M., and Piclke, J.D., 1984, Late Quaternary geomorphic history of the Silver Lake area, eastern Mojave Desert, California, *in* Dohrenwend, J.C., ed., Surficial geology of the eastern Mojave Desert, California: Geological Society of America 1984 Annual Meeting Field Trip 14 Guidebook, p. 69–87.

Wells, S.G., McFadden, L.D., and Dohrenwend, J.C., 1987, Influence of late Quaternary climatic changes on geomorphic processes on a desert piedmont, eastern Mojave Desert, California: Quaternary Research, v. 27, p. 130–146.

Wells, S.G., Anderson, R.Y., McFadden, L.D., Brown, W.J., Enzel, Y., and Miossec, J-L., 1989, Late Quaternary paleohydrology of the eastern Mojave River drainage basin, southern California: Quantitative assessment of the late Quaternary hydrologic cycle in a large arid watershed: New Mexico Water Resources Research Institute, Technical Report 242, 250 p.

Wohl, E.E., and Enzel, Y., 1995, Data for paleohydrology, *in* Gregory, K.J., et al., eds., Continental paleohydrology: New York, John Wiley and Sons, p. 23–59.

MANUSCRIPT ACCEPTED BY THE SOCIETY AUGUST 1, 2002

Latest Pleistocene lake highstands in Death Valley, California

Diana E. Anderson*
*Center for Environmental Sciences and Education and the Quaternary Sciences Program,
Box 5694, Northern Arizona University, Flagstaff, Arizona 86011, USA*

Stephen G. Wells
Desert Research Institute, 2215 Raggio Parkway, Reno, Nevada 89512, USA

ABSTRACT

The modern Amargosa River–Death Valley hydrogeologic system includes portions of arid southwestern Nevada and southeastern California. During the latest Quaternary pluvial Lake Manly and other lakes in Death Valley were the termini for surface waters from both the Amargosa River, which drained local areas in the Great Basin Desert, and the Mojave River, which flowed through the Mojave Desert with headwaters in the San Bernardino Mountains. Analysis of four radiocarbon-dated, shallow (<30 m) cores along a 75 km transect in southern Death Valley allow the reconstruction of three lake highstands during the latest Pleistocene. These lake highstands occurred at 26 ka, ca. 18 ka, and 12 ka (uncalibrated ages) suggesting that shallow, oscillating lakes occupied the valley during that period. The ca. 18 ka and 12 ka lake highstands included surface-water contributions from the Amargosa and Mojave rivers based on concurrent lacustrine sediments from Death Valley, Lake Dumont, and Lake Mojave. The 26 ka highstand included only local Amargosa River runoff since there is no evidence for Mojave River (or Owens River) overflow into Death Valley at that time.

INTRODUCTION

During the latest Quaternary, pluvial Lake Manly and other lake stands of Death Valley were the termini for surface waters from at least two rivers emanating from diverse hydrogeologic settings: the Mojave River with headwaters at elevations >3000 m in the Transverse Ranges of southern California, and the Amargosa River draining highlands with elevations up to 2400 m in the Great Basin Desert. After emanating from headwater regions, both river systems flow through areas of regionally low relief including some of the driest regions of the western United States (Fig. 1). In addition to surface flow, lake stands in Death Valley received groundwater discharged from a regional carbonate aquifer with a recharge area that included the Spring Mountains (with elevations up to 3635 m) near Las Vegas, Nevada (Hunt, 1975).

Elsewhere in the Mojave Desert and Great Basin, studies of interior continental basin lake sediments and surficial features have led to the reconstruction of paleoenvironmental changes during the late Pleistocene (R. Smith, 1978; G. Smith, 1984, 1991; Mifflin and Wheat, 1979; Winograd and Doty, 1980; Benson, 1981; Smith and Street-Perrott, 1983; Benson and Paillet, 1989; Benson et al., 1990, 1995; Oviatt, 1990; Oviatt et al., 1990; Wells et al., 1997, also this volume). These studies may be used in conjunction with other paleohydrologic indicators, such as oxygen isotopes from vein and cave calcite (Winograd et al., 1985, 1992), to elucidate the hydrologic response to climate change during the latest Quaternary.

The goal of this paper is to present a reconstruction of the latest Pleistocene (35–10 ka) lake highstand history in Death Valley using radiocarbon-dated sediments from shallow cores taken from the basin floor. The chronology of lake highstand events is com-

---

*E-mail: diana.anderson@nau.edu

Anderson, D.E., and Wells, S.G., 2003, Latest Pleistocene lake highstands in Death Valley, California, *in* Enzel, Y., Wells, S.G., and Lancaster, N., eds., Paleoenvironments and paleohydrology of the Mojave and southern Great Basin Deserts: Boulder, Colorado, Geological Society of America Special Paper 368, p. 115–128. © 2003 Geological Society of America.

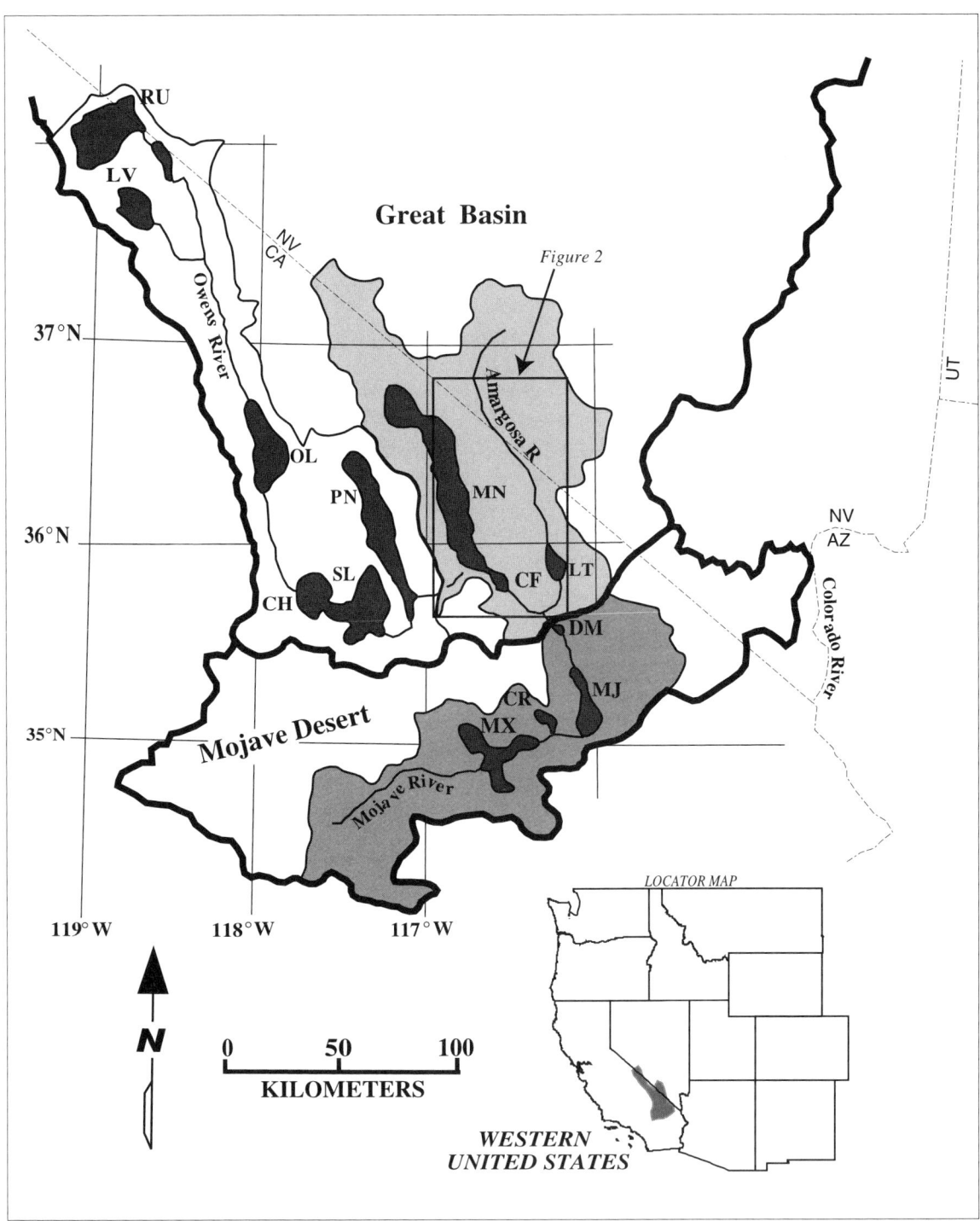

Figure 1. Pluvial lakes in basins of the Owens, Mojave, and Amargosa Rivers. RU—Lake Russell; LV—Long Valley; OL—Owens Lake; CH—China Lake; SL—Searles Lake; PN—Panamint Lake; MN—Lake Manly; LT—Lake Tecopa; DM—Lake Dumont; MJ—Lake Mojave; CR—Cronese Lakes; MX—Lake Manix. CF denotes Confidence Flats subbasin. Figure modified from Morrison (1991).

pared to existing chronologies from the basin and to chronologies from the Mojave and Amargosa drainage basins.

BACKGROUND

In their seminal work on the geology and hydrology of the Death Valley region, Hunt et al. (1966), Hunt and Mabey (1966), and Hunt (1975) found evidence for late Pleistocene and Holocene lakes based on shoreline features, basin stratigraphy, and archaeological evidence. Hooke (1972) recovered several cores from Badwater Basin and described lake sediments dating from 26 to 10.5 ka based on radiocarbon dating. Hooke (1972) also identified a terminal highstand (Blackwelder Stand) that is thought to have occurred during an early lake phase of Illinoian age (Hooke and Lively, 1979; Hooke and Dorn, 1992).

A recent study in Death Valley by Lowenstein et al. (1998, 1999), Li et al. (1996, 1997), Ku et al. (1998), Spencer et al. (1996), and Roberts et al. (1996) recovered a single, 185-m core from Badwater Basin, core DV93–1 (Fig. 2). Based on this well-studied core, these researchers suggested the presence of a perennial saline lake from ca. 35 to 10 ka based on ostracodes in mud layers interbedded with halite and suggested that while salinities and lake depths fluctuated during this interval, the overall climate was relatively wet. Core DV93–1 was dated using U-series ages on halite that were extrapolated by facies-specific sedimentation rates (Ku et al., 1998). An older perennial lake phase, existing from ca. 186 to 128 ka, was also identified (Lowenstein et al., 1998, 1999; Ku et al., 1998). The presence of two perennial lakes was further substantiated by Forester (1996), who found ostracodes in siliciclastic muds in core DV93–1, which suggests the existence of wetter climates manifested in deeper lakes from 35 to 10 ka and from 186 to 128 ka.

Preliminary field geomorphic assessment of the Wingate Pass area by the authors did not reveal young fluvial features, with ages <30 ka. The model of surface flow into Death Valley during the past 30 ka therefore includes only the Amargosa and Mojave Rivers (Fig. 1), in addition to local runoff. Based on stable isotopes and other geochemical analyses from the lower portion of core DV93–1, Spencer et al. (1996) suggested that inflow from the west (Owens River via Wingate Pass) was not significant and occurred only briefly prior to desiccation of the 186–128 ka lake. Forester (1996) also suggested, based on estimates of solute composition and total dissolved solids from ostracode samples, that the 35–10 ka lake probably had an Amargosa River source, whereas the penultimate (186–128 ka) lake may have included Amargosa, Mojave, or Owens River inputs.

GEOMORPHIC SETTING OF DEATH VALLEY AND THE TERMINUS OF THE AMARGOSA RIVER

Rosen (1994) presented a classification of playas based on the hydrology of a basin that included the location of the capillary fringe, and whether the basin is open or closed. Under this designation, Death Valley may be considered a hydrologically closed playa with regional groundwater discharge (shallow groundwater) where evaporites accumulate. The Confidence Flats portion of the study area (CF in Fig. 1; location of cores 1, 2, 3, and 4 in Fig. 2) is a through-flow playa that has only local discharge with groundwater moving through the subbasin to discharge in Death Valley. These differences (through-flow versus a closed system) are reflected in the surface of the modern playas, saline pan in Death Valley, and mudflat in Confidence Flats.

The components of a classic playa complex (Kendall, 1984) include (1) alluvial fan, (2) sandflat, (3) mudflat, (4) ephemeral saline lake, (5) perennial saline lake, (6) dune field, (7) perennial stream floodplain, (8) springs, and (9) shoreline features. The Death Valley playa complex has three components: (1) bordering alluvial fans, (2) mudflats, and (3) perennial or ephemeral lakes, or saline pan. Ephemeral lakes are present when there is surface flow into the basin and a deep groundwater table; perennial lakes are present when surface flow and groundwater discharge are high, and saline pans are present when the capillary fringe is shallow. There are no appreciable dune forms in the portion of the basin studied because the saline pan captures available sand, which is then indurated by salt. Springs are currently found in the basin and were certainly present during the wetter late Pleistocene.

Previous research on core DV93–1 has yielded evaporite mineral assemblages characteristic of certain regional moisture conditions and inflow sources. Li et al. (1997) suggested that the mineralogy within core DV93–1 indicates an evaporite geochemistry that varied with inflow source (continental surface and groundwater). Calcium sulfate–poor mineral assemblages formed during inflow from the Owens or Amargosa systems during relatively wet periods. During relatively dry periods, Ca-rich minerals dominated (gypsum, $CaSO_4 \times 2H_2O$, and glauberite, $CaSO_4 \times Na_2SO_4$) as spring inflow produces brines with Ca > HCO_3. Li et al. (1997) discussed changes in the source waters to Death Valley during the past 100 k.y. (k.y. = thousand years) during dry and wet periods as represented by core DV93–1, several important points are summarized below.

Glauberite and gypsum precipitated during dry periods in saline pan and deposits where Ca-rich spring discharge initially removed HCO_3 as calcite (Li et al., 1997). As evaporative concentration continued, glauberite and gypsum formed from the remaining Ca. This type of $CaSO_4$-rich environment existed during 0–10 and 60–100 ka. The modern analog for this situation in Death Valley involves the formation of Na-Cl-SO_4 brines from a mixture of the following:

1. Na-HCO_3-rich and Na-Cl-SO_4-HCO_3-rich meteoric water from the Amargosa River and springs from central and northern Death Valley (Travertine, Nevares, and Tule Springs).

2. Na-Ca-Cl spring and ground waters from southern Death Valley (possibly related to volcanism, hydrothermal activity, and a midcrustal magma chamber).

$CaSO_4$-bearing minerals are rare during wet periods. During these times, increased discharge of Amargosa River water introduces abundant HCO_3, which removes Ca from the brine by precipitation of calcite; the brines lack Ca and do not form $CaSO_4$

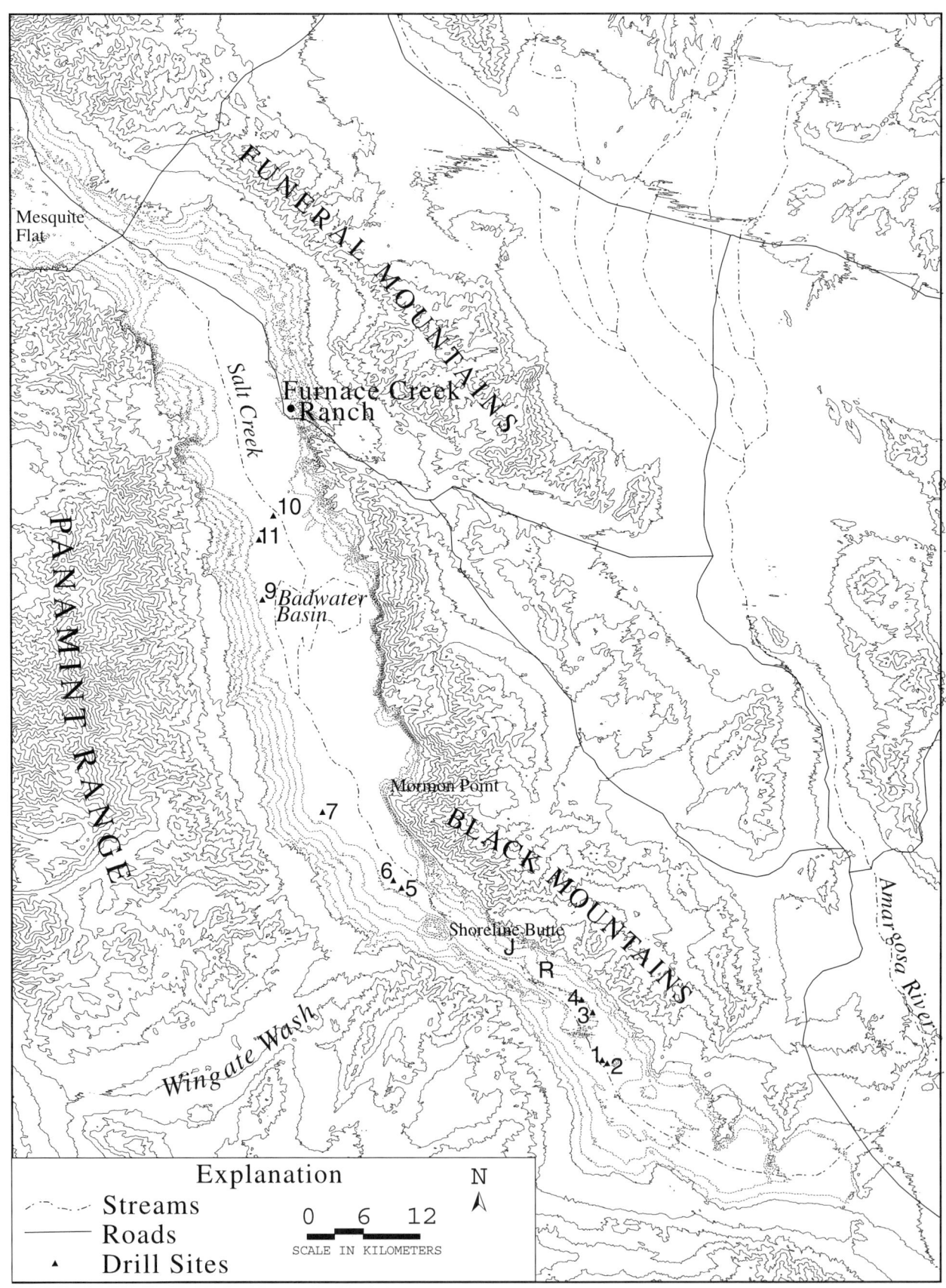

Figure 2. Location map for Death Valley cores DVDP96-1 through 11. Solid lines indicate 200-m contour intervals. Below 200 m, dashed contours are at 50-m intervals. Sites 1 through 4 are located in the Confidence Flats subbasin. See Table 1 for core site surface elevations. Deep core DV93–1 (Li et al., 1996; Lowenstein et al., 1998, 1999) is located at Badwater Basin.

minerals during further evaporative concentration (Li et al., 1997). These brines may precipitate thenardite (Na_2SO_4). This $CaSO_4$-poor environment existed from 10 to 60 ka in Death Valley.

RESEARCH APPROACH

Core acquisition

Ten shallow (<30 m) sediment cores were taken from southern Death Valley (Anderson and Wells, 1997a, 1997b; Anderson, 1999; Table 1; Fig. 2). Site selection was based on two criteria: (1) distribution of sites to encompass a range from the basin depocenter to the area of the delta of the Amargosa River as it enters the basin, and (2) locations allowable by National Park Service Wilderness Area policies and procedures. A permit was granted for drilling 11 sites, although core site 8 at Eagle Borax Spring was not drilled due to potential artesian conditions. Cores were acquired using a truck-mounted, hollow-stem auger; segments were drilled in 5-ft sections, containing two 2′ long 4″ diameter acrylic liners. Total core length for all 10 sites is 185.2 m.

Core recovery ranged from 50 to 92% (Table 1), and the stratigraphy was very well preserved. Fluvial, alluvial fan, mudflat, saline pan, and lake sediments, along with one paleosol, have been identified based on apparent texture, bedding, presence of evaporites, and Munsell color. Selected strata were sampled for dating by radiocarbon methods and for reconnaissance paleoecological analyses including extractions for pollen, diatoms, and ostracodes.

Sediment descriptions and interpretations

Following a modified set of criteria from Smith (1991), Li et al. (1996), and Lowenstein et al. (1999), sediments were classified as lake deposits where sediment hues were green, yellow, olive-brown, or black (Munsell colors 5GY, 10Y, 5Y, 2.5Y), clasts were well sorted and ranged in texture from clay to medium sand, and bedding was distinct, thin, laminar, or massive.

Since the overall goal of this research is to identify lake *highstands*, two more criteria were added to isolate the significant deposits. Li et al. (1996), Lowenstein et al. (1999), and Kendall (1984) noted the formation of subaqueous halite in saturated brines of perennial lakes. The lake deposits in this study were therefore simply classified as "shallow" (saturated, potentially low inflow rates) or "deep" (undersaturated, potentially high inflow rates) depending on the presence or absence, respectively, of subaqueous halite deposits. Lake "highstands" in this paper refer to these relatively deep, latest Pleistocene lakes.

As a further criteria, deposits >25 cm in thickness were highlighted for regional paleohydrologic correlation. Lowenstein et al. (1999) found sedimentation rates in the muds of core DV93–1 to be on the order of 0.4 to 1.0 m/k.y. Sedimentary units >25 cm in thickness should therefore represent events on the order of hundreds to thousands of years duration. This report focuses on long-lived, deep lake events (by the definitions given above) identified in the four cores that have been radiocarbon dated: cores 2, 6, 9, and 10. Further analysis of the cores may continue to resolve the paleohydrology of the basin by including shorter-lived and shallower lake events.

Other depositional environments were interpreted from the core sediments based on combined criteria from Reading (1978), Kendall (1984), Birkeland (1984), Collinson and Thompson (1982), Rosen (1994), and Li et al. (1996). Mudflat facies were represented by well-sorted, oxidized clay to sand strata or laminae with abrupt to clear smooth to wavy boundaries with or without desiccation structures such as mud cracks. Saline pan facies were characterized by moderately sorted sand to silt with evaporite minerals present. Fluvial deposits included well-sorted sand to subrounded gravel. Alluvial fan deposits were represented by subangular to subrounded gravels in a clayey to sandy matrix or poorly sorted to well-sorted sand.

Chronologic analyses

Radiocarbon dating of the bulk organic fraction was used to date the deposition of lake sediments in Death Valley during the latest Quaternary. Ranges of probable age for undated lacustrine strata were estimated by interpolation between, or extrapolation from, dated sample sites. Radiocarbon dating of lacustrine

TABLE 1. GPS-DETERMINED LOCATIONS, ELEVATIONS, AND RECOVERY OF CORES 1 THROUGH 11

Core	Access road	Universal Transverse Mercator northing	Universal Transverse Mercator easting	Surface elevation (m)	Core depth (m)	Recovery (%)	Geomorphic setting
1	Harry Wade	3958190	542797	10	8.98	50	Terrace
2	Harry Wade	3958138	542739	9	20.67	60	Channel floor
3	Harry Wade	3963256	541911	–2.0	22.04	85	Fan
4	Harry Wade	3965354	540749	–3.0	16.54	62	Fan
5	West Side	3977435	524152	–61	19	84	Fan
6	West Side	3978643	522977	–67	26.03	92	Fan
7	West Side	3986492	516310	–74	22.04	88	Fan
9	West Side	4010770	510703	–76	15.05	89	Fan
10	Devil's Seedway	4022029	512316	–84	18.74	92	Salt pan
11	Devil's Speedway	4019354	510057	–76	16.11	62	Fan

Note: Recoveries below 80% due to interval sampling (0.6 m preserved per 1.5 m drilled in unconsolidated gravels).

sediments is problematic because the carbon system may remain open after deposition. Thompson et al. (1990) dated dispersed organic matter and biogenic carbonate from sediment cores by accelerator-mass spectrometer (AMS) methods and compared the results to ages from shoreline features in the Bonneville, Franklin, and Lahontan Basins. Their results suggest internally consistent ages for the past 20 ka, whereas older ages were not stratigraphically consistent. Inconsistencies were more likely in shallow saline facies, while dispersed organics yielded reasonable ages during fresher water intervals (Thompson et al., 1990).

The findings of Thompson et al. (1990) have implications for chronologic control in Death Valley. The sediment from the DVDP96 cores is very low in identifiable carbon, and previous work (i.e., Lowenstein et al., 1999) has suggested that shallow saline lakes were likely. This study includes the time during which ages began to show inconsistencies (>20 ka) in the Bonneville, Franklin, and Lahontan Basins (Thompson et al., 1990).

A preliminary attempt was made to discriminate the types of organics being used for radiocarbon dating by running two sets of sample splits on the northernmost and southernmost cores, core 10 and core 2, respectively (Fig. 2, Table 2). All radiocarbon samples received acid pretreatments, but these additional sample splits were pretreated using acid/alkali/acid in an effort to differentiate between humic and fulvic acids and humin compounds as sources for the age determination (Abbott and Stafford, 1996), and for a preliminary assessment of potential contamination by modern material or ancient groundwater. Ages derived from the splits are similar, but not within 1 sigma (Table 2). All ages reported herein are uncalibrated years B.P. (years before present).

Paleoecological analyses

Extractions, identification, and interpretations for the pollen samples were conducted by P. Wigand, Desert Research Institute. Development of custom extraction procedures (P. Wigand, 1997, personal commun.) allowed the preservation of pollen that is normally lost during the HF phase from alkaline oxidized sediments.

R. Forester, U.S. Geological Survey, conducted the extraction, identification, and interpretation of ostracodes and other identifiable organic material. None of the six samples submitted to P. Bradbury, U.S. Geological Survey, contained diatoms.

DVDP96 CORE ANALYSIS RESULTS AND INTERPRETATIONS

Core 10

Description. The site drilled for core 10 is located immediately off the Devil's Speedway east of the modern drainage of Salt Creek (Figs. 2 and 3A). The top of the core hole is at –84 m elevation The core is 18.74 m in length and contains sediments from five different depositional environments. Two radiocarbon ages were obtained from the core at 3.0 m depth and 14.4 m depth (Table 2). The upper sample was split; the split that was pretreated with acid yielded an age of 12,160 ± 80 yr B.P., and the split that was pretreated with acid/alkali/acid yielded an age of 12,420 ± 60 yr B.P. They overlap with a 2-sigma error and the age discrepancy may reflect slight contamination from younger material (R. Hatfield, Beta Analytic, Inc., 1996, personal commun.). The lower radiocarbon sample had a resultant age of 45,800 ± 1300 yr B.P. Since this age is near the limit of modern radiocarbon dating techniques, little confidence was placed in this latter age.

The lowermost depositional unit, Unit 10A, is located from 16.5 to 18.9 m depth. Unit 10A is characterized by greenish gray to black, massive, poorly sorted clay to gravel with clasts up to 5 cm in diameter with some coarse sand to gravel fining upward sequences. Within Unit 10A, from 17.4 to 17.5 m depth, is dark brown, massive, poorly sorted clay to gravel with $CaCO_3$ nodules up to 2 cm in diameter. This unit was distinct from the rest of Unit 10A due to the oxidized color and presence of $CaCO_3$ nodules. Unit 10B is characterized by light gray, massive clay extending from 15.6 to 16.5 m depth. Sediments in Unit 10C extend from 14.0 to 15.6 m and consist of light gray to olive gray sandy clay to sand interbedded with evaporites in

TABLE 2. ACCELERATOR MASS SPECTROMETRY RADIOCARBON DATES FROM CORES 2, 6, 9, AND 10 FROM BULK ORGANIC FRACTION

Core	Sample elevation (m)	Sample depth in core (m)	Laboratory number	$\delta^{13}C$ (‰)	Uncalibrated ages (yr B.P.)
2	–6.78	15.8	Beta-97591	–25.5	20,020 ± 80
2	–6.78	15.8	Beta-97592	–25.6	19,530 ± 80
2	–7.90	17.0	Beta-93412	–25.2	19,710 ± 70
6	–84.1	17.6	Beta-97797	–21.7	17,550 ± 80
6	–91.9	25.4	Beta-94206	–20.6	26,200 ± 150
9	–84.2	8.0	Beta-97593	–27.3	9780 ± 60
9	–86.9	10.7	Beta-94205	–21.1	14,450 ± 60
10	–87.0	3.0	Beta-97594	–22.0	12,160 ± 80
10	–87.0	3.0	Beta-97595	–22.5	12,420 ± 60
10	–98.2	14.4	Beta-93413	–25.3	45,800 ± 1300

Note: Samples underwent acid wash pretreatments. Cores 2 and 10 include splits using acid/alkali/acid pretreatments (second date listed).

planar parallel to wavy parallel laminae. Unit 10D extends from 5.0 to 14.0 m depth and consists of yellowish brown to dark brown well-sorted clay in the upper portion of the unit, and gray to olive brown well-sorted clay to sand in the lower portion. Strata are 2 mm to 11 cm thick with wavy to planar discontinuous boundaries with mud ripples, sand drapes, and mud cracks found throughout. Unit 10E is found from 3.3 m to 5.0 m depth and is characterized by brown to dark brown clay and sand, often overlain by light gray evaporite crystals. The evaporites may also be disseminated throughout the units as soft nodules. Unit 10F is found from 2.6 to 3.3 m depth and is represented by grayish brown to light brownish gray clay with weak laminae and occasional evaporites. Unit 10G extends from the surface to 2.6 m depth and consists of grayish brown to dark brown massive to horizontally bedded clay to silt with sand-sized evaporite crystals.

Interpretation. Unit 10A represents alluvial fan deposition with a hiatus represented by the buried paleosol in the middle of the alluvial fan deposit. The boundary between the paleosol and overlying alluvial fan is erosional, very abrupt, and irregular. The parent material of the paleosol is massive with matrix-supported clasts and is possibly a debris flow facies of the alluvial fan. If this is indeed a paleosol, it is significant because it represents a period of stability on the alluvial fan prior to the lake-filling event evidenced by the subsequent Unit 10B. Bracketing ages for the Unit 10B deep lake event are difficult to calculate. If an average sedimentation rate of 0.3 m/k.y. is used (based on the radiocarbon ages of 12,160 yr B.P. at 3.0 m depth and 45,800 yr B.P. at 14.4 m depth), the estimated age for the lower deep lake event is ca. 51 ka. If the 45,800 yr B.P. age is deemed unreliable for the reasons discussed above, and the average sedimentation rate of 1 m/k.y. from core DV93–1 (Lowenstein et al., 1999) is applied using the radiocarbon age from the top of the core, the estimated age for the lower deep lake event is ca. 25 ka. Further age control should resolve this problem.

Subsequent to the Unit 10B deep lake event, the site underwent a drying phase from shallow lake (Unit 10C) to a long-lived mudflat (Unit 10D). Pollen was recovered in concentrations of 25,018 grains/gram from the mudflat (Unit 10D) sediments at 12 m depth. The sample indicates a desert scrub type vegetation (P. Wigand, 1997, personal commun.), with the identifiable grains dominated by Cheno-Am (21.21%), *Tubuliflorae* (17.81%), Ambrosia-type (11.31%), and *Juniperus* (4.81%). Cheno-Am is an abbreviation used in palynology to describe a pollen type that subsumes several genera from the Chenopodiaceae family and species from the *Amaranthus* genus (S. Smith, 2001, personal commun.). Such high concentrations of identifiable pollen are unusual in these extremely arid and alkaline conditions, and basin floor records of paleobotanical assemblages are not common in southwest deserts.

Drying continued as the depositional environment changed to that of a saline pan, which continues to modern times. Deposition was interrupted once by the Unit 10F deep lake event that occurred at ca. 12 ka.

Core 9

Description. Core 9 was taken at Tule Springs off West Side Road in Death Valley (Figs. 2 and 3B). The top of the core hole is at –76 m elevation. Two AMS radiocarbon ages were acquired from the core: 9780 ± 60 yr B.P. at 8.0 m depth and 14,450 ± 60 yr B.P. at 10.7 m depth (Table 2). Core 9 is 15.05 m long and contains evidence of five different depositional environments. The average sedimentation rate based on the two radiocarbon ages is 0.6 m/k.y.

The base of the core is floored in a light gray, very well indurated sand >7 cm thick with violent effervescence when dilute HCl was applied. After drilling through this material, and leaving the drill site overnight, artesian flow was observed the next morning with a hydraulic head of ~1.2 m and a flow rate estimated to be 3 gpm. The hardpan appears to have acted as an aquiclude. This site was the only drill site where artesian conditions were encountered.

Unit 9A extends from 13.3 m to 15.0 m depth and consists of light gray to dark grayish brown, massive, poorly to well-sorted silt to gravel. Unit 9B extends from 9.8 m to 13.3 m depth and consists of white to very pale brown clay in horizontal lamellae with occasional sand or gravel. These carbonate muds effervesce strongly to dilute HCl. Unit 9C extends from 7.5 m to 9.8 m depth and is characterized by gray to black clay to silty clay or, rarely, sand that is either massive or has weak horizontal laminae or strata, often defined by color changes. The black color is due to the reduction of sulfur species (Kendall, 1984) although there are some organic fragments in the unit as well. Unit 9D extends from 1.72 m to 7.5 m depth and is characterized by light yellowish brown to brown clay, silt, and sand. The uppermost unit, Unit 9E, which underlies the road fill, is predominantly fluvial with pale brown to dark yellowish brown fine to medium sand.

Interpretation. Units 9A and 9E appear to be dominated by fluvial deposition, from streams related to the alluvial fans on the western margin of the valley (Fig. 2). The carbonate muds of Unit 9B may have been deposited in deeper waters but, adhering to the specifications described above (undersaturated vs. saturated), they have been labeled as "shallow lake" for this analysis. Pollen recovered from 10.5 m depth in Unit 9B had a concentration of 2916 grains/gram. The assemblage suggests a transitional environment between semiarid woodland and desert scrub vegetation (P. Wigand, 1992, personal commun.), with the identifiable grains dominated by total *Pinus* (30.23%), *Tubuliflorae* (16.28%), *Ephedra nevadensis* (9.30%), Cheno-Am (9.30%), and *Juniperus* (6.98%). Ostracodes were found at a depth of 10.3 m in Unit 9B, revealing evidence of a regional low Alk-Ca saline spring with an estuarine ostracode, *Cyprideis beaconensis*. Thus, Unit 9B shows some evidence of ancient spring activity at Tule Springs. Unit 9C is interpreted as deep lake with bracketing ages of 7 ka to 13 ka using a sedimentation rate based on the two radiocarbon ages from the core. Framboidal pyrite root tubes were recovered from 8.9 m depth and are thought to indicate a wetland-marsh complex similar

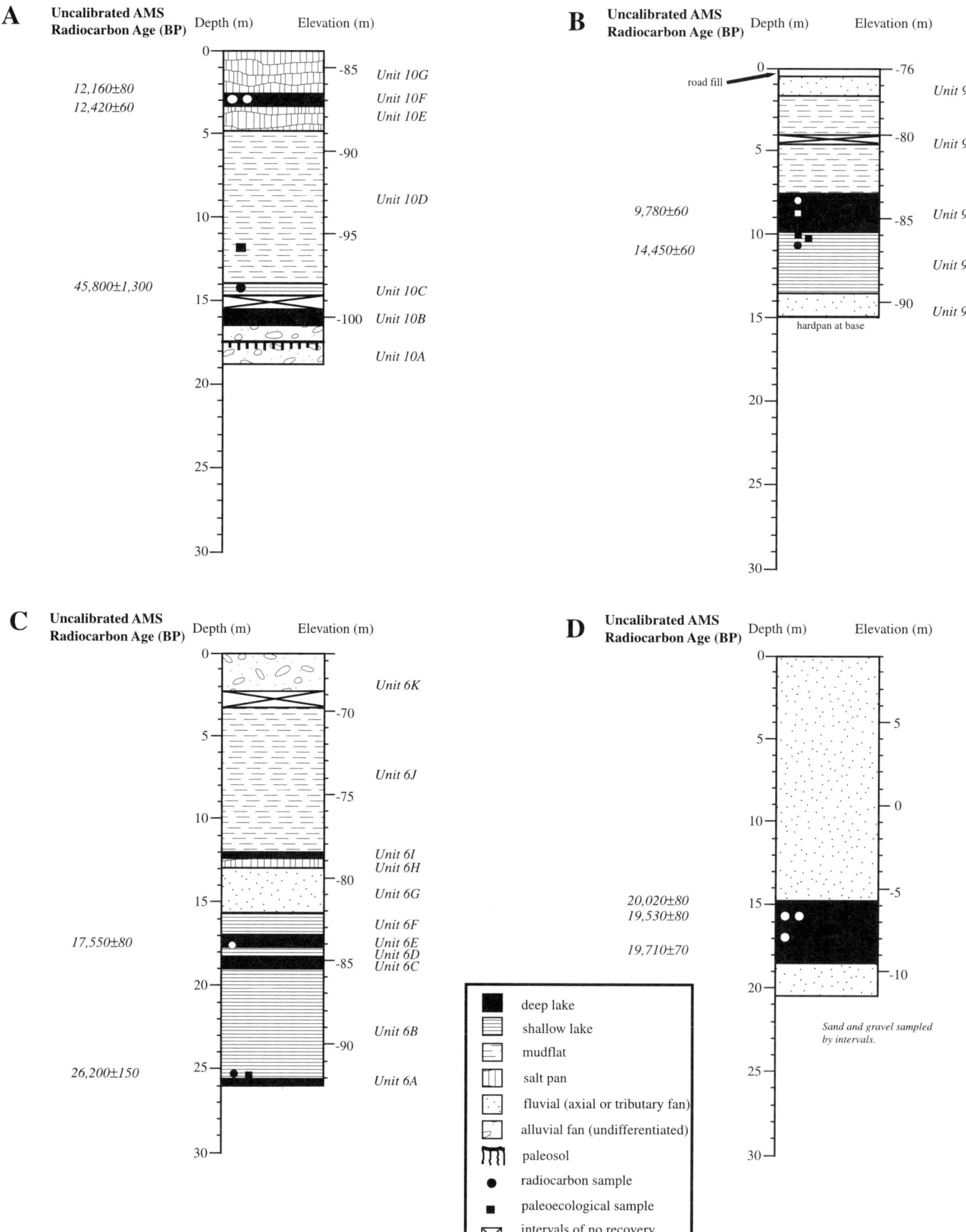

to local *Phragmites* wetlands (R. Forester, 1997, personal commun.). Unit 9D is interpreted to represent a mudflat environment.

Core 6

Description. Core 6 was taken 5 km north of Shoreline Butte in southern Death Valley (Figs. 2 and 3C), downstream from Wingate Wash. The top of the core hole is at –67 m el. The core is 26.03 m long and contains evidence of six different depositional environments. Two AMS radiocarbon ages were acquired from the core: 26,200 ± 150 yr B.P. at 25.4 m depth and 17,550 ± 80 yr B.P. at 17.6 m depth (Table 2). An average sedimentation rate for this core based on the two radiocarbon ages is 0.9 m/k.y.

A thick section comprising the lower portion of the core includes Units 6B, 6D, and 6F extends from 15.6 m to 26.0 m depth, and is characterized by light gray to dark brown well-sorted clay, silt and fine sand, massive or with planar parallel or wavy laminae. This depositional environment is interspersed with Unit 6A (25.6–26.0 m depth), 6C (18.2–19.0 m depth), and 6E (170.0–17.8 m depth) which consist of massive gray to greenish gray clay or fine sand in planar parallel laminae. Unit 6G extends from 12.9 m to 15.6 m depth and consists of poorly sorted, olive gray to dark gray sand to gravel with occasional clay drapes. Unit 6H is found from 12.4 m to 12.9 m depth and is characterized by very pale brown clay and sand with light gray evaporites in diffuse layers >1.5 cm thick. Unit 6I extends from 12.0 m to 12.4 m depth and is characterized by light brownish gray massive silty clay. Unit 6J extends from 3.0 m to 12 m depth and consists of yellowish brown to dark brown well-sorted clay, silt, and sand laminae with mud cracks, ripples, and rip-ups throughout. Unit 6K extends from the surface to 2.3 m depth and is characterized by grayish brown, poorly sorted sand and gravel.

Interpretation. The lower portion of the core is dominated by shallow lake deposits (Units 6B, 6D, and 6F) interspersed with three deep lake events. Using a sedimentation rate of 0.9 m/k.y. the ages of the deep lake events in the lower part of the core are 27 ka (Unit 6A), 19 ka (Unit 6C), and 17 ka (Unit 6E; in previous papers these latter two ages were averaged to 18 ka). The site then underwent changes to a fluvially dominated system and then to a short-lived salt pan. The next deep lake event occurred at 12 ka (Unit 6I). The site was then dominated by mudflat and alluvial fan depositional environments to modern times.

Figure 3. Stratigraphy and interpreted facies for (A) core 10 on the Devil's Speedway within 100 m east of Salt Creek (B) core 9, located at Tule Springs on West Side Road (C) core 6 located at Butte Valley Road–West Side Road junction, and (D) core 2 located in the channel of the Amargosa River where it is crossed by the Harry Wade Road in southern Death Valley Basin southeast of the Confidence Hills. For all diagrams, circles indicate radiocarbon sample depths and squares indicate pollen, ostracode, and/or diatom sample depths. For paired ages, upper sample underwent acid pretreatment, lower sample underwent acid/alkali/acid pretreatment.

Core 6 was sampled for pollen at 25.7 m depth and yielded pollen concentrations of 10,574 grains/gram. The sample is considered to represent semiarid woodland (P. Wigand, personal commun.), although the presence of a great quantity of immature juniper may attest to the presence of a portion of a juniper stamen, which could skew the data. The dominant pollen types include the immature *Juniperus* (49.86%), mature *Juniperus* (15.32%), total *Pinus* (13.37%), Poaceae (5.01%), and Cheno-Am (4.46%).

Core 2

Description. Core 2 was extracted from the bed of the Amargosa River where it is crossed by Harry Wade Road in southern Death Valley (Figs. 2 and 3D), ~9 km southeast of the Confidence Hills. The location of core 2 (along with cores 1, 3, and 4) is considered to be in a separate subbasin, Confidence Flats, isolated behind a constriction created by the Confidence Hills and the Jubilee and Rhodes alluvial fans, indicated by the "J" and "R" in Figure 2. The surface elevation of the core hole is 9 m el., and the core is 20.67 m long. This core had the poorest recovery due to interval sampling techniques used to drill through sand and gravel.

Two radiocarbon ages were obtained from the core at 15.8 m depth and 17.0 m depth (Table 2). The upper sample was split; the split that was pretreated with acid yielded an age of 20,020 ± 80 yr B.P., and the split that was pretreated with acid/alkali/acid yielded an age of 19,530 ± 80 yr B.P. The resultant age of the splits suggests slight contamination from older groundwater, as there is an older relative age for the acid pretreatment split compared to the acid/alkali/acid pretreatment split of the sample (R. Hatfield, BETA Analytic, 1996, personal commun.). The lower sample from 17.0 m depth yielded a radiocarbon age of 19,710 ± 70 yr B.P.

The sediments throughout the core are dominated by sand and gravel with the exception of the interval between 15.4 m and 17.4 m depth, which contains light olive gray to olive gray sandy clay to clay. This predominantly clayey unit contains occasional <1 cm thick brown sandy strata.

Interpretation. Although the depositional environment at this site appears to have been similar to the modern fluvial environment through most of the core, the intervening clayey unit is thought to represent a deep lake with the sandy strata representing individual flood events. The deep lake event occurred at ca. 19 ka.

DISCUSSION

Death Valley cores

Sedimentology and radiocarbon ages of the four dated cores indicate the presence of more than one lake highstand between 35 and 10 ka. Figure 4 shows a longitudinal profile of the cores with deep lake deposits and ages highlighted. These lake highstands fall within the late Pleistocene perennial lake interval (35–10 ka) identified from the DV93-1 core (Lowenstein et al., 1998, 1999), Li et al. (1996, 1997), and Ku et al. (1998). Analyses of the DVDP96 cores from sites above the basin depocenter did not provide evidence of a large, long-lived, continuous deep lake between 35 and 10 ka; instead, there is evidence for three lake highstands when lake depths and concurrent sedimentation reached higher elevations. Cores 6, 9, and 10 show correlative deep lake deposits at 12 ka. Older deep lake deposits are found in core 10 at 25 ka (using a regional sedimentation rate), in core 6 at 19 ka, 17 ka, and 26 ka, and in core 2 at 19 ka. Note that ages are approximate due to errors associated with radiocarbon analyses (Table 2) and potential errors from using averaged sedimentation rates.

The lack of deep lake deposits dating ca. 19–17 ka in core DVDP96-10 may be attributable to (1) lack of good chronologic control in the base of core 10, (2) failure to recognize lacustrine strata in core 10 based on the established criteria, or (3) reworking of the strata in core 10 by Salt Creek, which is adjacent to the drill site (Fig. 2) and flows into Death Valley from the north.

Lake highstands in Death Valley during the latest Pleistocene appear to occur at 26 ka, ca. 18 ka, and 12 ka, based on the DVDP96 cores. Lowenstein et al. (1998, 1999) identified three "less saline brackish lake phases" during the late Pleistocene perennial lake interval at 34 ka, 25 ka, and 16 ka based on the occurrence of *Candona* and *Limnocythere ceriotuberosa* ostracode species in mud layers. Further details of the late Pleistocene perennial lake are unclear because muds from depths of 9–14 m (ca. 12–25 ka) were partially lost during coring (Lowenstein et al., 1999).

Correlation with regional paleohydrologic events

Lake-filling events in the Lake Mojave Basin (the modern Silver Lake and Soda Lake Playas) along the Mojave River represent possible overflow events when the Mojave River drainage basin could have contributed significant volumes of water to Death Valley lakes. Study of the Pleistocene Lake Mojave (Fig. 1) by Wells et al. (1989, this volume), Enzel et al. (1989, 1992, this volume), and Enzel (1992) revealed a record of hydrologic responses to climatic events in the Transverse Ranges during the past 22 k.y. Lake Mojave I was a prolonged lake stand with no drying events lasting from 18.4 to 16.6 ka, and was thought to be associated with more frequent overflow from Lake Manix (Wells et al., this volume). A second lake highstand, Lake Mojave II, occurred from 13.7 to 11.4 ka. These lake highstands are correlated to dated shoreline features, shorelines A and B. Water crossing the sill at the northern end of the basin flowed northward, ultimately to southern Death Valley (Fig. 1).

The Dumont Basin lies between the Lake Mojave Basin and Death Valley (Fig. 1), and contains sediments from at least two lake events during the past 30 ka (K. Anderson and Wells, 1996, 1997, this volume). The two phases of Lake Dumont include Lake Dumont I at ca. 30–25.3 ka and Lake Dumont II at ca. 19.4–18 ka. The sill at the northern end of Dumont Basin was breached after Lake Dumont II.

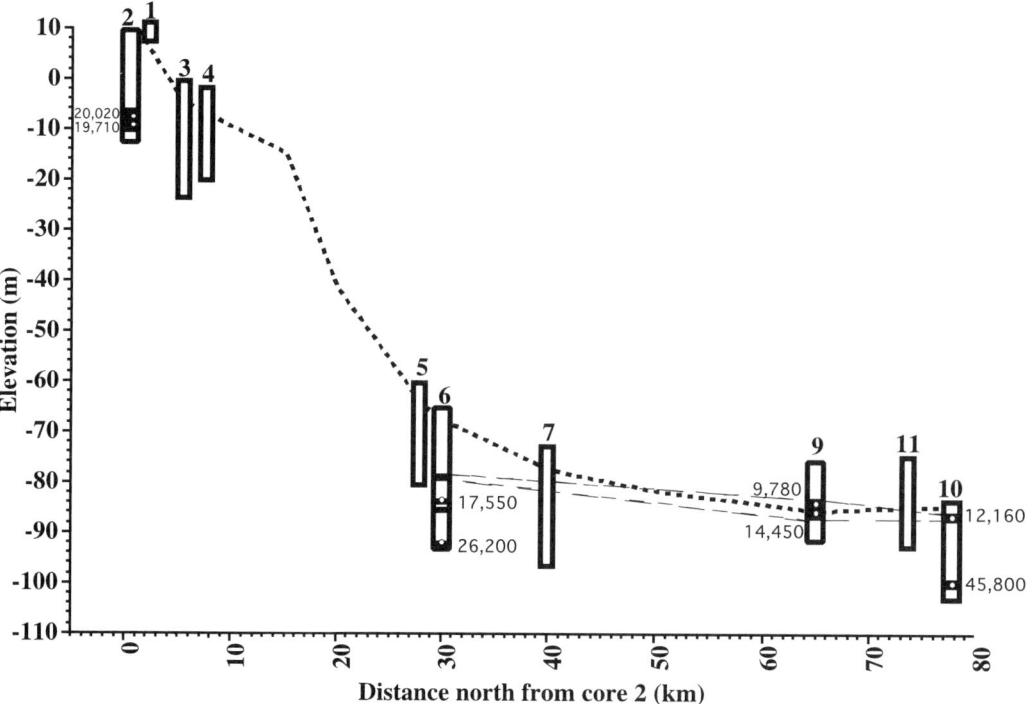

Figure 4. Longitudinal profile (see Figure 2 for map locations) showing sediments defined as deep lake in black with radiocarbon ages noted by the white circles. Thick dotted line is basin floor; note that the Confidence Flats subbasin (cores 1 through 4) is separate from Death Valley Basin proper. Dramatic change in gradient between cores 4 and 5 occurs near the constriction provided by the Confidence Hills and the Rhodes and Jubilee alluvial fans. Cores shown entirely in white have not yet been dated.

Figure 5 shows Lake Mojave, Lake Dumont, and Death Valley lake highstands during the past 35 ka. The 26 ka Death Valley lake highstand is contemporaneous with the end of Lake Dumont I in Dumont Basin. Since the deep lake deposit in core 6 is at the bottom of the core, it is possible that that lake phase began much earlier. There is currently no evidence that there was overflow from the Mojave River system into either the Dumont or Death Valley Basins at this time, suggesting that these two basins contained lakes that were maintained by local, interior continental, dominantly low-elevation drainages (Fig. 6). The hypsometric curve for the present upper and lower Amargosa River drainage basins, which includes the modern Dumont Basin, shows that about half of the area (~11,313 km^2) is below 1000 m, with <3% of the drainage basin above 2000 m, where precipitation is greatest.

The Death Valley lake highstand at ca. 18 ka correlates with the beginning of Lake Mojave I and the end of Lake Dumont II, suggesting that at this time, the Mojave River was integrated with Death Valley (Fig. 7). The Death Valley lake highstand at 12 ka correlates with Lake Mojave II. At this time the Dumont Basin had been breached and was a through-flowing basin (K. Anderson and Wells, this volume). Again, based on the presence of an overflowing Lake Mojave II, exotic runoff likely reached Death Valley (Fig. 8).

SUMMARY

Li et al. (1996, 1997) and Lowenstein et al. (1998, 1999) reported the presence of a saline perennial lake at the basin depocenter at Badwater Basin from 35 to 10 ka. The four dated DVDP96 cores, from elevations above that of the depocenter, show evidence of lake fluctuations within this time interval and provide a more sensitive indicator of lake highstands. Lake highstands were identified based on sediments >25 cm in thickness that were reduced, massive or finely laminated clays that did not contain significant evaporite minerals. Three lake highstands were identified at 26 ka, ca. 18 ka, and 12 ka. These highstands correlate well with lake highstands of both Lake Mojave and Lake Dumont. The three "less saline brackish lake phases" identified in the DV93-1 core (Lowenstein et al., 1998, 1999) correlate less well; the youngest lake highstand (12 ka) is not represented and an older lake is present (34 ka). In addition, the 18 ka lake highstand found in the DVDP96 cores is not represented but a 15 ka lake is noted. This lack of correspondence may be due in part to only partial recovery of sediments dating from ca. 12 to 25 ka in the upper portion of core DV93-1 (Lowenstein et al., 1999) or from age interpolation problems in the DVDP96 cores.

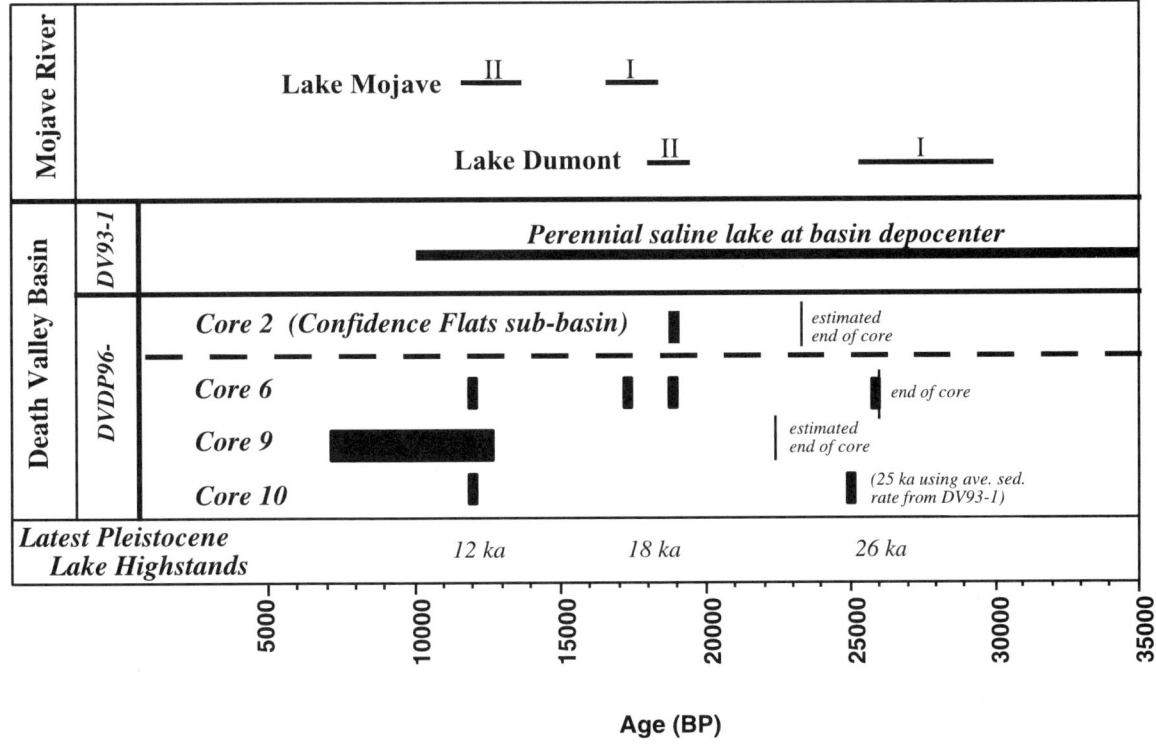

Figure 5. Composite of 10–35 ka lacustrine chronologies for the Mojave River drainage basin in uncalibrated years B.P. Lakes Manix (Dohrenwend et al., 1991; Enzel et al., this volume), Mojave (Wells et al., 1989, this volume; Enzel, 1992; Enzel et al., 1989, 1992, this volume), and Dumont (K. Anderson and Wells, 1996, 1997, this volume) and the chronologies for Death Valley Basin (Hunt et al., 1966; Hunt and Mabey, 1966; Hooke, 1972; Hunt, 1975; Li et al., 1996; Lowenstein et al., 1999) with the results of this study.

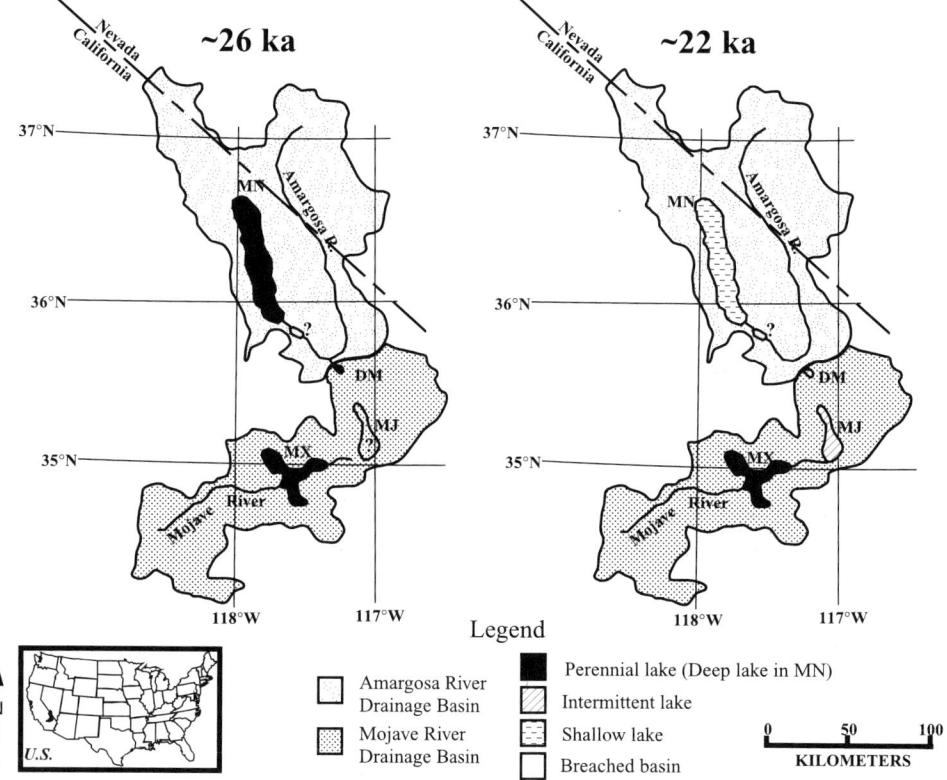

Figure 6. Proposed paleogeography of pluvial lakes of the Mojave and Amargosa rivers for periods the period ca. 26 ka and 22 ka. Note that lake areas are arbitrary and not based on dated shorelines. See Figure 1 for lake and basin names.

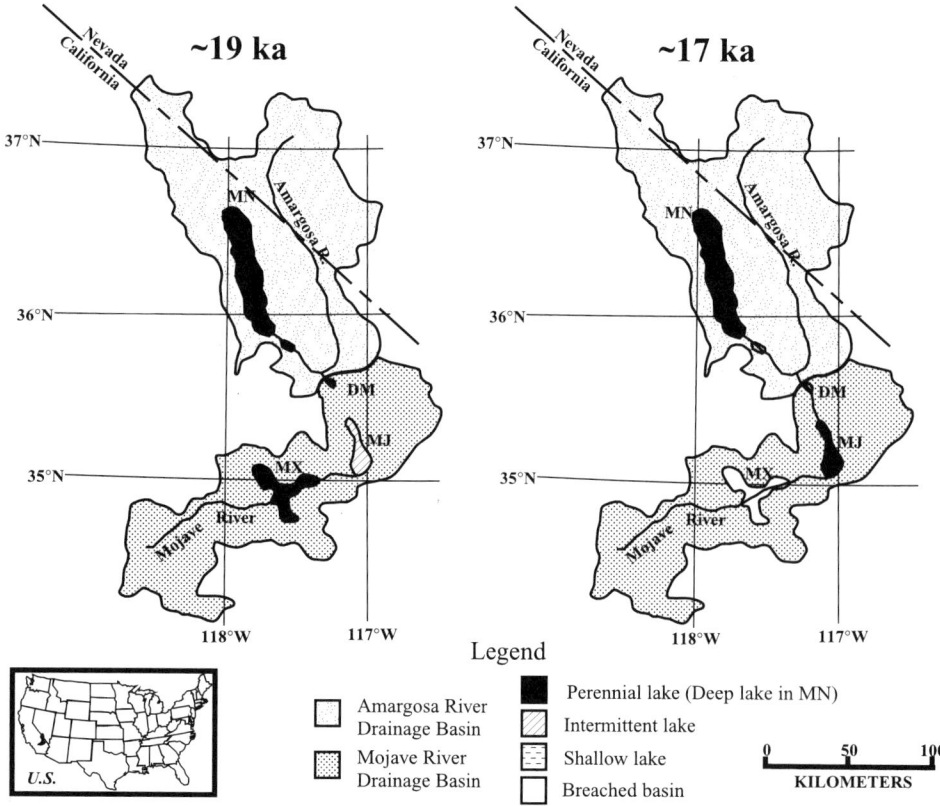

Figure 7. Proposed paleogeography of pluvial lakes of the Mojave and Amargosa rivers for periods the period ca. 19 ka and 17 ka. Note that lake areas are arbitrary and not based on dated shorelines. See Figure 1 for lake and basin names.

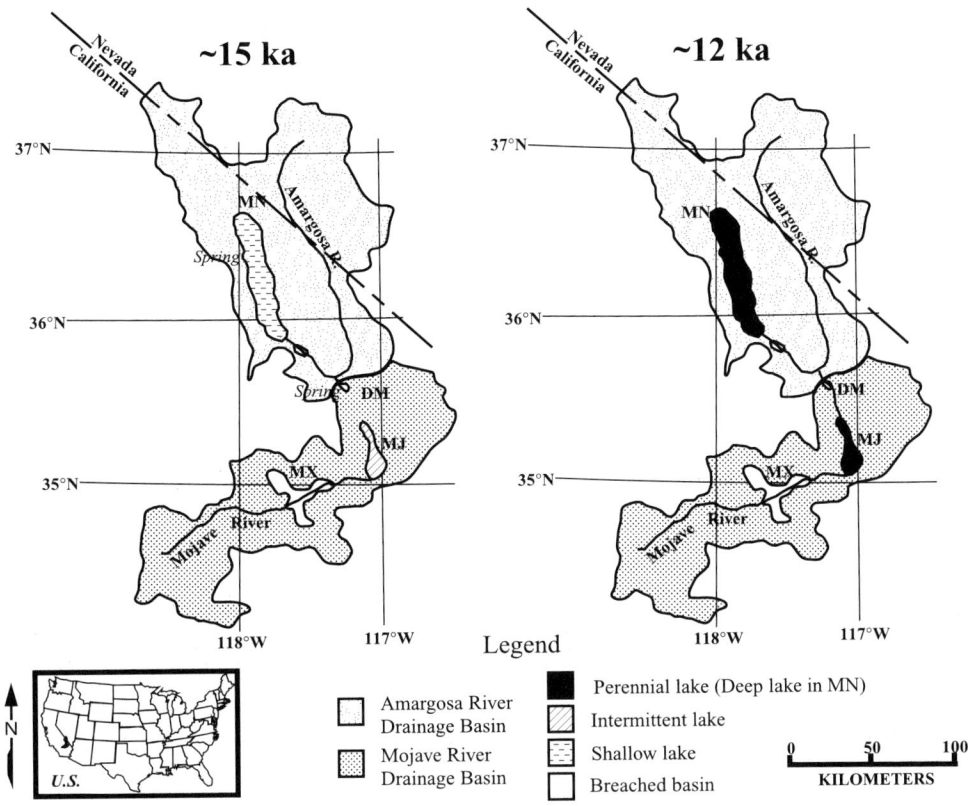

Figure 8. Proposed paleogeography of pluvial lakes of the Mojave and Amargosa rivers for periods the period ca. 15 ka and 12 ka. Note that lake areas are arbitrary and not based on dated shorelines. See Figure 1 for lake and basin names.

The DVDP96 cores suggest that while there was a long-lived late Pleistocene perennial lake on the floor of the basin, sediments from within several meters elevation along the edges of the basin show evidence of fluctuating lake levels. Lakes during this time period were likely shallow and dynamic, responding not only to changes in climate but also to changes in the storage capacity of basins upstream (Wells et al., this volume; Enzel et al., this volume; K. Anderson and Wells, this volume). Additional evidence for the fluctuating nature of lakes during the latest Pleistocene include the paucity of shoreline features and the pollen recovered from the DVDP96 cores, which suggest oligotropic conditions, or short-lived lakes.

The ca. 18 ka and 12 ka lake highstands included both Amargosa River and Mojave River waters in addition to local runoff. The 26 ka lake highstand appears to have been maintained solely from Amargosa River flow and local runoff, despite the generally low elevation and interior continental position of the watershed.

ACKNOWLEDGMENTS

This work was supported by National Science Foundation Hydrological Sciences Program Grant #EAR 9696058 to S.G. Wells. We thank U.S. Department of the Interior National Park Service staff at Death Valley for access and support during the drilling. We also thank P. Wigand (Desert Research Institute), R. Forester and P. Bradbury (U.S. Geological Survey) for paleoecological analyses and interpretations. We appreciate the thoughtful manuscript reviews by N. Lancaster, C. Oviatt, and T. Lowenstein; their efforts greatly improved the quality and organization of this manuscript. Finally, D.E.A. thanks K. Anderson and J. Knott for insightful and animated discussions on the paleohydrology of this extraordinary basin.

REFERENCES CITED

Abbott, M.B., and Stafford, T.W., Jr., 1996, Radiocarbon geochemistry of modern and ancient lake systems, Baffin Island, Canada: Quaternary Research, v. 45, p. 300–311.

Anderson, D.E., 1999, Latest Quaternary (<30 ka) lake highstand fluctuations and evolving paleohydrology of Death Valley, in Slate, J., ed., Proceedings of Conference on Status of Geologic Research and Mapping in Death Valley National Park, April 1999, Las Vegas, Nevada: U.S. Geological Survey Open-File Report 99-153, p. 124–131.

Anderson, D.E., and Wells, S.G., 1997a, Late Pleistocene Death Valley lakes: Subsurface records of changing paleoenvironments, in Reynolds, R.E., and Reynolds, J., eds., Death Valley: The Amargosa Route: San Bernardino County Museum Association, San Bernardino, California, p. 89–92.

Anderson, D.E., and Wells, S.G., 1997b, Latest Quaternary lakes in Death Valley, California, USA: Paleogeography and highstands, in Federici, P.R., ed., Abstracts, International Association of Geomorphologists' Fourth International Conference on Geomorphology: Comitato Glaciologico Italiano, Torino, Bologna, Italy, p. 51.

Anderson, K.C., and Wells, S.G., 1996, Late Pleistocene lacustrine record of Lake Dumont and the relationship to pluvial lakes Mojave and Manly: Geological Society of America Abstracts with Programs, v. 28, no. 7, p. A458.

Anderson, K.C., and Wells, S.G., 1997, Late Pleistocene and Holocene valley-fill deposits of Lake Dumont, in Reynolds, R.E., and Reynolds, J., eds., Death Valley: The Amargosa Route: San Bernardino County Museum Association, San Bernardino, California, p. 29–32.

Benson, L.V., 1981, Paleoclimatic significance of lake-level fluctuations in the Lahontan Basin: Quaternary Research, v. 16, p. 390–403.

Benson, L.V., and Paillet, F.L., 1989, The use of total lake-surface area as an indicator of climatic change: Examples from the Lahontan Basin: Quaternary Research, v. 32, p. 262–275.

Benson, L.V., Currey, D.R., Dorn, R.I., Lajoie, K.R., Oviatt, C.G., Robinson, S.W., Smith, G.I., and Stine, S., 1990, Chronology of expansion and contraction of four Great Basin lake systems during the past 35,000 yr: Paleogeography, Paleoclimatology, Paleoecology, v. 78, p. 241–286.

Benson, L., Kashgarian, M., and Rubin, M., 1995, Carbonate deposition, Pyramid Lake subbasin, Nevada, 2: Lake levels and polar jet stream positions reconstructed from radiocarbon ages and elevations of carbonates (tufas) deposited in the Lahontan Basin: Paleogeography, Paleoclimatology, Paleoecology, v. 117, p. 1–30.

Birkeland, P.W., 1984, Soils and geomorphology: New York, Oxford University Press, 372 p.

Collinson, J.D., and Thompson, D.B., 1982, Sedimentary structures: London, George Allen and Unwin, 194 p.

Dohrenwend, J.C., Bull, W.B., McFadden, L.D., Smith, G.I., Smith, R.S.U., and Wells, S.G., 1991, Quaternary geology of the Basin and Range province in California, in Morrison, R., ed., Quaternary nonglacial geology: Conterminous U.S.: Boulder, Colorado, Geological Society of America, The Geology of North America, v. K-2, p. 321–352.

Enzel, Y., 1992, Flood frequency of the Mojave River and the formation of late Holocene playa lakes, southern California, USA: The Holocene, v. 2, p. 11–18.

Enzel, Y., Cayan, D.R., Anderson, R.Y., and Wells, S.G., 1989, Atmospheric circulation during Holocene lake stands in the Mojave Desert: Evidence of regional climate change: Nature, v. 341, p. 44–48.

Enzel, Y., Brown, W.J., Anderson, R.Y., McFadden, L.D., and Wells, S.G., 1992, Short-duration Holocene lakes in the Mohave River drainage basin, southern California: Quaternary Research, v. 38, p. 60–73.

Forester, R., 1996, A Death Valley ostracode glacial lake hydrochemical history: Geological Society of America Abstracts with Programs, v. 28, no. 7, p. A457.

Hooke, R.L., 1972, Geomorphic evidence for late-Wisconsin and Holocene tectonic deformation, Death Valley, California: Geological Society of America Bulletin, v. 83, p. 2073–2098.

Hooke, R.L., and Dorn, R., 1992, Segmentation of alluvial fans in Death Valley, California: New insights from surface exposure dating and laboratory modeling: Earth Surface Processes and Landforms, v. 17, p. 557–574.

Hooke, R.LeB., and Lively, R.S., 1979, Dating of late Quaternary deposits and associated Tectonic events by U/Th methods, Death Valley, California: Final Report for NSF Grant EAR-7919999.

Hunt, C.B., 1975, Death Valley geology, ecology, archaeology: Berkeley, University of California Press, 234 p.

Hunt, C.B., and Mabey, D.R., 1966, Stratigraphy and structure Death Valley, California: U.S. Geological Survey Professional Paper 494-A, 162 p.

Hunt, C.B., Robinson, T.W., Bowles, W.A., and Washburn, A.L., 1966, Hydrologic basin Death Valley California: U.S. Geological Survey Professional Paper 494-B, 138 p.

Kendall, A., 1984, Evaporites, in Walker, R.G., ed., Facies models: Toronto, Geological Association of Canada, 317 p.

Ku, T.-L., Luo, S., Lowenstein, T.K., Li, J., and Spencer, R.J., 1998, U-series chronology of lacustrine deposits in Death Valley, California: Quaternary Research, v. 50, p. 261–275.

Li, J., Lowenstein, T., Brown, C.B., Ku, T.-L., and Luo, S., 1996, A 100 ka record of water tables and paleoclimates from salt cores, Death Valley, California: Paleogeography, Paleoclimatology, Paleoecology, v. 123, p. 179–203.

Li, J., Lowenstein, T., and Blackburn, I., 1997, Responses of evaporite mineralogy to inflow water sources and climate during the past 100 k.y. in Death Valley, California: Geological Society of America Bulletin, v. 109, p. 1361–1371.

Lowenstein, T.K., Li, J., and Brown, C.B., 1998, Paleotempertures from fluid inclusions in halite: Method verification and a 100,000 year paleotemperature record, Death Valley, CA: Chemical Geology, v. 150, p. 223–245.

Lowenstein, T.K., Li, J., Brown, C., Roberts, S.M., Ku, T.-L., Luo, S., and Yang, W., 1999, 200 k.y. paleoclimate record from Death Valley salt core: Geology, v. 27, p. 3–6.

Mifflin, M.D., and Wheat, M.M., 1979, Pluvial lakes and estimated pluvial climates of Nevada: Reno, Nevada, Nevada Bureau of Mines and Geology, Bulletin 94, 57 p.

Morrison, R.B., 1991, Quaternary stratigraphic, hydrologic, and climatic history of the Great Basin, with emphasis on Lakes Lahontan, Bonneville, and Tecopa, in Morrison, R., ed., Quaternary nonglacial geology: Conterminous U.S.: Boulder, Colorado, Geological Society of America, The Geology of North America, v. K-2, p. 283–320.

Oviatt, C.G., 1990, Late Pleistocene and Holocene lake fluctuations in the Sevier Lake Basin, Utah, USA, in Davis, R.B., ed., Paleolimnology and the reconstruction of ancient environments: Boston, Kluwer Academic Publishers, p. 9–21.

Oviatt, C.G., Currey, D.R., and Miller, D.M., 1990, Age and paleoclimatic significance of the Stansbury Shoreline of Lake Bonneville, northeastern Great Basin: Quaternary Research, v. 33, p. 291–305.

Reading, H.G., editor, 1978, Sedimentary environments and facies: New York, Elsevier, 557 p.

Roberts, S., Spencer, R., and Yang, W., 1996, Climatic response in Death Valley, California, during Glacial Terminiation II: Geological Society of America Abstracts with Programs, v. 28, no. 7, p. A459.

Rosen, M., 1994, The importance of groundwater in playas: A review of playa classification and the sedimentology and hydrology of playas, in Rosen, M., ed., Paleoclimate and basin evolution of playa systems: Boulder, Colorado, Geological Society of America Special Paper 289, p. 1–18.

Smith, G.I., 1984, Paleohydrologic regimes in the southwestern Great Basin, 0–3.2 m.y. ago, compared with other long records of "global" climate: Quaternary Research, v. 22, p. 1–17.

Smith, G.I., 1991, Stratigraphy and chronology of Quaternary-age lacustrine deposits, in Morrison, R., ed., Quaternary nonglacial geology: Conterminous U.S.: Geological Society of America, The Geology of North America, v. K-2, p. 339–352.

Smith, G.I., and Street-Perrott, F.A., 1983, Pluvial lakes in the western United States, in Wright, H.E., and Porter, S.C., eds., Late Quaternary environments of the United States, Volume 1: The late Pleistocene: Minneapolis, University of Minnesota Press, p. 190–214.

Smith, R.S.U., 1978, Pluvial history of Panamint Valley, California: A guidebook for the friends of the Pleistocene, Pacific Cell: Geology Department, University of Houston, Texas, 36 p.

Spencer, R., Yang, W., Krouse, H.R., and Roberts, S., 1996, Hydrology of Death Valley between 180 and 100 ka: Geological Society of America Abstracts with Programs, v. 28, no. 7, p. A457.

Thompson, R.S., Toolin, L.J., Forester, R.M., and Spencer, R.J., 1990, Accelerator-mass spectrometer (AMS) radiocarbon dating of Pleistocene lake sediments in the Great Basin: Paleogeography, Paleoclimatology, Paleoecology, v. 78, p. 301–313.

Wells, S.G., Anderson, R.Y., McFadden, L.D., Brown, W.J., Enzel, Y., and Miossec, J., 1989, Late Quaternary paleohydrology of the eastern Mojave River drainage, southern California: Quantitative assessment of the late Quaternary hydrologic cycle in large arid watersheds: New Mexico Water Resources Research Institute Technical Completion Report (14-08-0001-G1312), New Mexico State University, Las Cruces, 253 p.

Wells, S.G., Anderson, K.C., Anderson, D.E., Williamson, T., and Enzel, Y., 1997, Geomorphic and sedimentologic responses to late Quaternary hydrologic events: Implications for paleo-precipitation regimes across the hyperarid Mojave River drainage basin of southern California, USA, in Federici, P.R., ed, Abstracts, International Association of Geomorphologists' Fourth International Conference on Geomorphology: Comitato Glaciologico Italiano, Torino, Bologna, Italy, p. 402.

Winograd, I.J., and Doty, G.C., 1980, Paleohydrology of the southern Great Basin, with special reference to water table fluctuations beneath the Nevada Test Site during the late(?) Pleistocene: U.S. Geological Survey Open-File Report 80-569, 91 p.

Winograd, I.J., Szabo, B.J., Coplen, T.B., Riggs, A.C., and Kolesar, P.T., 1985, Two-million-year record of deuterium depletion in Great Basin groundwaters: Science, v. 227, p. 519–522.

Winograd, I.J., Coplen, T.B., Landwehr, J.M., Szabo, B.J., Kolesar, P.T., and Revesz, K.M., 1992, Continuous 500,000-year climate record from vein calcite in Devils Hole, Nevada: Science, v. 258, p. 255–260.

MANUSCRIPT ACCEPTED BY THE SOCIETY AUGUST 1, 2002

Latest Quaternary paleohydrology of Silurian Lake and Salt Spring basin, Silurian Valley, California

Kirk C. Anderson*
Quaternary Studies Program, Northern Arizona University, Flagstaff, Arizona 86001, USA

Stephen G. Wells
Desert Research Institute, 2215 Raggio Parkway, Reno, Nevada 89512-1095, USA

ABSTRACT

Three shallow sediment cores taken from two basins in Silurian Valley, California, contain evidence related to the latest Pleistocene paleohydrology of the Mojave River and, by inference, the paleohydrology of the latest Pleistocene Death Valley lakes. The core from Silurian Lake contained no deposits that could be interpreted as lacustrine or spring discharge. The entire 33 m core is characterized by brown playa silt and clay, and distal alluvial fan sand. Rounded pebbles in two stratigraphic locations may, however, represent a through-flowing Mojave River during overflow episodes from Pleistocene Lake Mojave. Two radiocarbon ages are in reverse stratigraphic order and therefore temporal associations are not possible. Two cores from the Salt Spring basin represent latest Pleistocene lacustrine and spring activity, and exposed basin-fill deposits from the basin center represent three periods of aggradation during the Holocene. Core DU-2 from Salt Spring basin has two radiocarbon ages—18.1 ka at 0.8 m depth, and 27.5 ka at 10.8 m depth, giving an estimated sedimentation rate of 1.07 m/1000 yr. Using the sedimentation rate and fossil ostracod analysis, the following relationships are inferred for Lake Dumont, which formed in the Salt Spring basin for a period of time: (1) shallow lakes and spring activity from 30 ka to 25.3 ka, (2) desiccation from 25.3 ka to 23.4 ka, (3) shallow lakes or springs at 23.4 ka, (4) a drier period between 23.4 ka and 19.4 ka, (5) lakes and possible spring activity from 19.4 ka to 18 ka, and (6) breaching after 18 ka as a result of increased discharge of the Mojave River down Silurian Valley. This chronology correlates well with lake activity in Death Valley (30–18 ka), and Lake Mojave (18 ka), possibly constraining Mojave River contributions to Death Valley lakes to the period of overflow and eventual breaching of the Salt Spring basin. Lake Dumont was probably fed by local streams and springs from the Avawatz Mountains and Kingston Wash prior to Lake Mojave overflow. These earlier lakes indicate moister climatic conditions that also allowed lakes to form in Death Valley from increased discharge down the Amargosa River. The prominent shoreline along the colluvial apron of the Salt Spring Hills and cutting into a late Pleistocene alluvial fan is correlated with the ca. 18 ka (OIS 2) high lake stand.

---

*E-mail: kirk.anderson@nau.edu

Anderson, K.C., and Wells, S.G., 2003, Latest Quaternary paleohydrology of Silurian Lake and Salt Spring basin, Silurian Valley, California, *in* Enzel, Y., Wells, S.G., and Lancaster, N., eds., Paleoenvironments and paleohydrology of the Mojave and southern Great Basin Deserts: Boulder, Colorado, Geological Society of America Special Paper 368, p. 129–141. © 2003 Geological Society of America.

INTRODUCTION

Silurian Valley connects Silver Lake to the south with Death Valley to the north (Fig. 1). Salt Creek currently flows intermittently through the ~30 km long, northwest-trending valley, though during the late Pleistocene the Mojave River flowed through Silurian Valley, influencing the hydrology and geomorphology of two basins along this reach, Silurian Lake and Salt Spring basin (Wells et al., this volume). The latest Quaternary paleohydrology of these two basins is the focus of this chapter.

The lacustrine records from Silver Lake and Death Valley are well documented (D. Anderson and Wells, this volume; Enzel et al., this volume; Wells et al., this volume) and although they represent a relatively detailed paleohydrologic history for the Mojave River system, several important questions can be addressed through a better understanding of the paleohydrology of Silurian Valley. For example, when did the Mojave River flow through Silurian Valley on its way toward a confluence with the Amargosa River and into Death Valley? What record, if any, did the Mojave River leave in Silurian Valley and how significant was the contribution of Mojave River waters to Death Valley lakes?

How do the ages of lacustrine deposits in the Salt Spring basin correspond to the lacustrine records of Pleistocene Lake Mojave and the Death Valley lakes? Was Lake Dumont (the name given to the Pleistocene lake in the Salt Spring basin) fed only by Mojave River waters, or did lake and spring activity originate from more localized sources? And finally, what can we infer about the past climatic conditions of the area? Silurian Valley is uniquely situated to evaluate these questions as it provides a hydrologic link between dated lacustrine sequences of pluvial Lake Mojave (Silver and Soda Lake) and Lake Manly (Death Valley; see Enzel et al., this volume).

Silurian Lake is actually an undissected playa rather than a lake, whereas the heavily dissected Salt Spring basin contains exposures of lacustrine deposits and a prominent shoreline. Silurian Lake was chosen because of its position within Silurian Valley and the potential for preservation of subsurface deposits to aid in the interpretation of the hydrologic history of the Mojave River. Salt Spring basin was chosen because of (1) previous research in the area on latest Quaternary stratigraphy and soils (Ritter, 1987; McFadden et al., 1989), (2) discernible geomorphic relationships between basin-fill and piedmont fans, (3) good potential

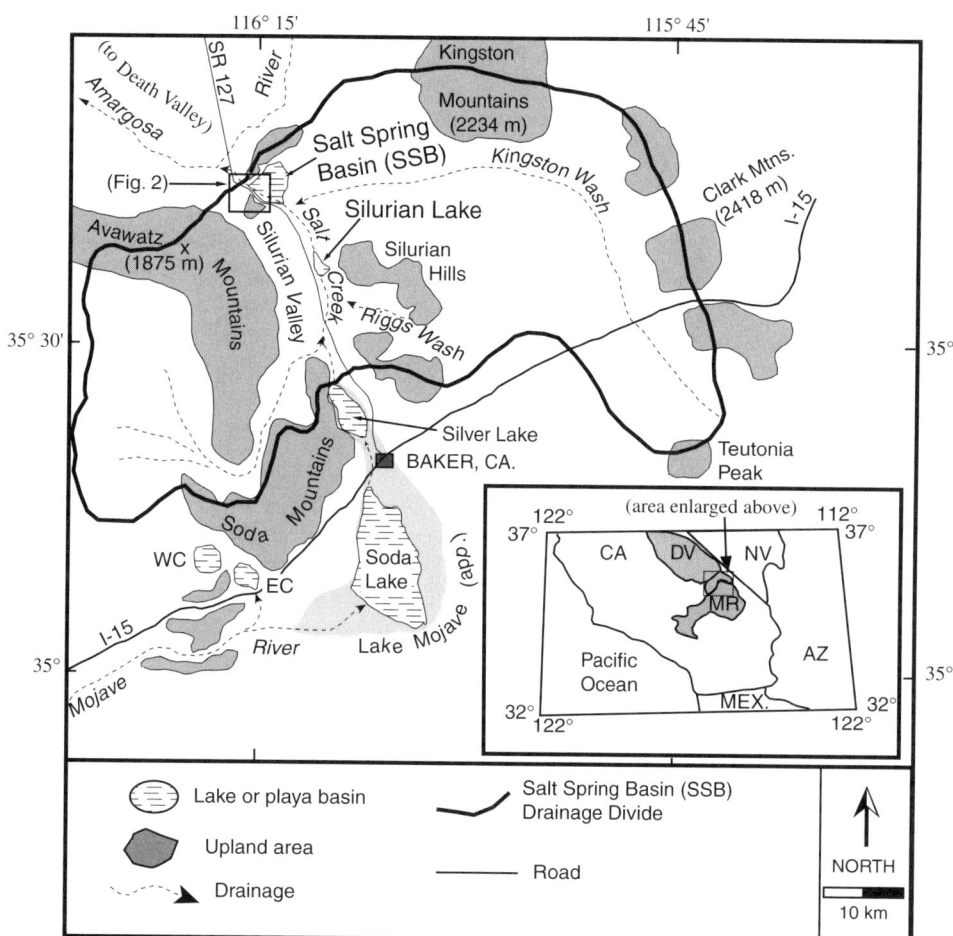

Figure 1. Map showing location of features discussed in the text. Note particularly Salt Spring basin (SSB) and Silurian Lake.

for radiocarbon control on basin-fill depositional units, and (4) its geographic location linking the Mojave and Amargosa Rivers.

PREVIOUS WORK

There has been limited geologic investigation of Silurian Lake, although the U.S. Geological Survey (USGS) is currently mapping the area. Evidence for a lake, such as lacustrine deposits, is sparse, and the absence of extensive archaeological sites, such as those found in areas of perennial water, suggests that there has not been a reliable water supply in Silurian Lake for much of the Holocene (Byrd, 1998).

The lake history of the Salt Spring basin was first studied by Malcolm Rogers (1939) in connection with archaeological investigations. He termed the area "Salt Spring basin" and inferred that a rather complex lake and erosion history had occurred. Although his notes are somewhat ambiguous, he determined that, based primarily on projectile point types and an understanding of glacial/interglacial periods, a Pleistocene lakebed dried out ca. 27 ka, and underwent a postglacial erosive episode. Between ca. 6 ka and 3.5 ka a wetter period created a lake or playa followed by a dry period from ca. 3 ka to the present. He also stated that this interpretation was entirely wrong, but did not provide a newer interpretation. As shall be seen, this first interpretation contains some surprisingly accurate inferences. Hooke (1999) discussed the prominent shoreline along the Salt Spring Hills on the southwest edge of the basin, and questioned whether the shoreline was attributed to a very large oxygen isotope stage (OIS) 6/5e lake or a more restricted OIS 2 lake.

In Silver Lake basin, which is a subbasin of Lake Mojave basin, Wells et al. (1987, 1989, 1990, this volume) identified a period of high lake stands between ca. 22 ka and 8.7 ka. Based on uncalibrated radiocarbon ages on tufa, mollusk shells, and bulk sediment, and calculated sedimentation rates, two full-lake phases range from 22 ka to 17 ka, and from 14 ka to 12 ka (Wells et al., this volume). Intermittent and ephemeral lakes continued until ca. 8.7 ka. The period from 16 ka to 14 ka was drier, with eolian deposits and desiccation cracks forming (Wells et al., this volume). Enzel (1992) identified Holocene lakes in Silver Lake at ca. 3.6 ka and 0.39 ka. In Death Valley, pluvial Lake Manly existed from ~30,000 to 10,000 yr ago (Hunt, 1975; Hooke, 1972; Lowenstein et al., 1999). Anderson (1998) and D. Anderson and Wells (this volume) further subdivided lacustrine events in Death Valley into three periods: >26 ka, 18 ka, and 12 ka.

GEOMORPHIC SETTING

The study area is dominated by the northwest-southeast–trending Avawatz Mountains, where the highest point, Avawatz Peak, rises to an elevation of 1875 m above sea level. Salt Creek flows northward through Silurian Valley from Silver Lake to its confluence with the Amargosa River, northwest of Salt Spring Hills (Fig. 1). Salt Creek's potential drainage area is currently ~1400 km^2, which includes Kingston Wash, Riggs Wash, and the western slopes of the Avawatz Mountains. The headwaters of Kingston Wash are in the Cima Dome–Teutonia Peak area, though it also drains Kingston (2234 m) and Clark (2418 m) mountains. The Mojave River presently terminates in Silver Lake basin near Baker, California, and there is no evidence that the Mojave River overflowed the Silver Lake basin or flowed through Silurian Valley since earliest Holocene or latest Pleistocene time (Enzel et al., 1989).

Silurian Lake is ~4.5 km long and 2.25 km wide, with the widest portion at the southern end; it narrows to the north to form the main Salt Creek channel flowing northward to Salt Spring basin (Fig. 1). Where Salt Creek enters Silurian Lake in the south, an alluvial fan has prograded over the playa surface. The playa is generally undissected although there are a series of deep desiccation fissures in the southern end. Alluvial fans from the Avawatz Mountains form the western piedmont, and distal portions of these fans form the western edge of the playa. To the east, fans from Valjean Valley form the northeastern periphery of the playa and fans from Riggs Wash form the southeastern periphery (Fig. 1). The east side of the Silurian Lake playa extends eastward toward the Silurian Hills. Surface runoff from the local piedmonts supplies water during wet years, and standing water may be present for several weeks during these times.

The present-day geomorphic setting of the Salt Spring basin is characterized by erosional topography (Fig. 1). The brown fine-grained Holocene deposits in the center of the basin are incised, scoured, and deflated creating a broadly undulating surface broken by a few isolated dune mounds held in place by mesquite trees. Salt Creek, which flows along the southwestern basin periphery, and a small tributary informally named Dumont Wash along the northern basin periphery, have incised up to four meters into the latest Quaternary deposits (Fig. 2). Numerous tributary gullies dissect the basin-fill material. Salt Creek, which flows perennially for ~1.5 km in Salt Spring basin, has cut a channel through the granitic northern Salt Spring Hills at a present elevation of 150 m and flows northward toward the Amargosa River and eventually into Death Valley. The elevation of the top of the basin-fill deposits, toward the center of the basin, is ~170 m. Toward the basin edges active channels have left isolated gravel remnants on top of the dissected brown and gray, fine-grained silts. A prominent shoreline on the southwestern portion of the basin has beveled quartzite colluvial material on the apron of the southern Salt Spring Hills (Fig. 3) and is cut into a late Pleistocene alluvial fan at an elevation of 177–178 m (Fig. 2). The tread of the shoreline cut into the colluvium is generally <0.5 m-wide, and generally <0.75 m-wide where it is cut into the alluvial fan.

METHODS

Surficial geologic mapping of the piedmont was conducted on aerial photographs at scales ranging from 1:30,000 to 1:12,000. The mapped units from the aerial photographs were

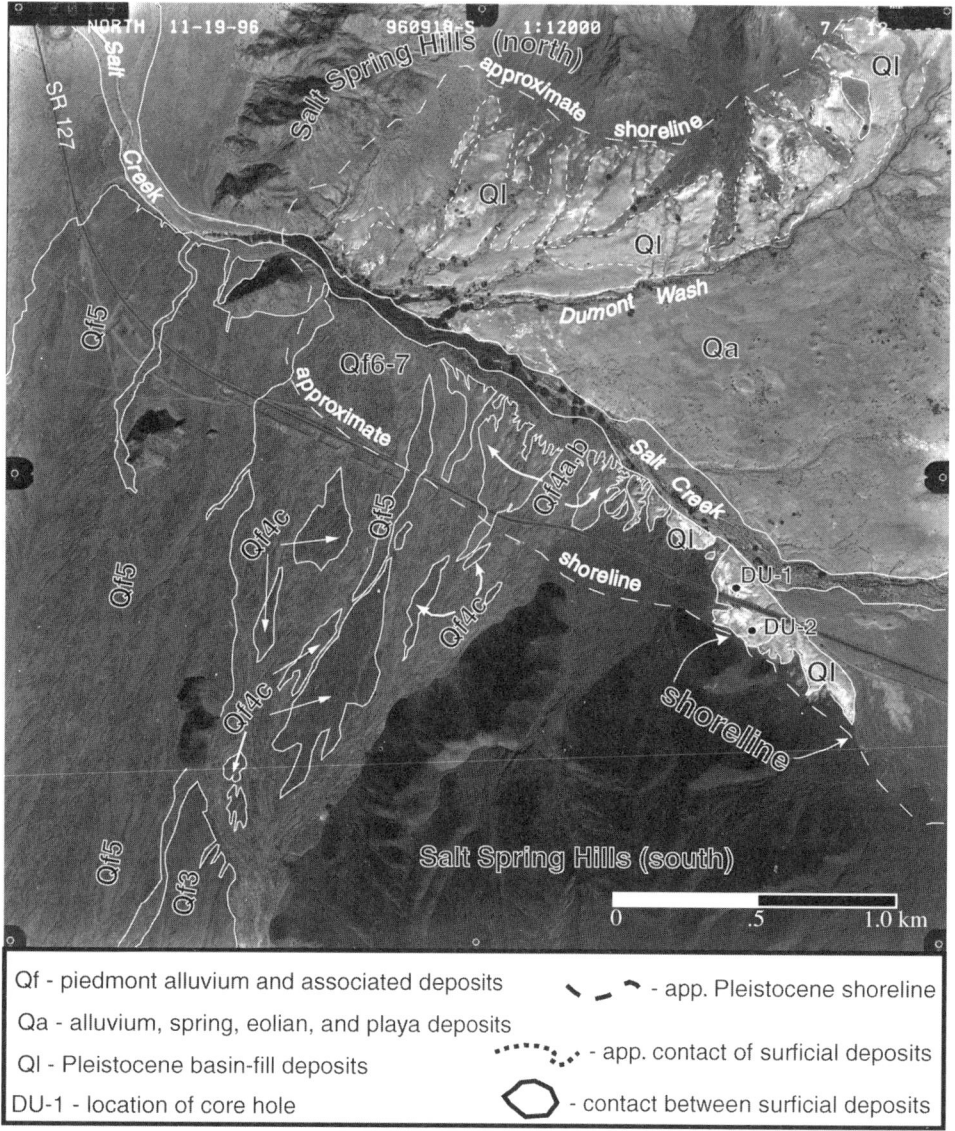

Figure 2. Aerial photograph of Salt Spring basin showing geologic contacts of surficial deposits discussed in the text. Note also the estimated location of the shoreline for late Pleistocene Lake Dumont.

transferred to USGS 7.5-minute topographic maps using a Bausch and Lomb zoom transfer scope.

Two sediment cores were collected from Salt Spring basin and one from Silurian Lake using a truck-mounted, hollow-stem auger coring rig. The maximum depth obtainable by the coring equipment was ~35 m. The two Salt Spring basin cores are each 15 m deep, and the Silurian Valley core is ~33 m deep. Deposits were described according to color, texture, sedimentary structures, and thickness following Smith (1991). The drill hole location for the Silurian Lake core was positioned toward the southern end of the playa, where Salt Creek enters the basin, to increase the possibility of collecting sediment representative of both stream and lake environments. Drill hole locations for Salt Spring basin were selected based on surface characteristics and the potential for obtaining sediments representative of lacustrine environments, which basically consisted of drilling into the grayish green silts along the southwestern basin periphery. Coring locations were restricted to within ~30 m of the road due to the designation of Salt Springs as an Area of Critical Environmental Concern by the Bureau of Land Management.

Age estimates for the deposits are based on numerical and relative dating techniques. Numeric age control is provided by accelerator mass spectrometry (AMS) radiocarbon analyses of bulk organic carbon extracted from the sediment cores (Table 1). Beta Analytic, Inc., performed the pretreatment of the carbon, and Lawrence Livermore National Laboratory performed the accelerator analysis. Pretreatment included acid-alkali-acid washes to remove carbonate and organic acid contaminants.

Pollen analysis and ostracode data were used to interpret depositional settings and inferred paleoenvironmental condi-

Figure 3. Photograph of shoreline (white arrows) cut into colluvium at the base of the southern Salt Spring Hills. Shoreline elevation is 177–178 m above sea level, and the shoreline is generally <50 cm wide. Light colored deposits in foreground are the upper section of core DU-2, which is dominated by lacustrine and spring deposits.

TABLE 1. RADIOCARBON AGES DISCUSSED IN THE TEXT

Sample no.	Beta lab no.	Depth (m)	δ^{13}C (‰)	Setting	Material	Measured (yr B.P.)	Convent. (yr B.P.)	2σ calibrated (yr B.P.)
Silurian Lake								
Si-1-I	85541	5.8	–23.5	Playa	OC	13,430 ± 60	13,450 ± 60	Reversed
Si-1-II	85542	18	–22.1	Playa	OC	9210 ± 70	9250 ± 70	ages
Lake Dumont								
DU-1097	104954	0.3	–21.6	Alluvium	Charcoal	1720 ± 50	1780 ± 50	1555–1820
DU-5-Qal	93386	0.6	–13.9	Alluvium	Charcoal	2330 ± 50	2510 ± 50	2365–2750
KCADC-1	119674	3	–26.6	Spring	OC	5200 ± 70	5180 ± 70	5750–6160
DU-3-Qal	94111	0.6	–18.5	Alluvium	Charcoal	9200 ± 60	9300 ± 60	10,080–10,390
DU-1-I	93120	6.6	–23	Spring	OC	14,870 ± 60	14,910 ± 60	–
DU-2-I	85538	0.76	–22.6	Lake	OC	18,110 ± 80	18,150 ± 80	–
DU-2-IV	88136	10.8	–23.2	Lake	OC	27,470 ± 360	27,500 ± 360	–

Note: OC—bulk organic carbon.

tions. Ostracodes were analyzed by Rick Forester, USGS, Denver (1997, written commun.). Pollen analysis was undertaken by Peter Wigand, Desert Research Institute, Reno (1998, written commun.).

RESULTS

Basin-fill investigations: Silurian Lake

The Silurian Lake core represents 33 m of sediment from the southern edge of the playa. Silurian Lake deposits are dominated by yellowish brown (10YR) sand, silt, and clay throughout the core, although rounded pebble layers were recorded at 4 and 19 m depth. Both of the pebble layers are 76 cm thick. From ~9 to 14 m depth, the sediments consist of light yellowish brown (10YR 6/4) to brown (10YR 5/3) banded sand, silt, and clay. From 19.5 to 25.5 m the core is dominated by brown (10YR 4/4) clay. Two AMS samples, from 6 and 17 m, produced uncalibrated ages of 13,450 ± 60 yr B.P. and 9250 ± 70 yr B.P., respectively. These ages are in reverse stratigraphic order. There are no grayish or green deposits throughout the entire 33 m of the core.

Basin-fill investigations: Salt Spring basin

Two cores collected from Salt Spring basin, DU-1 and DU-2, have distinctive sedimentologic characteristics, although they were collected from nearly the same elevation, ~100 m apart in the Ql mapped unit (Fig. 2). Core DU-1 is characterized by yellowish brown (10YR) gravel, sand, silt, and clay, with several

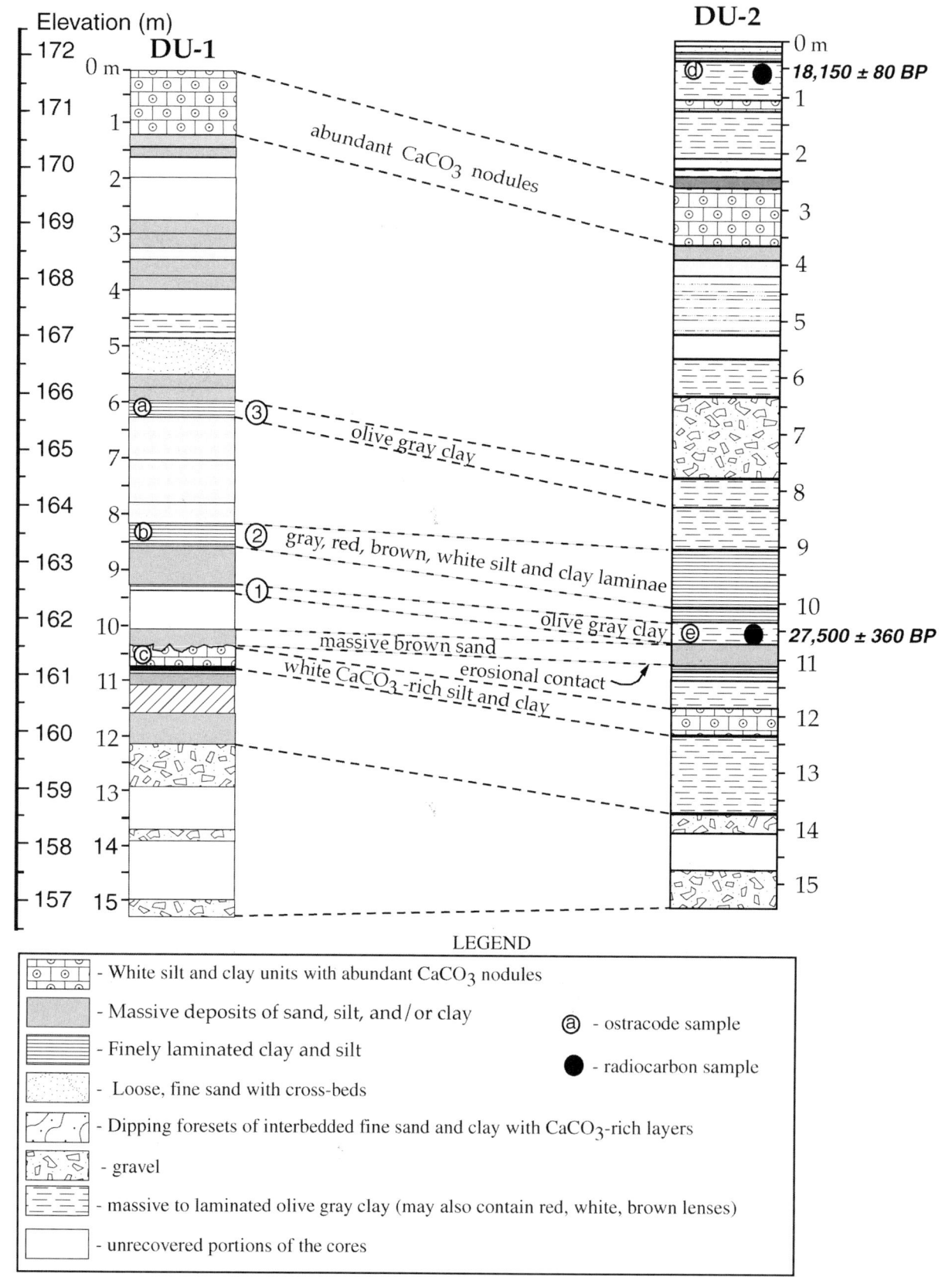

Figure 4. Stratigraphic section of cores DU-1 and DU-2 from the Salt Spring basin showing sedimentary features, location of ostracod samples, radiocarbon ages, and stratigraphic relationships between the two cores. See text for a more complete discussion.

white, salt-rich deposits (Fig. 4). Three relatively thin olive-gray (5YR) clay layers in core DU-1 are finely laminated and occur with white, brown, and reddish clay and silt layers. DU-2 is characterized by olive-gray (5YR) clay and silt that is generally finely laminated, but may be unstratified (Anderson, 1999).

In core DU-1, basal gravel gives way to sand with basinward-dipping foresets and Fe and Mn oxidized mottles (Fig. 4). Overlying the sand is a brown clay and black, sandy, organic-rich clay deposit that is overlain by 35 cm of a very finely laminated white $CaCO_3$. The top of this unit, at 161.4 m, is an erosional boundary that abruptly truncates the $CaCO_3$ layer and is overlain by massive, dark yellowish brown fine sand. Further up the core, three olive-gray clay units are banded with brown, red (2.5YR 5/2), and white (10YR 8/2) laminae. Sediment samples from two of these olive-gray clay units contain the following ostracodes: *Limnocythere staplini*, *Candona rawsonoid*, and *Heterocypric* sp. (R. Forester, 1997, personal commun.). Fine silts and sands are present for the remainder of the core, with the exception of well-sorted, very fine sand at 166.33–166.56 m. The uppermost 160 cm consists of white $CaCO_3$-rich deposits with abundant $CaCO_3$ nodules. A radiocarbon age estimate of 14,910 ± 60 yr B.P. was obtained on bulk organic sediment at a depth of 6 m. However, because this date does not agree with the reconstructed core stratigraphy, we believe it is invalid and do not include it in our interpretations (see below).

Core DU-2 is dominated by olive-gray (5Y 5/2) silt and clay deposits. Nearly 5 m of olive-gray silt and clay overlie basal gravel (Fig. 4). A 1.5 m thick, poorly sorted, matrix-supported unit occurs at 165–166.5 m, above the dominantly olive-gray clays. This massive unit, in turn, underlies another 75 cm thick olive-gray clay layer. Brown sand and clay deposits occur between 167.33 and 170.43 m, followed by another succession of olive-gray, red, and brown clay lenses. The ostracodes *Candona rawsonoid* and *C. caudata* were found at 172 m. Two AMS radiocarbon samples from 172.2 m and 161.8 m in elevation provided uncalibrated ages of 18,150 ± 80 yr B.P. and 27,500 ± 360 yr B.P., respectively (Table 1).

Younger deposits that fill the center of the basin are broadly designated Qa and consist of alluvial, playa, eolian, and spring deposits (Fig. 2). These deposits, inset below the olive-gray silts along the periphery (Ql), are subdivided into depositional units 1, 2, and 3. Unit 1 consists of a fining upward sequence of thinly bedded sand, silt, and clay, with ripple laminae and <5 cm thick trough cross-bedding preserved. A 9300 ± 60 yr B.P. age was obtained from the upper 60 cm of an ~3 m thick alluvial section of unit 1 (Table 1). The sample was collected from a 1 cm thick dark gray, ashy disseminated charcoal layer overlying a 1 cm thick oxidized red zone, possibly a burn layer.

Unit 2 consists of decimeter-thick bedded sands with minor, discontinuous clay lenses. Approximately 1 km downstream from the 9.3 ka section is a channel-fill sequence of yellowish, oxidized sand underlying olive-gray, finely laminated, reduced clays, which in turn underlie a dark brown, organic-rich sandy clay, a stratigraphic sequence similar to black mat–paleospring deposits identified in southern Nevada (Quade et al., 1995). An AMS radiocarbon age of 5180 ± 70 yr B.P. was obtained on the organic-rich layer of unit 2 (Fig. 5).

Unit 3, the upper 0.5–1.5 m of Qa, is variable, but generally consists of numerous thin black organic or charcoal bands, which are small channels filled with bedded sand cemented with gypsum and $CaCO_3$. The upper 0.5 m yielded an AMS radiocarbon age of 2510 ± 50 yr B.P. (Fig. 6).

Figure 5. Photograph of middle Holocene deposits of the interior of Salt Spring basin. The green clay underlying the dark organic layer is reminiscent of late Pleistocene black mat/paleospring deposits found elsewhere in the southwest.

Inset into the early to middle Holocene deposits discussed above are sandy deposits overlain by alluvial fan gravels of the Avawatz Mountains fan sequence. Charcoal within the upper 2 cm of this alluvium yielded an AMS radiocarbon age of 1780 ± 50 yr B.P. It is clear from the above chronostratigraphy that late Pleistocene deposits occur along the margins of the basin and a remarkably complete Holocene section is preserved in the basin interior.

Piedmont stratigraphy related to basin-fill deposits of Salt Spring basin

Alluvial fans along the eastern and northeastern piedmonts contain seven general units, designated oldest to youngest, Quaternary fans (Qf) 1 through 7 (Fig. 2). Alluvial fans on the piedmont were investigated with regard to geomorphic position and soil development for relative age control. Fan units Qf1 and 2 were not investigated because they occur only along the mountain front and in mountain drainages. The following discussion focuses on units Qf3 through Qf7.

Qf3 is the oldest fan unit investigated, with dark varnish covering nearly all of the clast surfaces, subdued bar-and-swale topography, and Bt and stage II+ Bk soil horizons. Calcium carbonate nodules are common with some whitening of the matrix (Gile et al., 1966; Birkeland, 1999). Many of the plutonic clasts within the Qf3 fan are highly weathered, fractured, and grussified.

The Qf4 fan complex is subdivided into three stratigraphic units: Qf4a, b, and c (Fig. 2). Qf4a and b are buried by the younger Qf4c unit (Fig. 7). Qf4 is highly dissected and best observed in exposed section, where it is characterized by a smooth, gray, thick, very hard gypsic crust on eroded slopes. It has a gently dipping, planar surface that contrasts with the higher gradients of the overlying alluvial fan deposits. Qf4a consists of the following deposits: (1) a series of massive, muddy to sandy, matrix-supported layers; and (2) well-sorted fine to coarse sand lenses and channels commonly cemented with $CaCO_3$. The upper sandy lenses grade into the Qf4b unit. The Qf4b unit consists of horizontally bedded, $CaCO_3$-cemented, sand and gravel layers, and shallow channels (<1 m deep). Abundant, vermicular, 1 cm diameter and decimeter-long trace fossils can be found on bedding planes of the cemented sandstone. It seems that the same distinct, soft, chalky, white $CaCO_3$-rich zone that occurs at the top of cores DU-1 and DU-2 also occurs at the top of Qf4b. The uppermost layers of Qf4b are truncated by the overlying Qf4c fan unit. Qf4c is a relatively thin (<2 m) alluvial fan unit characterized by angular fan gravels that unconformably overly Qf4b deposits and thicken toward the piedmont. Soils developed in the Qf4c unit consist of a reddish brown Bw horizon. Qf4c consists of a clast-supported and minor matrix-supported unit, and a well-sorted sandy facies.

Qf5 occurs as a broad fan sequence extending from the mountain front to the basin floor (Fig. 2). Qf5 grades to the approximate level of the dated Holocene deposits in the center of the basin and exhibits a similar degree of dissection as well. Also characteristic of the Qf5 fan complex are occasional very large (>2.5 m diameter) boulders. Soils have slight reddening and thin stage I carbonate accumulation on clast bottoms. The vesicular A horizon averages 3–5 cm thick. Primary depositional sedimentary structures have not been disturbed by pedogenic processes.

The Qf6a and Qf6b units consist of thin fan gravels that grade to gravels on strath terraces along Salt Creek. In one location, the Qf6 fan overlies Salt Creek alluvium radiocarbon dated to 1780 ± 50 yr B.P. Qf7, the modern stream deposits, is characterized by (1) boulder levees and (2) extensive, fine-grained deposits with smooth surfaces and desiccation cracks.

DISCUSSION

Cores collected from Silurian Lake and Salt Spring basin have recorded sedimentation of generally fine textured deposits for the latest Pleistocene. The surficial deposits preserved in the Salt Spring basin represent latest Pleistocene and early, middle, and late Holocene deposits that provide information on the basin paleohydrology, and constraints on the paleohydrology of the Mojave River, allowing inferences to be made about regional lacustrine correlations with Lake Mojave and the latest Pleistocene Death Valley lakes. In addition, an estimated age of the prominent shoreline along the Salt Spring Hills is inferred.

Paleohydrology of Silurian Lake

The 33 m deep Silurian Lake core represents exclusively subaerial deposition as commonly found in playa and distal alluvial fan settings. The 10YR brown and yellowish brown banded sand, silt, and clay probably represent runoff from the surrounding piedmont, ponding, and settling in a playa environment. The rounded pebble layers do not represent local runoff events, but rather may indicate streamflow conditions, perhaps from higher energy flows such as would be expected if the Mojave River flowed through Silurian Valley. The reverse order of the radiocarbon samples is unfortunate and precludes age interpretations. Even if we assume that the ages are correct, and the older age represents erosion and redeposition of older material over younger deposits, as appears common in some settings (Stanley and Hait, 2000), more dates are necessary for proper evaluation. Perhaps the only definite inferences to be made from the core stratigraphy are that there appears to be no evidence for lacustrine phases throughout the entire length of the 33 m core, and that there does seem to be evidence for through-flowing streams, though their timing and significance is unclear.

Also, the question arises, why is the playa here? There may be a tectonic component, although there is no evidence for extensional faulting in the immediate area. It may be that a persistent lake has formed and lakeshore action trims the fan edges, although there is little evidence of shorelines. The preferred hypothesis is that the playa occurs because of the proximity of the Silurian Hills to the eastern playa boundary, thereby decreasing the drainage area and source area for supplying sediment to

Figure 6. Photograph of Unit 3, late Holocene deposits of Salt Spring basin. These deposits are characterized by channels cut into Unit 2, commonly filled with either sand and silt, or gravel deposits.

Figure 7. Photograph of late Pleistocene alluvial fan (Qf4c) overlying the fan delta (Qf4a, b) deposits of Salt Spring basin.

the valley center. That is, if the Silurian Hills were not present, the Silurian Lake playa would not be present.

Late Pleistocene paleohydrology of Salt Spring basin

Sediment cores DU-1 and DU-2, and units Qf4a and b represent different facies during the same period of basin-fill activity. We believe that Qf4a and b represent a fan-delta sequence from the Avawatz Mountains, that DU-1 represents a distal, generally subaerial to subaqueous sequence, and that DU-2 represents the dominantly subaqueous deeper water facies that possibly received water from both the Avawatz Mountains and Silurian Valley sources, such as spring discharge and streamflow. In our interpretation, we follow Smith (1991) and identify finely laminated to massive bedded 5Y 2/1, olive-gray clay as shallow lakes. Areas of olive-gray clay interlaminated with pink and white clays are tentatively identified as shallow saline lakes or possibly paleospring deposits. Ostracod ecology helps in the interpretation of depositional environments for selected sediment samples.

A comparison of cores DU-1 and DU-2 reveals significant sedimentologic and stratigraphic similarities that help interpret depositional environments (Fig. 4). Because core DU-1 represents a shallow, dominantly subaerial facies and core DU-2 represents a dominantly subaqueous facies, elevational differences between correlated units may be due to a variety of reasons, such as original depositional topography or erosion. Both cores met refusal at nearly the same 156 m elevation, where gravels were encountered. A calcium carbonate–rich zone occurs in both cores in about the same stratigraphic location, in core DU-1 between 161 and 161.4 m and in DU-2 at 159.6 and 160.8 m. At 161.4 m in both cores there is an erosional boundary buried by massive flood sand. At three elevations, designated (1), (2), and (3) in Figure 4, laminated olive-gray (5Y) clays are in the same stratigraphic position in both cores. A particularly diagnostic sedimentary sequence, designated (2) in Figure 4, consists of red, gray, and brown laminated clays that can be correlated from DU-1 (163.3–163.6 m) to DU-2 (162.4–163.6 m). Subaerial deposits that can be correlated include the eolian deposits of DU-1 at 166.5 m with the alluvial fan gravels in DU-2 at 166 m. Near the top of each core, sediments containing $CaCO_3$ nodules are also correlated. Stratigraphic and sedimentologic evidence indicates that cores DU-1 and DU-2 represent the same interval of depositional time. However, radiocarbon ages suggest that they represent a different period of basin-fill activity. We accept the facies interpretations and therefore the $14,900 \pm 60$ yr B.P. age from 165.8 m elevation in core DU-1 is not used in the chronostratigraphic reconstructions.

The Qf4a and Qf4b depositional units represent a fan-delta sequence. Qf4a is dominated by matrix-supported debris flows 20–40 cm thick, with 10 cm thick sandy lenses increasing in frequency toward the top of the section. Qf4b is a sand and gravel channel facies of a braided distributary stream across the fan-delta complex. The sand lenses are commonly cemented and form distinctive, resistant channel lenses, often with trace fossils on bedding planes. The upper portion of the Qf4b unit consists of a white $CaCO_3$-rich horizon that correlates stratigraphically, elevationally, and in relative thickness to the white zone at the top of cores DU-1 and DU-2 that we believe represents a significant desiccation event (Fig. 4).

Depositional units Qf4a and b represent facies changes of the same aggradational episode as the two sediment cores DU-1 and DU-2. Field relationships indicate interfingering of yellowish brown subaerial sands with olive-gray subaqueous silts and clays, suggesting rapid textural and color changes over short distances (cm), which further indicates fan-delta conditions. Properties common to the fan-delta–lacustrine deposits include (1) distinct color change over only a few decimeters, with reddish brown sands grading to green and olive gray sand and clay; (2) cemented sand lenses dipping at ~10° basinward in the fluvial facies; (3) massive, poorly sorted debris flow facies; and (4) marginal facies containing trace fossils, eolian and playa deposits, desiccation cracks, and $CaCO_3$-cemented deposits. These properties, identified in the deposits of Salt Spring basin, are generally accepted as indicative of arid zone fan-delta deposits (Wescott and Etheridge, 1990; Etheridge, 1985).

Ostracodes were found in three of the five samples investigated. Samples (see Figure 4, a and b) yielded three ostracod types that are diagnostic of shallow groundwater and spring-supported wetlands: *Limnocythere staplini*, *Canadona rawsonoid*, and *Heterocyprix* sp. Sediments associated with sample (a) are light olive-gray (5Y 6/2) clay banded with white and brown laminae, and for sample (b) are olive-gray (5Y 5/2) clay finely banded with white, brown, and red silt and clay. The other productive sample, d, contained ostracodes diagnostic of lake conditions that receive the majority of their water from streams: *Canadona rawsonoid* and *Canadona caudata* (R. Forester, 1997, personal commun.). Sediments associated with these deposits are massive olive-gray clay (5Y 5/2) with a few brown lenses. The brown lenses are interpreted as flooding events.

Interpretations from the pollen fraction are somewhat tentative due to the limited pollen data. Abundant pine pollen, possibly derived from local high-elevation ranges such as the Kingston or Avawatz Mountains, and local desert scrub pollen, such as *Ambrosia*, *Poacea*, and *Ephedra*, were found in the 18 ka lake. The presence of *Ephedra* suggests that the local vegetation was at the mesic end of the desert scrub community during that time (P. Wigand, 1997, personal commun.).

Lacustrine chronostratigraphy of Lake Dumont: Core DU-2

One of the main goals of these basin investigations is to understand the sequence of lacustrine events of Lake Dumont, which is best represented in core DU-2. Because core DU-1 represents a shallow water to subaerial facies, and because of the apparent discrepancy in a radiocarbon age estimate from that core, the following discussion focuses on core DU-2 (Fig. 4).

Bulk sediment samples collected from 0.8 and 10.8 m in depth from DU-2 provided uncalibrated AMS radiocarbon ages of $18,150 \pm 80$ yr B.P. and $27,500 \pm 360$ yr B.P., respectively. Using these ages and depths, an inferred sedimentation rate for DU-2 is 1.07 m per 1000 yr. Using this sedimentation rate, the ages of wetter (intermittent lake, perennial lake, spring) and drier periods (alluvial fan, playa) are estimated. The oldest, most persistent wet period lasted from ca. 30 ka to 25.3 ka. Following this period, alluvial fan progradation onto a dry basin deposited ~1.5 m of sand and gravel. This drier period occurred ca. 24.7 ka, followed by another wetter period with 0.75 m of gray clay ca. 23.4 ka. Between ca. 23.4 and 19.4 ka the basin experienced periods of intermittent saline lake–playa and distal fan activity, which deposited nearly 3.0 m of brown sand and sandy clay, culminating in nodular carbonate and white clay deposits thought to represent desiccation. Lake conditions returned between ca. 19.4 and 18 ka, represented by nearly 2 m of olive-gray clay and brown sandy clay deposits indicating moister conditions. Again, evidence for a lake at 18 ka is provided by ostracode analysis; the presence of *Candona rawsonoid* and *C. caudata* indicates a system dominated by streamflow rather than spring activity (R. Forester, 1997, personal commun.).

Using the above sequence for interpreting wetter and drier periods in the Salt Spring basin, we present the following scenario of the latest Pleistocene paleohydrology for the Mojave River, from Silver Lake through Silurian Valley to Death Valley. The two ages of 27.5 ka and 18.1 ka from core DU-2 correspond to a period of lacustrine activity in the Death Valley lake system when lakes have been recorded at 26 ka and 18 ka (Anderson, 1998). In addition, the period 22–17 ka approximates the Lake Mojave I phase in the Silver–Soda Lake basins (Wells et al., this volume). During this time, at ca. 18 ka, water from the Mojave River drainage overflowed from Lake Mojave, flowed through Silurian Valley (and Silurian Lake), and reached the Salt Spring basin, adding significant amounts of water to Lake Dumont. Prior to the input of Mojave River waters (>22 ka), Lake Dumont was fed by spring and surface flow from Silurian Valley, Kingston Wash, and the Avawatz Mountains, presumably under the same generally moister climatic conditions that allowed the Amargosa River to be the only contributor to Death Valley lakes during this period (Anderson, 1998). However, after the overflow of Lake Mojave the addition of significant discharge down Silurian Valley and into Salt Spring basin would have filled the basin with water, creating the highstand of Lake Dumont at approximately 19.4–18 ka. Because the Salt Spring basin is a relatively small basin (<10 km^2), such a postulated increase in discharge probably overwhelmed the basin, and the sill was eventually breached. How long the lake lasted before filling the basin and breaching the sill is unclear, but the ages based on sedimentation rates from core DU-2 suggest that the lake persisted nearly continuously from ca. 19.4 ka to shortly after 18 ka. We postulate that the Salt Spring basin sill was breached sometime after ca. 18 ka, although overflow may have occurred for some time prior to that. We also suggest this is the most likely time for the formation of the prominent shoreline cut into the colluvial material along the southern Salt Spring Hills and into the late Pleistocene alluvial fan. It is possible that the shoreline was formed during this 19.4–18 ka interval when waters persisted at the 178 m elevation during periods of overflow and prior to breaching of the basin. After 18 ka, the breached basin continued to experience episodic flow from the Mojave River through Silurian Valley, as well as from the Avawatz Mountains. The continued discharge scoured the center of the basin sometime after 18 ka and prior to 9.3 ka.

Holocene basin-fill deposits of Salt Spring basin

A period of alluviation began sometime prior to the early Holocene, as shown with the 9300 yr B.P. age sampled from the upper 60 cm of a 3 m section (Fig. 8). The late Pleistocene to early Holocene period of aggradation in Salt Spring basin could have resulted from increased discharge along the Mojave River through Silurian Valley during Lake Mojave II time, ca. 13– 8.7 ka (Wells et al., 1989). The increase in discharge may have been accompanied by an increase in sediment supply during the Pleistocene to Holocene transition, perhaps resulting from a decrease in vegetative cover and landscape instability (Bull, 1991; Wells et al., 1987). The Salt Spring basin may have acted as a sediment sink where material was trapped at the bedrock-Qf5 fan constriction at a much lower elevation (159 m) than the Pleistocene level (175 m). The 9.3 ka unit 1 sediments are characterized by streamflow and overbank deposition along Salt Creek.

During the middle Holocene, Salt Spring basin experienced an increase in moisture represented by a dark gray, organic-rich deposit overlying an olive-green layer. This couplet is common for black mat spring-fed deposits common in the latest Pleistocene but is rare in the middle Holocene (Quade et al., 1995). An AMS radiocarbon age of 5180 ± 70 yr B.P. on the dark layer approximates the same time as two dates from East Cronese Lake: 5375 ± 180 yr B.P. on lake tufa (Drover, 1979) and 5770 ± 60 yr B.P. on bulk organic sediment (Anderson and Wells, 1996), as well as 5.3–4.4 ka-old peat deposits at Ash Meadows (Mehringer and Warren, 1976). These correlative ages may suggest a regional increase in moisture during this generally arid period (Spaulding, 1991).

Although paleospring deposits associated with discharge from Salt Springs and groundwater discharge from Silurian Valley are present, they are a minor contribution in the deposits investigated. Clearly, spring discharge from Silurian Valley and the Avawatz Mountains contributed to the late Pleistocene and Holocene basin-fill record, but much of the record represents surface flow attributed to stream activity during the late Pleistocene from Kingston Wash, Silurian Valley, and the Mojave River, and during the Holocene from Kingston Wash and Salt Creek.

The sediments that were deposited ca. 2.5 ka are dominated by alluvial sand and gypsum-encrusted playa sediments toward the top of the basin-fill sequence. Stratigraphic relationships and the location of radiocarbon ages near the top of the basin-fill sections that span the entire Holocene (9.3 and 2.5 ka) indicate reworking of the basin-fill deposits. However, archaeological evidence indicates that prehistoric sites located on the surface of the basin-fill deposits are intact, suggesting that at least portions of these deposits have remained relatively stable for thousands of years. The inset well-sorted, medium to very coarse, clean sands represent a minor aggradation of Salt Creek ca. 1.7 ka.

CONCLUSIONS

Based on radiocarbon ages, calculated sedimentation rates, and geomorphic relationships, Lake Dumont lacustrine activity includes (1) an early phase of shallow lakes and spring activity from ca. 30 ka to 25.3 ka, (2) desiccation from 25.3 ka to 23.4 ka, (3) a series of shallow lakes centered ca. 23.4 ka, (4) a drier period between 23.4 ka and 19.4 ka, (5) lakes and possible spring activity from 19.4 ka to 18 ka, and (6) probable basin breaching after 18 ka as a result of increased discharge of the Mojave River down Silurian Valley from overflow of Lake

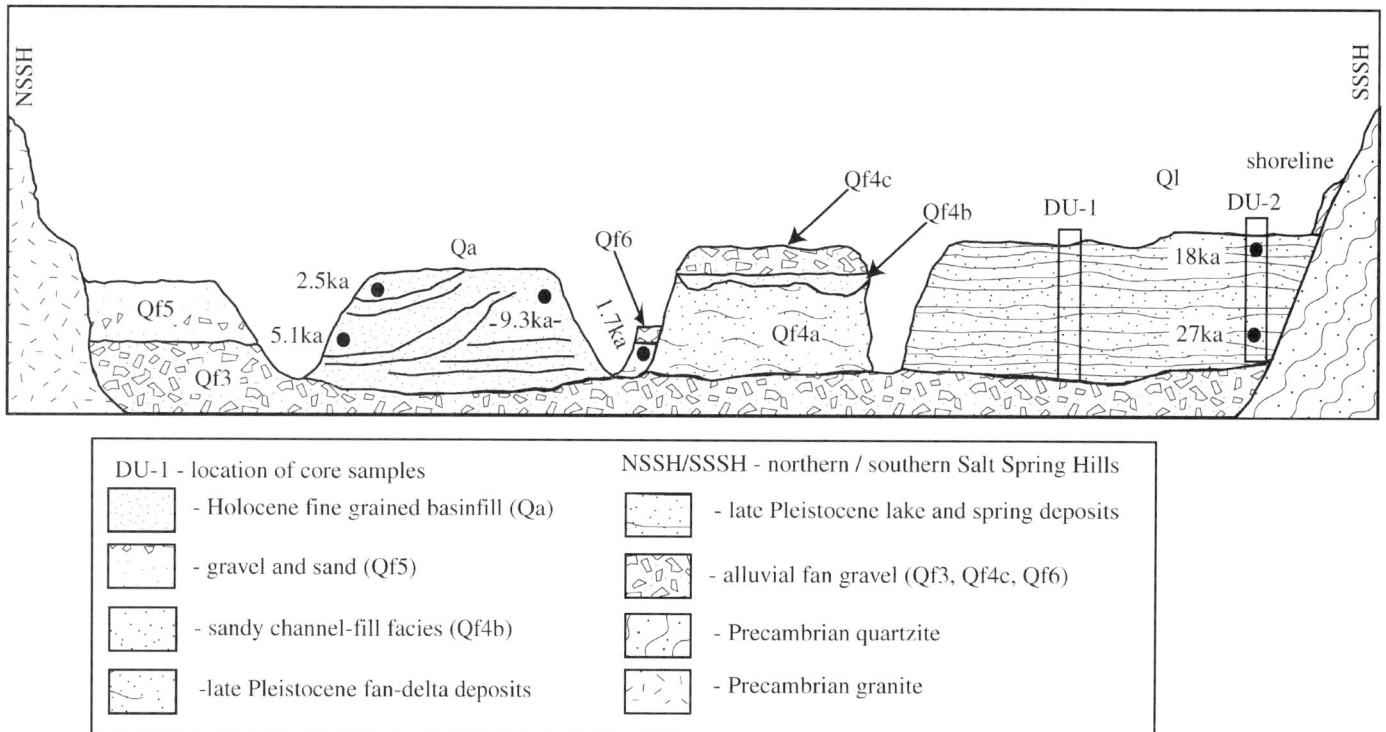

Figure 8. Schematic cross section of Salt Spring basin showing geomorphic relationships and age ranges of basin-fill deposits.

Mojave. The Lake Dumont chronology correlates with lake activity in Death Valley for the 30 ka to 18 ka period, and Lake Mojave for the 18 ka period, providing constraints on the timing of Mojave River contributions to Death Valley lakes. Lake Dumont was probably fed by local streams and springs from the Avawatz Mountains and Kingston Wash for the earlier phases prior to the Lake Mojave overflow. These earlier lakes are therefore indicative of moister climatic conditions within the Mojave Desert, which also allowed lakes to form in Death Valley from increased discharge down the Amargosa River. The prominent 178 m shoreline along the colluvial apron of the Salt Spring Hills that cuts into a late Pleistocene alluvial fan is correlated with the ca. 18 ka (OIS 2) high lake stand. Holocene valley-fill alluvium was intermittent, with several cut and fill episodes, but in general aggradation began prior to 9.3 ka, coincident with Lake Mojave II, and ended at ca. 2.5 ka. The middle Holocene experienced increased spring discharge at ca. 5.1 ka, represented by black mat paleospring deposits. A small terrace underlying a latest Holocene alluvial fan is dated to 1.7 ka. There is no evidence for perennial lakes throughout the 33 m deep Silurian Lake core. Although age control is lacking, the absence of evidence for lacustrine deposits suggests that there were no lakes filling Silurian Valley during the late Pleistocene or Holocene. Rounded pebbles indicate higher energy flows that may represent discharge of the Mojave River through Silurian Valley.

ACKNOWLEDGMENTS

Anderson would like to thank Diana Anderson, Jeff Knott, George Elder, Tim Dolan and Yvonne Wood for useful discussions and help in the field; Pete Sadler and Stephen Wells for providing insights into lacustrine sedimentation; Rick Forester for ostracod analysis and interpretation; Peter Wigand for pollen analysis and interpretation; Rob Fulton at the Desert Studies Center, Zzyzx, for maintaining the research station at Sheep Creek Springs; Mary Elder and Francis Kirk for keeping the field camp together; and Louella Holter for editing the manuscript. Jeff Knott and John Ritter provided very helpful comments that greatly improved the original manuscript. Funding and support was provided by ASM Affiliates, San Diego; Desert Research Institute, Reno; Bilby Research Center at Northern Arizona University; and the University of California, Riverside.

REFERENCES CITED

Anderson, D.E., 1998, Late Quaternary paleohydrology, lacustrine stratigraphy, fluvial geomorphology, and modern hydroclimatology of the Amargosa River/Death Valley hydrologic system, California and Nevada [Ph.D. thesis]: Riverside, University of California, 521 p.

Anderson, K.C., 1999, Processes of vesicular horizon development and desert pavement formation on basalt flows of the Cima volcanic field and alluvial fans of the Avawatz Mountains piedmont, Mojave Desert, California [Ph.D. thesis]: Riverside, University of California, 191 p.

Anderson, K.C., and Wells, S.G., 1996, Late Pleistocene lacustrine record of Lake Dumont and the relationships to pluvial Lakes Mojave and Manly: Geological Society of America Abstracts with Programs, v. 28, no. 7. p. A-458.

Birkeland, P.W., 1999, Soils and geomorphology (3rd edition): New York, Oxford University Press, 430 p.

Bull, W.B., 1991, Geomorphic responses to climatic change: New York, Oxford University Press, 326 p.

Byrd, B., editor, 1998, Springs and lakes in a desert landscape: Archaeological and paleoenvironmental investigations in the Silurian Valley and adjacent areas of southeastern California: Encinitas, California, ASM Affiliates, 848 p.

Drover, C.E., 1979, The late prehistoric human ecology of the Northern Mohave Sink, San Bernardino County, California [Ph.D. thesis]: Riverside, University of California, 257 p.

Enzel, Y., 1992, Flood frequency of the Mojave River and the formation of late Holocene playa lakes, southern California, USA: The Holocene, v. 2, p. 11–18.

Enzel, Y., Cayan, R.D., Anderson, R.Y., and Wells, S.G., 1989, Atmospheric circulation during Holocene lake stands in the Mojave Desert: Evidence of a regional climatic change: Nature, v. 341, p. 44–48.

Etheridge, F.G., 1985, Modern alluvial fans and fan deltas: Lecture notes for short course No. 19: Society of Economic Paleontologists and Mineralogists, p. 101–126.

Gile, L.H., Peterson, F.F., and Grossman, R.B., 1966, Morphological and genetic sequences of carbonate accumulation ion desert soils: Soil Science, v. 101, p. 347–360.

Hooke, R.L., 1972, Geomorphic evidence for late Wisconsin and Holocene tectonic deformation, Death Valley, California: Geological Society of America Bulletin, v. 83, p. 2073–2098.

Hooke, R.L., 1999, Lake Manley (?) shorelines in the eastern Mojave Desert, California: Quaternary Research, v. 52, p. 328–336.

Hunt, C.B., 1975, Death Valley geology, ecology, archaeology: Berkeley, University of California Press, 234 p.

Lowenstein, T., Jianren, L., Brown, C.B., Roberts, S.M., Lung, K.T., Shangde, L., and Wenbo, Y., 1999, 200,000 year paleoclimate record from Death Valley salt core: Geology, v. 27, p. 3–6.

McFadden, L.D., Ritter, J.B., and Wells, S.G., 1989, Use of multiparameter relative-age methods for age estimation and correlation of alluvial fan surfaces on a desert piedmont, eastern Mojave Desert, California: Quaternary Research, v. 32, p. 267–290.

Mehringer, P.J., Jr., and Warren, C.N., 1976, Marsh, dune and archaeological chronology, Ash Meadows, Amargosa Desert, Nevada, in Elston, R., and Headrick, P., eds., Holocene environmental change in the Great Basin: Nevada Archaeological Survey Research Paper 6, p. 120–150.

Quade, J., Mifflin, M.D., Pratt, W.L., McCoy, W., and Burckle, L., 1995, Fossil spring deposits in the southern Great Basin and their implications for changes in water-table levels near Yucca Mountain, Nevada, during Quaternary time: Geological Society of America Bulletin, v. 107, p. 213–230.

Ritter, J.B., 1987, The response of alluvial-fan systems to late Quaternary climate change and local base-level change, eastern Mojave Desert, California [M.S. thesis]: Albuquerque, University of New Mexico.

Rogers, M., 1939, Early lithic industries of the lower basin of the Colorado River and adjacent areas: San Diego Museum Papers No. 3.

Smith, G.I., 1991, Stratigraphy and chronology of Quaternary-age lacustrine deposits, in Morrison, R.B., ed., Quaternary nonglacial geology of the conterminous United States: Boulder, Colorado, Geological Society of America, The Geology of North America, v. K-2, p. 339–345.

Spaulding, W.G., 1991, A middle Holocene vegetation record from the Mojave Desert of North America and its paleoclimatic significance: Quaternary Research, v. 35, p. 427–437.

Stanley, D.J., and Hait, A.K., 2000, Deltas, radiocarbon dating, and measurements of sediment storage and subsidence: Geology, v. 28, p. 295–298.

Wells, S.G., McFadden, L.D., and Dohrenwend, J.C., 1987, Influence of late Quaternary climatic change on geomorphic and pedogenic processes on a desert piedmont, eastern Mojave Desert, California: Quaternary Research, v. 27, p. 130–146.

Wells, S.G., Anderson, R.Y., McFadden, L.D., Brown, W.J., Enzel, Y., and Miossec, J.L., 1989, Late Quaternary paleohydrology of the eastern Mojave River drainage basin, southern California: Quantitative assessment of the late Quaternary hydrologic cycle on a large arid watershed: Las Cruces, New Mexico, New Mexico Water Resources Research Institute Report 242, 250 p.

Wells, S.G., McFadden, L.G., and Harden, J., 1990, Preliminary results of age estimations and regional correlations of Quaternary alluvial fans within the Mojave Desert region of southern California, in Reynolds, R.E., et al., eds., At the end of the Mojave: Quaternary studies in the eastern Mojave Desert: Redlands, California, Special Publication of the San Bernardino County Museum Association, 1990 Mojave Desert Quaternary Research Center Symposium, May 18–21, 1990, p. 45–54.

Wescott, W.A., and Etheridge, F.G., 1990, Fan deltas: Alluvial fans in coastal settings, in Rachocki, A.H., and Church, M., eds., Alluvial fans: A field approach: New York, John Wiley and Sons, p. 195–211.

MANUSCRIPT ACCEPTED BY THE SOCIETY AUGUST 1, 2002

Geological Society of America
Special Paper 368
2003

Isotopic and geochemical evidence for Holocene-age groundwater in regional flow systems of south-central Nevada

Timothy P. Rose
M. Lee Davisson
Lawrence Livermore National Laboratory, Livermore, California 94550, USA

ABSTRACT

Regional groundwaters in the north-central Death Valley flow system exhibit $\delta^{18}O$ values that are 1‰–2‰ depleted relative to modern local recharge, and were previously interpreted to have originated from Pleistocene-age recharge. New evidence presented in this study is consistent with Holocene ages for these groundwaters. Isotopic, chemical, and hydrogeologic data suggest the Death Valley groundwaters are transported southward up to 300 km through regional carbonate aquifers of the Railroad Valley flow system. Groundwaters in the northern part of the flow system have $\delta^{18}O$ values nearly identical to recharge derived from winter precipitation in central Nevada. Isotopic analyses of snow cores revealed kinetic isotope enrichments in the snowpack *prior to melting* may account for observed isotopic enrichments in groundwaters relative to the global meteoric water line. Geographically consistent trends in $\delta^{18}O$ values, water chemistry, and carbon isotopes are observed along the entire 300 km regional flow path. Dissolved inorganic carbon quickly reacts with the carbonate aquifer due to elevated subsurface temperatures and cation exchange processes, yielding ^{14}C values <10 pmc within 30 km of the northernmost recharge area. Groundwater mixing along the flow path maintains uniformly low nonzero ^{14}C values. The complexities of the flow system preclude the application of ^{14}C age correction models. Hydrologic mass balance calculations imply turnover rates are too high to store significant amounts of pluvial recharge in the regional carbonate aquifer. Closed alluvial basins provide more favorable geologic environments for remnant accumulations of Pleistocene groundwater, as indicated by rare, isolated pockets of unusually low $\delta^{18}O$ groundwater.

INTRODUCTION

The eastern half of the Great Basin is underlain by an immense carbonate aquifer system that channels regional groundwater flow into southern Nevada from topographically high recharge areas in eastern and central Nevada. Rapid population growth in southern Nevada has motivated a number of studies aimed at determining the feasibility of tapping these vast groundwater reserves (e.g., Mifflin and Hess, 1979; Prudic et al., 1995; Dettinger et al., 1995; Thomas et al., 1996). In southwestern Nevada, similar studies are under way to assess the long-term risk of contaminant transport from underground nuclear test cavities at the Nevada Test Site (e.g., Laczniak et al., 1996; D'Agnese et al., 1997; U.S. Department of Energy, 1997). A key issue in each of these investigations is the rate at which the carbonate aquifers are replenished, and the concomitant rate at which water is transmitted through the system.

Part of the difficulty in assessing recharge and flow rates in this region is a lack of understanding regarding where the groundwater originates, and how disparate basins are interconnected. For example, only one regional flow system is believed to extend all the way from the recharge areas of east-central Nevada into

Rose, T.P., and Davisson, M.L., 2003, Isotopic and geochemical evidence for Holocene-age groundwater in regional flow systems of south-central Nevada, *in* Enzel, Y., Wells, S.G., and Lancaster, N., eds., Paleoenvironments and paleohydrology of the Mojave and southern Great Basin Deserts: Boulder, Colorado, Geological Society of America Special Paper 368, p. 143–164.

the low desert discharge areas of southern Nevada—the White River flow system in eastern Nevada (Eakin, 1966; Harrill et al., 1988). The possibility that regional flow systems of similar magnitude may transect central Nevada was not previously examined in detail because surface evidence for such flow systems is lacking. In this paper, regional inflow from central Nevada is investigated as a possible means of interpreting environmental isotope data from the northwestern Nevada Test Site (Fig. 1).

Environmental isotope studies (e.g., studies using δD, $\delta^{18}O$, $\delta^{13}C$, ^{14}C, ^{3}H data) can help to constrain regional flow systems by providing evidence for the origin and transport rate of groundwater. Many groundwater isotope studies have already been conducted in southern Nevada (e.g., Winograd and Friedman, 1972; Winograd and Pearson, 1976; Claassen, 1986; White and Chuma, 1987; Benson and Kleiforth, 1989; Thomas et al., 1996; Hershey and Acheampong, 1997; Davisson et al., 1999). However, most of these studies focused on the region south of latitude 38°N. Relatively little work has been done to assess the isotope hydrology of the recharge areas in eastern and central Nevada, or to evaluate possible links between these recharge areas and major discharge areas in southern Nevada.

In this paper, we examine the patterns and variations of groundwater chemistry and isotope data in the Railroad Valley flow system (central Nevada), and investigate the possibility that this region is hydraulically connected to the Death Valley regional flow system in southern Nevada (Fig. 1). The process of evaluating this flow path has permitted a more rigorous test of the concepts developed in our previous study of regional flow in Nevada (Davisson et al., 1999). In general, the results presented here validate our earlier conclusions that groundwater mixing and water-rock interaction are the major controls on oxygen and carbon isotope variations, respectively. In addition, we bring new evidence to bear on the stable isotope signatures inherited during recharge, regional variations in groundwater solute chemistry, and the distribution of paleo-groundwater in central and southern Nevada. With regard to the latter, our findings indicate that the paleowater distribution depends on the hydrodynamics of individual flow systems, and that several of the effects that have been widely cited as evidence for pluvial-age recharge can simply be attributed to Holocene-age processes.

BACKGROUND

Hydrogeologic setting of the Great Basin

The Great Basin province encompasses a 360,000 km² area, largely in Nevada and western Utah, characterized by subparallel north-south–trending mountain ranges separated by alluvial and fluviolacustrine basins (Harrill et al., 1988). The climate of the region is strongly influenced by the rain shadow effect of the Sierra Nevada and Cascade Ranges, which diminishes the moisture content of prevailing westerly winds originating from the Pacific Ocean (Houghton et al., 1975; Lamb et al., 1976). As a result, Nevada receives the lowest amount of annual precipitation of the 50 United States. Winter frontal systems deposit >50% of the annual precipitation as snowfall, producing relatively large accumulations in alpine areas. Rapid melting of the mountain snowpack during the spring and early summer probably accounts for most of the annual groundwater recharge budget (Benson and Klieforth, 1989; Winograd et al., 1998; Rose et al., 1999a).

The principal aquifers in the region consist of basin-fill deposits and fractured carbonate rock, although fractured volcanic rocks form locally important aquifers (Winograd and Thordarson, 1975; Thomas et al., 1996). Much of the eastern half of the Great Basin is underlain by Paleozoic carbonate rocks with depositional thicknesses as great as 9000 m (Stewart, 1980; Dettinger et al., 1995). Within this carbonate province, groundwater is transported along deep regional flow paths that transcend local topographic boundaries, driven by hydraulic gradients that may be laterally continuous over hundreds of kilometers (Eakin, 1966; Winograd and Thordarson, 1975; Mifflin and Hess, 1979; Harrill et al., 1988; Dettinger et al., 1995). Mesozoic thrust faulting and Cenozoic normal faulting have juxtaposed rocks of different ages and lithologies, locally compartmentalizing the aquifers, and creating hydraulic gradients that tend to be step-like rather than smooth (Winograd and Thordarson, 1975). Hence, several areas of spring discharge may occur along a single regional flow path, as in the White River flow system (Maxey and Eakin, 1949; Eakin, 1966).

Basin-fill deposits generally consist of unconsolidated or semiconsolidated sediments ranging from 500 to 1500 m in thickness, with local sediment accumulations sometimes reaching 3000 m depth (Harrill et al., 1988). The basin-fill aquifers may be hydraulically connected with adjacent and underlying bedrock aquifers, especially where carbonate rock is prevalent (Thomas et al., 1986). The degree of continuity depends in part on the vertical hydraulic conductivity of the deep basin fill (Dettinger et al., 1995). Where low-permeability rocks and sediments surround the basin-fill aquifers, hydrologically closed systems may develop. Closed basins are most common in the western part of the Great Basin, where low-permeability volcanic bedrock is widespread (Thomas et al., 1989, 1996).

Current models for regional groundwater flow in southwestern Nevada

The Death Valley regional flow system covers an area of ~40,000 km² in southwestern Nevada and adjacent eastern California (Fig. 1), and is characterized by carbonate-hosted interbasin flow associated with large regional springs (Winograd and Thordarson, 1975; Harrill et al., 1988; Laczniak et al., 1996). It is

Figure 1. Map of central and southern Nevada showing the approximate boundaries of major regional groundwater flow systems discussed in the text (flow boundaries from Harrill et al., 1988, and Eakin, 1966). Arrows denote the general direction of regional groundwater flow. New evidence presented in this study suggests the Railroad Valley and Death Valley flow systems are linked.

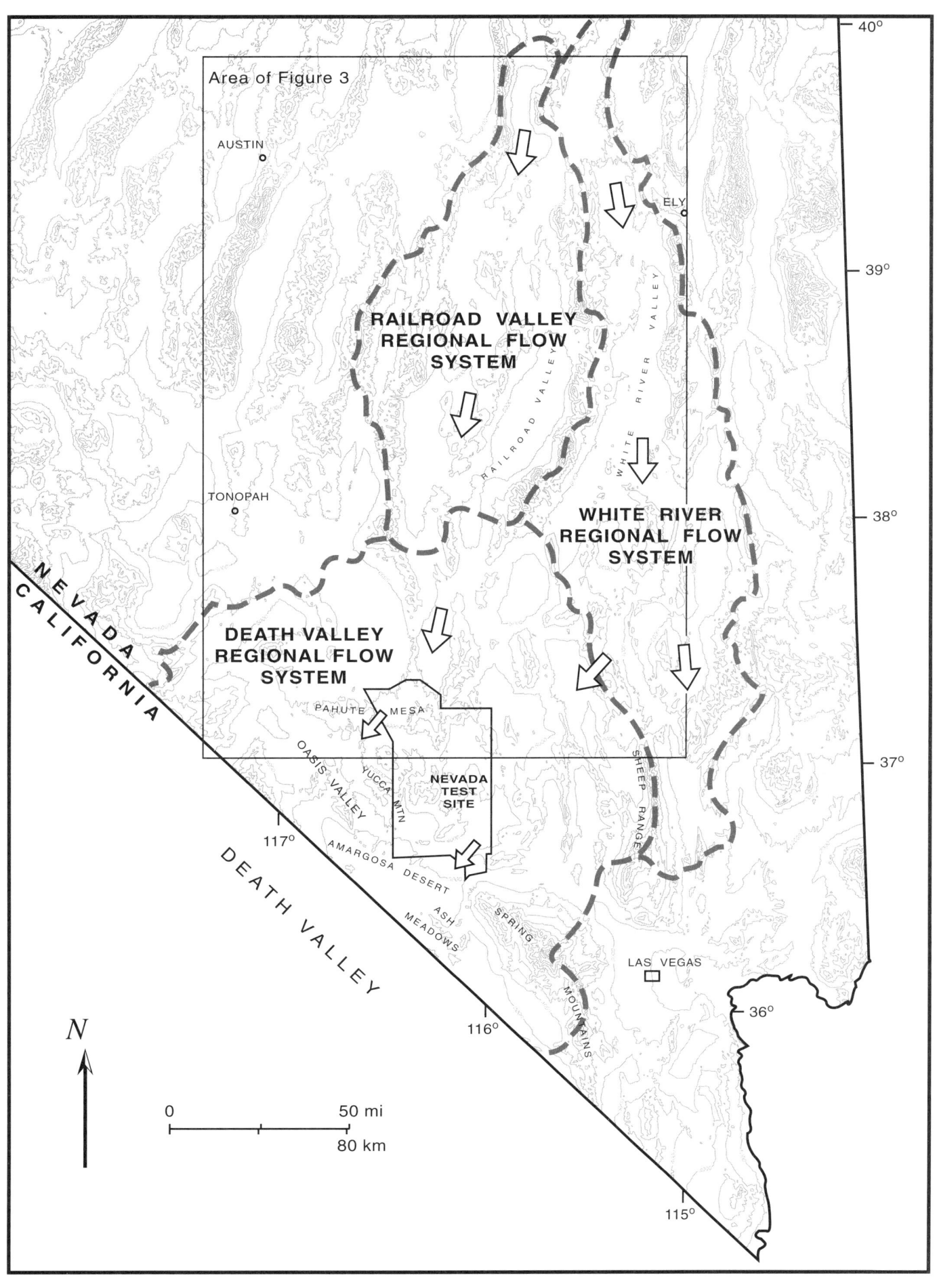

perhaps the best studied of the regional flow systems in the Great Basin because it contains the U.S. Department of Energy's Nevada Test Site and Yucca Mountain project. The Death Valley playa is the terminus of the system, although several subsystems discharge at intermediate points along the flow path (Harrill et al., 1988; Laczniak et al., 1996). The most prominent of these is the Ash Meadows spring system (Winograd and Thordarson, 1975; Winograd and Pearson, 1976). Recharge to the Death Valley flow system is assumed to be small for three reasons: (1) precipitation rates are generally low; (2) there is a paucity of high elevation catchments; and (3) observed discharge rates and totals are small. Significant recharge does occur in the Sheep Range and Spring Mountains, the latter being an important source of groundwater for both the Las Vegas Valley (Malmberg, 1965; Morgan and Dettinger, 1996) and the Ash Meadows spring system (Winograd and Friedman, 1972; Winograd and Thordarson, 1975).

For the present study, we will focus on the Pahute Mesa–Oasis Valley subsystem, in the north-central part of the Death Valley flow system (Fig. 1), and its possible linkage to regional flow to the north. Pahute Mesa is underlain by a thick sequence of rhyolitic volcanic rocks through which groundwater flow occurs principally along interconnected faults and fractures (Blankennagel and Weir, 1973). This subsystem is of particular interest because Pahute Mesa was the location of nearly 90 underground nuclear tests conducted between 1965 and 1992, the majority of which were detonated below the water table (Laczniak et al., 1996). The underground test area is situated ~40 km northeast of Oasis Valley, where ~8000 ac-ft yr^{-1} (27,000 m^3 d^{-1}) of groundwater discharges via evapotranspiration and spring flow (Reiner et al., 1999). Groundwater stable isotope data (δ^{18}O and δD) is consistent with a direct flow path between Pahute Mesa and Oasis Valley (White and Chuma, 1987; Rose et al., 1999b). However, the origin of the groundwater underlying Pahute Mesa remains poorly constrained.

Cool season precipitation (October–April) collected at 2145 m elevation on Pahute Mesa has a weighted mean δ^{18}O value significantly higher (less negative) than groundwater sampled below the Pahute Mesa water table (>600 m depth; see Figure 2; precipitation data from Milne et al., 1987; Benson and Klieforth, 1989). Perched groundwater sampled from nearby Rainier Mesa is also isotopically enriched relative to the deeper groundwater (Fig. 2). These data indicate that modern local recharge is not the principal source of the deeper groundwater. Two hypotheses have been proposed to explain these observations: (1) groundwater was recharged locally during a cooler climate episode, or (2) recharge is Holocene and occurred in higher elevation regions of central Nevada, and flowed into Pahute Mesa from the north. Each of these hypotheses merit further examination.

Paleoclimate studies

Proxy paleoclimate records are consistent with greater effective moisture in the Great Basin during the late Pleistocene. For example, nearly 100 basins in the northern Great Basin contain

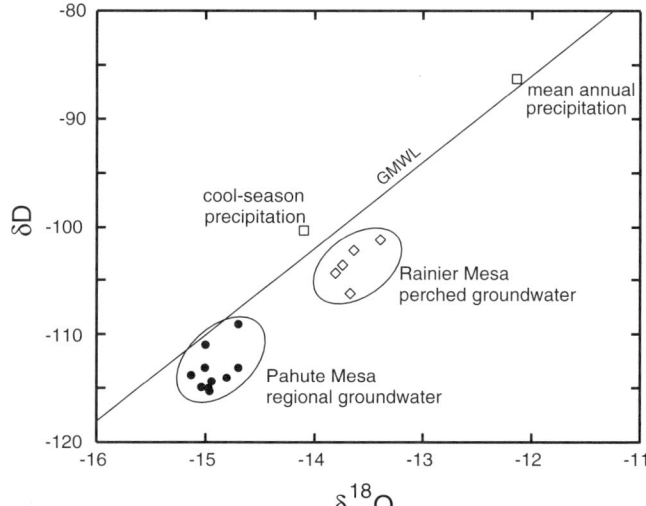

Figure 2. Plot of δD vs. δ^{18}O values for water samples from the northwestern Nevada Test Site. Regional groundwater flow beneath Pahute Mesa (filled circles) is isotopically depleted relative to both mean annual precipitation on Pahute Mesa (open squares; Milne et al., 1987), and perched groundwater from nearby Rainier Mesa (open diamonds). These data suggest the regional groundwater cannot have a local origin under present climate conditions. GMWL—global meteoric water line (Craig, 1961).

evidence of large Pleistocene lakes that oscillated in size in response to changes in climate (Mifflin and Wheat, 1979; Benson et al., 1990). Variations in oxygen isotopes and total organic carbon in cored sediments from these paleolake basins tend to correlate with North Atlantic ice-core and marine paleoclimate records (e.g., Benson et al., 1996, 1998). In the southern Great Basin, the widespread occurrence of paleospring deposits has been interpreted as evidence for increased recharge and higher water tables during the late Pleistocene (Quade et al., 1995, 1998). However, with few exceptions, the effective moisture was insufficient to support large lakes in the southern Great Basin (Quade et al., 1995), and the full pluvial climate in south-central Nevada may have been similar to the modern climate in extreme northwestern Nevada (Mifflin and Wheat, 1979; Spaulding and Graumlich, 1986).

Increased groundwater recharge rates during the late Pleistocene imply a significant volume of groundwater was added to the regional aquifers at that time. The extent to which this water persists today is largely a function of groundwater flow velocities and turnover rates for individual subbasins. Pleistocene-age groundwater is most likely to be preserved within hydrologically closed or semiclosed alluvial basins. In some cases, this water may mix into the regional flow systems via downward leakage into the underlying carbonate aquifers (e.g., Winograd and Thordarson, 1975). However, this is likely to be a slow process since vertical conductivities are typically 1 to 2 orders of magnitude lower than horizontal conductivities in stratified alluvium (Domenico and Schwartz, 1990).

Previous isotope hydrology studies in the Oasis Valley–Yucca Mountain–Amargosa Desert region (central Death Valley regional flow system) concluded that groundwater in volcanic and alluvial aquifers in south-central Nevada was probably recharged during the last major pluvial cycle (Claassen, 1986; White and Chuma, 1987; Benson and Klieforth, 1989). Cited evidence includes the stable isotope depletions noted above, low deuterium excess values relative to modern precipitation (cf. Dansgaard, 1964; Merlivat and Jouzel, 1979), and unadjusted groundwater ^{14}C ages that yield predominantly late Pleistocene dates (9000–18,000 yr B.P.). Whereas paleoclimate models provide one possible interpretation for these data, an alternative hypothesis is developed in the following sections that potentially links the central Death Valley groundwaters to regional inflow from central Nevada.

Revised model for regional groundwater flow into southwestern Nevada

Blankennagel and Weir (1973) inferred that up to 70% of the groundwater underlying Pahute Mesa originates as regional flow from the north, via Kawich Valley and Gold Flat. Precipitation patterns suggest that significant recharge should occur between latitudes 38° and 40°N, where a number of mountain ranges crest well above 3000 m, and precipitation rates can exceed 50 cm yr^{-1} (e.g., Lamke and Moore, 1965). Although the boundary of the Death Valley flow system, as delineated by Harrill et al. (1988), does not extend into this area, regional groundwater flow southward from central Nevada is a plausible source of the low $\delta^{18}O$ groundwater observed beneath Pahute Mesa.

Davisson et al. (1999) noted that regional groundwater $\delta^{18}O$ values show a systematic increase of nearly 5‰ going from 39° to 36°N latitude in Nevada. This variation is consistent with higher latitude recharge following continuous flow paths along north-south–trending graben valleys and progressively mixing with more ^{18}O-enriched recharge at lower latitudes. New isotopic evidence presented in this paper is used to further test this model. These data suggest groundwater in the Pahute Mesa–Oasis Valley area is linked to a regional flow system with recharge areas as far north as 40°N latitude. We infer that much of this groundwater originates from the Railroad Valley regional flow system, located directly north of the central Death Valley flow system (Fig. 1).

Hydrogeology of the Railroad Valley region, central Nevada

The 14,500 km^2 Railroad Valley flow system extends north-south almost 200 km from the Newark Valley to the central Railroad Valley playa, and includes the Little Smoky, Hot Creek, and Big Sand Springs Valleys (see Figure 3; Rush and Everett, 1966; Van Denburgh and Rush, 1974; Harrill et al., 1988). The flow system is generally considered to terminate at the Railroad Valley playa, which is located within a deep structural depression that hosts a shallow geothermal system associated with productive oil fields (Hulen et al., 1994; Lund et al., 1993). Potentiometric data from deep volcanic aquifers indicate eastward groundwater flow from the northern Hot Creek Valley toward the central Railroad Valley (Dinwiddie and Schroder, 1971). However, water-level data for carbonate and volcanic rock aquifers are lacking for southern Railroad and Hot Creek Valleys, as well as Reveille and Kawich Valleys.

The Railroad Valley flow system approximately coincides with the western extent of the regional carbonate rock province (Dettinger et al., 1995). Carbonate rock exposures occur in many of the ranges that border the flow system (e.g., Stewart and Carlson, 1978) representing potential recharge connections to the carbonate aquifers underlying the alluvial and volcanic deposits in this region. Approximately 15 springs with discharge rates >100 gpm (545 m^{3}d^{-1}) occur within the Railroad Valley system, nearly all of which are associated with the regional carbonate aquifer. The largest of these (Big Warm Spring, near Duckwater) has an average discharge rate >6000 gpm (32,700 m^{3}d^{-1}) (Van Denburgh and Rush, 1974). Little detailed isotope hydrology work has been done in central Nevada, despite the potential significance of this region as a groundwater recharge area. Previous studies that report isotopic data for the Railroad Valley region include Roth and Campana (1989) and Hulen et al. (1994).

Although the Railroad Valley and White River flow systems share a common geographic boundary (Fig. 1), they exhibit some notable differences in hydrogeology. The White River flow system displays a continuous north to south water-level gradient in both the basin-fill and carbonate-rock aquifers (Thomas et al., 1986), and is underlain by a thick, laterally continuous carbonate aquifer along the entire flow path length (Dettinger and Schaefer, 1996). In contrast, basin-fill aquifers in both the Newark and Railroad Valleys contain closed water-level contours in the central part of their basins, rather than a north-south gradient. Moreover, the distribution and lateral continuity of carbonate-rock aquifers is poorly constrained beneath much of the Railroad Valley flow system. Water-level data for the Railroad Valley flow system are examined in more detail later in this paper.

SAMPLING AND ANALYTICAL METHODS

Water samples were collected from >70 springs, creeks, and wells in central Nevada between 1997 and 1999 within an area approximately bounded by latitudes 37°30′ to 40°N and longitudes 115° to 117°30′W (Fig. 3). Analytical results are presented in Table 1. Most samples collected south of latitude 37°30′ are from an earlier data set (Rose et al., 1997), as discussed in Davisson et al. (1999). Water temperature, pH, and conductivity values were measured in the field at the time of sample acquisition. Cation samples were typically passed through a 0.45 μm filter and acidified with HNO_3 in the field; anion samples were untreated. Water chemistry was measured by ICP-AES for cations and ion chromatography for anions.

Carbon isotope samples were collected in glass bottles with teflon-lined septa caps, treated with $HgCl_2$ to prevent biological fractionation effects, and kept refrigerated until analysis. Dissolved inorganic carbon (DIC) was extracted by acidifying the sample under vacuum and cryogenically trapping the evolved CO_2 gas

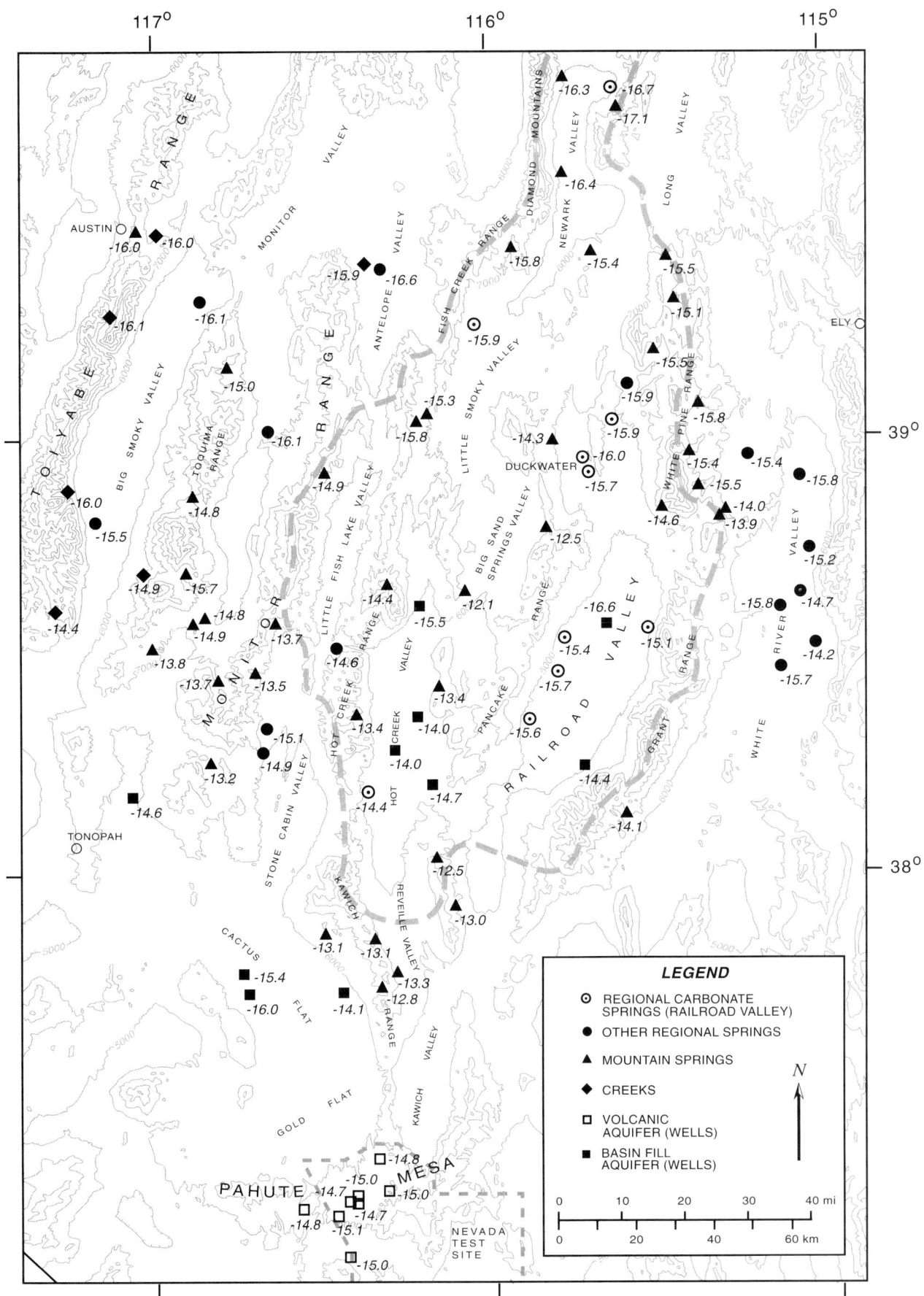

Figure 3. Map of central and southwestern Nevada showing the distribution of samples described in this study, the sample type (see legend), and the corresponding $\delta^{18}O$ values. Geographic locations discussed in the text are also shown. The approximate outline of the Railroad Valley regional flow system is shown as a heavy dashed line (from Harrill et al., 1988).

TABLE 1. WATER SAMPLE LOCATIONS AND CHEMICAL AND ISOTOPIC DATA

Sample name	Geographic location	Latitude	Longitude	Temp (°C)	pH	Cond. (µS/cm)	HCO_3^- (mg/L)	F^- (mg/L)	Cl^- (mg/L)	SO_4^{2-} (mg/L)	Na^+ (mg/L)	K^+ (mg/L)	Ca^{2+} (mg/L)	Mg^{2+} (mg/L)	$\delta^{18}O$ (‰ V-SMOW)	δD (‰ V-SMOW)	$\delta^{13}C$ (‰ PDB)	^{14}C (pmc)
UE-18r Well, NTS	Pahute Mesa	37°08'05"	116°26'41"	32	8.1	394	160	3	6.3	23	73	2.0	15.6	0.3	-15.0	-113	-1.4	8.2
ER-20-5 #3 Well, NTS	Pahute Mesa	37°13'13"	116°28'39"	34	8.6	302	105	3.3	17	35	70	3.1	3.2	0.1	-15.1	-114	-5.7	–
ER-20-5 #1 Well, NTS	Pahute Mesa	37°13'14"	116°28'39"	34	8.2	621	178	10.1	23	39	104	4.5	6.6	0.3	-14.9	-114	-2.3	–
Pahute Mesa #3 Well *	Pahute Mesa	37°14'21"	116°33'37"	–	8.3	748	158	2.5	84.2	92	124	12.3	18.9	4.0	-14.8	-115	–	–
UE-20bh #1 Well, NTS *	Pahute Mesa	37°14'42"	116°24'33"	26	8.3	398	214	4.3	4.7	14	88	8.7	3.1	0.6	-14.7	-109	-9.2	21.0
U-20 Water Well, NTS *	Pahute Mesa	37°15'05"	116°25'45"	37	8.2	309	107	2.2	12.1	32	59	1.6	7.8	0.3	-14.7	-113	-6.2	8.6
ER-20-6 #3 Well, NTS	Pahute Mesa	37°15'33"	116°25'16"	28	8.4	310	102	2.5	13.6	32	56	3.6	10.1	0.8	-15.0	-115	-7.2	16.3
ER-20-6 #2 Well, NTS	Pahute Mesa	37°15'35"	116°25'16"	26	8.2	308	105	3.8	11.6	32	61	3.1	8.3	0.7	-15.0	-115	-7.3	–
ER-20-6 #1 Well, NTS	Pahute Mesa	37°15'36"	116°25'15"	32	8.2	318	103	2.5	12.6	32	60	2.1	5.6	0.4	-15.0	-115	-5.8	–
UE-19c Well, NTS	Pahute Mesa	37°16'08"	116°19'10"	37	7.7	174	65	–	3.1	6	36	0.5	1.4	0.2	-15.0	-111	-5.3	8.1
U-20al Well, NTS *	Pahute Mesa	37°16'12"	116°29'51"	–	8.4	675	250	–	32.8	78	117	11.1	23.7	2.1	–	–	–	–
U-19ba #1 Well, NTS *	Pahute Mesa	37°17'46"	116°18'47"	–	8.1	622	189	–	40.9	10	79	5.5	20.8	1.2	–	–	–	–
UE-19gs Well, NTS *	Pahute Mesa	37°18'14"	116°21'53"	42	7.8	508	181	1.2	9.9	100	74	1	43	0.1	–	–	–	–
UE-19h Well, NTS	Pahute Mesa	37°20'34"	116°22'25"	28	8.3	415	144	–	8.5	–	64	4.0	14.9	1.5	-14.8	-112	-3.3	9.4
Roller Coaster Well	Cactus Flat	37°43'16"	116°44'11"	26	7.8	513	116	0.9	38	34	65	8.0	21.3	1.6	-16.0	-129	-6.6	13.2
Rose Spring	Kawich Range	37°44'46"	116°19'56"	18	7.4	668	322	0.3	23	49	44	1.9	82.4	11.2	-12.8	-104	-8.7	66.7
Cedar Creek Pass Well	Cactus Flat	37°44'48"	116°28'58"	28	7.7	288	100	0.5	15.2	25	35	9.4	20.7	0.5	-14.1	-111	-10.1	21.9
Summer Spring	Kawich Range	37°46'23"	116°17'25"	15	7.8	530	234	0.3	23.6	50	41	2.7	62.7	8.7	-13.3	-107	-8.6	79.1
Sandia Well #6	Cactus Flat	37°47'03"	116°44'59"	23	9.1	624	164	1.6	26.6	44	103	5.6	2.1	0.0	-15.4	-123	-6.7	13.5
Georges Water Spring	Kawich Range	37°51'36"	116°20'59"	10	7.1	179	72	0.3	4	9	14	1.6	16.7	2.7	-13.1	-98	-11.6	–
Silverbow Spring	Kawich Range	37°52'04"	116°30'23"	24	7.1	577	229	0.4	22.6	43	47	2.3	49.1	9.1	-13.1	-108	-13.7	110.9
Pyramid Spring	Reveille Range	37°55'55"	116°07'08"	13	7.3	495	226	0.3	11.7	30	46	1.4	53.1	5.6	-13.0	-100	-10.4	107.0
Reveille Spring	Reveille Range	38°01'54"	116°10'05"	11	7.3	404	191	–	–	–	–	–	–	–	-12.5	-95	-11.0	–
Adavan Spring	Quinn Canyon Range	38°08'19"	115°36'05"	10	7.1	500	290	–	–	–	–	–	–	–	-14.1	-108	-10.0	85.0
Warm Springs †	Jct. Hwy 6 and 375	38°11'16"	116°22'21"	59	6.6	1322	813	3.6	36	96	199	23.3	91.5	22.5	-14.4	-109	-2.8	–
Tonopah City Well	Ralston Valley	38°11'28"	117°04'41"	12	7.4	290	175	–	–	–	–	–	–	–	-14.6	-112	-13.6	–
Twin Springs Ranch Well	Pancake Range	38°12'13"	116°10'29"	12	7.6	570	374	0.3	14.5	24	84	11.8	35.6	4.4	-14.7	-121	-2.8	10.9
Saulsbury Ranch Spring	Monitor Range	38°14'16"	116°49'27"	21	7.8	1350	–	–	–	–	–	–	–	–	-13.2	-106	–	–
Sharp Ranch Well (Nyala)	Railroad Valley	38°14'56"	115°43'40"	11	7.2	1200	234	–	132	209	40	3.2	12	60.8	-14.4	-110	-7.1	39.0
Point of Rocks Spring	West Stone Cabin Valley	38°17'16"	116°40'07"	20	8.4	850	415	–	–	–	–	–	–	–	-14.9	-118	-2.3	–
Base Camp Well, DOD	Hot Creek Valley	38°18'38"	116°16'41"	18	6.5	362	122	–	9	23	25	4.3	34.3	10.0	-14.0	-109	-7.4	19.7
Warm Spring	West Stone Cabin Valley	38°20'23"	116°39'43"	23	7.6	700	335	10	14.8	39	161	2.3	8.3	0.7	-15.1	-122	-4.3	13.2
Abel Spring	Railroad Valley	38°21'53"	115°52'01"	44	6.6	1114	708	1.7	17.1	46	106	20.9	102	27.3	-15.6	-123	-1.9	1.8
Blue Jay Well	Hot Creek Valley	38°22'12"	116°13'39"	14	7.9	369	168	–	–	–	–	–	–	–	-14.0	-106	-6.9	–

TABLE 1. WATER SAMPLE LOCATIONS AND CHEMICAL AND ISOTOPIC DATA (continued)

Sample Name	Geographic Location	Latitude	Longitude	Temp (°C)	pH	Cond. (µS/cm)	HCO$_3^-$ (mg/L)	F$^-$ (mg/L)	Cl$^-$ (mg/L)	SO$_4^{2-}$ (mg/L)	Na$^+$ (mg/L)	K$^+$ (mg/L)	Ca^{2+} (mg/L)	Mg^{2+} (mg/L)	δ^{18}O (‰ V-SMOW)	δD (‰ V-SMOW)	δ^{13}C (‰ PDB)	^{14}C (pmc)
Mule Shoe Spring	Hot Creek Range	38°22'22"	116°25'46"	11	7.5	463	—	—	—	—	—	—	—	—	-13.4	-107	—	—
Hot Creek Spring	White River Valley	38°22'56"	115°09'11"	33	7.2	482	264	1.0	9.5	42	25	5.6	58.4	22.2	-15.7	-122	-3.2	—
Flag Springs (north spr.)	White River Valley	38°25'25"	115°01'14"	16	7.7	357	225	0.3	5.3	9	7	3.2	47	20.3	-14.2	-107	-6.8	—
Rattlesnake Spring	Pancake Range	38°26'59"	116°09'32"	17	7.4	300	—	—	—	—	—	—	—	—	-13.4	-107	—	—
Hunts Canyon Spring	Monitor Range	38°27'27"	116°48'57"	9	7.0	221	98	—	—	—	—	—	—	—	-13.7	-104	-14.5	—
Chimney Spring	Railroad Valley	38°27'46"	115°47'35"	67	6.7	663	519	1.3	12	47	72	19.0	98.6	16.7	-15.7	-123	-1.4	6.4
McCann Canyon Spring	Monitor Range	38°28'03"	116°41'01"	7	7.0	485	190	—	—	—	—	—	—	—	-13.5	-105	-14.1	—
Keller Spring	Toquima Range	38°31'26"	116°59'12"	10	7.3	1092	252	—	—	—	—	—	—	—	-13.8	-107	-8.9	—
Upper Warm Spring †	Hot Creek Range	38°31'57"	116°27'52"	35	8.1	201	100	0.4	7	19	38	0.8	4.7	0.1	-14.6	-113	-14.7	—
Lockes Big Spring †	Railroad Valley	38°33'21"	115°46'15"	37	7.7	662	450	1.2	10	59	52	10	66	21	-15.4	-124	-2.5	5.4
Blue Eagle Spring	Railroad Valley	38°33'43"	115°31'41"	27	7.0	640	409	0.4	10.1	30	36	5.3	73.3	23.2	-15.1	-116	-5.0	12.9
The Big Well (artesian)	Railroad Valley	38°34'50"	115°38'15"	22	7.8	443	208	0.6	25	19	63	10	12	9.6	-16.6	-125	-6.0	0.7
Mexican Spring	Toquima Range	38°35'20"	116°52'55"	10	7.2	509	261	—	—	—	—	—	—	—	-14.9	-112	-12.0	—
East Barley Creek Summit Spring	Monitor Range	38°35'44"	116°39'10"	6	7.0	323	65	—	—	—	—	—	—	—	-13.7	-103	-11.1	—
Moorman Spring	White River Valley	38°35'44"	115°08'20"	36	7.2	495	257	1.3	9.6	42	26	5.9	62.5	20.3	-15.8	-122	-1.9	—
Combination Spring	Toquima Range	38°35'47"	116°51'26"	11	7.2	446	175	—	—	—	—	—	—	—	-14.8	-111	-8.3	—
Peavine Creek	Toiyabe Range	38°36'59"	117°18'07"	12	—	239	—	—	—	—	—	—	—	—	-14.4	-108	-13.6	—
HTH-1 Well (853 m) §	N. Hot Ck. Valley	38°37'34"	116°12'45"	41	8.2	588	261	10	21.4	39	134	2.2	3.0	0.1	-15.5	-118	-2.8	0.8
HTH-2 Well (174 m) §	N. Hot Ck. Valley	38°37'39"	116°12'48"	19	8.0	287	177	—	4.1	0.7	19	1.4	36.9	5.2	-14.1	-107	-10.3	63.7
UC-1-P-1S Well (150 m) §	N. Hot Ck. Valley	38°37'54"	116°12'42"	18	8.2	217	134	—	2.6	0.6	23	1.4	23	1.7	-14.1	-104	-9.2	44.6
Hardy Springs	White River Valley	38°38'12"	115°04'27"	14	7.5	420	286	0.2	2.8	13	7	3.0	58.7	26.2	-14.7	-112	-7.6	—
Needles Spring	Big Sand Springs Valley	38°39'44"	116°04'10"	15	7.4	565	285	—	—	—	—	—	—	—	-12.1	-98	-11.3	—
North Sixmile Canyon Spring	Hot Creek Range	38°40'46"	116°18'13"	8	7.1	386	246	0.1	2.3	6	6	1.7	42.6	25.2	-14.4	-113	-11.0	97.6
Meadow Canyon Spring	Toquima Range	38°41'38"	116°55'10"	9	7.5	198	95	0.2	4.4	11	21	4.5	14.3	2.8	-15.7	-119	-13.4	—
Shoshone Creek	Toquima Range	38°42'41"	117°02'45"	9	—	161	—	—	—	—	—	—	—	—	-14.9	-112	-7.6	—
Indian Spring	White River Valley	38°45'19"	115°03'03"	13	7.5	444	278	0.2	3.1	20	6	1.6	60.3	23.6	-15.1	-117	-10.7	—
Martilletti Spring	Pancake Range	38°47'59"	115°49'57"	10	7.3	430	219	—	—	—	—	—	—	—	-12.5	-102	-10.3	—
Darroughs Hot Spring †	Big Smoky Valley	38°49'16"	117°10'48"	92	9.1	496	153	14	12	53	110	2.6	1.3	0.1	-15.5	-121	-11.5	—
Summit Spring	Horse Range	38°49'23"	115°17'59"	13	7.7	381	232	0.4	13.8	15	27	2.7	51.3	11.1	-13.9	-108	-10.8	—
Secret Spring	Horse Range	38°50'24"	115°17'20"	12	7.2	403	192	—	—	—	—	—	—	—	-14.0	-110	-9.1	—
Silver Spring	White Pine Range	38°50'39"	115°29'02"	10	7.8	715	411	0.9	13.3	42	40	2.6	94.6	17.6	-14.6	-112	-12.6	107.5
Logan Spring	Toquima Range	38°52'08"	116°53'05"	8	7.4	487	212	0.4	15.5	44	45	1.8	49.4	6.5	-14.8	-116	-10.9	84.6
Little Currant Ck Springs	White Pine Range	38°53'12"	115°22'08"	9	8.2	331	265	—	1.4	3	5	0.8	54.0	11.7	-15.5	-119	-12.3	—
North Fork Twin River	Toiyabe Range	38°53'38"	117°15'14"	9	7.9	86	82	—	—	—	—	—	—	—	-16.0	-125	-5.2	—
Arnoldson Spring	White River Valley	38°54'35"	115°03'49"	23	7.7	408	184	0.4	15	37	12	3.4	41.1	19.5	-15.8	-122		

TABLE 1. WATER SAMPLE LOCATIONS AND CHEMICAL AND ISOTOPIC DATA (continued)

Sample Name	Geographic Location	Latitude	Longitude	Temp (°C)	pH	Cond. (µS/cm)	HCO$_3^-$ (mg/L)	F$^-$ (mg/L)	Cl$^-$ (mg/L)	SO$_4^{2-}$ (mg/L)	Na$^+$ (mg/L)	K$^+$ (mg/L)	Ca^{2+} (mg/L)	Mg^{2+} (mg/L)	δ^{18}O (‰ V-SMOW)	δD (‰ V-SMOW)	δ^{13}C (‰ PDB)	^{14}C (pmc)
East Dobbin Summit Spring	Monitor Range	38°55'26"	116°30'35"	–	–	–	–	–	–	–	–	–	–	–	-14.9	-119	–	–
Spring, east of Duckwater	Railroad Valley	38°55'41"	115°42'12"	19	7.4	665	350	0.3	9	67	30	8.1	62.0	26.1	-15.7	-122	-6.3	29.6
Big Warm Spring	Railroad Valley	38°56'59"	115°42'00"	34	7.1	597	291	0.6	7.5	43	29	7.1	63.6	22.9	-16.0	-125	-2.9	3.0
Williams Hot Spring	West of White River Valley	38°57'07"	115°14'01"	52	9.2	295	56	13	8.9	14	64	0.4	1.6	0.0	-15.4	-123	-7.0	–
Saddle Spring	White Pine Range	38°58'36"	115°23'58"	5	7.3	186	81	0.2	5.5	7	12	1.5	18.5	4.2	-15.4	-118	-10.9	106.6
Young Florio Spring	Pancake Range	38°58'46"	115°48'44"	12	7.0	447	186	0.8	12.2	21	46	6.4	39.4	6.8	-14.3	-111	-11.5	89.9
Dianas Hot Spring †	Monitor Valley	39°01'47"	116°40'02"	47	6.9	608	285	2.8	8	59	57	15	47	11	-16.1	-128	-2.8	–
Snowball Ranch Spring	Antelope Range	39°02'23"	116°13'00"	15	6.9	279	157	0.2	4.5	18	10	3.3	32.5	9.3	-15.8	-123	-9.8	19.3
Bull Creek Spring	Railroad Valley	39°03'01"	115°37'35"	13	7.3	361	194	0.1	12	26	17	3.5	33.1	17.7	-15.9	-123	-7.1	15.4
Willow Spring	Antelope Range	39°04'10"	116°10'28"	16	7.6	205	–	–	–	–	–	–	–	–	-15.3	-123	–	–
Tom Plain Spring	White Pine Range	39°05'15"	115°22'33"	6	7.5	364	167	0.2	11.2	15	10	3.1	56.8	4.0	-15.8	-124	-9.3	50.7
Green Springs	Railroad Valley	39°06'57"	115°34'06"	19	7.1	432	–	–	–	–	–	–	–	–	-15.9	-117	–	–
Sam's Spring	Toquima Range	39°10'13"	116°46'49"	6	7.1	663	297	–	–	–	–	–	–	–	-15.0	-118	-12.0	–
Spring, southeast of Mount Hamilton	White Pine Range	39°12'41"	115°30'40"	9	7.1	473	301	–	5	5	10	1.4	61.5	22.1	-15.5	-115	-11.4	91.9
Kingston Creek	Toiyabe Range	39°16'17"	117°09'28"	8	6.6	481	208	–	–	–	–	–	–	–	-16.1	-125	-14.5	–
Fish Creek Springs **	Fish Creek Range	39°16'37"	116°02'18"	17	8.2	444	267	–	11	37	38	–	28	29	-15.9	-124	–	–
Spencer Hot Spring †	Big Smoky Valley	39°19'37"	116°51'17"	60	7.0	1199	710	5.2	26	47	198	34	51	9.4	-16.1	-135	-2.9	–
Sand Spring	White Pine Range	39°19'52"	115°27'15"	12	7.6	884	–	–	–	–	–	–	–	–	-15.1	-119	–	–
Klobe Hot Spring †	Antelope Valley	39°24'18"	116°20'50"	65	9.0	305	162	4.8	11	16	66	1.0	10.3	0.0	-16.6	-130	-12.7	–
Faulkner Creek	Monitor Range	39°24'41"	116°22'33"	14	6.3	106	–	–	–	–	–	–	–	–	-15.9	-123	–	–
Little Antelope Spring	North White Pine Range	39°24'57"	115°27'34"	9	7.3	1421	–	–	–	–	–	–	–	–	-15.5	-122	–	–
Sulphur Spring	Pancake Range	39°25'24"	115°40'49"	13	7.0	552	–	–	–	–	–	–	–	–	-15.4	-119	–	–
Lucky Springs	Diamond Range	39°27'30"	115°56'29"	9	6.4	341	–	–	–	–	–	–	–	–	-15.8	-122	–	–
Bade Creek	Toiyabe Range	39°28'27"	117°00'05"	6	7.3	195	–	–	–	–	–	–	–	–	-16.0	-121	–	–
Austin Summit Spring	Toiyabe Range	39°28'49"	117°02'36"	9	8.0	239	–	–	–	–	–	–	–	–	-16.0	-121	–	–
DeBernardi Ranch Spring	Newark Valley	39°37'28"	115°45'53"	16	7.6	288	–	–	–	–	–	–	–	–	-16.4	-124	–	–
Cottonwood Spring (runoff)	Buck Mountain	39°46'08"	115°35'45"	–	–	–	–	–	–	–	–	–	–	–	-17.1	-132	–	–
Simonson Warm Spring	Newark Valley	39°48'41"	115°36'30"	25	7.4	543	294	0.7	6.3	30	18	6.3	61.6	23.6	-16.7	-129	-2.5	7.9
Cold Spring	Newark Valley	39°50'23"	115°45'10"	11	7.3	320	154	0.2	3.8	10	8	1.1	44.4	8.2	-16.3	-123	-10.3	79.6

* Water chemistry measured by the Desert Research Institute, University of Nevada (unpublished data).
† Water chemistry data from Garside and Schilling (1979).
§ Water chemistry, δ^{18}O, δD data from Chapman et al. (1994).
** Water chemistry data from Rush and Everett (1966).

(McNichol et al., 1994). An aliquot of CO_2 is reduced to graphite and analyzed for ^{14}C on an accelerator mass spectrometer; the remaining CO_2 is analyzed for its $^{13}C/^{12}C$ ratio on an isotope ratio mass spectrometer. The ^{14}C results are reported as percent modern carbon (pmc) relative to a National Institute of Standards and Technology oxalic acid standard (Stuiver and Polach, 1977). All data presented in subsequent figures are uncorrected pmc values.

Oxygen and hydrogen stable isotope samples were collected in glass bottles with air tight seals to prevent atmospheric exchange. Water samples were prepared for $^{18}O/^{16}O$ and $^{2}H/^{1}H$ (or D/H) ratio measurements by the CO_2 equilibration (Epstein and Mayeda, 1953) and zinc-reduction (Coleman et al., 1982) methods, followed by analysis on an isotope ratio mass spectrometer. Oxygen, hydrogen and carbon stable isotope ratios are reported in the standard delta (δ) notation as per mil (‰) deviations from Vienna standard mean ocean water (V-SMOW; oxygen and hydrogen) or Peedee belemnite (PDB; carbon) reference standards.

RESULTS

Regional distribution of oxygen isotopes in central Nevada groundwater

The geographic distribution of groundwater $\delta^{18}O$ values in central Nevada is shown in Figure 3. Groundwater samples were collected over a broad region that includes the Railroad Valley flow system (the approximate boundaries of which are shown in Figure 3). Individual samples are identified with respect to sample type, and include springs associated with regional flow systems (principally the carbonate aquifer), samples from wells perforating basin-fill aquifers, and mountain springs associated with perched aquifers (Fig. 3). The latter are especially useful for characterizing the isotopic composition of local groundwater recharge. In general, the term "regional" groundwater will be used to refer to the carbonate aquifer system. However, we recognize that the basin-fill and carbonate aquifers are sometimes interconnected, and that some large warm springs related to "regional" flow are not associated with the carbonate aquifer.

Groundwater $\delta^{18}O$ values are generally lower (more negative) at higher latitudes, or in proximity to high elevation recharge areas, in accordance with the strong temperature dependence of the isotope fractionation factor between water and water vapor (see Criss, 1999, for review). The physiography of Nevada tends to accentuate this effect. Relative to southern Nevada, the average surface elevation in central Nevada is ~1 km higher, and the mean annual temperature is 5–10 °C cooler (Houghton et al., 1975). The $\delta^{18}O$ values of mountain springs clearly reflect these differences, averaging around –15.5‰ near latitude 39°N, and increasing to around –13‰ near latitude 38°N. In contrast, regional groundwaters have $\delta^{18}O$ values that average around –16‰ at latitude 39°N, decreasing to only about –15‰ at latitude 37°15′N (beneath Pahute Mesa). Two details are worth emphasizing: (1) the $\delta^{18}O$ values of regional groundwaters are similar to those of local mountain springs in the upper part of the regional flow systems; and (2) the $\delta^{18}O$ values of the regional groundwaters increase much more slowly with decreasing latitude compared to local mountain springs.

In the upper part of the Railroad Valley flow system (northeast section of Figure 3), local mountain springs have $\delta^{18}O$ values within ~0.5‰ of the regional carbonate springs. One possible interpretation of these data is that modern recharge provides a significant fraction of the regional flow in this area. However, regional springs near latitude 39°N are invariably slightly lower in $\delta^{18}O$ (more negative) than springs in adjacent mountain ranges. This ^{18}O-depletion may suggest that local recharge originates predominantly from the highest-elevation parts of the ranges (up to 3500 m altitude), where winter snowfall amounts are greatest, and $\delta^{18}O$ values are the most depleted. Alpine spring samples collected above 2500 m are very poorly represented in the data set.

It should be noted that most of the mountain springs used to delineate recharge areas were sampled only once. The $\delta^{18}O$ values of these springs are generally thought to reflect the mean isotopic value of recharge for a local area, particularly for those with perennial discharge. However, Ingraham et al. (1991) showed that the $\delta^{18}O$ values of small springs in southern Nevada may fluctuate by several per mil in response to large precipitation events with "anomalous" isotopic values. Hence, a multiyear isotopic record for small-volume mountain springs is ultimately needed to constrain $\delta^{18}O$ variations in key recharge areas.

Whereas both regional and local (perched) groundwater $\delta^{18}O$ values generally become more ^{18}O-enriched (less negative) at lower latitudes, the latitudinal ^{18}O gradient in the regional carbonate flow system is quite gradual. For example, carbonate springs discharging along the Railroad Valley system vary from –16.7‰ in Newark Valley to –15.6‰ in the western Railroad Valley—a lateral distance of >150 km. This variation is consistent with a southward-moving regional flow system that progressively mixes with local recharge along the length of its flow path. A convincing argument can be made for this type of process in the White River flow system (cf. Davisson et al., 1999). The continuity in regional groundwater $\delta^{18}O$ values between the western Railroad Valley and Pahute Mesa suggests a possible southward extension of this flow system via Hot Creek, Reveille, and Kawich Valleys, and possibly Gold Flat. This would require additional mixing along the flow path to increase the regional groundwater to $\delta^{18}O$ values around –15‰ beneath Pahute Mesa. Further evidence for this model is discussed in subsequent sections.

The relationship of the geothermal oil fields in central Railroad Valley to the regional flow system is uncertain, although there is some indication that central Railroad Valley is separated from regional flow to the west. Blue Eagle Spring in eastern Railroad Valley (38°34′N, 115°32′W) has a $\delta^{18}O$ value of –15.1‰, and is isotopically similar to thermal waters associated with Railroad Valley oil fields (Hulen et al., 1994). In comparison, the carbonate warm springs emerging along the western side of Railroad Valley are notably lower (more negative) in their $\delta^{18}O$ and δD values (up to 0.6‰ lower in $\delta^{18}O$ and 8‰ lower in δD).

Rush and Everett (1966) proposed that spring discharge along the western side of Railroad Valley originates from the southwestern part of adjacent Big Sand Springs Valley (see Figure 3). This is in accord with potentiometric data for deep boreholes in northern Hot Creek and Big Sand Springs Valleys (Dinwiddie and Schroder, 1971). The line of springs in western Railroad Valley may therefore indicate the presence of a structural barrier to eastward groundwater flow into central Railroad Valley. In general, these springs have isotopic values that are consistent with a southward continuation of flow from the vicinity of Big Warm Spring (38°57′N, 115°46′W; $\delta^{18}O$ = –16‰) and Fish Creek Springs (39°17′N, 116°02′W; $\delta^{18}O$ = –15.9‰). Some of this groundwater may also originate in the ranges northwest of Big Sand Springs Valley.

Figure 4 is a map identical to that of Figure 3, except that each sample point is now labeled with respect to water level (in feet above mean sea level). For springs and creeks, the water level is simply the local surface elevation; for wells, it is the static water level in the borehole. Recharge in the mountains bordering the flow system provides the hydrodynamic potential necessary to drive the regional flow system. Arrows show the inferred direction of water flow in the carbonate aquifers or deep fractured volcanic aquifers. Note that carbonate aquifer water levels are constrained by only 10 carbonate springs in the Railroad Valley flow system. Assuming the entire water budget of the aquifer is not expended at these springs, the difference in the deep potentiometric head between this area and Pahute Mesa may favor continued southward transport. Although water-level data for the carbonate flow system are limited in extent, it does not appear to preclude the interpretations developed on the basis of geochemical evidence.

Finally, it is notable that anomalous pockets of low $\delta^{18}O$ groundwaters are observed in two locations within the study area. Groundwater samples with $\delta^{18}O$ values less than –16‰ were observed from a single well in the central Railroad Valley and a single well in Cactus Flat. The lack of isotopic continuity between these groundwaters and nearby regional carbonate aquifer groundwaters suggests the low-^{18}O waters may represent small, isolated accumulations of pluvial-age groundwater "trapped" in basin-fill aquifers that are not connected to regional flow systems. The $\delta^{13}C$ and ^{14}C results for these samples generally support this conclusion (see discussion).

Kinetic isotope effects during recharge

Craig (1961) observed that the $\delta^{18}O$ and δD values of worldwide precipitation samples conform to the empirical linear relationship $\delta D = 8\delta^{18}O + 10$, known as the global meteoric water line (GMWL). The y-intercept of the GMWL was subsequently defined as the deuterium excess, or d-value, where $d = \delta D - 8\delta^{18}O$ (Dansgaard, 1964). The d-value is inherited from the initial isotopic composition of an air mass, as determined by the moisture deficit above the air-sea interface where the air mass originates (Craig and Gordon, 1965; Merlivat and Contiac, 1975). Merlivat and Jouzel (1979) suggested the d-value may decrease to values <10 during maximum glaciation, and that the meteoric water line described by Pleistocene-age groundwater should reflect this shift. This model has been proposed to account for the conspicuous difference between the d-values of regional groundwaters and local precipitation in southern Nevada (e.g., White and Chuma, 1987; Benson and Klieforth, 1989). However, this is not the only process that can result in isotopic enrichments of $\delta^{18}O$ and δD values relative to the GMWL.

Kinetic isotope effects during evaporation also cause heavy isotope enrichments that may not be readily distinguishable from the d-value effect in regional groundwater systems. Evaporation effects are well documented in desert environments, and may occur both during the descent of rain droplets, and during runoff and infiltration processes (e.g., Gat and Dansgaard, 1972; Barnes and Allison, 1988; Friedman et al., 1992). In addition, isotope enrichment effects occur during snowpack aging due to mass transport processes within the snowpack (e.g., Stichler et al., 1981; Friedman et al., 1991), and have been widely observed in central Nevada snowpacks (Rose et al., 1999a).

Figure 5 is a plot of $\delta^{18}O$ versus δD values for groundwater samples from the Railroad Valley regional flow system. This plot excludes samples collected in other parts of central Nevada although similar relationships are observed throughout the region. Most samples have isotopic values that plot to the right of the GMWL, with d-values for individual samples ranging from +8 to –3, and an average d-value near +4. Note that the regional groundwaters exhibit a shift to the right of the GMWL that is comparable to that of the mountain springs in the recharge areas. This is consistent with a link between the regional groundwaters and *modern* recharge in central Nevada, and implies the observed shift off the GMWL is inherited during recharge. Note also that the isotopic values of the different aquifers are indistinguishable from one another, implying no net isotopic effect due to past climate.

A recent study comparing the stable isotope values of spring waters and precipitation in the Spring Mountains near Las Vegas concluded that springtime snowmelt accounts for up to 90% of all groundwater recharge in southern Nevada (Winograd et al., 1998). To test this idea for central Nevada, precipitation gauges were installed at four separate locations at elevations between 2130 and 2280 m, and integrated samples of cool and warm season precipitation were collected during 1999. Evaporation effects were minimized by adding mineral oil to the bottom of the gauges. Small mountain springs located near each of the gauge sites were also sampled at the conclusion of the summer precipitation cycle (end of October). The stable isotope results for these samples are shown on a plot of $\delta^{18}O$ versus δD (Fig. 6). In each case, the local springs have stable isotope values that are similar to the integrated value for winter precipitation, which is dominated by snowfall. Summer rainfall accounts for ~10–30% of the annual precipitation total, but apparently accounts for a nominal amount of the total recharge budget. However, whereas the winter precipitation samples plot very close to the GMWL, the spring waters are all shifted to the right of the line with d-values similar to the regional groundwater. An extended study is under way to

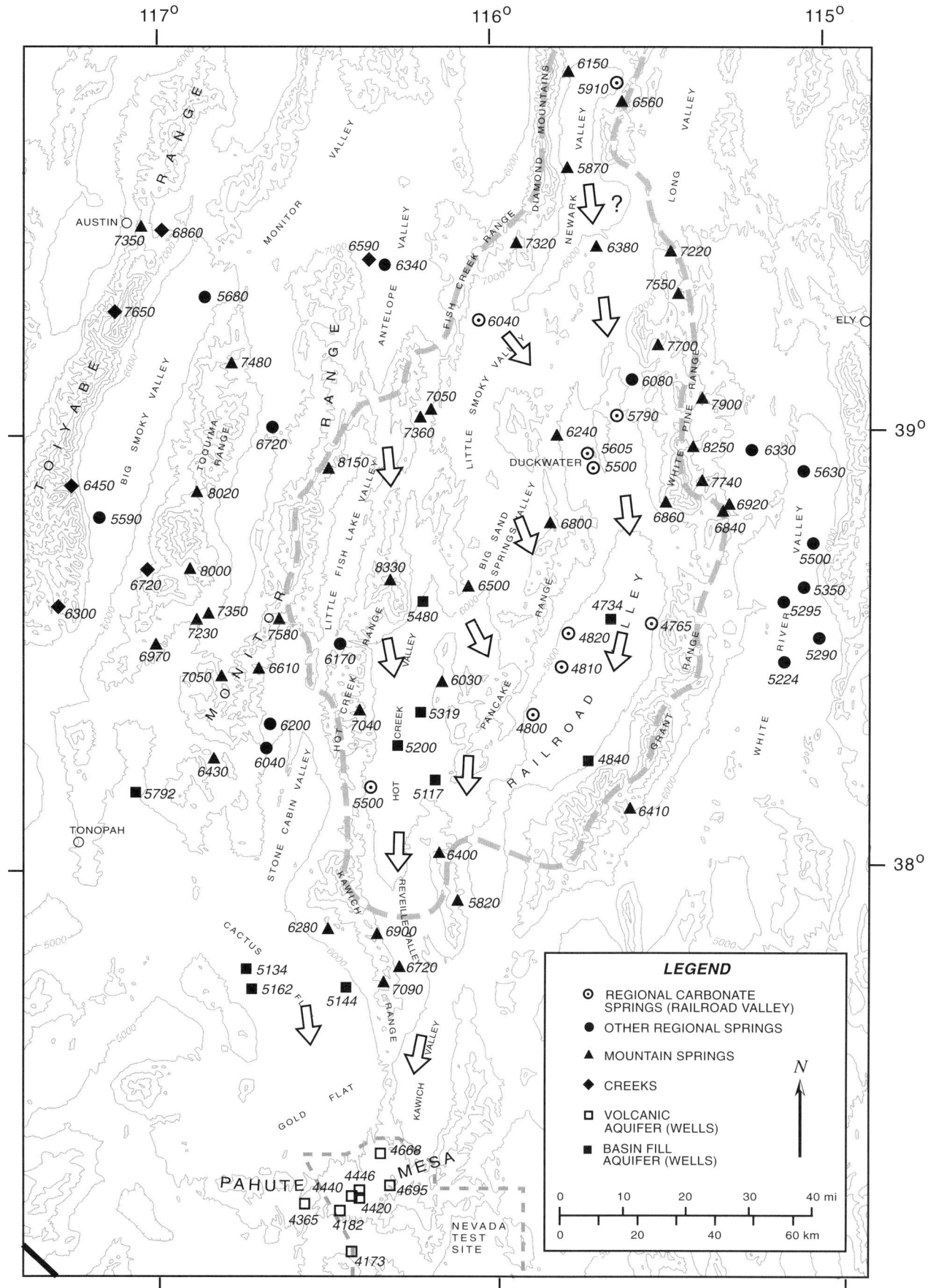

Figure 4. Map of central and southwestern Nevada showing water-level elevations (in feet above mean sea level) for sample locations shown in Figure 3. Arrows indicate the inferred direction of water flow in the carbonate aquifers or deep fractured volcanic aquifers.

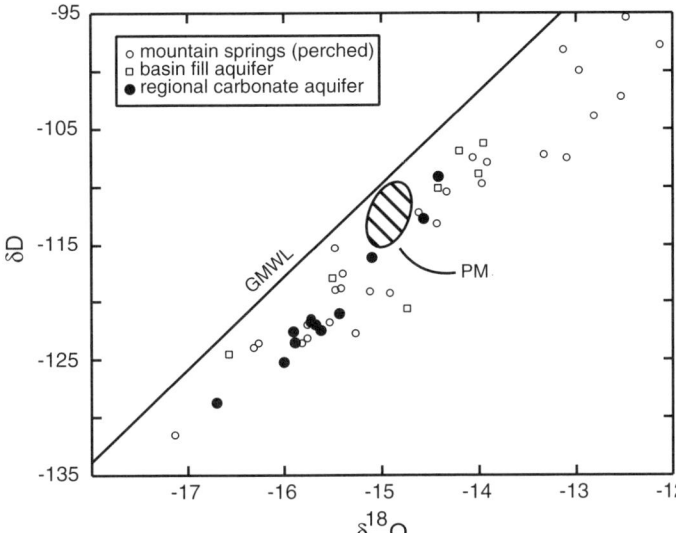

Figure 5. Plot of δD vs. $δ^{18}O$ values for groundwater samples from the Railroad Valley regional flow system. Groundwater samples are shifted to the right of the global meteoric water line (GMWL) with an average deuterium excess (d-value) near +4. The oval area marked "PM" denotes the Pahute Mesa regional groundwater sample field (from Figure 2).

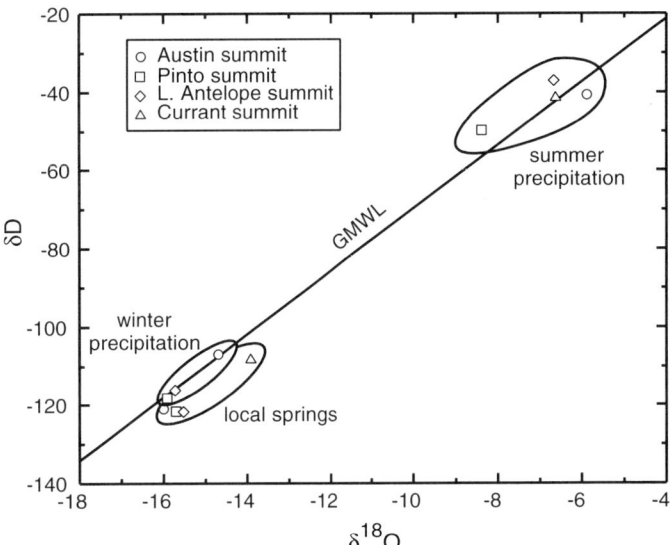

Figure 6. Plot of δD vs. $δ^{18}O$ values for seasonal accumulations of precipitation and small springs collected at four different mountain sites in central Nevada during 1999. Summer precipitation is highly enriched in heavy isotopes relative to winter precipitation. The spring waters are isotopically similar to winter precipitation, implying that most groundwater recharge is derived from snowmelt. Spring samples show a characteristic shift to the right of the global meteoric water line (GMWL).

determine the long-term variations in precipitation amount and isotopic values at these gauging sites.

Rose et al. (1999a) investigated isotopic variations in the winter snowpack in central Nevada to determine whether the observed enrichments in spring waters are related to processes during snowpack aging and ablation. At the time of peak accumulation (early March), most snowpacks showed physical evidence of snow metamorphism, including the development of depth hoar and grain clusters (Colbeck, 1987). Heat transfer from the ground to the overlying snowpack creates temperature and vapor pressure gradients within the snowpack (Benson and Trabant, 1973). Under these conditions, mass transport and recrystallization processes are accompanied by isotopic fractionation effects (Friedman et al., 1991), and the initial isotopic variability of individual snow layers is strongly attenuated (Judy et al., 1970).

Figure 7 shows a $δ^{18}O$-δD plot for 28 bulk snow cores collected throughout central Nevada in March 1998 (Rose et al., 1999a). Many of the data points are shifted to the right of the GMWL, with an average d-value near +5. This value is essentially the same as for the groundwaters plotted in Figure 5. These results indicate that kinetic isotope effects related to vapor loss during snow metamorphism cause isotopic enrichments in the snowpack that are not observed when the snowfall is simply accumulated in a precipitation gauge (as shown in Figure 6). Although subsequent processes occurring during late stage melting of the snowpack can further modify the isotopic composition of recharge, this initial vapor loss prior to melting may be a key factor determining the widely observed isotopic shift off the GMWL in Nevada groundwaters.

Carbon isotope data

The dissolved inorganic carbon (DIC) in groundwater is commonly derived from two main sources: the biochemical production of CO_2 gas in the soil zone, and the chemical dissolution of carbonate minerals. Differences in the $δ^{13}C$ values of these source materials provide insight into their relative contributions to the total DIC inventory. Biogenic CO_2 has $δ^{13}C$ values that range from approximately –27 to –12‰, depending on the type of plant material (Deines, 1980; Cerling, 1984), whereas marine carbonate rocks in Nevada have $δ^{13}C$ values between –2 and +3‰ (Thomas et al., 1996). Deep pedogenic carbonates (>50 cm depth) in southern Nevada vary in $δ^{13}C$ from about –9 to +4‰ and show a systematic ^{13}C depletion with increasing elevation (Quade et al., 1989).

The application of ^{14}C measurements to dating groundwater requires the careful evaluation of processes influencing the evolution of DIC along a flow path (see Mook, 1980, for review). A number of different models have been proposed to quantify these processes (e.g., Fontes, 1983). In practice, however, it is difficult to account for all the chemical processes in complex regional flow systems. A large amount of ^{14}C dilution (up to 50%) can occur in the recharge zone by the neutralization of dissolved soil CO_2 (^{14}C ~100 pmc) with ^{14}C-absent carbonate minerals (^{14}C ~0 pmc). Further reaction with carbonate minerals may occur along the flow path due to processes such as ion exchange or the addition of deep (magmatic) CO_2 to the system (e.g., Andrews et al.,

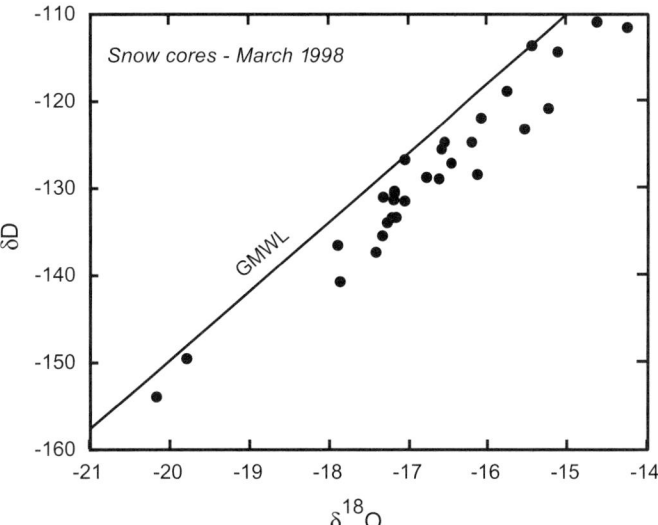

Figure 7. Plot of δD vs. δ^{18}O values for snow core samples collected throughout central Nevada during March 1998. Many of the data points are shifted to the right of the global meteoric water line (GMWL) due to kinetic isotope fractionation effects that occur during the aging of the snowpack. The average d-value for these samples (+5) is nearly the same as for the groundwaters shown in Figure 5. This comparison suggests the observed isotopic shifts in groundwaters are predominantly inherited from the snowpack.

1994; Rose and Davisson, 1996). Oxidation of organic material can add bicarbonate to the system that contains no ^{14}C but has a δ^{13}C value similar to the soil gas (Pearson and Hanshaw, 1970). Multiple recharge sources and mixing along the flow path further complicate ^{14}C age correction models (Davisson et al., 1999). For the present study, these various processes are not sufficiently well constrained to permit the calculation of meaningful ^{14}C ages. Nevertheless, it is possible to gain insight into some of the more important water-rock processes by examining regional variations in carbon isotopes and water chemistry.

Figure 8 shows a plot of δ^{13}C versus ^{14}C data for the Railroad Valley flow system. The results are separated into three groups representing samples from the regional carbonate aquifer, basin-fill aquifers, and mountain springs (perched aquifers). The data show a general trend toward decreasing ^{14}C with increasing δ^{13}C values that is consistent with chemical dissolution or isotopic exchange with carbonate rocks along the flow path. Mountain spring samples have δ^{13}C values between about –8 and –12‰, and most have ^{14}C values >75 pmc. We assume most of the mountain spring samples have undergone very little radioactive decay of ^{14}C, and the observed decreases in ^{14}C (to values <100 pmc) mostly reflect dilution from carbonate mineral dissolution during infiltration. One mountain spring (Snowball Ranch Spring) has a rather low ^{14}C value (19 pmc) and Ca-HCO$_3$ water chemistry that suggest substantial carbonate reaction, but a light δ^{13}C value (–9.8‰) and relatively low pH value (6.9) that suggest recent infiltration. These data may indicate dissolution/exchange with carbonate under partially open system conditions, although other interpretations are also possible.

Groundwater samples from the basin-fill aquifers are generally lower in ^{14}C than the mountain springs, and in some cases have carbon isotope signatures similar to the carbonate aquifer springs (Fig. 8). The two basin-fill samples with the highest δ^{13}C values are both from Hot Creek Valley. One is from a deep well in northern Hot Creek Valley (HTH-1, total depth = 1129 m) that perforates tuffaceous sediments (Chapman et al., 1994). The other is from a shallow well in a groundwater discharge area at Twin Springs Ranch, at the southern end of the Pancake Range. The low ^{14}C and high δ^{13}C values for these samples (see Table 1) suggest a possible hydrologic link to the regional carbonate aquifer.

Groundwater from an artesian well in the central Railroad Valley (total depth = 367 m) has a δ^{13}C value of –6‰ and a ^{14}C value <1 pmc (Fig. 8). The well perforates alluvium that overlies productive carbonate-hosted oil fields. Borehole temperature versus depth profiles and mineralogical studies of drill core samples indicate the carbonate aquifer is isolated from the overlying alluvial aquifer in this part of Railroad Valley (Hulen et al., 1994). Compared to other groundwaters in the central Railroad Valley, this sample has an unusually low δ^{18}O value (–16.6‰) that is similar to groundwater from Newark Valley. However, available evidence suggests the basin-fill aquifers are not interconnected between the Newark and Railroad Valleys (Harrill et al., 1988) and this sample may therefore represent "paleowater." Its low ^{14}C value suggests either older Pleistocene recharge, modification of the ^{14}C value by water-rock interaction, or introduction of ^{14}C-depleted CO$_2$ from the underlying oil field.

Springs associated with the regional carbonate aquifer vary in δ^{13}C from about –7.1 to –1.5‰ with ^{14}C values between 30 and 2 pmc (Fig. 8). The carbonate springs do not show a simple progression toward lower ^{14}C values in a southward direction, probably because they do not occur along a single flow path. For example, three carbonate springs located near Duckwater (Fig. 3) show a significant variation in ^{14}C (between 3 and 30 pmc), in δ^{13}C (between –7.1 and –2.9‰) and in temperature (13–34 °C). These springs occur in an area where Paleozoic carbonate rocks are exposed in low hills along the valley floor, and likely represent groundwater that is forced to the surface by structural barriers. The data variations imply the presence of multiple flow paths at different depths within the carbonate rocks in this area.

Simonson Warm Spring in Newark Valley is located in the uppermost part of the Railroad Valley flow system, but has a ^{14}C value of 8 pmc, and δ^{13}C value of –2.5‰. The low δ^{18}O value for this spring (–16.7‰) may indicate recharge in the southern Ruby Mountains. Even so, it is difficult to envision a flow path length much greater than 30 or 40 km between the recharge area and the spring. This suggests a significant amount of carbon isotope exchange occurs between the DIC and the carbonate rock along a relatively short path length. Davisson et al. (1999) reached the same conclusion for the springs in the upper part of the White River flow system.

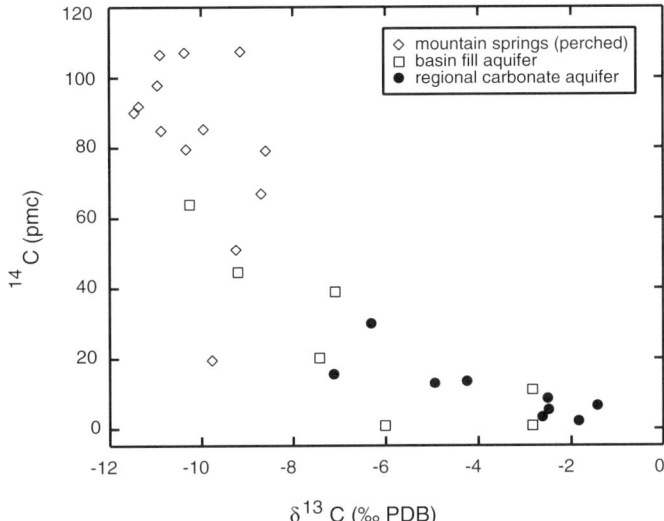

Figure 8. Plot of ^{14}C vs. δ^{13}C values for dissolved inorganic carbon in groundwater from the Railroad Valley regional flow system. Carbon isotope values for mountain springs are generally consistent with soil CO_2 gas dissolution and reaction with carbonate minerals in the recharge zone. Regional carbonate aquifer groundwaters have relatively high δ^{13}C and low ^{14}C values that indicate water-rock interaction with the host aquifer.

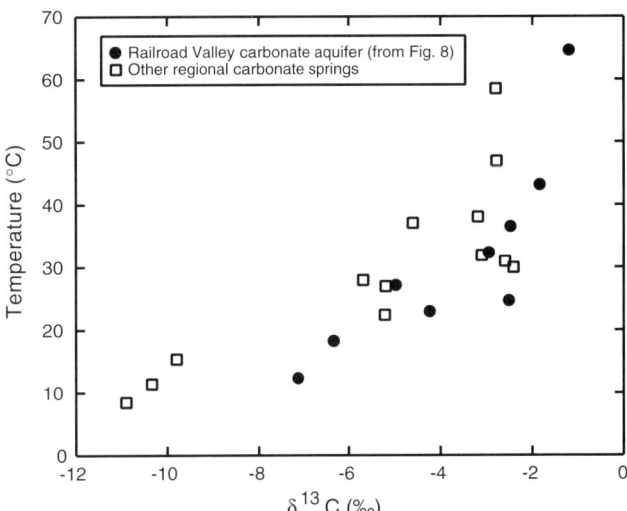

Figure 9. Temperature vs. δ^{13}C plot for samples collected from regional carbonate aquifer springs throughout central and southern Nevada. The trend toward higher δ^{13}C values with increasing temperature is interpreted to indicate increasing water-rock interaction, and may be driven in part by cation exchange processes with clay minerals in the aquifer (see discussion in text). Plot includes additional data from Rose et al. (1997).

There is a tendency for the carbonate springs with higher δ^{13}C values to be associated with higher water temperatures. Springs from carbonate aquifers throughout central and southern Nevada show a nonlinear increasing trend in δ^{13}C with temperature (Fig. 9; data from this study and Rose et al., 1997). The increasing δ^{13}C values reflect a shift toward isotopic equilibrium between the DIC and the carbonate rock. At 35 °C, dissolved HCO_3 is depleted in ^{13}C by ~1.4‰ relative to $CaCO_3$ under equilibrium conditions (Salomons and Mook, 1986). Thus, many of the warm springs are probably close to equilibrium with the carbonate rock, assuming a δ^{13}C value near 0‰ for the rock. The springs with high δ^{13}C values tend to also have higher DIC concentrations, implying greater amounts of carbonate rock dissolution at higher temperatures. This is actually *opposite* to the solubility-temperature relationship for calcite (e.g., Stumm and Morgan, 1981). Water chemistry data (discussed below) suggests that cation exchange with clay minerals may help drive this process. In addition, it is notable that all four springs with water temperatures >40 °C in Figure 9 have pH values <7, implying high levels of dissolved CO_2.

DISCUSSION

Chemical and isotopic evidence for a Railroad Valley–Pahute Mesa regional flow path

The isotopic results presented up to this point are generally consistent with a continuous regional flow path extending from the upper Railroad Valley flow system (Newark Valley) southward into western Railroad Valley and Hot Creek Valley. The regional carbonate springs have δ^{18}O and δD values that can be reasonably derived from modern snowmelt recharge in high elevation areas of central Nevada. These data do not appear to be consistent with pluvial period recharge processes. Carbon isotope data indicate rapid reaction of the DIC in groundwater with the carbonate host rock, and relatively low ^{14}C values (<10 pmc) are found in regional warm springs within the northernmost part of the flow system (Newark Valley). Regional carbonate springs in the western Railroad Valley, 150 km south of Newark Valley, show similarly low ^{14}C values between 2 and 6 pmc. The comparatively small ^{14}C difference of over more than 150 km of regional flow implies mixing with young groundwater recharge along the flow path in order to maintain a nonzero ^{14}C content (see Davisson et al., 1999). Gradual increases in carbonate spring δ^{18}O values along the flow path are consistent with this mixing argument.

If regional groundwater beneath Pahute Mesa originated from the western part of the Railroad Valley flow system, then mixing along the flow path can readily account for the increase in δ^{18}O values from approximately –15.5‰ in the western Railroad Valley to near –15.0‰ beneath Pahute Mesa. Groundwater discharge at Oasis Valley constrains the approximate volume of water flowing *out* of Pahute Mesa (8000 ac-ft/yr; Reiner et al., 1999). Regional inflow to Pahute Mesa is ~5500 ac-ft/yr (Blankennagel and Weir, 1973) implying a "local" recharge amount of ~2500 ac-ft/yr. For simple two-component mixing:

$$\delta^{18}O_{mix} = \delta_1 X_1 + \delta_2 X_2 \qquad (1)$$

where δ_1 and δ_2 are the isotopic values of the two components, X_1 and X_2 are their fractional mixing proportions, and $X_1 + X_2 = 1$. Assuming the regional inflow has a $\delta^{18}O$ value of –15.5‰, and "local" recharge has a $\delta^{18}O$ value of –13.5‰ (e.g., Figure 2, perched aquifer data), then:

$$\delta^{18}O_{mix} = (-15.5)(5500/8000) + (-13.5)(2500/8000) = -14.88‰. \quad (2)$$

This is similar to $\delta^{18}O$ values observed in wells beneath Pahute Mesa (Fig. 3), and in the Oasis Valley discharge area (White and Chuma, 1987; Rose et al., 1999b).

Hydrogeologic and isotopic data that would confirm the proposed regional flow path is presently unavailable between 38° and 37°15′N latitude. Geologic evidence suggests that carbonate rocks may underlie basin-fill deposits as far south as Kawich Valley (Ekren et al., 1971). However, all of the deep wells on Pahute Mesa are completed in fractured volcanic rocks, and the link to the regional carbonate flow system is by inference. Groundwater beneath Pahute Mesa also has a different chemistry and carbon isotope content than the regional carbonate aquifer to the north. The chemistry of Pahute Mesa groundwater is dominated by Na^+ and HCO_3^- ions (Blankennagel and Weir, 1973), and is lower in Ca^{2+} and HCO_3^- compared to the carbonate aquifer groundwaters. In addition, Pahute Mesa groundwaters have ^{14}C values ranging from 8 to 21 pmc, and $\delta^{13}C$ values between –1.4 and –9.2‰ (Table 1). The increase in ^{14}C values relative to the Railroad Valley carbonate aquifer requires an added component of "younger" water.

Comparison of water chemistry data for the Railroad Valley and Pahute Mesa regional aquifers provides insight into possible processes relating these flow systems. Figure 10 is a plot of HCO_3^- versus Na^+ ion concentrations for both aquifer systems. In general, the Railroad Valley carbonate springs show a systematic increase in HCO_3^- with Na^+ that is closely related to distance along the proposed regional flow path. One possible mechanism to account for this variation is a cation exchange process wherein Ca^{2+} is removed from solution by ion exchange on clay minerals, according to the exchange reaction:

$$2NaX + Ca^{2+} = CaX_2 + 2Na^+ \quad (3)$$

where X denotes an exchange site on the clay minerals (Pearson and Swarzenki, 1974; Andrews et al., 1994). During this process, HCO_3^- will increase even in calcite-saturated waters, because the removal of Ca^{2+} drives the solution toward calcite undersaturation, allowing further calcite dissolution. This process is therefore consistent with the observed increases in groundwater $\delta^{13}C$ values. It should be noted that while the highest ion concentrations occur in the southernmost carbonate spring (Warm Springs at 38°11′N), the $\delta^{18}O$ value of this spring (–14.4‰) indicates a different origin from all central Railroad Valley samples. The high Na^+ concentration in this sample may imply additional water-rock interaction with volcanic rocks located near the spring.

Repetition of the ion exchange process (equation 3) along a carbonate aquifer flow path will cause a correlated increase in

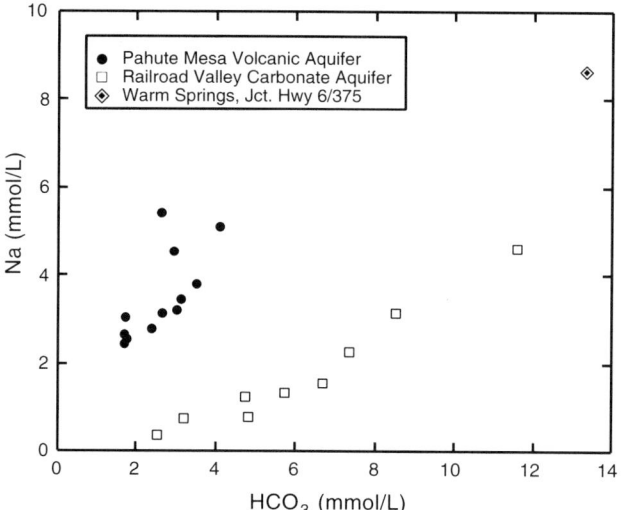

Figure 10. Plot of Na^+ vs. HCO_3^- ion concentrations in groundwater samples from the Railroad Valley carbonate and Pahute Mesa volcanic aquifers. The Railroad Valley data show an increase in HCO_3^- with Na^+ that may indicate calcium-sodium ion exchange on clay minerals. Ion exchange would drive the solution toward calcite undersaturation, allowing further calcite dissolution, and causing a progressive increase in HCO_3^- along the regional flow path. Pahute Mesa groundwaters have Na^+ ion concentrations similar to chemically evolved carbonate aquifer groundwaters, but significantly lower HCO_3^-, implying carbonate mineral precipitation (see Figure 11). The Warm Springs sample (Jct. Hwy 6/375) is from a travertine spring in the western Hot Creek Valley, but it has a $\delta^{18}O$ value that is isotopically distinct from all other Railroad Valley carbonate springs.

Na^+ with HCO_3^- (Fig. 10), while Ca^{2+} should at most remain constant. However, on a plot of Na^+ versus Ca^{2+} concentrations (Fig. 11), the carbonate springs show an increase in Na^+ with Ca^{2+} along a least-squares regression line of slope 2.18 (r = 0.90). This slope is in close agreement with the stoichiometry of the cation exchange reaction. We propose that an independent process may cause additional calcite to dissolve while the exchange reaction controls the observed Na^+ to Ca^{2+} ratio. Some of the carbonate springs, especially those with higher temperatures, exhibit relatively low pH values (<7). One possible explanation is that CO_2 is added to the system, possibly from a deep source. Low pH values associated with high CO_2 concentrations will enhance calcite dissolution rates, and may account for the increasing Ca^{2+} concentrations in the carbonate groundwaters (Fig. 11).

Variations in chemical and isotopic data suggest that two processes may occur during the proposed transition from carbonate to volcanic aquifer lithologies: (1) calcite precipitation, and (2) mixing with local recharge. Groundwater from the Pahute Mesa volcanic aquifer is significantly lower in both HCO_3^- and Ca^{2+} compared to most carbonate springs, suggesting that calcite is precipitated. This process would be driven by the equilibrium state of the reaction

$$Ca^{2+} + CO_3^{2-} = CaCO_3. \quad (4)$$

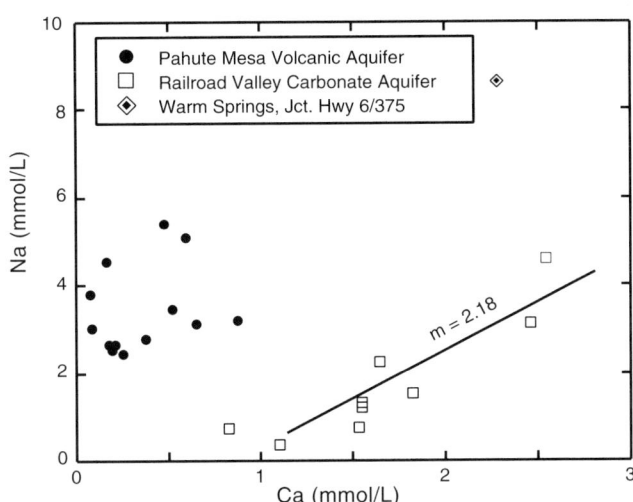

Figure 11. Plot of Na⁺ vs. Ca²⁺ ion concentrations for the same samples as shown in Figure 10. The Railroad Valley carbonate groundwaters show a correlated increase in Na⁺ with Ca²⁺ along a line with a slope of ~2, in accord with the proposed ion exchange model. The difference in Ca²⁺ concentrations between the two different aquifers likely reflects calcite precipitation. Scatter in the volcanic aquifer data may reflect variable amounts of mixing with dilute local recharge.

If the groundwater is no longer in contact with a large calcite reservoir, then this reaction will be driven toward the right, according to LeChatelier's principle. Calcite saturation index values indicate that most carbonate aquifer groundwaters are calcite saturated whereas most volcanic aquifer groundwaters are not. Benedict et al. (1999) examined calcite veins and fracture linings from deep boreholes on Pahute Mesa and found the calcite $\delta^{13}C$ and $\delta^{18}O$ values were consistent with deposition from thermal waters originating from the carbonate aquifer.

Groundwater mixing may account for variations in Na⁺ versus Ca²⁺ and HCO_3^- in the volcanic aquifer (Figs. 10 and 11) following calcite deposition. Na⁺ is conservative during groundwater flow, and is not readily removed once it is in solution. Continuity along a flow path between the carbonate and volcanic aquifers therefore requires that the Na⁺ concentration cannot decrease, except by mixing with a more dilute groundwater. Mixing with chemically dilute recharge may account for the range in Na⁺ concentrations in Pahute Mesa groundwater. Observed increases in $\delta^{18}O$ and ^{14}C values between Railroad Valley and Pahute Mesa are consistent with mixing of local recharge and regional groundwater. Although the water chemistry data does not prove a flow path linkage between the Railroad Valley and Pahute Mesa aquifer systems, the data are generally consistent with the proposed flow path.

A possible analog for the chemical processes described above is found near Yucca Mountain, ~40 km south of Pahute Mesa. In the Yucca Mountain area, the hydraulic head measured in the deep carbonate aquifer is 20 m higher than the head in the overlying volcanic aquifer, indicating an upward groundwater flow potential (Bredehoeft, 1997). Groundwater temperature anomalies in the volcanic aquifer are consistent with upwelling of warm (57 °C) carbonate aquifer water along fault zones (Fridrich et al., 1994). Groundwater from the carbonate aquifer is high in Ca²⁺, Na⁺ and HCO_3^- compared to groundwater in the overlying volcanic aquifer (e.g., Benson and McKinley, 1985). This *may* suggest the occurrence of calcite precipitation and mixing processes similar to those inferred beneath Pahute Mesa, provided the source of groundwater in the volcanic aquifer is from the underlying carbonate aquifer.

Hydrologic mass balance

Estimated groundwater budgets for the Railroad Valley flow system range from ~75,000 to 134,000 ac-ft yr⁻¹ (Rush and Everett, 1966; Van Denburgh and Rush, 1974; Roth and Campana, 1989; Prudic et al., 1995). A useful summary is provided in Roth and Campana (1989), who estimate an annual hydrologic mass balance of 115,000 ac-ft yr⁻¹ (1.4×10^8 m³yr⁻¹). In general, these models predict that recharge from precipitation and subsurface inflow is balanced by discharge from evapotranspiration and spring flow. It is notable that Van Denburgh and Rush (1974) estimated that only 4% of the annual precipitation budget recharges the groundwater system. This value may significantly underestimate the actual recharge rate, given that previous models assumed regional flow did not extend beyond the boundaries of the central Railroad Valley subbasin. Blankennagel and Weir (1973) estimated that ~5500 ac-ft (6.8×10^6 m³) of groundwater enters Pahute Mesa annually from the north. Uncertainties in the Railroad Valley mass balance far exceed the amount necessary to account for this underflow.

It is instructive to evaluate the volume of water stored within the carbonate aquifer in Railroad Valley in order to estimate the length of time required for the system to turnover (assuming a well-connected fracture network). The moderately high permeabilities in the carbonate aquifer are the result of fracturing during extensional tectonism and solution enlargement. Effective fracture porosities for the carbonate aquifer are typically ≤0.5%, whereas the effective intercrystalline porosity averages ~1% (Winograd and Thordarson, 1975; U.S. Department of Energy, 1997). The simplest approach is to assume that the entire Railroad Valley flow system is uniformly underlain by a certain thickness of permeable carbonate rock. If we infer that most groundwater flow occurs in the upper 2000 m of the aquifer, and apply an effective porosity of 1%, then the effective volume of water in storage is 14,500 km² × 2 km × 0.01 = 290 km³ (or 2.9×10^{11} m³).

We will assume a hydrologic mass balance of 1.4×10^8 m³yr⁻¹ for the Railroad Valley system (Roth and Campana, 1989). In the process of calibrating a regional-scale conceptual model, Prudic et al. (1995) achieved a reasonable water balance by assuming that ~32% of the total influx enters the carbonate aquifer system, with the remainder going to the alluvial aquifers. Using this value, an annual recharge budget of 4.5×10^7 m³yr⁻¹ is obtained for the carbonate aquifer. Dividing this number into the estimated volume of water in storage gives a mean turnover rate of 6400 yr. This suggests the effective volume of water within interconnected fractures could have turned over in Holocene time. This value is consistent with the range of flow rates predicted by permeability and hydraulic gradient data. In reality, groundwater flow is not homogeneously distributed throughout the carbonate rock, but occurs preferentially along a few discrete fractures in more permeable layers (e.g., Mifflin and Hess, 1979). Hence, groundwater turnover rates could be higher than implied by this simple calculation. For example, decreasing the fracture porosity to 0.1% causes a corresponding order-of-magnitude increase in the estimated turnover rate (to 640 yr).

Carbonate springs in the Railroad Valley flow system have water temperatures as high as 67 °C. Assuming a maximum flow depth of ~3000 m (e.g., 2000 m of carbonate rock overlain by ~1000 m of basin fill) and a geothermal gradient of 30 °C/km, groundwater temperatures ≥100 °C are expected. The lack of higher spring temperatures may suggest circulation depths <3000 m, or it may indicate the spring waters are not from the maximum depth, or that they cooled during ascent. Sass et al. (1971) report a relatively wide range in measured vertical temperature gradients in central Nevada, ranging from 7 to 52 °C/km. They also note a large region of low heat flow (<1.5 μcal cm^{-2} s^{-1}) in central Nevada (the "Eureka low") that is centered on the Railroad Valley flow system. Sass et al. (1971) propose the Eureka low is related to interbasin groundwater flow "with appreciable vertical velocity components to depths of ~3 km." This suggests that regional flow may advectively transport a sizable fraction of the heat budget in this region.

Implications for pluvial recharge and the distribution of Pleistocene-age groundwater

There is good evidence to suggest that the effective moisture in the Great Basin increased during the last glacial-interglacial transition, between ~18,000 and 9000 yr before present time. We could infer that groundwater recharge rates increased in central Nevada during that time. Estimates of the full glacial mean-annual temperature depression in the Great Basin range from ~3 to 8 °C (Dohrenwend, 1984; Claassen, 1986; Spaulding and Graumlich, 1986; Benson and Klieforth, 1989). From these estimates, we can calculate the corresponding depletion in the $\delta^{18}O$ value of late Pleistocene recharge. Precipitation $\delta^{18}O$ values vary as a function of mean monthly surface temperature along a line with a slope between ~0.52 and 0.57 (Yurtsever, 1975; Van der Straaten and Mook, 1983; Claassen, 1986). Hence, an average temperature decrease of 5 °C at full glacial maximum would presumably cause precipitation $\delta^{18}O$ values to decrease by around –2.6 to –2.8‰. This decrease is comparable to that observed in other basins at similar latitudes in the western U.S. (cf. Phillips et al., 1986). If we assume that recharge from winter precipitation currently averages around –15.5‰ at latitude 39°N, we would predict that similar recharge at glacial maximum would have a $\delta^{18}O$ value < –18‰. Even assuming a conservative temperature shift of only 3 °C, we still arrive at a $\delta^{18}O$ value < –17‰. Isotopic depletions of this magnitude are not observed in the carbonate aquifer groundwaters at this latitude, nor is there evidence for similarly depleted groundwater in carbonate aquifers at lower latitudes.

Throughout this paper, we have argued that the observed variations in regional groundwater $\delta^{18}O$ and ^{14}C values can be explained by a combination of Holocene-age recharge, groundwater mixing, and water-rock interaction processes. The continuity in regional groundwater $\delta^{18}O$ values over distances up to 300 km (Newark Valley to Pahute Mesa) suggests a well-connected flow path and relatively rapid transport through the system. In general, the data suggest that pluvial-age groundwater either is no longer represented in most regional carbonate and basin-fill aquifers, or it is "lost" in the mixture and difficult to unequivocally identify from existing evidence. Mass balance calculations further indicate that regional groundwater transport rates are too high to store significant amounts of pluvial recharge.

Closed basins are more favorable locations for trapping pluvial-age groundwater, particularly in areas that presently receive relatively little recharge. We have noted two locations within the study area where the isotopic data suggest this may have occurred: central Railroad Valley and Cactus Flat. The latter case was briefly discussed in Davisson et al. (1999) who noted the unusually low $\delta^{18}O$ value of groundwater from a well completed in this alluvial basin. Roller Coaster Well in Cactus Flat (37°43′ N) has a $\delta^{18}O$ value of –16‰, and a low d-value of –1. Springs associated with local perched aquifers in the Kawich and Monitor Ranges have present-day $\delta^{18}O$ values near –13.5‰, implying that recharge during the glacial maximum could have been near –16‰, assuming a 5 °C paleo-temperature shift. In this case, the stable isotope data fit nicely with the inferred pluvial-age climate conditions.

Further illustrations of pluvial-age groundwater stored in closed basins are found in the literature. For example, Thomas et al. (1989) studied the geochemical and isotopic evolution of groundwater from a hydrologically closed basin in west-central Nevada, and found that less evaporated basinal groundwaters are significantly depleted in their $\delta^{18}O$-δD values relative to modern recharge. Correcting these data to the GMWL along the observed evaporation trajectory gives an initial $\delta^{18}O$ value near –19‰, which is 3–4‰ lighter than modern recharge in this area. In another set of studies, Davisson and Criss (1993, 1996) observed that deep Pleistocene-age groundwaters in the closed basin of Sacramento Valley, California have $\delta^{18}O$ values ~2.5‰ depleted relative to shallow, modern recharge. Corrected ^{14}C ages for Sacramento Valley groundwaters suggest turnover rates for this alluvial basin are at least a factor of 2–3 slower than in the Railroad Valley regional carbonate aquifer.

CONCLUSIONS

Isotopic and chemical data presented in this study are consistent with the hypothesis that predominantly Holocene-age groundwater enters the north-central Death Valley flow system along regional flow paths originating from central Nevada. We have proposed that groundwater originates from recharge areas in the Railroad Valley flow system, moves southward along regional carbonate aquifers, and then transitions to fractured volcanic aquifers in the vicinity of the Nevada Test Site (Fig. 1). Supporting evidence for this model includes the following:

1. The Railroad Valley flow system is rimmed by high elevation recharge areas containing significant exposures of carbonate bedrock. These areas receive relatively large amounts of winter snowfall, which is rapidly recharged in the spring. Water-level data for regional carbonate springs in the Railroad Valley flow system and wells on Pahute Mesa (Fig. 4) are generally consistent with a continuous regional flow path.

2. Groundwater $\delta^{18}O$ values for carbonate springs of the Railroad Valley flow system are similar to $\delta^{18}O$ values of modern precipitation and mountain springs in central Nevada. The regional distribution of $\delta^{18}O$ values defines a continuous north-to-south flow path that extends into the Pahute Mesa–Oasis Valley region of the Death Valley flow system. Modest increases in $\delta^{18}O$ along this flow path are inferred to indicate mixing with ^{18}O-enriched "local" recharge at lower latitudes.

3. Carbon isotope and major ion data indicate extensive water-rock interaction along the carbonate aquifer flow paths, with higher solute concentrations and greater ^{13}C enrichments generally occurring in warmer springs at lower latitudes. Inferred reaction processes include cation exchange on clay minerals coupled with calcite dissolution. A "deep" CO_2 source may be required to account for relatively low pH values (<7) observed in some high temperature, chemically evolved carbonate springs. Removal of calcite from chemically evolved carbonate aquifer groundwater can yield a modified groundwater chemistry similar to that observed in volcanic aquifers beneath Pahute Mesa. Subsequent mixing with local recharge may account for variations in the chemistry and carbon isotope content of the Pahute Mesa groundwater.

4. Hydrologic mass balance considerations suggest that regional groundwater transport rates are too high to store significant amounts of pluvial recharge in the regional carbonate aquifer. Pleistocene-age water remaining in the regional flow system is probably highly mixed, and indistinguishable from Holocene-age groundwater. Limited evidence for pristine Pleistocene-age groundwater is found within two closed alluvial basins located within the study area. In each case, the groundwater $\delta^{18}O$ values are significantly depleted relative to local recharge and regional flow.

Although the evidence outlined above is consistent with the proposed regional flow system, we cannot preclude other possible flow path models. For example, some (or all) of the regional flow entering Pahute Mesa could originate from the region just to the west of the Railroad Valley flow system (via Stone Cabin Valley and Cactus Flat; see Figure 3). However, few of the springs in this area have $\delta^{18}O$ values less than the Pahute Mesa groundwaters, suggesting there could be little mixing with isotopically heavier groundwater along this flow path. In addition, the extent to which carbonate rocks underlie the region west of the Railroad Valley flow system is poorly determined. A chemical analysis of the water sampled from a warm spring in Stone Cabin Valley does not suggest interaction with carbonate rocks. Hence, if regional flow does occur in this area, it may be hosted entirely in fractured volcanic aquifers. At this time, there is not a compelling reason to believe that such a flow path is more plausible than the proposed Railroad Valley flow path, but it clearly warrants further consideration.

In order to more fully test the ideas proposed in this study, a broader range of evidence is required, particularly in terms of hydrogeologic data (e.g., geologic and hydraulic data from boreholes) and geochemical interpretations (e.g., flow path modeling of chemical data). The development of deep boreholes in the region between the southern Railroad Valley and Pahute Mesa (e.g., Kawich Valley) would be particularly beneficial for obtaining hydraulic, geologic, and geochemical data within the inferred transition zone between carbonate and volcanic aquifers. These evaluations are beyond the scope of this study, although we hope that this paper may provide a framework for developing future studies of regional groundwater flow paths in central and southern Nevada.

ACKNOWLEDGMENTS

This manuscript was significantly improved by thoughtful reviews provided by Jim Thomas, Brent Newman, and one anonymous reviewer. We thank Dave Smith, Ron Hershey, Chris Benedict, and Bob Criss for fruitful discussions regarding this work. Gail Eaton contributed valuable assistance in map preparation, field work, and laboratory analyses. Additional field assistance was provided by Jackie Kenneally, Brenda Ekwurzel, and Ross Williams. Funding for this project was provided by the Hydrologic Resources Management Program and Underground Test Area Project of the U.S. Department of Energy, Nevada Operations Office. This work was performed under the auspices of the U.S. Department of Energy by Lawrence Livermore National Laboratory under contract number W-7405-Eng-48.

REFERENCES CITED

Andrews, J.N., Fontes, J.Ch., Aranyossy, J.F., Dodo, A., Edmunds, W.M., Joseph, A., and Travi, Y., 1994, The evolution of alkaline groundwaters in the continental intercalaire aquifer of the Irhazer Plain, Niger: Water Resources Research, v. 30, p. 45–61.

Barnes, C.J., and Allison, G.B., 1988, Tracing of water movement in the unsaturated zone using stable isotopes of hydrogen and oxygen: Journal of Hydrology, v. 100, p. 143–176.

Benedict, F.C., Rose, T.P., and Eaton, G.F., 1999, Fracture coating mineral phase characterization in the Pahute Mesa–Oasis Valley flow system [abs.]: Eos (Transactions, American Geophysical Union), v. 80, p. 327.

Benson, C.S., and Trabant, D.C., 1973, Field measurements on the flux of water vapour through dry snow, in Proceedings, Symposia on the Role of Snow and Ice in Hydrology, Banff, September 1972, Vol. 1: Geneva, Unesco-WMO-IAHS Publication no. 107, p. 291–298.

Benson, L.V., and Klieforth, H., 1989, Stable isotopes in precipitation and groundwater in the Yucca Mountain region, southern Nevada: Paleoclimatic implications, in Peterson, D.H., ed., Aspects of climate variability in the

Pacific and western Americas: American Geophysical Union, Geophysical Monograph Series, v. 55, p. 41–59.

Benson, L.V., and McKinley, P.W., 1985, Chemical composition of groundwater in the Yucca Mountain area, Nevada, 1971–1984: U.S. Geological Survey Open-File Report 85-484, 10 p.

Benson, L.V., Currey, D.R., Dorn, R.I., Lajoie, K.R., Oviatt, C.G., Robinson, S.W., Smith, G.I., and Stine, S., 1990, Chronology of expansion and contraction of four Great Basin lake systems during the past 35,000 yr: Palaeogeography, Palaeoclimatology, Palaeoecology, v. 78, p. 241–286.

Benson, L.V., Burdett, J.W., Kashgarian, M., Lund, S.P., Phillips, F.M., and Rye, R.O., 1996, Climatic and hydrologic oscillations in the Owens Lake Basin and adjacent Sierra Nevada, California: Science, v. 274, p. 746–749.

Benson, L.V., Lund, S.P., Burdett, J.W., Kashgarian, M., Rose, T.P., Smoot, J.P., and Schwartz, M., 1998, Correlation of late-Pleistocene lake-level oscillations in Mono Lake, California, with North Atlantic climate events: Quaternary Research, v. 49, p. 1–10.

Blankennagel, R.K., and Weir, J.E., Jr., 1973, Geohydrology of the eastern part of Pahute Mesa, Nevada Test Site, Nye County, Nevada: U.S. Geological Survey Professional Paper 712-B, 35 p.

Bredehoeft, J.D., 1997, Fault permeability near Yucca Mountain: Water Resources Research, v. 33, p. 2459–2463.

Cerling, T.E., 1984, The stable isotopic composition of modern soil carbonate and its relationship to climate: Earth and Planetary Science Letters, v. 71, p. 229–240.

Chapman, J.B., Mihevc, T.M., and Lyles, B.F., 1994, The application of borehole logging to characterize the hydrogeology of the Faultless site, Central Nevada Test Area: Desert Research Institute, Water Resources Center Publication No. 45119, 36 p.

Claassen, H.C., 1986, Late-Wisconsin paleohydrology of the west-central Amargosa Desert, Nevada, USA: Chemical Geology, v. 58, p. 311–323.

Colbeck, S.C., 1987, Snow metamorphism and classification, in Jones, H.G., and Orville-Thomas, W.J., eds., Seasonal snowcovers: Physics, chemistry, hydrology: Dordrecht, Reidel Publishing, p. 1–35.

Coleman, M.L., Sheperd, T.J., Durham, J.J., Rouse, J.E., and Moore, G.R., 1982, Reduction of water with zinc for hydrogen isotope analysis: Analytical Chemistry, v. 54, p. 993–995.

Craig, H., 1961, Isotopic variations in meteoric waters: Science, v. 133, p. 1702–1703.

Craig, H., and Gordon, L.I., 1965, Deuterium and oxygen-18 variations in the ocean and the marine atmosphere, in Schink, D.R., and Corless, J.T., eds., Proceedings, Symposium on Marine Geochemistry, 29–30 October 1964: Kingston, Rhode Island, Narraganset Marine Laboratory, University of Rhode Island Publication, v. 3, p. 277–374.

Criss, R.E., 1999, Principles of stable isotope distribution: New York, Oxford University Press, 254 p.

D'Agnese, F.A., Faunt, C.C., Turner, A.K., and Hill, M.C., 1997, Hydrogeologic evaluation and numerical simulation of the Death Valley regional groundwater flow system, Nevada and California: U.S. Geological Survey Water-Resources Investigations Report 96-4300, 124 p.

Dansgaard, W., 1964, Stable isotopes in precipitation: Tellus, v. 16, p. 436–468.

Davisson, M.L., and Criss, R.E., 1993, Stable isotope imaging of a dynamic groundwater system in the southwestern Sacramento Valley, California, USA: Journal of Hydrology, v. 144, p. 213–246.

Davisson, M.L., and Criss, R.E., 1996, Stable isotope and groundwater flow dynamics of agricultural irrigation recharge into groundwater resources of the Central Valley, California, in Proceedings, Symposium on Isotopes in Water Resources Management, Vienna, 20–24 March 1995, Volume 1: International Atomic Energy Agency, p. 405–418.

Davisson, M.L., Smith, D.K., Kenneally, J., and Rose, T.P., 1999, Isotope hydrology of southern Nevada groundwater: Stable isotopes and radiocarbon: Water Resources Research, v. 35, p. 279–294.

Deines, P., 1980, The isotopic composition of reduced organic carbon, in Fritz, P., and Fontes, J.Ch., eds., Handbook of environmental isotope geochemistry, Volume 1: The terrestrial environment, A: Amsterdam, Elsevier, p. 329–406.

Dettinger, M.D., and Schaefer, D.H., 1996, Hydrogeology of structurally extended terrain in the eastern Great Basin of Nevada, Utah, and adjacent states from geologic and geophysical models: U.S. Geological Survey Hydrologic Investigations Atlas HA-694-D, 1 sheet.

Dettinger, M.D., Harrill, J.R., Schmidt, D.L., and Hess, J.W., 1995, Distribution of carbonate-rock aquifers and the potential for their development, southern Nevada and adjacent parts of California, Arizona and Utah: U.S. Geological Survey Water-Resources Investigations Report 91-4146, 100 p.

Dinwiddie, G.A., and Schroder, L.J., 1971, Summary of hydraulic testing in and chemical analyses of water samples from deep exploratory holes in Little Fish Lake, Monitor, Hot Creek, and Little Smoky Valleys, Nevada: U.S. Geological Survey Report USGS-474-90, 70 p.

Dohrenwend, J.C., 1984, Nivation landforms in the western Great Basin and their paleoclimatic significance: Quaternary Research, v. 22, p. 275–288.

Domenico, P.A., and Schwartz, F.W., 1990, Physical and chemical hydrogeology: New York, John Wiley and Sons, 824 p.

Eakin, T.E., 1966, A regional interbasin groundwater system in the White River area, southeastern Nevada: Water Resources Research, v. 2, p. 251–271.

Ekren, E.B., Anderson, R.E., Rogers, C.L., and Noble, D.C., 1971, Geology of northern Nellis Air Force Base bombing and gunnery range, Nye County, Nevada: U.S. Geological Survey Professional Paper 651, 91 p.

Epstein, S., and Mayeda, T.K., 1953, Variation of O^{18} content of waters from natural sources: Geochimica et Cosmochimica Acta, v. 4, p. 213–224.

Fontes, J.Ch., 1983, Dating of groundwater, in Guidebook on nuclear techniques in hydrology: Vienna, International Atomic Energy Agency Technical Reports Series, no. 91, p. 285–317.

Fridrich, C.J., Dudley, W.W., and Stuckless, J.S., 1994, Hydrogeologic analysis of the saturated-zone groundwater system under Yucca Mountain, Nevada: Journal of Hydrology, v. 154, p. 133–168.

Friedman, I., Benson, C., and Gleason, J., 1991, Isotopic changes during snow metamorphism, in Taylor, H.P., Jr., et al., eds., Stable isotope geochemistry: A tribute to Samuel Epstein: The Geochemical Society, Special Publication No. 3, p. 211–221.

Friedman, I., Smith, G.I., Gleason, J.D., Warden, A., and Harris, J.M., 1992, Stable isotope composition of waters in southeastern California, 1: Modern precipitation: Journal of Geophysical Research, v. 97, p. 5795–5812.

Garside, L.J., and Schilling, J.H., 1979, Thermal waters of Nevada: Nevada Bureau of Mines and Geology, Bulletin 91, 163 p.

Gat, J.R., and Dansgaard, W., 1972, Stable isotope survey of the fresh water occurrences in Israel and the northern Jordan Rift Valley: Journal of Hydrology, v. 16, p. 177–212.

Harrill, J.R., Gates, J.S., and Thomas, J.M., 1988, Major groundwater flow systems in the Great Basin region of Nevada, Utah, and adjacent states: U.S. Geological Survey Hydrologic Investigations Atlas HA-694-C, scale 1:1,000,000, 2 sheets.

Hershey, R.L., and Acheampong, S.Y., 1997, Estimation of groundwater velocities from Yucca Flat to the Amargosa Desert using geochemistry and environmental isotopes: Desert Research Institute, Water Resources Center Publication No. 45157, 49 p.

Houghton, J.G., Sakamoto, C.M., and Gifford, R.O., 1975, Nevada's weather and climate: Nevada Bureau of Mines and Geology, Special Publication 2, 78 p.

Hulen, J.B., Goff, F., Ross, J.R., Bortz, L.C., and Bereskin, S.R., 1994, Geology and geothermal origin of Grant Canyon and Bacon Flat oil fields, Railroad Valley, Nevada: American Association of Petroleum Geologists Bulletin, v. 78, p. 596–623.

Ingraham, N.L., Lyles, B.F., Jacobson, R.L., and Hess, J.W., 1991, Stable isotopic study of precipitation and spring discharge in southern Nevada: Journal of Hydrology, v. 125, p. 243–258.

Judy, C., Meiman, J.R., and Friedman, I., 1970, Deuterium variations in an annual snowpack: Water Resources Research, v. 6, p. 125–129.

Laczniak, R.J., Cole, J.C., Sawyer, D.A., and Trudeau, D.A., 1996, Summary of hydrogeologic controls on groundwater flow at the Nevada Test Site, Nye County, Nevada: U.S. Geological Survey Water-Resources Investigations Report 96-4109, 59 p.

Lamb, D., Nielsen, K.W., Klieforth, H.E., and Hallett, J., 1976, Measurements of liquid water content in winter cloud systems over the Sierra Nevada: Journal of Applied Meteorology, v. 15, p. 763–775.

Lamke, R.D., and Moore, D.O., 1965, Interim inventory of surface-water resources of Nevada: Nevada Department of Conservation and Natural Resources, Water Resources Bulletin 30, 38 p.

Lund, K., Beard, L.S., and Perry, W.J., Jr., 1993, Relation between extensional geometry of the northern Grant Range and oil occurrences in Railroad Valley, east-central Nevada: American Association of Petroleum Geologists Bulletin, v. 77, p. 945–962.

Malmberg, G.T., 1965, Available water supply of the Las Vegas groundwater basin, Nevada: U.S. Geological Survey Water-Supply Paper 1780, 116 p.

Maxey, G.B., and Eakin, T.E., 1949, Groundwater in White River Valley, White Pine, Nye, and Lincoln Counties, Nevada: Nevada Department of Conservation and Natural Resources, Water Resources Bulletin 8, 59 p.

McNichol, A.P., Jones, G.A., Hutton, D.L., Gagnon, A.R., and Key, R.M., 1994, The rapid preparation of seawater ΣCO_2 for radiocarbon analysis at the National Ocean Sciences AMS facility: Radiocarbon, v. 36, p. 237–246.

Merlivat, L., and Contiac, M., 1975, Study of mass transfer at the air-water interface by an isotopic method: Journal of Geophysical Research, v. 80, p. 3455–3464.

Merlivat, L., and Jouzel, J., 1979, Global climatic interpretation of the deuterium-oxygen 18 relationship for precipitation: Journal of Geophysical Research, v. 84, p. 5029–5033.

Mifflin, M.D., and Hess, J.W., 1979, Regional carbonate flow systems in Nevada: Journal of Hydrology, v. 43, p. 217–237.

Mifflin, M.D., and Wheat, M.M., 1979, Pluvial lakes and estimated pluvial climates of Nevada: Nevada Bureau of Mines and Geology, Bulletin 94, 57 p.

Milne, W.K., Benson, L.V., and McKinley, P.W., 1987, Isotope content and temperature of precipitation in southern Nevada, August 1983–August 1986: U.S. Geological Survey Open-File Report 87-463, 32 p.

Mook, W.G., 1980, Carbon-14 in hydrogeological studies, in Fritz, P., and Fontes, J.Ch., eds., Handbook of environmental isotope geochemistry, Volume 1: The Terrestrial Environment, A: Amsterdam, Elsevier, p. 49–74.

Morgan, D.S., and Dettinger, M.D., 1996, Groundwater conditions in Las Vegas Valley, Clark County, Nevada, Part II: Hydrogeology and simulation of groundwater flow: U.S. Geological Survey Water-Supply Paper 2320-B, 124 p.

Pearson, F.J., and Hanshaw, B.B., 1970, Sources of dissolved carbonate species in groundwater and their effects on carbon-14 dating, in Isotope Hydrology: Vienna, International Atomic Energy Agency, p. 271–286.

Pearson, F.J., and Swarzenki, W.V., 1974, ^{14}C evidence for the origin of arid region groundwater, Northeastern province, Kenya, in Proceedings, Isotope Techniques in Groundwater Hydrology, 11–15 March 1974, Volume 2: Vienna, International Atomic Energy Agency, p. 95–109.

Phillips, F.M., Peeters, L.A., Tansey, M.K., and Davis, S.N., 1986, Paleoclimatic inferences from an isotopic investigation of groundwater in the central San Juan Basin, New Mexico: Quaternary Research, v. 26, p. 179–193.

Prudic, D.E., Harrill, J.R., and Burbey, T.J., 1995, Conceptual evaluation of regional groundwater flow in the carbonate-rock province of the Great Basin, Nevada, Utah, and adjacent states: U.S. Geological Survey Professional Paper 1409-D, 102 p.

Quade, J., Cerling, T.E., and Bowman, J.R., 1989, Systematic variations in the carbon and oxygen isotopic composition of pedogenic carbonate along elevation transects in the southern Great Basin, United States: Geological Society of America Bulletin, v. 101, p. 464–475.

Quade, J., Mifflin, M.D., Pratt, W.L., McCoy, W., and Burckle, L., 1995, Fossil spring deposits in the southern Great Basin and their implications for changes in water-table levels near Yucca Mountain, Nevada, during Quaternary time: Geological Society of America Bulletin, v. 107, p. 213–230.

Quade, J., Forester, R.M., Pratt, W.L., and Carter, C., 1998, Black mats, spring-fed streams, and late-glacial-age recharge in the southern Great Basin: Quaternary Research, v. 49, p. 129–148.

Reiner, S.R., Laczniak, R.J., DeMeo, G.A., Smith, J.L., Nylund, W.E., and Elliott, P.E., 1999, Preliminary estimate of mean annual evapotranspiration, Oasis Valley, Nye County, Nevada [abs.]: Eos (Transactions, American Geophysical Union), v. 80, p. 328.

Rose, T.P., and Davisson, M.L., 1996, Radiocarbon in hydrologic systems containing dissolved magmatic carbon dioxide: Science, v. 273, p. 1367–1370.

Rose, T.P., Kenneally, J.M., Smith, D.K., Davisson, M.L., Hudson, G.B., and Rego, J.H., 1997, Chemical and isotopic data for groundwater in southern Nevada: Lawrence Livermore National Laboratory Report UCRL-ID-128000, 35 p.

Rose, T.P., Davisson, M.L., Criss, R.E., and Smith, D.K., 1999a, Isotopic investigation of recharge to a regional groundwater flow system, Great Basin, Nevada, USA, in Proceedings, International Symposium on Isotope Techniques in Water Resources Development and Management, Vienna, 10–14 May 1999: International Atomic Energy Agency, IAEA-CSP-2/C, session 2, p. 63–72.

Rose, T.P., Hershey, R.L., Thomas, J.M., Benedict, F.C., Hudson, G.B., Eaton, G.F., and Kenneally, J.M., 1999b, Environmental isotopes in the Pahute Mesa–Oasis Valley groundwater flow system [abs.]: Eos (Transactions, American Geophysical Union), v. 80, p. 328.

Roth, J.G., and Campana, M.E., 1989, A mixing-cell model of the Railroad Valley regional groundwater flow system, central Nevada: Desert Research Institute, Water Resources Center Publication # 41123, 175 p.

Rush, F.E., and Everett, D.E., 1966, Water-resources appraisal of Little Fish Lake, Hot Creek, and Little Smoky Valleys, Nevada: Nevada Department of Conservation and Natural Resources, Water Resources, Reconnaissance Series Report 38, 38 p.

Salomons, W., and Mook, W.G., 1986, Isotope geochemistry of carbonates in the weathering zone, in Fritz, P., and Fontes, J.Ch., eds., Handbook of environmental isotope geochemistry, Volume 2: The terrestrial environment, B: Amsterdam, Elsevier, p. 239–269.

Sass, J.H., Lachenbruch, A.H., Munroe, R.J., Greene, G.W., and Moses, T.H., Jr., 1971, Heat flow in the western United States: Journal of Geophysical Research, v. 76, p. 6376–6413.

Spaulding, G.W., and Graumlich, L.J., 1986, The last pluvial climatic episodes in the deserts of southwestern North America: Nature, v. 320, p. 441–444.

Stewart, J.H., 1980, Geology of Nevada: A discussion to accompany the geologic map of Nevada: Nevada Bureau of Mines and Geology, Special Publication 4, 136 p.

Stewart, J.H., and Carlson, J.E., 1978, Geologic map of Nevada: U.S. Geological Survey, scale 1:500,000, 2 sheets.

Stichler, W., Rauert, W., and Martinec, J., 1981, Environmental isotope studies of an alpine snowpack: Nordic Hydrology, v. 12, p. 297–308.

Stuiver, M., and Polach, H., 1977, Reporting of ^{14}C data: Radiocarbon, v. 19, p. 355–363.

Stumm, W., and Morgan, J.J., 1981, Aquatic chemistry (2nd edition): New York, Wiley, 780 p.

Thomas, J.M., Mason, J.L., and Crabtree, J.D., 1986, Groundwater levels in the Great Basin region of Nevada, Utah, and adjacent states: U.S. Geological Survey Hydrologic Investigations Atlas HA-694-B, scale 1:1,000,000, 2 sheets.

Thomas, J.M., Welch, A.H., and Preissler, A.M., 1989, Geochemical evolution of groundwater in Smith Creek Valley: A hydrologically closed basin in central Nevada, USA: Applied Geochemistry, v. 4, p. 493–510.

Thomas, J.M., Welch, A.H., and Dettinger, M.D., 1996, Geochemistry and isotope hydrology of representative aquifers in the Great Basin region of Nevada, Utah, and adjacent states: U.S. Geological Survey Professional Paper 1409-C, 100 p.

U.S. Department of Energy, 1997, Regional groundwater flow and tritium transport monitoring and risk assessment of the underground test area, Nevada Test Site, Nevada: U.S. Department of Energy, Nevada Operations Office, Environmental Restoration Division Report DOE/NV-477, 396 p.

Van Denburgh, A.S., and Rush, F.E., 1974, Water-resources appraisal of Railroad and Penoyer Valleys, east-central Nevada: Nevada Department of Conservation and Natural Resources, Water Resources, Reconnaissance Series Report 60, 61 p.

Van der Straaten, C.M., and Mook, W.G., 1983, Stable isotopic composition of precipitation and climatic variability, in Proceedings, Symposium on Paleoclimates and Paleowaters, 25–28 November 1980: A Collection of Environmental Isotope Studies: Vienna, International Atomic Energy Agency, p. 53–64.

White, A.F., and Chuma, N.J., 1987, Carbon and isotope mass balance models of Oasis Valley–Fortymile Canyon groundwater basin, southern Nevada: Water Resources Research, v. 23, p. 571–582.

Winograd, I.J., and Friedman, I., 1972, Deuterium as a tracer of regional groundwater flow, southern Great Basin, Nevada and California: Geological Society of America Bulletin, v. 83, p. 3691–3708.

Winograd, I.J., and Pearson, F.J., Jr., 1976, Major carbon 14 anomaly in a regional carbonate aquifer: Possible evidence for megascale channeling, south-central Great Basin: Water Resources Research, v. 12, p. 1125–1143.

Winograd, I.J., and Thordarson, W., 1975, Hydrogeologic and hydrochemical framework, south-central Great Basin, Nevada-California, with special reference to the Nevada Test Site: U.S. Geological Survey Professional Paper 712-C, 126 p.

Winograd, I.J., Riggs, A.C., and Coplen, T.B., 1998, The relative contributions of summer and cool-season precipitation to groundwater recharge, Spring Mountains, Nevada, USA: Hydrogeology Journal, v. 6, p. 77–93.

Yurtsever, Y., 1975, Worldwide survey of stable isotopes in precipitation: International Atomic Energy Agency, Isotope Hydrology Section Report, November, 1975, 40 p.

MANUSCRIPT ACCEPTED BY THE SOCIETY AUGUST 1, 2002

Late Quaternary paleohydrologic and paleotemperature change in southern Nevada

Jay Quade
Desert Laboratory/Department of Geosciences, University of Arizona, Tucson, Arizona 85721, USA

Richard M. Forester
U.S. Geological Survey, MS 980, Box 25046, Denver, Colorado 80225-0046, USA

Joseph F. Whelan
U.S. Geological Survey, MS 963, Box 25046, Denver, Colorado 80225-0046, USA

ABSTRACT

Paleo-spring discharge activity in the southern Great Basin responded to changes in recharge, hence climate changes, in high mountain areas during the late Quaternary. In our study, we examined four stratigraphic sections in southern Nevada in order to reconstruct paleohydrologic change spanning the last two major discharge cycles. The largest discharge event in those sections is expressed as extensive wetland deposits (Unit B) that fall beyond the range of radiocarbon dating (>41 ka). We tentatively correlate this event with marine isotope stage 6, which is so conspicuously represented in cores from Death Valley and Owens Lake. Major wetlands were also present during last glacial maximum (Unit D) deposited between 16.4 and <26.3 ^{14}C ka. The absence of any dates between 16.4 and ca. 14.5 ^{14}C ka may indicate a period of relative aridity. Wetlands are also strongly expressed between ca. 13.9 and 13.5 ^{14}C ka in several sections, followed by contraction beginning between 12.9 and 12.8 ^{14}C ka. The region witnessed a modest resurgence of spring activity, expressed as black mats and spring-fed channels, starting at 11.6 ^{14}C ka, and peaking between 11 and 9.5 ^{14}C ka, followed by desiccation of most springs between 9.5 and 7 ^{14}C ka.

Detailed analysis of ostracode taxa from three stratigraphic sections shows that a complex depositional mosaic composed of wet meadows, seeps, flowing springs, streams, and wetlands covered the valley bottom during the last two glacial periods. Differences in ostracode species assemblages suggest that climate associated with the earlier discharge cycle (Unit B) was colder and perhaps wetter than that of the younger cycle (Units D and E). δ^{18}O values from >400 ostracode shells vary by ~5‰, and there is no consistent, section-wide, difference in isotopic values between standing water and spring taxa. This pattern strongly suggests short residence times for water in local basins, due to loss of water from basins by outflow as groundwater or overflow, rather than by evaporation.

We used the δ^{18}O value of fossil ostracodes to place constraints on paleotemperature in the valley bottoms during glacial periods. This analysis entails at least three key assumptions: no vital or evaporation effects during valve formation of the ostracode *Cypridopsis vidua*, short transit times in the aquifer, and the basic relationship between modern air and spring water temperature holds for the past. If these and other assumptions are satisfied, we estimate that mean annual air temperature during the penultimate wet period (Unit B_2) in the valley bottom was at least 10.8 °C colder than today, and at least 5.6 °C colder during the last glacial maximum (Unit D). If a vital effect of 0.8–1‰ is assumed using $\delta^{18}O$ values from groundwater candonids, then the above estimates of maximum valley-bottom temperatures during Unit D time increase by ~2–3 °C.

INTRODUCTION

Major basins in the southern Great Basin are largely dry today because regional water tables remain below the surface and stream flow supported by snow-pack and high-mountain spring discharge is restricted. During the late Pleistocene, climate was wetter and cooler, and regional water tables were higher across the southern Great Basin, producing seeps, flowing springs and associated wet meadows, streams, and wetlands in the centers of the now largely dry basins (Quade, 1986; Quade and Pratt, 1989). As climate became drier and warmer during the Holocene, water tables fell, valley fill aquifers dried, and stream flow ceased, leading to dissection and extensive exposure of these spring-related deposits. The stratigraphy, paleontology, paleohydrologic, and paleoclimatic implications of the deposits have been the focus of many papers and maps (Haynes, 1967; Quade, 1986; Quade and Pratt, 1989; Quade et al., 1995, 1998; Bell et al., 1998, 1999). These studies document the repeated change in regional hydrology over the past ~500,000 yr in response to regional climate change. Relative age estimates of these paleo-discharge periods are based largely on amino acid racemization/epimerization dating of fossil mollusks, while ^{14}C dates, largely from carbonized wood and organic mats, provide absolute age control on the younger deposits.

Geologists rely heavily on paleolake deposits and plant macrofossils from packrat middens to document changes in paleoclimate in the Great Basin. Spring-related deposits, however, offer another perspective that enjoys several advantages over the paleoenvironmental information available from lake deposits. Young (<15 ka) spring and wetland deposits can often be reliably radiocarbon dated from the remains of vascular plants, unlike lake sediments where carbon reservoir effects (e.g., Bischoff et al., 1997) or contamination problems associated with carbonates and organic matter make accurate age determination uncertain. Spring discharge and wetland deposits can also be studied in extensive natural exposures, providing the equivalent to multiple exposed "cores" in which the lateral continuity of a hydrologic event can be traced. Although caution must be exercised, the $\delta^{18}O$ values of carbonates from ostracodes and mollusks in spring deposits offer insights into the $\delta^{18}O$ values of the paleo-recharge. Because spring discharge represents flowing water, it is less likely to become evaporatively enriched than lake water and should have a $\delta^{18}O$ value close to the recharge value, commonly from winter precipitation (Winograd et al., 1998).

A key limitation of the spring-related sedimentary records is that the magnitude of hydrologic changes can only be semiquantitatively estimated. Springs produce a complex mosaic of aquatic environments including the flowing spring or seep areas themselves, wet meadows, streams, and wetlands. The aquatic environments responsible for the deposits do not leave bathtub rings as lakes do. Thus, they offer no means of quantifying the precipitation/evaporation balance related only to climate. In the case of the spring deposits we studied, isotopic and taxonomic evidence does show that evaporation was minor, but the causes are probably both climatic and topographic (i.e., the system overflowed). In contrast to lake hydrologic budgets, flow-dominated environments such as springs only provide insights into the limiting hydrologic conditions necessary for their genesis. So, one can only estimate what reduction in temperature and (or) rise in precipitation is necessary to raise the regional water tables to ground surface, recharge the valley fill aquifers, and produce stream flow. Actual conditions may greatly exceed the minimum needed to initiate surface discharge.

In this paper we synthesize new results from four stratigraphic sections with previous results from across the region. The new study sections include two (LPM-34 and OCI-11) from Corn Creek in the northern Las Vegas Valley, one (LPM-35) from near Cactus Springs, Nevada, and one (LWDD 6/8N) through diatomaceous deposits next to Highway 95 in the eastern Amargosa Valley, informally termed the "Lathrop Wells Diatomite" (Fig. 1). The new paleoenvironmental data includes tabulation of ostracode species by stratigraphic level and environmental preference, and stable isotope analyses of both ostracodes and aquatic mollusks, all placed in the context of new and extensive radiocarbon age control of the deposits. The ostracode data in particular provide an important new perspective on the sources, temperature, solute-composition, and solute-concentration (total dissolved solids, TDS) of the water that discharged in valley bottoms during wet periods. Finally, we combine our isotopic evidence with data on modern springs to place constraints on the maximum mean annual temperatures in valleys during wet periods.

LABORATORY METHODS

The sediment samples were processed for ostracodes according to technical procedure HP-78, R1, as described in Forester (1991a). Sections OCI-11 and LPM-35 were sampled continuously in 5-cm thick intervals, and each sample was divided in four parts. Ostracodes were picked from about a 15-g subsample of the 5-cm thick field sample. A second subsample of ~50 g was also prepared for multiple uses, such as stable isotope

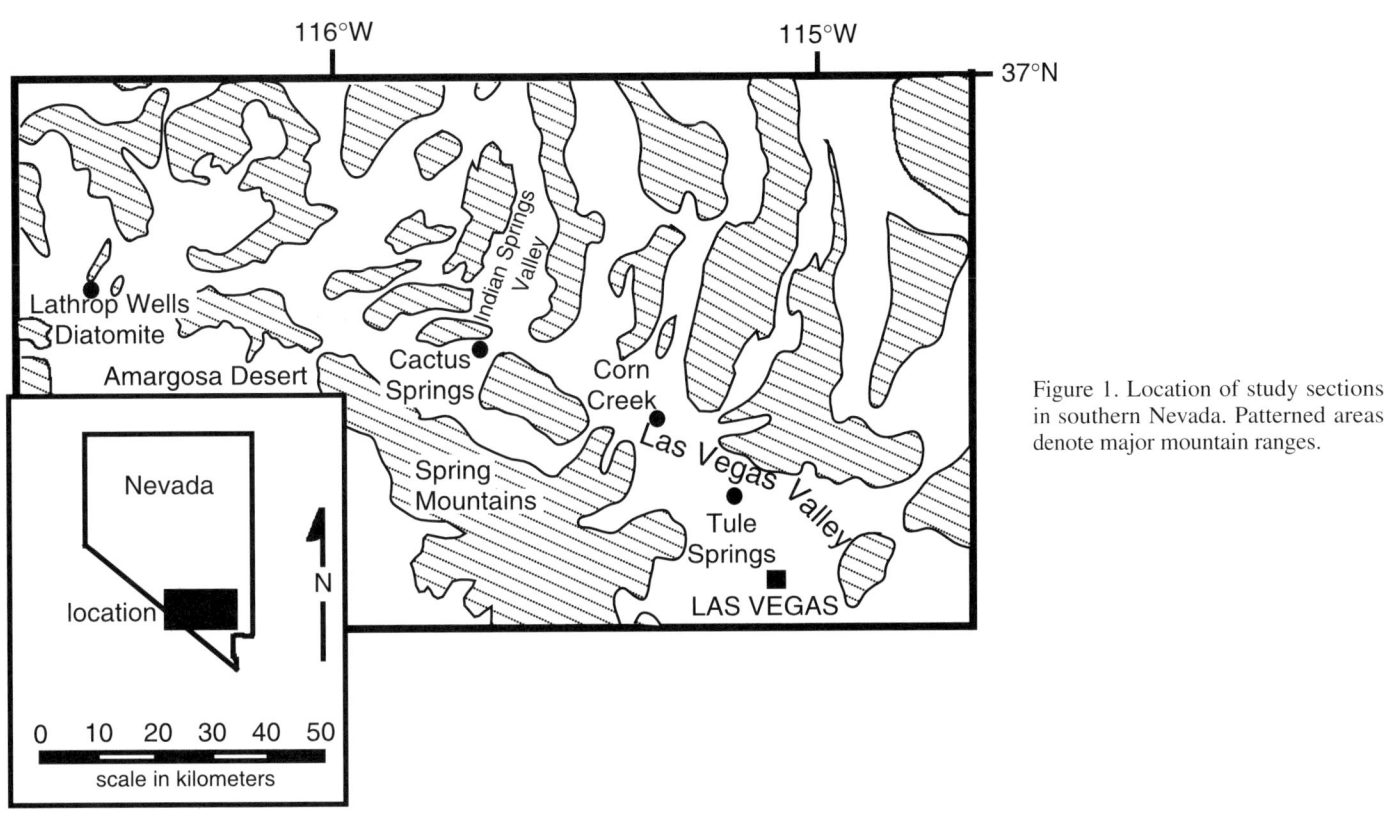

Figure 1. Location of study sections in southern Nevada. Patterned areas denote major mountain ranges.

or radiocarbon analyses. A third subsample of about five grams was taken for pollen extraction, but examination of a few of these samples suggests pollen is not commonly preserved. A fourth subsample was archived.

Fossil mollusks were handpicked and identified at the University of Arizona and at the U.S. Geological Survey, Denver. Specimens were washed repeatedly in an ultrasonic bath of distilled water until free of any adhering detritus, soaked in 2% H_2O_2, washed, dried, and then separated for ^{14}C dating by taxa. Coiled taxa such as *Vallonia* were crushed and the shell fragments viewed under a microscope to insure no detritus was entrapped within. One to five shells were required for a ^{14}C date. One radiocarbon date also was obtained from ostracodes. All dating was by accelerator mass spectrometry through Beta Analytic, Inc.

Carbonized wood was used for dating of some horizons. Though identical in appearance to charcoal, carbonized wood differs by being much more soluble in base treatment. All samples were digested in 2–3N HCl for at least one hour, until effervescence stopped but pH remained <2, and then thoroughly washed in distilled H_2O. Base treatment consisted of 2–3 soakings in 2–4% NaOH, followed by digestion and filtration, which generally resulted in near complete dissolution of the organic material. Humate solids were precipitated from the supernatant by acidification, followed by filtration, washing to a pH of ~3 and drying. Except where noted, the humate fraction was used for dating. Previous studies have shown this fraction to yield very reliable dates (Quade et al., 1998). All dates are reported in ^{14}C years B.P., unless otherwise noted.

Ostracode and mollusk shell carbonate CO_2 was extracted with 100% H_3PO_4 at 75 °C in a Kiel automated extraction device, and $\delta^{13}C$ and $\delta^{18}O$ values were determined using a Finnigan MAT 251 mass spectrometer. For larger ostracode species, such as *Strandesia meadensis*, a single valve was sufficient for analysis, whereas for the smaller species, such as the groundwater taxa, an average of 8 valves were loaded. Mass spectrometer analyses were corrected to calcite or aragonite, as appropriate, by three-point calibration using calcite standards NBS-19 ($\delta^{13}C_{PDB}$ = 1.92‰; $\delta^{18}O_{smow}$ = 28.65‰), NBS-18 ($\delta^{13}C$ = –5.0‰; $\delta^{18}O$ = 7.20‰), and NBS-20 ($\delta^{13}C$ = –1.06‰; $\delta^{18}O$ = 26.64‰). The $\delta^{18}O_{smow}$ values of water samples were determined by CO_2-H_2O equilibration according to procedures developed by Epstein and Mayeda (1953) and Kishima and Sakai (1980). Isotopic compositions were determined on a Finnigan MAT 252 mass spectrometer and the measured values corrected using a two point correction based on Standard Mean Ocean Water (δ^2H and $\delta^{18}O$ = 0‰) and Standard Light Antarctic Precipitation (δ^2H = –428‰; $\delta^{18}O$ = –55.5‰) (Coplen, 1994).

RADIOCARBON DATING OF MOLLUSKS

Aquatic and terrestrial mollusks are locally abundant in spring deposits of all ages, and were extensively radiocarbon dated in this study. Previously, most of the older (>14,000) spring-related deposits were undatable due to a lack of organic carbon. This motivated Brennan and Quade (1997) to explore the potential of mollusk aragonite for dating purposes. They found that small land snails in the deposits yield reliable ^{14}C ages, based on the similarity of ages between land snails and associated vascular plant remains. However, the more abundant aquatic snails yield ages that are often too old, due incorporation of C from variably ^{14}C-deficient surface waters. The extent of this deficiency tends to be much less in glacial-age specimens than in Holocene ones. Brennan and Quade (1997) found that aquatic mollusks in this system are 0–3000 ^{14}C years older than the true stratigraphic age, less than the much larger (>10,000 yr) deficiencies observed by Riggs (1984) in the Ash Meadows system. Thus, the ages from aquatic snails provide maximum ages (by <3000 yr) for the deposits containing them. Exceptions to this generalization can occur when mollusks are reworked, such as within channels cutting through older deposits. However, deposits in our study sections formed mostly in low-energy wetlands and springs. Except at lower Corn Creek Flat, deep channel cutting and reworking of mollusks is rarely in evidence, and then only locally. Moreover, with few exceptions, fossil ostracodes and mollusks that we analyzed are fresh, without any signs of significant abrasion.

OSTRACODES IN SPRING-DISCHARGE AND WETLAND ENVIRONMENTS

Ostracodes are microscopic aquatic crustaceans with calcitic bivalved shells that are easily identified to genus and species. Particular ostracode species often live in limited hydrologic settings such as springs, streams, lakes, wetlands, and groundwater. Within such settings, the species are further limited by having life cycles that are dependent on certain physical and chemical parameters, including: (1) total dissolved solids (TDS), (2) major dissolved-ion composition and especially the total-alkalinity/calcium (alk/Ca) ratio, (3) water-temperature, (4) the daily to annual variability of all of the latter parameters, and (5) the permanence of the environment (Delorme, 1969; DeDeckker, 1981; Forester, 1983, 1987, 1991a). Thus, the hydrologic setting, and the chemical and physical characteristics of the environment determine the species composition of ostracode assemblages. Identifying those assemblages in a stratigraphic sequence allows one to reconstruct past environments.

Lake and wetland species typically have biogeographic distributions that are related to climate. That is, the physical and chemical properties of the water in which those ostracodes live are often fully or partially determined by regional climate. Because these ostracodes routinely survive transportation from one place to another, their biogeographic ranges rapidly shift with climate-driven change of their environment.

Ostracodes that live only in springs also survive transportation. Unlike wetland or lake ostracodes, however, spring-discharge ostracodes may or may not reflect local climate conditions, because the physical and chemical characteristics of spring discharge may not be directly linked to local climate (Forester, 1991a).

Conversely, geologically long-lived and environmentally stable aquatic settings such as Lake Baikal, or large regional aquifers such as the lower carbonate aquifer in southern Nevada, often support endemic ostracode species swarms. These endemic surface- or groundwater species either cannot survive transportation or, if they do, cannot establish populations in new environments (Forester, 1991b). The presence of endemic species indicates long-term stability of their environment.

The basis for reconstructing paleoenvironments using fossil ostracode data comes from ostracode species occurrence and associated environmental data in the literature (e.g., Delorme, 1989) and from an unpublished ostracode species environmental database. The unpublished database contains ~800 sites, primarily lakes, wetlands, and springs, from throughout the United States and ~30 sites from Mexico. The data were collected by various U.S. Geological Survey personnel, by Alison Smith and Don Palmer at Kent State University, and by Brandon Curry at the Illinois State Geological Survey. The ostracode database is used here to both evaluate the most likely hydrologic habitat(s), such as spring or wetland, and the chemical characteristics of the water from those habitats.

HYDROLOGICALLY COMPLEX DEPOSITIONAL ENVIRONMENTS

The heterogeneous distribution of seeps, springs, streams, and wetlands along a valley floor results in environmentally complex depositional environments. Consequently, sediment accumulation at different sites along a valley floor may reflect different environmental settings, somewhat analogous to sedimentary facies in large lakes or the ocean. Similarly, because various ostracode species live in different hydrological environments, the stratigraphic distribution of ostracode species assemblages will also vary with location along a valley floor.

The potential or likelihood of different horizons of the same age to record different hydrological histories must then be taken into account in an environmental interpretation. There are several ways in which to evaluate spatial-temporal heterogeneity: (1) study several stratigraphic sections from different sites in a valley to capture the complex spatial-temporal relations of the sedimentary environments, (2) focus on a more homogeneous environmental setting such as in the valley center, analogous to taking a core from the center of a lake, (3) study sections from different valleys to identify synchronous environmental change, which should reflect a regional response to climate, and (4) use the hydrological sensitivity of ostracodes or other proxies to filter spatial from temporal changes in stratigraphic sections. So, for example, a change from stream to spring ostracodes might reflect a facies change, whereas a change from cold to warm wetland taxa might reflect climate change. All four approaches were employed in this study, but with an emphasis on numbers 2 and 4.

MODERN AND FOSSIL OSTRACODES FROM SOUTHERN NEVADA AND EASTERN CALIFORNIA

Springs are the most common natural aquatic environment in southern Nevada today. Large high-flow warm springs discharge from the lower carbonate aquifer, whereas typically low-flow cool springs discharge from valley fill or other local aquifers (see Winograd and Thordarsen, 1975). Overall, aquatic environments in the region today are very restricted when compared to the extensive valley bottom wetlands that were common in the past. Modern-day sites somewhat analogous to the paleo-hydrology of southern Nevada do exist in northeast Nevada, in, for example, the Ruby and Steptoe Valleys (Quade and Pratt, 1989; Quade et al., 1995).

Ostracodes were collected from a number of the extant springs and wetlands in southern Nevada and eastern California, providing a basis for a region-specific comparison of modern to past taxa. Ostracodes were also collected from the Ruby Valley wetlands and in springs in the Ruby Mountains, allowing for a comparison of this modern area to the paleoenvironments from southern Nevada. Because the ostracodes serve as hydrologic proxies, comparison of modern and fossil ostracodes provide a way to compare modern and past hydrology.

A diverse group of species composes the fossil ostracode assemblages found in the deposits on the valley floors in southern Nevada. The high diversity (5–10 or more species) reflects the heterogeneity of aquatic environments that contributed to the setting in which the sediments were deposited. Twenty-eight endemic fossil groundwater species were also found in Pleistocene deposits throughout the region. The groundwater ostracodes probably discharged to the surface, where their shells were deposited with surface-water taxa.

The four sections (OCI-11, LPM 34, 35, and LLDW-6/8N) discussed in this paper contain several common surface-water and groundwater ostracode species. Common surface-water species are defined as those occurring in more than five samples, whereas the groundwater species abundances are summed together and reported as a single entity. Rare surface-water taxa are omitted from discussion, because their limited stratigraphic distribution does not offer any insight into the general characteristics of the past environment. Groundwater taxa are summed together to signify groundwater discharge from the regional aquifer, which is believed to be the source of these taxa. No other environmental information can be derived from the groundwater taxa, because they are endemic, mostly known as fossils, and, with the exception of springs in the area, no environmental data is available for them. Although the discussion treats the taxa from all four sections, the data from section OCI-11 is emphasized, because that section contains both the longest sedimentary record and the most extensive set of samples.

The ostracode species from OCI-11 are further sorted and presented according to one of three key environmental settings: springs, wetlands, and groundwater (Fig. 2). The spring category includes ostracodes known to live in seeps, flowing springs, and spring-fed pools and streams. The wetland category includes taxa found in more extensive standing water bodies (marshes, ponds, littoral zones of small lakes) also chemically and perhaps thermally dominated by groundwater. No lacustrine taxa, in the sense of taxa found in standing water-bodies not dominated by groundwater, were found in these deposits. In some instances, species that commonly live in both wetlands and springs were placed in one environmental category in which they are believed to be more common. In other cases, such as with *Limnocythere paraornata*, abundance was divided equally between the spring and wetland environmental categories, following its modern-day distribution. Creating ostracode stratigraphic profiles for both general habitat types (Fig. 2) and for particular taxa (Fig. 3) provides a way to illustrate general paleohydrologic change and the components of the general change.

STRATIGRAPHY, SEDIMENTARY UNITS, AND OSTRACODE PROFILES

In this paper, we follow the basic stratigraphic nomenclature for the region developed by Haynes (1967) from the Tule Springs archeological site northwest of Las Vegas (Fig. 1). The oldest deposits at Tule Springs are alluvial and designated Unit A, but are not recognized from other areas. Unit B is more widespread and consists of alluvial silt and sand, as well as green muds (Unit B_2) representing very extensive paleo-spring discharge. Unit B yielded mostly infinite ^{14}C dates at Tule Springs, and is therefore thought to be >40,000 ^{14}C yr B.P. A well-developed paleosol is often preserved at the top of Unit B. Unit C is entirely alluvial and not widely exposed. Unit C marks a dry episode between Units B_2 and D, both of which are derived from spring discharge.

Extensive green muds compose overlying Unit D and represent spring discharge and wetland development across the area, greater in extent than in Unit B_2 at Tule Springs. Previous dating control on Unit D comes from Tule Springs on carbonized wood (25,300 ± 2500 ^{14}C yr B.P.; UCLA-539; Haynes, 1967) from the base of a spring feeder, and on large aquatic shells of *Planorbella subcrenata* (22,600 ± 550 ^{14}C yr B.P.; UCLA-536) from the middle of Unit D. The shell date should represent a maximum age for middle Unit D deposition. Spaulding (1995), citing McVickar and Spaulding (1993), describe dates on charcoal of 26,800 ± 700 and 27,580 ± 650 ^{14}C yr B.P. from somewhere in Unit D near Tule Springs. A thick marly caprock developed in the upper ~1 m of Unit D, probably as a result of the desiccation of Unit D wetlands prior to Unit E deposition.

Unit E caps Unit D, and like Unit D was deposited in spring-fed channels and wetlands, but over a smaller area. Unit E is further subdivided into Units E_1 (14.5–11 ^{14}C ka), characterized primarily by deposits derived from spring-fed channels, and Unit E_2 (11.3 to ca. 7.2 ^{14}C ka), characterized by black mats and some channel deposits, except for the uppermost part of the unit. Uppermost Unit E_2 consists of brown silt capped by gravel that together are typically 0.5–1 m thick. Complete desiccation of permanent surface moisture associated with Unit E_2 occurred in most sections by ca. 8–7 ^{14}C ka.

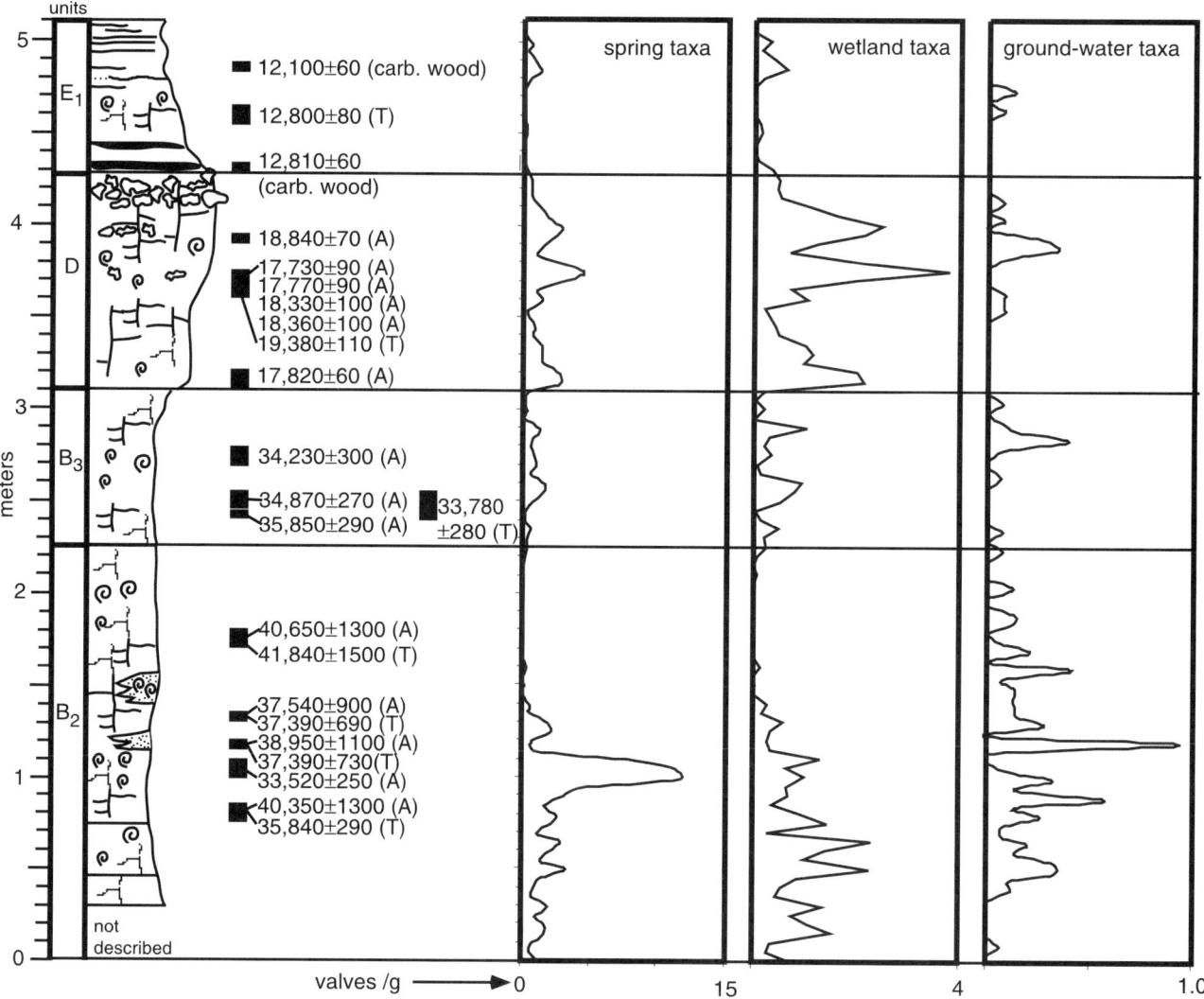

Figure 2. Stratigraphic section, units, and radiocarbon dates from OCI-11 in lower Corn Creek Flat (115°20′48″, 36°23′21″). See Figure 4 for explanation of lithologic symbols. Columns to right depict ostracode valves/gram of sediment, divided on the basis of ostracode environmental preferences. See text for explanation. A—aquatic mollusks, T—terrestrial mollusks.

In our study sections, Unit B and Unit D crop out only on lower Corn Creek Flat and probably in the eastern Amargosa section, whereas Unit E is present in all sections.

Corn Creek Flat sections OCI-11 and LPM-34

Two sections, OCI-11 and LPM-34, were described and sampled from Corn Creek Flat. These sections are located ~200 m apart in a small cluster of badlands near the bottom (southeast) end of the flat just before entering the narrows that connect to Tule Springs Flat to the south (see Quade et al., 1995, Figure 6A, or Bell et al., 1999, for a detailed map of the area). These sections were selected for their valley center location, which was thought would provide the most aquatic setting, and for their stratigraphic thickness of 3–5 m.

Section OCI-11

Unit B_2 (0–225 cm). The lower 225 cm is mostly pale green (5Y 6/3d) to brown (10 YR 6/3d) silt with little primary bedding, cut by a few small, sandy partings (Fig. 2). The locally reduced color, fine-grain size, and lack of bedding are typical of wetland deposits in the region, a characterization confirmed by the faunal assemblages discussed below.

The nine ^{14}C dates (five on aquatic and four on terrestrial snails) from five horizons fall between 33,520 ± 250 and 41,840 ± 1500 ^{14}C yr B.P. (Table 1; Fig. 2). Two lines of evidence suggest that the dates represent minima. First, Brennan and Quade (1997) obtained a date of 40,310 ± 310 on pre-Wisconsin age snails from spring deposits in the Pahrump Valley, showing that a demonstrably infinite-age shell is slightly contaminated with

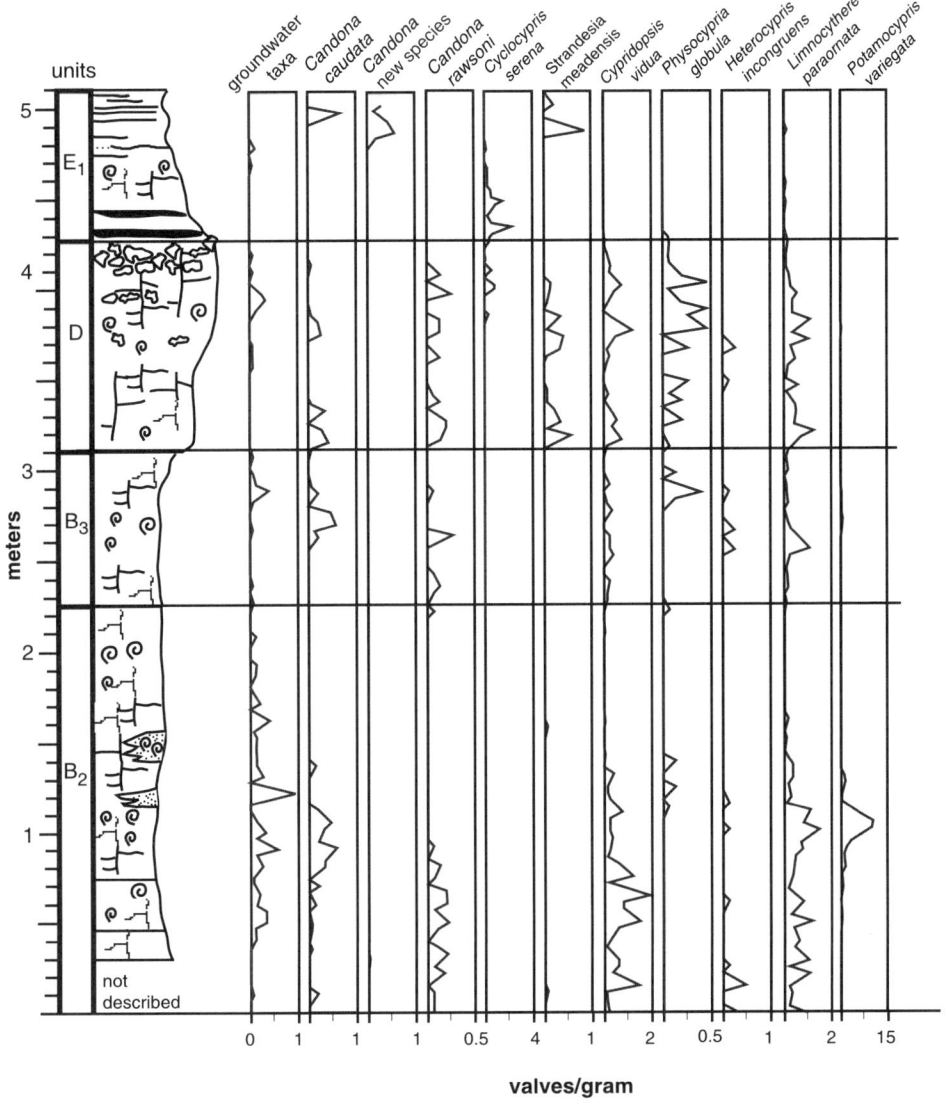

Figure 3. Stratigraphic section, units, and ostracode taxa from OCI-11 in lower Corn Creek Flat (115°20′48″, 36°23′21″). See Figure 4 for explanation of lithologic symbols.

young ^{14}C. Second, the dates follow no stratigraphic order, suggesting that the age differences arise from variable but slight (<1.5% modern) contamination.

We assign sediments in this interval to Unit B_2. This is based on the ^{14}C results and position at the base of the section, the same basal position occupied by Unit B_2 at Tule Springs.

Unit B_2 was deposited in three basic environmental settings based on the ostracode data: a wetland, a stream and flowing spring, and a wet-meadow (Fig. 2). The basal sediments in Unit B_2 (0 to ~75 cm) contain predominately wetland taxa, whereas the middle of Unit B_2 (~75 to ~125 cm) contains predominately spring/stream taxa. The upper part of Unit B_2 (~125 cm to ~225 cm) contains rare groundwater and scattered surface-water ostracode taxa (Fig. 2) and abundant aquatic gastropods. Groundwater taxa are most abundant in the basal through the middle part of unit B_2 compared to the rest of the section. This implies that the regional water table was both higher than today, and was discharging at a higher rate relative to the rest of the record in this section.

The ostracodes from basal Unit B_2 (0 to ~75 cm) provide additional insights into the chemical and temperature characteristics of the water, and from that, an indication of how climate in this period differed from the later periods. *Cypridopsis vidua*, *Candona rawsoni*, and to a lesser extent *Limnocythere paraornata* are wetland taxa in the basal part of unit B_2 (Fig. 3). Conversely, other wetland taxa, such as *Cyclocypris serena* and *Physocypria globula*, are common in the younger part of the section, principally Unit D, but are absent from the basal part of Unit B_2. All of these wetland species co-occur in the younger part of this section, and are known to live together at many, but not all, sites today. Therefore, the absence of *C. serena* and *P. globula* in the older part of the section is probably not related to differences in water-chemistry, but instead to differences in wetland hydrology and perhaps to seasonal water temperatures in the case of *P. globula*.

TABLE 1. ^{14}C DATES FROM STUDY SECTIONS IN SOUTHERN NEVADA

Sample	Beta #	Sampling interval (cm)	^{14}C date	δ^{13}C (PDB)	Material
Section 186/187					
CSCarb. 27b	73966		12,400 ± 60	−26.5	Carbonized wood
CSCarb. 28a	73967		12,410 ± 60	−27.0	Carbonized wood
CSCarb. 28b	73968		12,490 ± 50	−26.6	Carbonized wood
CSCarb. 30b	73969		12,180 ± 110	−28.7	Carbonized wood
Section OCI-11					
CSC-29b	84316	490	12,100 ± 60	−26.7	Carbonized wood
NV94RMF86CL+CM	84781	445–450	12,800 ± 80	−8.3	Terrestrial: *Vallonia cyclophorella*
CSC-27b	84315	430	12,810 ± 60	−27.2	Dispersed organic matter
NV94RMF86BY	91926	390–395	18,840 ± 70	−7.3	Aquatic: *Pisidium* sp.
NV94RMF86BU	84780	370–375	17,730 ± 90	−6.0	Aquatic: *Pisidium* sp.
NV94RMF86BT	84776	365–370	17,770 ± 90	−6.9	Aquatic: *Pisidium* sp.
NV94RMF86BT	84777	365–370	18,330 ± 100	−6.7	Aquatic: *Pisidium* sp.
NV94RMF86BT	84778	365–370	18,360 ± 100	−5.2	Aquatic: *Strandesia meadensis*
NV94RMF86BS+BT+BU	84779	360–375	19,380 ± 110	−8.9	Terrestrial: *Vallonia* sp. + Pupillid
CSC-25	86430	310–320	17,820 ± 60	−10.5	Semi-aquatic: *Stagnicola* sp.
CSC-14	86425	260–270	34,230 ± 300	−9.6	Aquatic: *Gyraulus parvus*
NV94RMF86AU+AT	85985	240–250	34,870 ± 270	−6.9	Aquatic: *Pisidium* sp.
NV94RMF86AS	85984	235–240	35,850 ± 290	−6.7	Aquatic: *Pisidium* sp.
NV94RMF86AU+AT+AS	85981	235–250	33,780 ± 280	−9.2	Terrestrial: *Vallonia* sp. + Pupillid
NV94RMF86AE+AF	84774	160–170	40,650 ± 1300	−7.2	Aquatic: *Pisidium* sp.
NV94RMF86AE+AF	84775	170–180	41,840 ± 1500	−8.1	Terrestrial: *Vertigo berryi* + *Vallonia* sp.
NV94RMF86W	84784	130–135	37,540 ± 900	−7.0	Aquatic: *Pisidium* sp.
NV94RMF86W	83993	130–135	37,390 ± 690	−9.5	Terrestrial: *Vallonia* sp.
NV94RMF86T	84783	120–125	38,950 ± 1100	−7.0	Aquatic: *Pisidium* sp.
NV94RMF86T	83994	120–125	37,390 ± 730	−9.4	Terrestrial: *Vallonia* sp.
CSC-16	86426	100–110	33,520 ± 250	−10.7	Aquatic: *Gyraulus parvus*
NV94RMF86M+L	84782	85–90	40,350 ± 1300	−6.5	Aquatic: *Pisidium* sp.
NV94RMF86M+L	85986	85–90	35,840 ± 290	−8.3	Terrestrial: *Vallonia* sp. + Pupillid
LPM-34 (floating section)					
RMF93NV71d, -0.25m	74329	75–80	16,390 ± 70	−7.3	Aquatic: *Pisidium* sp.
RMF93NV71d, -0.25m	76433	75–80	18,850 ± 100	−8.7	Terrestrial: *Vallonia* sp.
RMF93NV71g, -0.75m	76434	25–30	25,340 ± 180	−7.7	Terrestrial: *Vallonia* sp.
RMF93NV71f, -1.00m	74330	0–5	26,620 ± 160	−7.2	Aquatic: *Pisidium* sp.
LPM-35					
Cac.Spr.Carb.8	86427	330	10,030 ± 60	−26.7	Carbonized wood
Cac.Spr.QUADE90-100	86429	185–195	12,300 ± 60	−7.4	Terrestrial: Succineidae
Cac.Spr.QUADE45-55	86428	140–150	12,890 ± 60	−9.2	Aquatic: *Gyraulus circumstriatus*
NV94RMF84O	83992	137–142	13,350 ± 60	−8.2	Terrestrial: *Pupilla muscorum*
NV94RMF84N	83995	130–137	13,690 ± 80	−7.9	Terrestrial: *Euconulus fulvus*
NV94RMF84N	83991	130–137	13,270 ± 260	−7.5	Terrestrial: *Pupilla muscorum*
NV94RMF84N	83988	130–137	13,420 ± 60	−8.1	Terrestrial: *Vertigo berryi*
NV94RMF84M	83989	125–135	13,630 ± 60	−8.4	Terrestrial: *Euconulus fulvus*
NV94RMF84M	83990	125–135	13,560 ± 60	−8.5	Terrestrial: *Pupilla muscorum*
Lathrop Wells Diatomite					
LWD-7 (lower)	84448	490–500	13,970 ± 140	−8.3	Aquatic: *Gyraulus circumstriatus*
LWD-7	83708	490–500	13,480 ± 120	−7.4	Terrestrial: Succineidae
LWD-7	83709	490–500	13,510 ± 120	−8.7	Terrestrial: *Vertigo berryi*
LWD-7	83710	490–500	13,890 ± 140	−8.6	Terrestrial: *Vallonia* sp.
LWD-7	83495	490–500	14,080 ± 120	−7.4	Aquatic: *Pisidium* sp.
LWD-7	83496	490–500	13,550 ± 140	−8.5	Terrestrial: Succineidae
LWD-12	90244	70–80	41,910 ± 1860	−8.3	Terrestrial: *Vertigo berryi*
LWD-12	91922	70–80	39,970 ± 1280	−9.0	Aquatic: Hydrobiidae
LWD-12	91923	70–80	35,120 ± 700	−7.7	Aquatic: Hydrobiidae
LWD-12	91924	70–80	38,240 ± 1020	−7.4	Aquatic: *Pisidium* sp.
LWD-12	91925	70–80	36,880 ± 900	−7.3	Aquatic: *Pisidium* sp.

The wetland species *C. vidua* and *C. rawsoni*, as well as the spring (seep) species *Heterocypris incongruens*, often live in ephemeral environments, whereas *C. serena* and *P. globula* are typically, but not exclusively, found in permanent (less ephemeral) wetlands and lakes. The ephemeral characteristic could be as frequent as annually, or as infrequent as several drought years within a decade. Because the sediments of Unit B_2 are widely distributed throughout the basin, the wetlands were extensive, but if the wetlands were also ephemeral, then either their source (input) waters were also ephemeral (seasonal?) or evaporation (output) increased seasonally for the case of a high-frequency ephemeral setting. Alternatively, the record may indicate that wet years were punctuated by drought years, in the case of a low-frequency ephemeral environment. The $\delta^{18}O$ results discussed below do not support seasonal or interannual increases in evaporation (drought).

Basal Unit B_2 also contains *Candona caudata*, but in lesser abundance than the collective wetland taxa (Fig. 3). It tends to live in permanent flowing water. Groundwater taxa, although not as common as the surface-water taxa, persist throughout the interval.

The complete species assemblage collectively indicates that this period was one in which the regional water table rose above ground surface on some basis, probably seasonally, resulting in the appearance of extensive wetlands with flowing springs or small streams among the wetlands. The ostracodes indicate the water had a low TDS, below ~1000 mg/L, and that the solutes never evolved to a saline-solute composition, and so was a water whose solutes were dominated by calcium bicarbonate (Forester, 1991a). Overall, the ostracode assemblage is composed of eurythermic species indicating that water temperatures varied during the year. Summer warmth-loving ostracodes such as *Physocypria globula* are absent or rare, implying summer water temperatures were typically below ~18 °C. Cold-requiring ostracodes such as *Prionocypris canadensis* and *Cavernocypris wardi* (Forester, 1991a), which live in high mountain springs in the region, are also absent or rare, implying that seasonal water temperatures were above ~12 °C.

Ostracodes from the predominately spring/stream environments of the next highest subunit (75 to 125 cm) of Unit B_2 also provide insight into the nature of that environment. The predominant spring/stream ostracode in this interval is *C. caudata*, whereas *Strandesia meadensis* is absent (Fig. 3). That occurrence pattern is probably significant in that *S. meadensis* lives in flowing springs, but is not common in streams, whereas *C. caudata* lives both in flowing springs and in streams. The presence of abundant *C. caudata* therefore implies that this interval of Unit B_2 was deposited in a stream-dominated environment or an environment with greater flow than is typical of a flowing spring.

The change from a wetland-dominated to a stream-dominated environment between the basal and middle subunits of B_2 may indicate increased discharge related to climate change, or it could simply reflect a local facies shift as a stream migrated across the valley floor. The ostracode *Potamocypris variegata* is extremely abundant in this subinterval (Fig. 3), but provides little environmental information, because it has not been found commonly in North America. In Europe, it is known from both stream and wetland environments. The groundwater taxa are common throughout unit B_2, but are especially common in the two subintervals (0–75 and 75–125 cm) discussed above. The abundance of the groundwater taxa in these two subintervals implies that the regional water table was near the earth surface and was actively discharging.

The assemblage of ostracode species from 75 to 125 cm suggests that the water chemistry was fresh and TDS was typically below ~750 mg/L and, as with the lower unit, the solutes were dominated by calcium bicarbonate. The water temperature, based on all the sites where *C. caudata* occurs, falls into a large range from ~6 to 34 °C, so water temperature is not defined by the most common species for which there is modern data. If, however, only the occurrences near the Ruby Marsh (the probable modern analog) are considered, then the water temperature was between ~10 and 14 °C.

The third and youngest subinterval of Unit B_2 (~125 cm to ~225 cm) contains few to no ostracodes (Fig. 3), but aquatic gastropods are common. The presence of aquatic gastropods demonstrates that water was present, but the general absence of surface-water ostracodes implies that surface water bodies do not persist long enough to have the site fully populated by ostracodes. The wet meadow environment is one such environment where aquatic gastropods are common, but ostracodes are rare to absent (Quade et al., 1998), so it is the likely environment of deposition for these sediments.

The rich mollusk assemblage from Unit B_2 provides a much less detailed but congruent picture of a wetland environment depicted by the ostracodes. Mollusks are dispersed throughout most of the unit, but particularly abundant at 65–90, 110–130, and 190–220 cm. Although no exact counts of shells were made, aquatic mollusks far outnumber terrestrial ones. The presence of many large shells of *Planorbella subcrenata*, as well as of *Physa* sp., *Valvata humeralis*, *Pisidium* sp., and *Gyraulus parvus*, attest to the permanently aquatic character of some of Unit B_2 deposits. *P. subcrenata*, in particular, is not common in spring deposits in the southern Great Basin and is largely confined to the pond facies of Units D and B (Taylor, 1967; Quade et al., 1995). The presence of a few terrestrial snails attests to the nearness of dry banks to the site.

Unit B_3 (225–310 cm). This unit consists entirely of massive, pale-green (5Y 7/2d), hard, fine to medium silt. The upper and lower contacts are sharp and smooth. Four ^{14}C dates from two horizons range from 33,780 ± 280 to 35,850 ± 290 yr B.P. (Table 1; Fig. 2). The dates do fall in stratigraphic order and the youngest date is from a terrestrial snail, which is to be expected of finite dates. However, these results overlap the youngest date from Unit B_2, suggesting that all the dates may be slightly contaminated infinite-age dates. The true age of this unit remains unclear; based on available evidence we view it as >33,780 ^{14}C yr B.P. Aquatic mollusks are dispersed throughout the unit but were not identified.

Sediments of Unit B_3 (Fig. 2) contain ostracode taxa that imply a variable environmental sequence. The basal sediments from ~225 cm to ~260 cm variously contains few to no ostracodes. The wetland ostracodes in this interval imply the wetland was ephemeral, like the wetlands from the basal Unit B_2 (0–75 cm). The presence of aquatic gastropods in intervals with few to no ostracodes implies a wet-meadow environment, with sufficient water to support gastropods, but not ostracodes. Ostracodes indicating flowing water, such as *Candona caudata*, appear at ~260 cm and persist until ~290 cm (Fig. 3). As with the flowing-water intervals in Unit B_2, this flow was more likely in a stream than in flowing springs, and could represent either a climate-driven change in discharge or a local facies shift unrelated to regional changes.

A wetland environment developed at this site in upper Unit B_3 (280–310 cm). This wetland, however, contains ostracode species that did not occur or were not common in the older wetland environments. Taxa such as *Cyclocypris serena* appear for the first time and *Physocypria globula* becomes common. *Cypridopsis vidua*, *Candona rawsoni*, and *Limnocythere paraornata* are also present. Groundwater taxa are also quite common and imply the regional water table was high and actively discharging, supporting the wetlands. The appearance of *C. serena* indicates, in part, a more permanent wetland that probably supported subaquatic macrophytes. The commonness of *P. globula* is also important, because it implies summer water temperatures were warmer than in the older wetlands. Permanence of the wetland during the summer is an alternative explanation to warmer summer water temperatures.

Ostracodes are only abundant in the uppermost (~280–310 cm) subunit of Unit B_3. The common species from this upper interval, when considered collectively rather than by sample content, imply a freshwater environment with TDS values largely below 750 mg/L and probably below ~500 mg/L. Solute composition is dominated by calcium bicarbonate and the alk/Ca ratio is low, reflecting carbonate solute-source rocks (Forester, 1987). Water temperatures vary with season and unlike the older intervals, summer water temperatures were warm and likely remain in the high teens and low twenties celsius for two or more months during the year. Winter water temperatures are cool and perhaps similar to those in the Ruby Valley (northeast Nevada), where many of the taxa found in this subinterval live today, and where winter water temperatures are below ~12 °C.

Unit D (310–428 cm). Unit D consists entirely of pale-green (5Y 8/1–8/2d), hard, fine to medium silt in the lower part and browner silt above 385 cm. Six ^{14}C dates from this unit range from 19,380 ± 110 to 17,730 ± 90 ^{14}C yr B.P. (Table 1; Fig. 2). The oldest date comes from land snails (four shells combined), while the other dates are from aquatic snails. This pattern of ages would imply some mixing or reworking of shells. As such, the middle of Unit D at this location is at least as old as 17,770 ± 90 but no older than 19,380 ± 110.

Support for this 19–17 ^{14}C ka age range for Unit D comes from dates on reworked land snails in a Unit E_1 (<14,530 ± 80 ^{14}C yr B.P. based on dates from carbonized wood) channel cut directly into Unit D, located ~20 m to the west of OCI-11 (profiled in Quade, 1986, Figure 6). The terrestrial mollusks from this channel fill returned ages of 16,660 ± 70, 16,895 ± 170, 16,895 ± 170, 18,180 ± 90, 18,820 ± 90, and 21,260 ± 90 ^{14}C yr B.P. (Brennan and Quade, 1997). This range of ages overlaps the dates from in situ mollusks described above, and may be the best representation of true age range of Unit D.

Abundant carbonate nodules appear for the first time in the section in the upper ~50 cm of Unit D. This carbonate is a conspicuous feature of Unit D in the described sections at Tule Springs (Haynes, 1967) and elsewhere (Quade et al., 1998). It probably represents capillary fringe/pedogenic cementation along this surface following the desiccation of Unit D wetlands, which occurred after 17,730 ± 90 (this paper) but before 14,450 ± 80 ^{14}C yr B.P. (Quade et al., 1995).

The taxa from Unit D indicate that the depositional setting was a complex of flowing springs and wetlands (Fig. 2). Taxa such as *Cyclocypris serena* and *Physocypria globula* are common (Fig. 3) and indicate a relatively permanent and seasonally warm wetland environment. Unit D also contains two prominent abundance peaks of *Candona caudata*, indicating flowing water, but unlike the older occurrences of this taxon, these occurrences are associated with *Strandesia meadensis*. The presence of *S. meadensis* probably indicates flowing springs rather than streams. A possible shift from stream flow to spring flow may represent a change in discharge intensity linked to climate or simply a shift in hydrofacies. We favor the climate explanation because of the similarity of the ostracode assemblage of Unit D at OCI-11 to that of section LPM-34, a nearby section discussed below.

The collective ostracode species found in the Unit D samples imply that the environment was similar to that described above for the upper subunit of Unit B_3. The water is fresh with a TDS generally below 750 mg/L and probably 500 mg/L. Solutes are dominated by calcium bicarbonate, and temperatures vary seasonally. Summer water temperatures are in the high teens through low twenties celsius and winter temperatures are below ~12 °C.

Unit E_1 (428–510 cm). Unit E_1 consists of interbedded brown (10 YR 8/1d) silt and thin organic-rich layers in the lower part, and interbedded fine sand and silt above 470 cm. Overall, Unit E is more bedded and coarser than underlying Unit D. The basal contact is sharp and smooth.

Three dates were obtained on the humate fraction of two organic layers, one from the base of Unit E_1 that yielded a date of 12,810 ± 60 and another from carbonized wood at ~465 cm that returned a date of 12,100 ± 60 ^{14}C yr B.P. The date of 12,800 ± 80 on terrestrial mollusks from 455 to 465 cm confirms the accuracy of ^{14}C dates on such samples. These Unit E_1 sediments represent onlap along the margins of a deeply incised channel of Unit E_1 that dates to 14,530 ± 80 ^{14}C yr B.P. at the base (Brennan and Quade, 1997).

Unit E_1 contains a few wetland ostracodes in the basal part of the unit, associated with the organic-rich layers (Fig. 2).

These taxa include *Cyclocypris serena* (Fig. 3) and *Limnocythere paraornata*, indicating some similarity with the Unit D wetlands. There are also samples scattered throughout Unit E_1 with aquatic mollusks but no ostracodes, implying a wet-meadow environment. There is also a number of species indicating flowing water, especially toward the top the Unit E_1 (Fig. 2). This includes *Candona* n. sp. 1, known to live in wetlands and springs in the region today, as well as *C. caudata* and other rare taxa that are not shown in Figure 3, as noted at the beginning of the ostracode discussion above. One of the taxa not shown in Figure 3 is believed to be a new genus and a new species and it is only known today from a high mountain stream above Cedar City, Utah. Its presence here may imply cold flowing water. Remarkably, even though Unit E represents deposition during a late glacial climate in transition toward interglacial climate, the setting is a cold-flowing stream.

Environmental information is not available for all of the ostracodes that are common in this unit. However, the apparent cold-flowing nature of the environment in which the upper part of Unit E_1 was deposited, together with taxa for which environmental information is available, indicates a fresh, probably dilute calcium bicarbonate water. The water may have had a TDS below 300 mg/L. Water temperatures probably did not vary a great deal seasonally due to flow and were probably <15 °C.

Section LPM-34

LPM-34 (Fig. 4) is located ~200 m downvalley from OCI-11, and provides a test of the lateral continuity of the major units seen in OCI-11. Only nine faunal samples were taken at LPM-34, and three horizons ^{14}C dated. Green silt similar in appearance to Units B_2 and D makes up the basal ~30 cm of the section. Mollusks yielded ages of 26,620 ± 160 (aquatic) at the base and 25,340 ± 180 ^{14}C yr B.P. (terrestrial) at the top of the exposed unit (Table 1; Fig. 4). These ages suggest a basal Unit D equivalence.

The silts are sharply truncated by channel sand and gravel containing mammalian bone and abundant mollusks. The sand fines upward gradually into pale-green silt. Mollusks from ~75 cm returned dates of 16,850 ± 100 (terrestrial) and 16,390 ± 100 ^{14}C yr B.P. (aquatic). The sediment and character of the mollusks strongly resemble those of Unit D in OCI-11, and these dates would support that correlation. This also supports evidence discussed previously that Unit D is as young as ca. 16.4 ^{14}C ka.

Unit E overlies Unit D along a sharp contact at 290 cm. Although we have no dates on this horizon, the presence of organic matter, bedded layers, and brown color all point to Unit E.

LPM-34 was coarsely sampled at 25 cm intervals to provide general information about the ostracodes in the section. The ostracode data is summarized in terms of wetland, spring, and

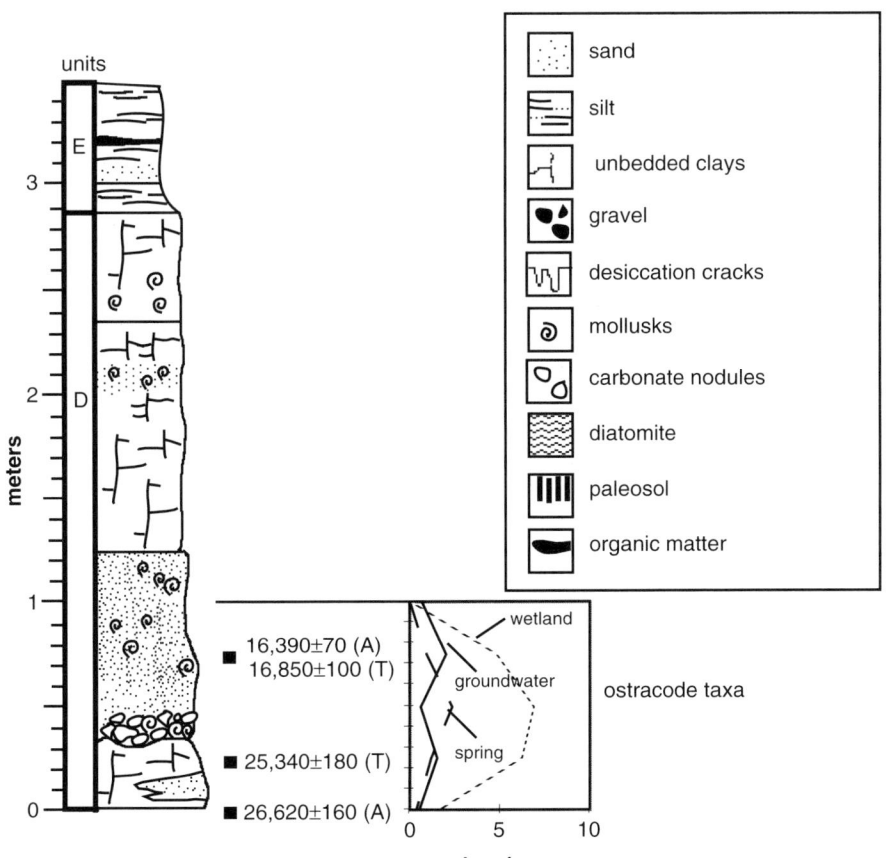

Figure 4. Stratigraphic section, units, and radiocarbon dates from LPM-34 in lower Corn Creek Flat (115°20′39″, 36°23′16″). A—aquatic mollusks, T—terrestrial mollusks.

groundwater taxa and then plotted against the stratigraphic section for lower Unit D (Fig. 4). Like section OCI-11, the ostracodes that commonly lived in both the wetlands (standing water), springs (flowing water), and groundwater are all represented.

Ostracodes indicating several different types of hydrologic environments are most abundant in the basal part of Unit D, especially in the lower 75 cm. At 100 cm, the abundance of surface-water ostracodes is lower and groundwater taxa are common. Above 100 cm, aquatic gastropods are common, but ostracodes are rare, indicating a transition from wetlands and flowing springs (lower 75 cm) to a wet-meadow environment.

The ostracode species that dominate the assemblage in the lower 75 cm of section LPM-34 include wetland taxa such as *Cyclocypris serena*, and *Physocypria globula*, as well as spring and stream taxa such as *Candona caudata* and *Strandesia meadensis* (see Fig. 4). This is the same species assemblage as that which lived in the Unit D wetlands described above for section OCI-11. These taxa indicate that the wetlands consisted of standing and flowing water with the flowing water coming primarily from springs rather than streams. The presence of *P. globula* implies that summer water temperatures were warm, in the high teens to low twenties celsius. As with Unit D in section OCI-11, the environment associated with this subunit is fresh water and the solutes are dominated by calcium bicarbonate.

The transition from environments with surface-water taxa to those with few ostracodes, but common aquatic gastropods, may depict a simple change in hydrofacies or a shift in toward a drier and/or warmer climate. In section OCI-11, the environment of Unit D remained a wetland until the transition to Unit E, whereas at this site (LPM-34) Unit D appears to be a wet-meadow at the Unit D to E transition. The differences in transitions may be due to erosion (prior to Unit E deposition) of the top of Unit D at OCI-11, which is supported by the radiocarbon ages. These two sections are ~200 m apart and yet preserve somewhat different histories, illustrating the need to examine several sections to acquire complete environmental histories.

Cactus Springs section LPM-35

LPM-35 was selected for study because of its basin center location near Cactus Springs (Fig. 1) and thickness of 4.3 m. Section LPM-35 is the same as locality 1 in Figure 3 of Quade and Pratt (1989). The lowest unit is 90 cm thick and is composed of pale-green sand alternating with unbedded silt (Fig. 5). No dates were obtained from this unit, but it probably represents either Unit D or more likely lower Unit E_1.

Clay, silty clay, and local sandy layers dominate the interval 90–230 cm. The lower (90–160 cm) portion is mostly pale-green (5Y 7/2d) clay and silty clay containing abundant ostracodes, and aquatic, semiaquatic, and terrestrial mollusks. Six ^{14}C dates on terrestrial snails range from 13,690 ± 80 in the middle of the unit to 13,350 ± 60 ^{14}C yr B.P. Semiaquatic mollusks sampled 10 cm below the top returned a date of 12,890 ± 60 ^{14}C yr B.P. (Fig. 5). A very weak soil appears to be developed on top of this interval at 150–160 cm, represented by Fe-staining, carbonate nodules, and a fine prismatic structure.

The upper 160–230 cm of the unit is composed of a basal sand capped by white (10YR 8/2d) silty clay. Terrestrial mollusks near the base of this interval gave a date of 12,300 ± 60 ^{14}C yr B.P. Mollusks are common in the lower 20 cm, then decrease upwards. Staining and extensive insect bioturbation probably denote weak pedogenesis of this unit prior to burial by brown silt.

The upper 230–430 cm is composed of finely bedded brown silt over most of the interval, capped by 30 cm of coarse gravel. Mollusks and ostracodes are entirely absent. A date of 10,030 ± 60 ^{14}C yr B.P. from dispersed carbonized wood fixes the age of the middle of the brown silt unit, and suggests that much if not all of this horizon belongs to Unit E_2.

The wettest part of the exposed record was being deposited no later than 13,560 ± 60 and continued through at least 12,890 ± 60 ^{14}C yr B.P. Persistent but declining moisture is indicated until shortly after 12,300 ± 60. Sometime between 12,300 ± 60 and 10,030 ± 60 ^{14}C yr B.P., standing surface moisture completely and permanently disappeared.

LPM-35 was sampled continuously at 5 cm-thick intervals from the base of the section to ~300 cm, and thereafter selectively to the top of the section. Ostracodes were restricted to a limited part of the section designated Unit E_1, while mollusks were found over a wider stratigraphic interval (Fig. 5).

The basal sediments in this section tentatively designated Units D-E (0 cm to 90 cm) and those assigned to Unit E_1 to ~120 cm contain no gastropods and rare ostracodes, indicating a wet-meadow environment. A similar environment typified upper Unit D and lower E at section LPM-34 (see previous discussion).

Mollusks from the two stratigraphic levels (1.4 and 2 m) that were collected indicate permanent water was present at both times. *Gyraulus circumstriatus*, *G. parvus*, as well as *Pisidium casertanum* are present in both horizons and require permanent water to survive. Terrestrial mollusks (e.g., *Pupilla muscorum*, *Vertigo berryi*) are also abundant at both levels, suggesting that standing water was not expansive and that moist banks were nearby.

The ostracode data is summarized with the stratigraphic profiles of four common species (Fig. 5). These species, but especially *Candona* n. sp., *Limnocythere paraornata*, and *Strandesia meadensis,* imply that the environment was dominated by flowing springs. The commonness of *Candona* n. sp. in these samples, as well as in the Unit E_1 in section OCI-11, provides a good example of the influence of regional climate over local hydrology. *Candona* n. sp. is a common species in springs at higher elevations in the region and north into central Nevada. It is rare to absent in the older record, so its abundance in these sediments implies climate change toward Holocene conditions. The ostracode assemblage from Unit E_1 in section OCI-11 also shows that flow dominated the environment. In the case of OCI-11, some of that flow appears to have been colder than is implied by the ostracodes in this section. The ostracodes at this section, with the exception of a single poorly preserved valve, are all surface-water species, implying that extensive discharge of regional groundwater did not occur here. Thus, much of the water at this site is likely derived from the valley-fill aquifer. The sediments in the basal part of Unit E_1 contained a poorly preserved valve of *Limnocythere staplini*, an ostracode that lives in saline water with low alk/Ca solute ratios. That ostracode was probably not living at this site, but instead was blown to this site from the nearby playa.

Amargosa Desert section LWDD 6/8N

This section (LWDD 6/8N) was measured and sampled through one of the thickest portions of the diatomaceous mound located just off Highway 95 north of Lathrop Wells (Fig. 1), informally referred to as the Lathrop Wells Diatomites (Paces et al., 1996). The composite section measured >7 m thick (Fig. 6). These deposits have been the focus of intense geological scrutiny, due to their proximity to the proposed High-Level Nuclear Waste Repository ~20 km away at Yucca Mountain (e.g., Paces et al., 1996). Here we report only on the stratigraphy, ^{14}C dating, and ostracode paleontology of the deposits, with primary attention given to the youngest part of the section. Samples were obtained only in selected intervals, generally where mollusks or ostracodes were apparent in the field.

The lower 2.7 m (LWD 8N) of section exposed in a shallow trench cut is composed mainly of pale green (2.5Y 8/2d) to pale brown (10 YR 8/2d) pebbly to silty sand. The beds are mostly massive with almost no preserved bedding. Mollusks were found in two layers but dated only from the 170 cm level. Terrestrial and aquatic mollusks yielded a range of ages from 35,120 ± 700 (B-91923) to 41,910 ± 1860 ^{14}C yr B.P. (B-90244) (Fig. 6; Table 1). As in the older deposits on Corn Creek Flat, we regard these as minimum ages.

The upper 5 m is exposed in one continuous section (LWD-station 6), and can be divided in three parts (Fig. 6). The lower 1.8 m consists of well-indurated greenish (5Y 7/2d) to white (10 YR 8/1d) sands and sandy loam. Volcanic pebbles and angular chips are dispersed throughout much of the unit, as are siliceous rhizoliths and carbonate nodules in the base. Several

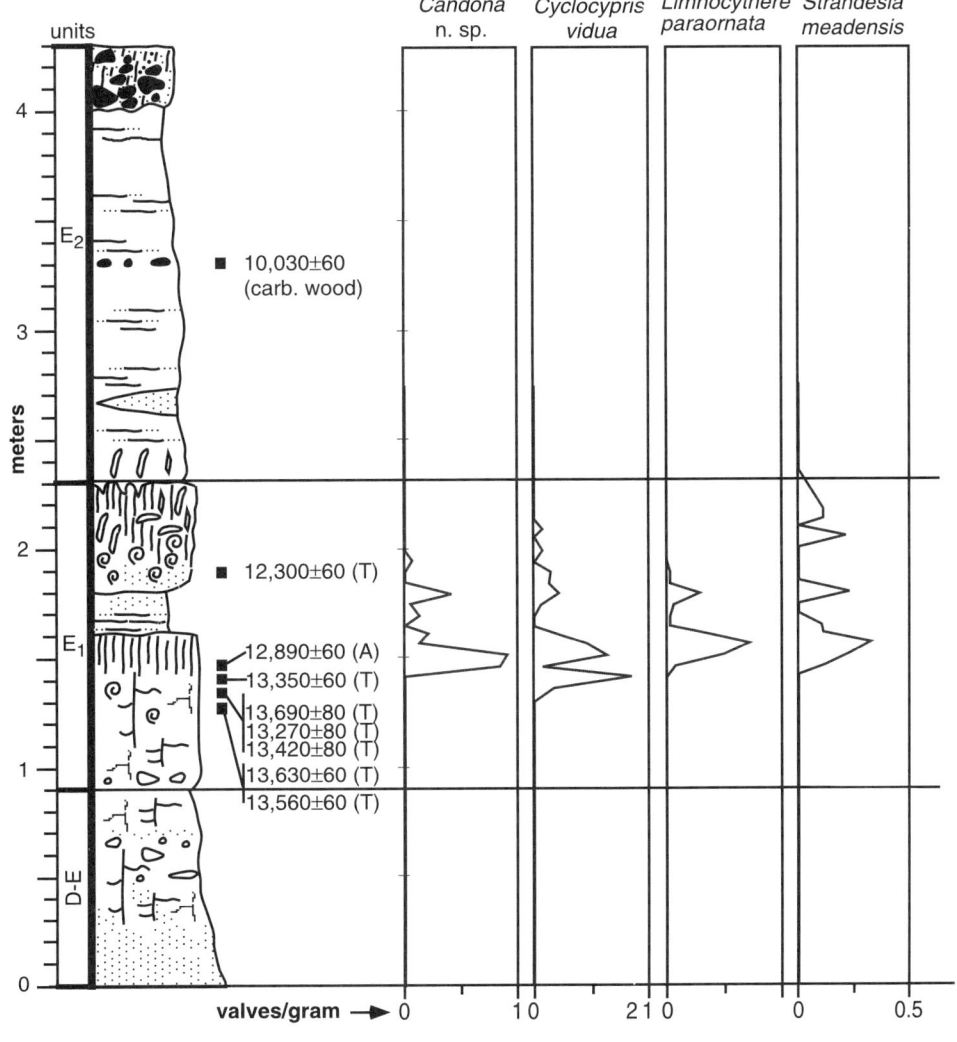

Figure 5. Stratigraphic section, units, and radiocarbon dates from LPM-35 west of Cactus Springs (115°48′03″, 36°34′11″). See Figure 4 for explanation of lithologic symbols. A—aquatic mollusks, T—terrestrial mollusks.

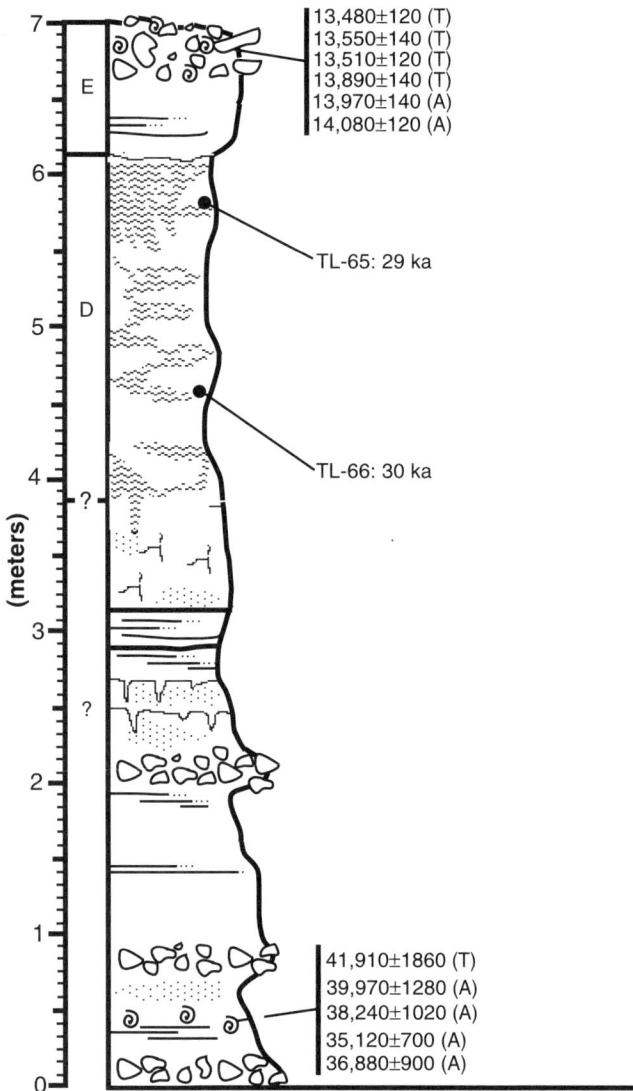

Figure 6. Stratigraphic section and dates from 6/8N at Lathrop Wells (116°35′14″, 36°42′28″). See Figure 4 for explanation of lithologic symbols. A—aquatic mollusks, T—terrestrial mollusks.

vertically cracked partings cut the unit, suggesting hiatuses in deposition. No ^{14}C ages were obtained on this unit, which is devoid of mollusks; Paces et al. (1996) report U-series ages ranging from 186 ± 3 to 115 ± 3 ka.

A hard, white diatomite (whiter than 2.5Y 8/0d) extends from 2.2 to 4.15 m. It is largely noncalcareous, and contains dispersed pebbles of tuff. This unit contains a proboscidean tusk, as well as mammoth and horse tooth fragments. No radiocarbon-datable material was found, although Paces et al. (1996) report TL ages of 29 ± 11 and $30 + 12/-17$ ka from this unit, and U-series ages of 42 ± 2 and 56 ± 6 ka on rhizoliths from the base of the diatomite.

The 5-m thick section is capped by 85 cm of very hard light orange to white marl. Dates from aquatic and terrestrial snails in the marl range from $13,480 \pm 120$ (B-83708) to $14,080 \pm 120$ ^{14}C yr B.P. (B-83495) (Table 1). As such, the marl is assigned to Unit E_1. The same ages within Unit E also show up at Cactus Springs (Fig. 5).

This section was sampled opportunistically for whatever fauna was locally present. All of the ostracodes recovered from samples of this section suggest a complex array of spring discharge environments, much like those at the other sections. Ostracodes typical of flowing spring and seep environments dominate over wetland environments, consistent with the location of the site on an alluvial fan, rather than on the valley floor. The marl belonging to Unit E_1 (6.15–7 m) contains *Candona* n. sp. (consistent with Unit E_1 elsewhere), implying flow, but lacks taxa such as *Candona caudata* or the new genus/new species, implying high and/or cold flow. *Gyraulus circumstriatus* and *Pisidium* sp. from the marl indicate permanent water, but the additional presence of terrestrial gastropods (Succinidae, *Vertigo berryi*) is consistent with a low-flow spring to seep environment.

The diatomite (2.2–6.15 m) is dominated by *Denticula valida* according to J. Platt Bradbury in Paces et al. (1996). In the latter, Bradbury states:

This species characterizes warm (usually <40 °C), alkaline and slightly saline springs that are related to volcanic hydrothermal processes. For example, it occurs in springs on Mount Rainier, Washington, and in warm water habitats in Yellowstone National Park. However, it does not necessarily imply warm, hydrothermal water. For example, the species is abundant in shallow, near-shore water and marginal spring seeps at Black Lake (Adobe Valley, Mono County, California). Here the water is near ambient air temperature and saline as a result of evaporation. This species has a wide salinity tolerance ranging from ~1 ppt (parts per thousand) to ~15 ppt. The other diatoms at the LWD [species?] suggest that salinity was low, probably ~1 ppt.

Thus, this interpretation of the diatoms suggests discharge of deep, possibly warm, groundwater at the site.

STABLE ISOTOPE RESULTS AND INTERPRETATION

δ^{18}O results

The δ^{18}O data set is composed of >400 analyses on the valves/shells of ostracodes and mollusks that were collected from all of the study sections except LWDD 6/8N. The analyses include species that lived variously in flowing, standing, and groundwater. Land snails were also analyzed, but those and some of the aquatic mollusk data will be presented elsewhere (also see Sharpe et al., 1994).

The δ^{18}O (VSMOW) values from ostracode and mollusk species from all sections and all time periods (Figs. 7–8; Tables DR1–DR3[1]) fall between about +16 and +22.2‰. In the largest data set from one location, section OCI-11, individual species show no clear clustering of δ^{18}O values according to habitat

---

[1]GSA Data Repository item 2003070, Tables DR1–DR3, is available on request from Documents Secretary, GSA, P.O. Box 9140, Boulder, CO 80301-9140, USA, editing@geosociety.org, or at www.geosociety.org/pubs/ft2003.htm.

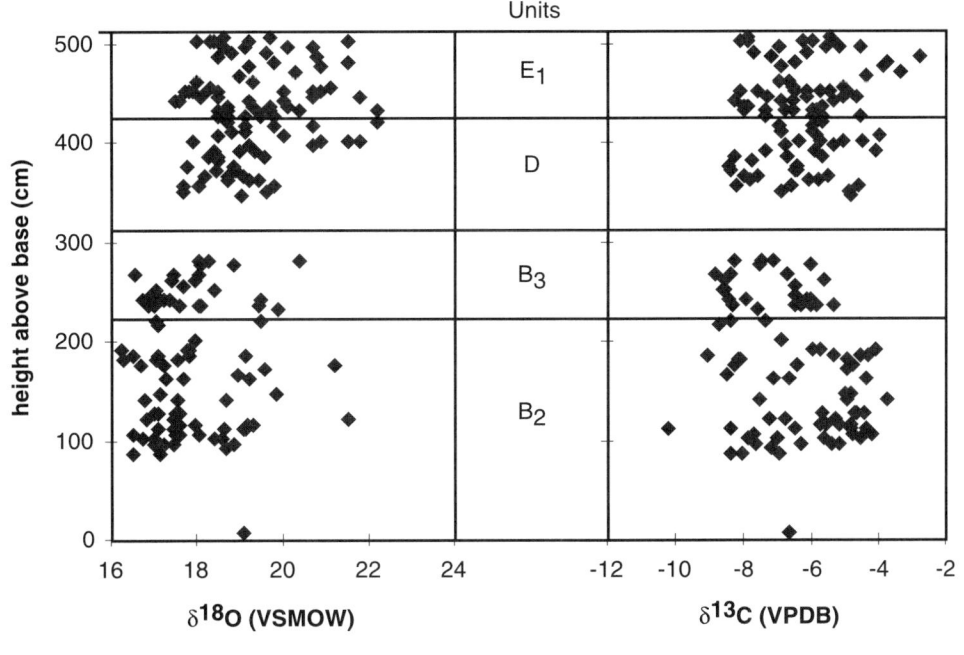

Figure 7. $\delta^{18}O$ (VSMOW) and $\delta^{13}C$ (VPDB) values of select ostracode taxa from OCI-11 on lower Corn Creek Flat.

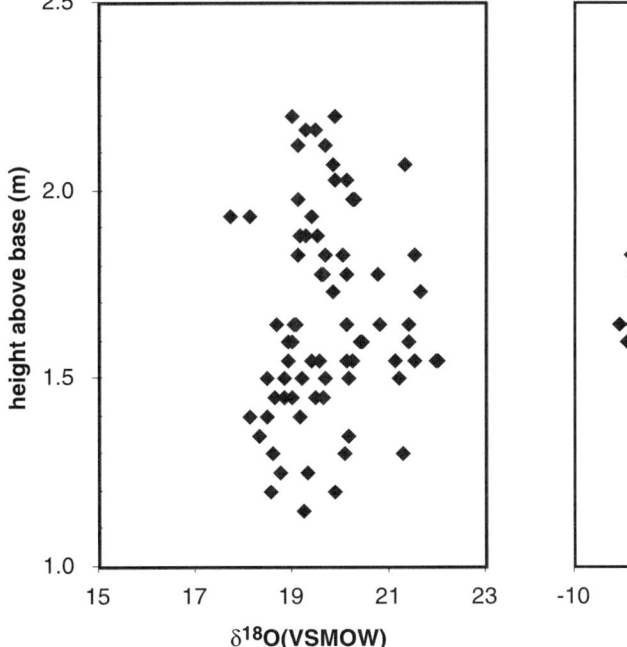
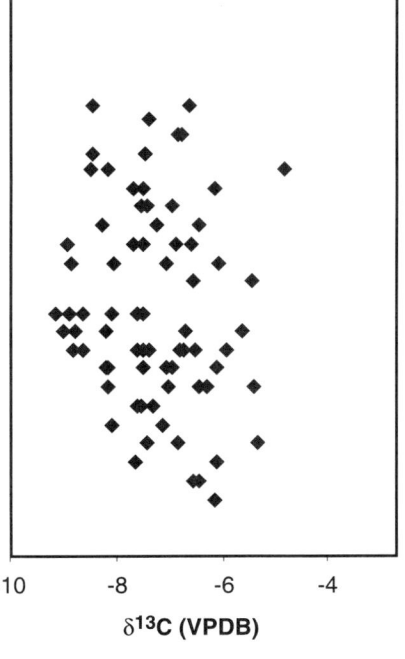

Figure 8. $\delta^{18}O$ (VSMOW) and $\delta^{13}C$ (VPDB) values of select ostracode and mollusk species from LPM-35 west of Cactus Springs.

preferences (Fig. 9A). For example, *Cypridopsis vidua*, a taxon with a wetland environmental preference, displays $\delta^{18}O$ values between +16.5 and +21.5‰ (n = 74) and *Candona rawsoni*, another wetland taxon, +17.3 to +22.0‰ (n = 36). Examples of flowing water taxa include *Candona caudata* (+16.7 to +22.2‰; n = 48) and *Strandesia meadensis* (+16.2 to +19.9‰; n = 19). Ostracodes that commonly live in and around the orifice of springs include *Candona accuminata* (+18.6 to +19.5‰; n = 4) and perhaps a new genus/new species (+19.3 to +21.7‰; n = 9). *Limnocythere paraornata*, which probably lived in both flowing and standing water settings, returned $\delta^{18}O$ values of +16.3 to +21.5‰ (n = 74). Groundwater taxa also returned a broad spread of values, from +17.5 to +22.2‰ (n = 6).

$\delta^{13}C$ (VPDB) values for all taxa range between –9 and –3‰ for all but a few results from OCI-11 (Fig. 7), and between –9 and –5‰ for LPM-35 (Fig. 8).

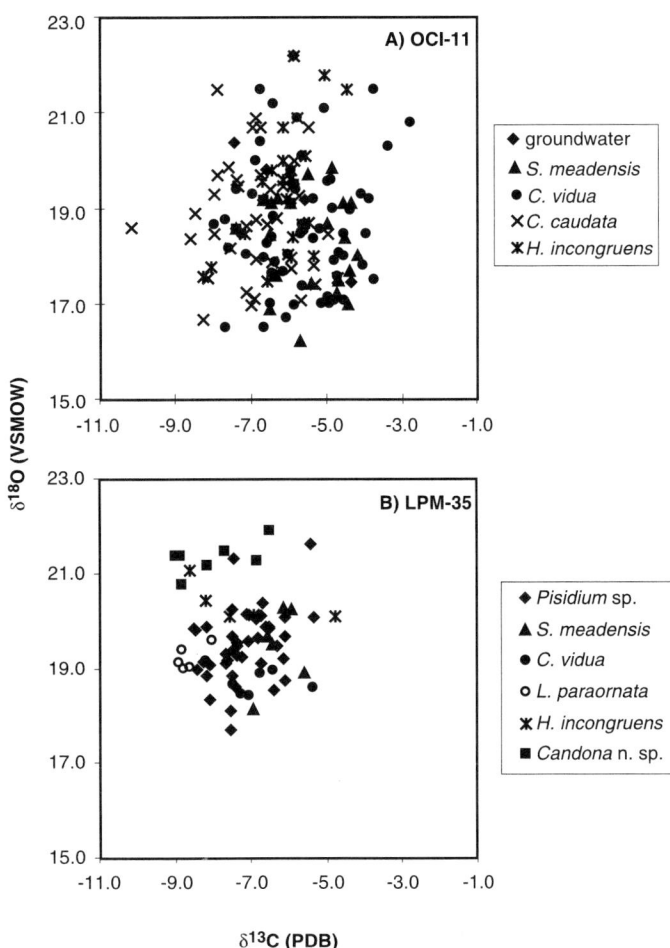

Figure 9. δ¹⁸O (VSMOW) vs. δ¹³C (PDB) of selected ostracodes. A: from OCI-11 on lower Corn Creek Flat. B: from LPM-35 west of Cactus Springs.

Interpretation

δ¹³C values. The δ¹³C value of ostracode shells is probably determined by the δ¹³C value of dissolved inorganic carbon (DIC), with some possible influence by dietary carbon, following the analogy of aquatic clams (Dettman et al., 1999). Natural waters in the area have a pH between 7.5 and 8 (Thomas et al., 1991), and therefore the dominant inorganic carbon species in water is bicarbonate. DIC δ¹³C values in groundwater are initially set by plant CO_2 at high elevations, which in the region is largely derived from C_3 plants (δ¹³C = ~ −24‰; Quade et al., 1989). Plant CO_2 with a δ¹³C value of −24‰ would produce a DIC δ¹³C value of ~ −11‰. However, reaction with limestone and dolostone along the flowpath should gradually increase the δ¹³C value of DIC. This probably explains δ¹³C values in springs in the area today, which vary from −10.9 to −7.8‰ (Table 2), and are higher at lower elevation. Average δ¹³C values of DIC in the past were the same or slightly more negative (Quade et al., 1989). Calcite formed from these waters should be 1.0‰ enriched in ¹³C (Romanek et al., 1992), if in isotopic equilibrium, and thus be in the range −8.5 to −6‰. This range in δ¹³C values overlaps that displayed by most fossil ostracodes, suggesting that water DIC was the dominant source of carbon for shell building. Some δ¹³C values in ostracodes are higher than −6‰, particularly from OCI-11, implying a food source for some shell carbon, or locally high productivity environments (MacKenzie, 1985).

δ¹⁸O values. δ¹⁸O values from biogenic carbonates provide important insights into the nature of the paleoenvironments reconstructed from the ostracodes. They also provide constraints on paleotemperature maxima for spring water and air, and on other aspects of the paleohydrology.

Our δ¹⁸O results show little discernible variation with depth in LPM-35, while there appears to be some small but probably significant variations in OCI-11. The overall lack of variation (Figs. 7–9) suggests that the hydrologic response in valley bottom environments to each major wet period, and even the later stages of the last wet period, was similar. Significant evaporation of streams and wetlands is also not in evidence. The similarity of mean δ¹⁸O values from flowing and standing water taxa points to aquatic systems dominated by rapid through flow of water, and to water loss by overflow rather than evaporation.

δ¹⁸O values in upper Unit D and Unit E display a modest (1–2‰) increase in mean values (Fig. 7). There are at least two possible explanations for this. The one favored by us is that this reflects an increase in the δ¹⁸O value of meteoric water, as climate warmed during that late glacial period in recharge areas, thus shifting the δ¹⁸O values of precipitation to higher values. Alternatively, some surface waters at this time may have experienced some evaporation prior to carbonate formation. However, the second explanation is inconsistent with the cold conditions implied by the ostracode taxa present, as discussed in a previous section.

Our extensive δ¹⁸O results provide a unique opportunity to reconstruct paleotemperatures for periods represented by our sections. To do this we must first constrain the variables that determine the δ¹⁸O value of associated carbonates in most aquatic settings. Other than temperature, these variables include the δ¹⁸O value of host water and the degree of attainment of temperature-controlled isotopic equilibrium during carbonate formation. In this section of the paper we will examine each of these variables in the wetlands and spring context. This analysis in turn provides some constraints of mean annual air temperature (MAT) in the region during the same periods.

Attainment of isotopic equilibrium. Some biologically precipitated calcites do not attain isotopic equilibrium with waters in which they form. Such "vital effects" may account for some of the range in δ¹⁸O values from various ostracodes from the same stratigraphic level. Xia et al. (1997) showed that *Candona rawsoni* exhibits a 0.8–1.0‰ enrichment over equilibrium δ¹⁸O values. Similarly, results from Bacon (1999) seem to suggest that some ostracodes (crawlers) make a shell with isotope values higher than for inorganic calcite in equilibrium with the same water. Conversely, other ostracodes (swimmers) make a shell that

TABLE 2. SPRING ELEVATIONS, TEMPERATURES, AND ISOTOPIC COMPOSITIONS
FROM THE SPRING AND SHEEP RANGES AND INTERVENING VALLEYS

Spring name	Spring elevation (m)	Spring temperature (°C)	$\delta^{18}O$ (VSMOW)	$\delta^{13}C$ (VPDB)
Peak Creek	3260	NA	−12.9 to −13.8	−9.5
Cave Creek	3048	5	−13.3	NA
Two Spring	3048	4	NA	NA
Lee Canyon Ski Spring	2967	7.7	−13.6	−9
Deer Creek #1	2870	4.3	−14 to −14.1	−8.2 to −8.5
West Spring	2804	5.5	NA	NA
Upper Macks Canyon	2710	9.2	NA	−9.1
Clark Spring	2648	10.4	−12.9	−8.6
Deer Creek #2	2646	7.8	−13.4	−9.6
Trough	2505	13.2	−14.1	−10.9
Lower Stanley B	2461	10	NA	NA
Trout Spring	2360	7.5 (6.9–8.5)	−12.9 to −14.1	−8.2
Fletcher Spring	2350	8.4	NA	NA
Fletcher Spring	2310	11.5	NA	NA
Mud	2280	11	−14.1	−8.7
Buck	2224	13.9	−13.6	−9.4
Cold Creek	1930	10 (9.5–12)	−13.3 to −14.2	−9.3
Willow Spring	1826	10.5	−13.4	−9.6
Harris Spring	1820	14	−13.7	−8.5
Wood Canyon Spring	1780	17.4	NA	NA
East House Spring	1620	18.9	NA	NA
Grassy Spring	1570	19.4	NA	NA
Grapevine Spring	1490	26.5	NA	NA
Cactus	987	20.6	NA	NA
Indian	969	26	−13.0 to −14.2	−7.8
Paiute Indian Reservation Well	927	18.7	−14	−8.7
Pahrump Spring Well	823	25	−13.6	−7.6
Corn Creek Spring	817	21	−12.9 to −13.0	−8.3
Tule Spring Park	750	21	NA	NA

Note: data from McKinley et al., 1991; Thomas et al., 1991; Winograd et al., 1998; and this study.

is in isotopic equilibrium or nearly in equilibrium with host water (Bacon, 1999). Vital effects therefore may account for the slightly lower $\delta^{18}O$ values among, for example, *Cypridopsis vidua* (aver. 18.6), a swimmer, compared to *Candona caudata* (aver. 19.1), a crawler (Fig. 9A). In our paleotemperature estimates, we use the isotopic results both from groundwater candonids, and assume 1.0‰ vital effect, and from *C. vidua*, for which we assume no vital effects during valve formation. It must be kept in mind, however, that neither the groundwater candonids nor *C. vidua* have been studied for vital effects in controlled laboratory settings.

$\delta^{18}O$ *value of paleowaters.* Study of the paleo-spring and wetland deposits in the valley bottoms shows them to be groundwater fed, with no evidence of any perennial flow from the surrounding fans. $\delta^{18}O$ value of standing and flowing water in valley wetlands is determined by the $\delta^{18}O$ value of meteoric water ($\delta^{18}O_{mw}$) feeding the groundwater system, and by the extent of subsequent evaporation. For the Corn Creek and Cactus Springs areas, the main recharge zones are the high peaks of the Spring Range; the rugged Sheep Range serves as in important additional source to the east side of Corn Creek Flat (Fig. 1). Most major springs in these mountains have been sampled, providing a detailed picture of the patterns of recharge. The $\delta^{18}O_{mw}$ values of these springs fall in a narrow range between −12.9 and −14.3‰ (Table 2). Winograd et al. (1998) demonstrated that this narrow isotopic range reflects the dominance of winter recharge (~90%) at all elevations, even though ~40% of precipitation falls in summer. Importantly, these values show up in valley-bottom springs fed by the deeper carbonate aquifer, and in much smaller mountain springs flowing along variable but often very short flow paths (Table 2). The abundance of groundwater (carbonate aquifer) taxa in some strata, but absence in others, suggests that both types of springs (long- and short-flow path) may have discharged into valley bottoms in the past.

Recharge from the Spring Mountains also flows through Devils Hole, where vein calcite deposits record fluctuations in the $\delta^{18}O_{mw}$ of groundwater over the past 500,000 yr (Winograd et al., 1992). Calcite formed at Devils Hole during glacial periods has $\delta^{18}O$ values that are 1.2–2.2‰ lower than interglacial-age calcite, with the smallest difference, 1.2‰, associated with the last deglaciation (Marine Isotope Stages I to II) (I.J. Winograd, 2000, personal commun.). These differences are thought to be caused mainly by colder conditions during glacial periods in recharge areas high in the Spring Mountains (Winograd et al., 1992). A variety of evidence presented by Winograd et al. (1992, 1997) and Thomas (1996) argues for very short (no more than a few thousand years) transit times in the carbonate aquifer from recharge areas, even though ^{14}C dates on associated groundwater yield apparent late-Pleistocene ages. Compared to Devils Hole, our study sites at Cactus Springs and Corn Creek are less than half the distance to recharge areas, and at least partly fed by the same carbonate aquifer. The vein calcite at Devils Hole therefore provides well-constrained $\delta^{18}O$ values for waters discharging in our valley bottom sites during glacial periods. For our paleotemperature estimates we assumed a $\delta^{18}O_{mw}$ value of –13.0‰ for Holocene-age groundwater. This assumption makes our paleotemperature estimates maxima, although we agree with Winograd et al. (1998) that the weighted average $\delta^{18}O_{mw}$ values (= –13.5‰) of three large-flow springs in the Spring Mountains is probably more representative of Holocene recharge. Using 1.2‰ for the glacial/interglacial shift in $\delta^{18}O_{mw}$ values, we estimate a $\delta^{18}O_{mw}$ value for glacial periods to be –14.2‰. Use of a 2.2‰ shift would render our paleotemperature estimates even colder.

The other important variable influencing the isotopic composition of paleo-waters is evaporation. The extent of evaporation can be quite variable, and depends on residence times of water, air and water temperature, windiness, and other factors. In many lakes, particularly in dry settings, evaporation is often by far the most important determinant of host water $\delta^{18}O$ values, so much so that in many lake studies the $\delta^{18}O$ value of carbonates is viewed as a proxy for evaporation-controlled lake volume.

In our paleo-spring systems, the situation is simpler than in most lakes, and appears to involve little or no evaporation. This feature of the spring/wetland system is not surprising given their physical character. At Corn Creek Flat, a shallow marsh, probably no more than a few meters deep, developed in Units B_2, B_3, and D time behind a shallow impoundment formed by encroachment of alluvial fans from both sides of the valley (Quade, 1986). The marsh deposits grade down valley into alluvial channel deposits containing abundant perennial water fauna, showing that the small marsh overflowed continuously through the narrows between the fans. This would have greatly reduced residence times of water in the marsh, and explains the total lack of primary salts in the deposits. In the case of the deposits at Cactus Springs and Lathrop Wells, there is no major topographic closure to impound discharge; springs waters must have flowed as "freeface" discharge (Quade et al., 1995), and ponded only locally (Quade and Pratt, 1989).

These inferences of short residence times for paleo-water are also strongly supported by isotopic evidence from both of the Corn Creek sections and the Cactus Springs section. The similarity in the means of $\delta^{18}O$ values between standing and flowing water taxa argues against large differences in average temperature between flowing and standing water, and for short residence times and little evaporation.

Estimates of water temperatures and mean annual temperatures. The $\delta^{18}O$ values of fossil ostracodes from our sections can provide constraints on *maximum* mean annual temperatures in the past, provided the list of assumptions discussed above is satisfied. For our paleotemperature reconstruction, we selected results from two types of ostracodes: groundwater candonids and *Cypridopsis vidua*. The advantages of using groundwater taxa are that (1) they grew within the groundwater, thus eliminating any possible influence of postdischarge evaporation on the $\delta^{18}O$ values, and (2) they would have grown at the same temperature as that of local spring water, temperatures that in turn relate to mean annual temperatures, as will be discussed below. The disadvantages of using the groundwater taxa are that (1) as candonids, they are probably subject to vital effects like their relative *C. rawsoni*, and (2) they are not common at all stratigraphic levels. The lowest $\delta^{18}O$ value (n = 15) among groundwater taxa in Unit D from OCI-11 and LPM-34 is +16.9, the value we adopt in our calculations below, thus ensuring that our paleotemperature estimates discussed are *maxima*. Too few analyses of groundwater taxa (n = 4) are available from Unit B for our analysis. Most $\delta^{18}O$ values from both ostracodes (spring and wetland) and aquatic clams are higher than 17‰. Use of the many higher $\delta^{18}O$ values would produce much colder paleotemperature estimates.

We use the following values, drawn from previous discussion, for reconstruction of paleo-water temperatures ($T_{watermax}$) from groundwater candonids at the point of discharge: $\delta^{18}O_{mw}$ (VSMOW) = –14.2 and $\delta^{18}O_{calcite}$ = +16.9‰ (VSMOW). Nearly all ostracode valves are composed of low-Mg calcite. The temperature control on fractionation for ^{18}O and ^{16}O between calcite and host water is (Kim and O'Neil, 1997):

$$1000\ln\alpha_{calcite\text{-}water} = (18.03 \times 10^3/T) - 32.42$$

where T is absolute temperature, and $\alpha_{calcite\text{-}water}$ is the calcite-water fractionation factor. These constraints produce an estimate of $T_{watermax}$ of +10.9 °C (Fig. 10) for Unit D. $T_{watermax}$ increases to ~15 °C using a +0.8 to +1‰ vital effect (Xia et al., 1997).

We can take a similar approach to paleotemperature reconstruction using other ostracode taxa. Our other choice is *C. vidua* because it is a swimmer, and therefore apparently not subject to vital effects (Bacon, 1999), and because we have >85 analyses spanning several time periods from several sections. We selected the lowest $\delta^{18}O$ values of *C. vidua* from each stratigraphic horizon, thereby making our paleotemperature estimates

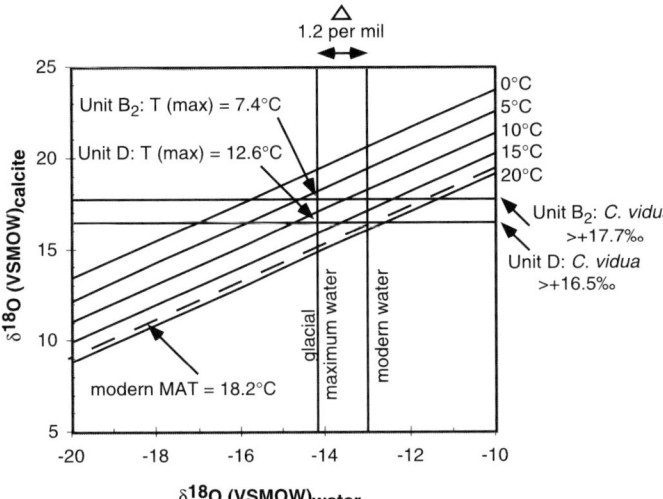

Figure 10. Graphical depiction of paleo-water temperature reconstruction using the limiting $\delta^{18}O$ (VSMOW) value from fossil ostracodes (*Cypridopsis vidua* only) of +16.5‰ for Unit D and B_3, +17.7‰ for Unit B_2, and assuming no vital effects during calcite shell growth. Lines represent predicted $\delta^{18}O$ (VSMOW) values of calcite in isotopic equilibrium with local spring water, using fractionation factors from 0 to 20 °C from Kim and O'Neil (1997). Constraints on the $\delta^{18}O$ (VSMOW) value of modern water are provided by Thomas et al. (1991), Winograd et al. (1998), and unpublished data of the authors, and for glacial-age water, by Winograd et al. (1992, 1997; 2000, personal commun.).

maxima, as well as minimizing any potential effects of evaporation. We adopt the following $\delta^{18}O$ values in our calculations below: +17.7‰ from Unit B_2 (OCI-11 only, n = 40), +16.5 for Unit B_3 (OCI-11 only, n = 7), and +16.5 from Unit D (OCI-11 and LPM-34, n = 30) (Fig. 10). Most $\delta^{18}O$ values from both ostracodes (spring and wetland) and aquatic clams are higher than 17‰. Use of the many higher $\delta^{18}O$ values would produce much colder paleotemperature estimates. Based on previous discussion, we assume $\delta^{18}O_{mw}$ (VSMOW) = −14.2‰ to reconstruct paleo–water temperatures ($T_{watermax}$) at the point of discharge. Using these constraints and the same equation as above from Kim and O'Neil (1997), we obtain an estimate of $T_{watermax}$ of +7.4 °C for Unit B_2, and +12.7 °C for both Unit B_3 and Unit D (Fig. 10), assuming no vital effects.

Two key issues surrounding the use of *C. vidua* are (1) post-discharge evaporation of waters could have occurred, and (2) many could and probably did live far from the discharge point, in places where spring water temperature may have shifted in the direction of air temperature. As for evaporation, we have already argued that this should have been minimal in this hydrologic system, and we selected the lowest $\delta^{18}O$ values to minimize any effects. As to temperature, *C. vidua* very likely built its valves in the summer half-year when air temperature exceeds mean annual temperature. Thus, use of $\delta^{18}O$ values from *C. vidua* should produce overestimates (*maxima*) of paleo-mean annual temperature.

Paleo-spring temperature calculated above provides constraints on mean annual air temperature (MAT) in the past, through the relation of spring and air temperature in the study area today. For MAT today, we used a linear regression of data from Kyle Canyon Ranger Station (2225 m) and Las Vegas (640 m). Data compiled for 28 springs in the Spring and Sheep Range shows that spring temperature generally increases with decreasing elevation (Fig. 11; Table 2). Only a few springs have seasonal data, which shows up to 3 °C annual variation, coldest in the early spring, and warmest in the late summer and fall (Winograd et al., 1998). The small amplitude results from the buffering effect on water temperature by bedrock. For this reason, annually averaged spring temperatures are generally close to mean annual air temperatures, at least for short-flow path springs (Mifflin, 1968). Our own plot of springs in nearby ranges shows that spring temperatures are similar to or higher than local MAT (Fig. 11). The difference between the two generally increases at lower elevation. This is to be expected, because water discharging in valleys will have traveled the furthest, and in the simplest case for most groundwaters, circulated the deepest. For these reasons, we view our estimates of paleo-water temperatures as maxima for local mean annual air temperature. Local MAT today is 18.2 °C at Corn Creek. Using the $\delta^{18}O_{mw}$ values from *C. vidua*, this implies that MAT was a *minimum* of 10.8 °C cooler during Unit B_2 deposition, and 5.6 °C cooler during Unit B_3 and Unit D deposition. If Unit B_2 proves to be correlative with Isotope Stage 6 at Devils Hole, then the large glacial-interglacial shift in $\delta^{18}O_{mw}$ of 2.2‰ for that transition produces an estimate of minimum cooling of 15.2 °C. Using $\delta^{18}O_{mw}$ values from groundwater candonids, MAT during Unit D deposition was *a minimum* of 2.9 °C colder than today, assuming a 1.0‰ vital effect.

We did not attempt a paleotemperature estimate for Unit E because of uncertainty over the value of $\delta^{18}O_{mw}$ during the late glacial period. At both OCI-11 and LPM-35, the average $\delta^{18}O$ value of ostracodes from uppermost Unit D and Unit E_1 (+18.7 ± 1.6; n = 64) is ~1‰ higher than for ostracodes in lower/middle Unit D levels (+17.6 ± 1.0; n = 53). We attribute this rise to probable increases in the $\delta^{18}O_{mw}$ value of local springs in this transition period from glacial to interglacial conditions. The full extent of this glacial-to-interglacial shift is 1.2‰, as previously noted. The exact composition of water discharging ca. 13–12 ^{14}C ka is hard to constrain. We will await documentation of calcite records from Devils Hole spanning the last 50 ka before attempting any late glacial temperature reconstructions.

Our paleotemperature estimates are in line with most other estimates from the southern Great Basin for glacial maxima. Roberts et al. (1997) estimated that paleotemperaures were 10–15 °C colder than today for the period 120–186 ka, using fluid inclusion homogenization temperatures in salts from Death Valley. This compares to our own estimate of at least 10.8 °C colder during Unit B_2 deposition. Dohrenwend (1984), using the distribution of paleo-nivation features with elevation in the Great Basin, estimated temperature changes of ~7 °C since the LGM. Spaulding (1984) and Thompson et al. (1999) relied on several

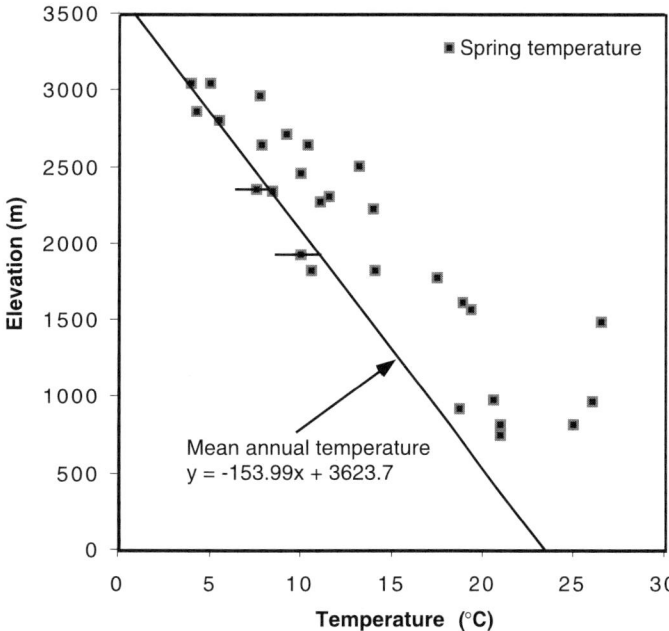

Figure 11. Mean annual air temperature and spring water temperature (using data from McKinley et al., 1991; Thomas et al., 1991) as a function of elevation for the Spring Mountains. A few springs have seasonal temperature data, also shown. This seasonal variation is ≤3 °C in springs at higher elevation, and negligible in low-elevation springs.

lines of evidence from fossil packrat middens for their estimates and arrived at changes of ~5.5–7 °C since the LGM. These estimates rely largely on climate reconstructions for conifers that exhibit down-elevation displacement of ~1000 m during the LGM, and on the absence of thermophilous shrubs such as creosote (*Larrea divaricata*) in the area of Yucca Mountain. The same studies call for paleotemperatures in the late glacial (post-12 ^{14}C ka) only 0–4 °C colder than today, coincident with the summer insolation maximum in the northern hemisphere. Unfortunately, we have no isotopic results nor constraints on the δ^{18}O value of paleo-water from that period (Unit E_2) upon which to base our own estimates of paleotemperature.

Several estimates of paleotemperature during the LGM center around the temperature changes required to restore pluvial lakes, mostly from the central and northern Great Basin. Estimated temperature changes since the LGM are generally <6 °C (e.g., Snyder and Langbein, 1962; Mifflin and Wheat, 1979). A good review of these estimates appears in Mifflin and Wheat (1979).

Finally, Benson and Klieforth (1989) derived paleotemperature estimates for the region based on the difference between the δ^{18}O value of modern and inferred full-glacial groundwater. This analysis assumed that changes in equatorial temperature since the LGM were small, and that ^{14}C dates on groundwater required no large correction. Evidence developed since 1992 (Winograd et al., 1992; Thomas, 1996; Stute et al., 1995) suggests that neither assumption is warranted, and thus we do not consider Benson's estimates further here.

SYNTHESIS OF CLIMATIC RECORDS IN THE REGION

The oldest spring deposits in our study sections belong to Unit B_2. These represent a large spring discharge event that flooded not only the bottom of Corn Creek Flat but portions of the Tule Springs area, and perhaps the spring cluster at the Lathrop Wells Diatomite. Amino acid ratios of mollusks collected from Unit B_2 in a previous study (Quade et al., 1995) are similar to those in deposits representing the second oldest major spring discharge event in the Pahrump Valley, west of the Spring Mountains. Unfortunately, we have no absolute age control on Unit B_2, other than the dates of >41 ^{14}C ka obtained in this study. For reasons discussed above, we regard these ages as infinite. A well-developed paleosol is preserved in some places (but not at OCI-11, where it was apparently eroded) on the top of Unit B_2 and below Unit D, demonstrating substantial age separation (>10^4 yr) between them. For this reason, we tentatively correlate Unit B_2 with the penultimate deep-lake episode in Death Valley (ca. 140 and ca. 190 ka; Ku et al., 1998; Lowenstein et al., 1999), and isotope Stage 6 at Devils Hole (186 to >130 ka; Winograd et al., 1992, 1997) and in the deep sea record (Imbrie et al., 1984). The wetlands of Unit B_2 are more extensive than Unit D at Corn Creek, but apparently less so than Unit D near Tule Springs (Haynes, 1967).

The types of ostracodes present during Unit B_2 time, combined with their shell δ^{18}O values, point to a climate much colder than today's, as well as that of the LGM (Unit D). Whereas the climate was very cold, the wetlands associated with Unit B_2 were also ephemeral. However, the δ^{18}O values from ostracodes in Unit B_2 do not show patterns of enrichment in ^{18}O indicative of evaporation of the waters. We suggest that periodic drying was caused by drops in the water table below the land surface rather than by evaporation, probably in the presence of a climate with cold, relatively dry winters, and cool summers with frontal-storm derived rainfall. This climate is somewhat analogous to that in the modern-day northern prairies of Canada. The continental ice sheet associated with isotope Stage 6 was very large, potentially displacing polar air masses much further south than in other glacial periods. As such, the cold dry arctic high pressure cells might have resided in southern Nevada during the winters of this period, whereas the polar lows would move northward to southern Nevada in the summers, much as both air masses do today in the northern prairies of Canada. This sort of climate is also consistent with the very large lake believed to have existed in Death Valley (Lowenstein et al., 1999), which would require a substantial reduction in air-temperature.

Unit D was deposited during the period of highest effective moisture in the last 30 ka, based on our evidence from Corn Creek Flat, Tule Spring, and perhaps at Lathrop Wells. Only the Corn Creek Flat sections contain well-dated evidence of this

<26.3–16.4 ^{14}C ka wet event. Dates falling in the period 16.4–14.5 ^{14}C ka are not known from any section that we have studied. This negative evidence suggests but does not prove that this was a relatively dry period.

Ostracodes show that paleo-wetlands associated with Unit D had greater permanence than those in Unit B_2, in settings that included flowing springs in addition to streams or perhaps instead of streams. The types of ostracodes in Unit D also imply greater summer temperatures than during the deposition of Unit B_2. Collectively, those conditions suggest a warmer and wetter climate compared to that of Unit B_2, perhaps consistent with atmospheric circulation associated with a smaller continental glaciation. During this time, the winter season may have been the wet season, with snow common at high and perhaps low elevation, whereas the summer season would have been relatively warm, but not hot like today, and dry. Such a climate is also consistent with the much smaller lakes in Death Valley discussed below, because a generally warmer climate with warm dry summers would be less likely to sustain large lakes. Wet-meadow environments appear to be common late in the deposition of Unit D and as with upper Unit B_2, possibly indicating climate change toward a warmer and drier climate.

The wet phase between 14.5 and 12.3 ^{14}C ka is well represented in several sections, including Corn Creek (OCI-11; previously published sections in Quade et al., 1995, and Brennan and Quade, 1997), the Tule Springs archeological site (Haynes, 1967), Cactus Springs (LPM-35), and at Lathrop Wells (LWD-6/8N). To this we can add new dates from Unit E_1 channels from Corn Creek Flat (Table 1, sections 186/187; Figure 12). This phase ended shortly after 12.3 ^{14}C ka, as indicated at LPM-35 at Cactus Springs, where no further discharge is recorded higher in the section. LPM-35 at Cactus Springs stands ~60 m above the current water table, and therefore probably records only the largest paleo-recharge events capable of producing that much rise in the water table. Taken together, this evidence suggests that the wet phase between 14.5 and 12.3 ^{14}C ka was the wettest post–Unit D period in the region, wetter than the period 11.6–9.8 ^{14}C ka when black mats formed all over the area (Quade et al., 1998).

This same chronology of paleohydrologic changes has been documented, although usually in less detail, in several lake and one other spring record from the region. The most robust records are those that involve U-series dating, because of the uncertainties surrounding reservoir corrections for ^{14}C dating. The chronology of lake fluctuations at Lake Manly in Death Valley and, to some extent Searles Lake, agree with our own. Paleo–Lake Manly was a saline but perennial lake between ca. 26 and 8.7 ^{14}C ka (30.1 ± 3.3 to 9.6 ± 3.3 cal. yr from U-series total-dissolution ages on cored halites) and attained its late Pleistocene highstand ca. 21–15.0 ^{14}C ka (24.7 ± 0.9 to 18 ± 1.6 cal. yr from U-series dates on shoreline tufas; Ku et al., 1998; Lowenstein et al., 1999). These dates overlap with the age range of Units D and E.

A U-series date on salt layer S-7 at Searles Lake further constrains the beginning of the same wet episode (called the "Parting Mud" in the core) to shortly after 23.7 ± 1.5 (calendar years) (Phillips et al., 1994). Curiously, dates from shoreline tufas at Searles Lake thought to be equivalent to the Parting Mud are no older than ca. 14 ^{14}C ka (Lin et al., 1997). U/Th dates from the tufas fall between 14.5 and 13.5 ka, while ^{14}C dates are ~1000 yr younger. This discrepancy remains unexplained. As it stands, the high shoreline in Searles Lake appears to correlate with Unit E_1, not D.

Extensive U-series dating of spring calcite in Browns Room constrains periods of higher water tables in the Devils Hole system to between 42–35.5 (44–38.3 cal. yr) and 25.5–17.6 (30–19.6 cal. yr) ^{14}C yr B.P. (Szabo et al., 1994). The

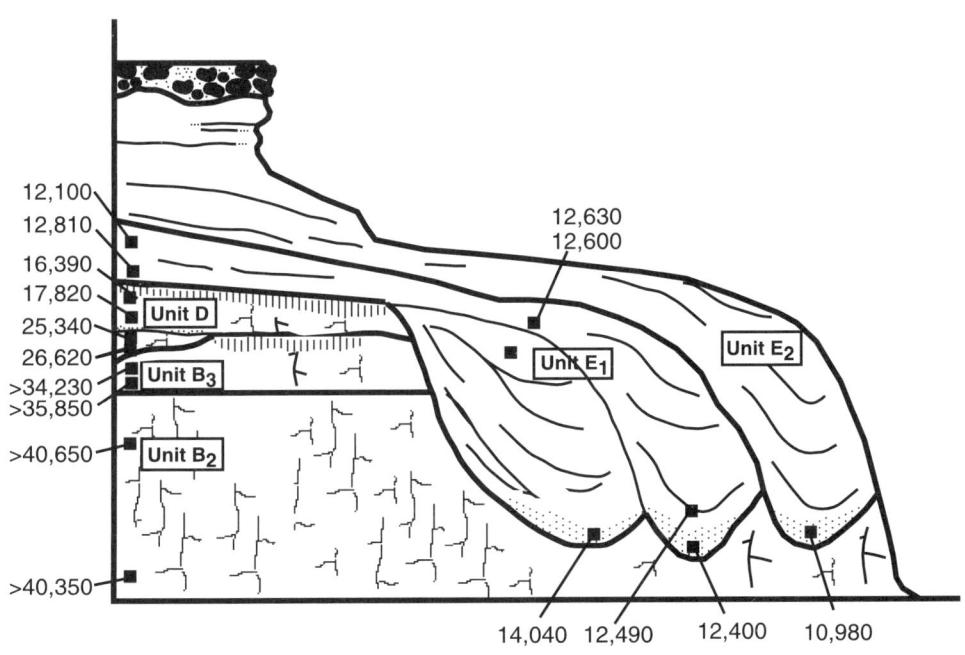

Figure 12. Composite stratigraphy of lower Corn Creek Flat, based on results from OCI-11, LPM-34, sections 186 and 187 (Table 1), and previous work.

older high-water episode partially overlaps the age of Unit B_3. The younger episode correlates almost perfectly with our estimated ages for Unit D. The water table was in decline but still higher than today's levels between 14.2 and 6 (17–6.8 cal. yr) ^{14}C yr B.P., correlative with Unit E.

The Silver Lake chronology is based on ^{14}C dates on both charcoal and lacustrine organic matter (Wells et al., 1989; Enzel et al., 1989; Wells et al., this volume), shells and tufa (Wells et al., 1987), and shells (Ore and Warren, 1971; Wells et al., 1989; Wells et al., this volume). The agreement of the charcoal dates with the other dates suggests that ^{14}C reservoir effects are minimal. Wells et al. (1989) places the termination of the earliest continuous high lake phase (Mojave Lake I) at ca. 16.5 ^{14}C ka, exactly the age we estimate for the end of Unit D deposition. Between 16.5 and ca. 13.7 ^{14}C ka, the lake is intermittent (Intermittent Lake II), then reverts to another continuously high lake condition between 13.7 and 11.3 ^{14}C ka (Mojave Lake II). This overlaps with the period of high discharge represented by Unit E_1 (14–12.3 ^{14}C ka). The lake then passes through a period of intermittent highstands (Intermittent Lake III) between 11.3 and 8.7 ^{14}C ka, almost exactly correlative with the modest resurgence in discharge represented by Unit E_2.

The paleoclimate record for Owens Lake is one of the most detailed in the region (e.g., Smith and Bischoff, 1997; Menking et al., 1997; Benson et al., 1996, 1997). However, it is premature to compare our records to that of Owens Lake until the chronology of Owens Lake is fully developed. At this point, at least two ^{14}C-based chronologies have been presented (Benson et al., 1996, 1997; Bischoff et al., 1997), but until there is consensus on the extent of the ^{14}C reservoir effect, we will refrain from comparison to our record.

Evidence from fossil packrat middens in the region and from glacial deposits in the Sierra Nevada provides further support for our chronology from spring deposits. Middens from the region contain very cold-tolerant vegetation during the LGM up until 16.5 ^{14}C ka. Glaciers on the nearby eastern slopes of the Sierra Nevada underwent several major expansions at the same time as Unit D deposition, termed the Tioga 2 (22 ± 1.5 ^{14}C ka) and 3 (16.5 ± 1.5 ^{14}C ka) advances, dated by ^{36}Cl (Phillips et al., 1996). After this time, thermophilous vegetation begins its slow movement both into the region and upwards in elevation. This date marks the drying of Unit D wetlands in our valley bottom records, and thus we take the change in vegetation also as a response to reduction in effective moisture. This reduction in effective moisture is also reflected in major contraction of Sierran glaciers prior to 14.5 ^{14}C ka. A brief readvance between ca. 14.5 and 13 ^{14}C ka (Tioga 4) correlates with the increase in discharge associated with Unit E_1, an event not remarked upon in the midden record. Unit E_2 (11.6 to ≤8 ^{14}C ka) time is characterized in the midden record by vegetation typical of moister and perhaps cooler conditions than today, particularly at lower elevations (Spaulding, 1995). No Sierran glacial advances are dated from this period; if present, they would be small advances below the headwall cirques.

CONCLUSIONS

Detailed analysis of four new stratigraphic sections in southern Nevada provides a wealth of paleoclimatic information on the last two glacial periods. Carbon-14 dates on land snails provide the geochronological framework for these deposits, augmented by a few dates on organic matter. The penultimate glacial period is probably represented by the pale green muds of Unit B_2. During this and the last glacial period, spring discharge supported extensive, but at times ephemeral, wet meadows, streams, and wetlands. These environments hosted a diverse assemblage of spring and wetland ostracodes and mollusks, as well as highly endemic ostracodes that lived in the groundwater. The types of ostracodes present, as well as isotopic evidence, point to colder conditions than during the LGM (Unit D). Unit B_2 is likely contemporaneous with the penultimate deep-lake episode in Death Valley (Lake Manly, Blackwelder Stand: ca. 140 and ca. 190 ka) and at Owens Lake (118–140 to >155 ka) and with isotope Stage 6 at Devils Hole (186 to >130 ka) and in the deep sea record.

Carbon-14 dates on land snails provide the first firm control on the age of Unit D in the region, and show that it spans much of the period <26.3–16.4 ^{14}C ka. As during Unit B_2 time, an extensive marsh covered the Tule Springs area, while Unit D on Corn Creek Flat is represented by both channel and marsh sediments. Evidence from ostracodes suggests that wetlands were more permanent during Unit D than Unit B_2 time. Unit D appears to be present as diatomites in localized spring deposits near Lathrop Wells. Unit D correlates with period of high lake–groundwater stands at Brown's Room (Devils Hole), Silver Lake and Lake Manly (Death Valley), and with glacial advances Tioga 2–3 in the eastern Sierra.

Unit E (14.5 to ca. 7.2 ^{14}C ka) is present at Corn Creek Flat, Cactus Springs, and in a marl above diatomites (Unit D?) near Lathrop Wells. Ostracodes from Unit E are similar to those of Unit D on Corn Creek Flat, except that wetland taxa are less abundant, and groundwater taxa are not present in the Cactus Spring section. Moderate highstands during this time are also evident in the Silver Lake record, and in an apparent highstand at Searles Lake as well as the Tioga 4 glacial advance. Comparison with the Owens Lake record awaits development of a firm geochronological framework for the cores. Deposition in Unit E_1 appears to represent largely a mix of cold-flowing springs and streams.

Over 400 isotopic analyses of ostracodes and mollusks from Units B_2, D and E yield δ^{18}O values ranging between +16 and +22.2‰. Mean δ^{18}O values for standing and flowing water are similar, suggesting that the water was lost from the wetlands and streams by overflow from the basin rather than by evaporation. In general, mean δ^{18}O values of all ostracodes in Units B and D are similar, while those from Unit E are slightly higher, perhaps the result of late glacial increases in air temperature in recharge areas. The δ^{18}O results from ostracodes provide a unique opportunity to constrain paleotemperatures in valleys during wet periods. Evidence from modern springs in the area and from nearby Devils Hole suggest that δ^{18}O values of paleo-waters during wet periods were no higher than −14.2‰. This and $\delta^{18}O$ values of +16.5 to +17.7‰ from *C. vidua* produce an estimated maximum paleotemperature for springs of 7.4 °C for Unit B_2 time, and 12.6 °C for Unit B_3 and Unit D time. Because spring temperature is always equal to or higher than MAT for air in the area today, MAT must have been ≥10.8 °C colder during Unit B_2 deposition than today at the sample sites, and ≥5.6 °C colder during Unit B_3 and Unit D deposition.

ACKNOWLEDGMENTS

We thank the U.S. Geological Survey for providing support for two of us (Forester, Whelan) as well as field support for Quade. The ostracode, stable isotope, and radiocarbon data were collected under the U.S. Geological Survey Yucca Mountain Program. We are grateful to Marti Mifflin for his review and for inspiring the paleotemperaure analysis, and to Saxon Sharpe and Ike Winograd for their thoughtful reviews. We also thank Kelly Conrad, Bob Brennan, Rick Moscati for all their field and laboratory support.

REFERENCES CITED

Bacon, S.W., 1999, Seasonal constraints on oxygen isotope values of living fresh water ostracodes [M.S. thesis]: Kent, Ohio, Kent State University, 61 p.

Bell, J.W., Ramelli, A.R., and Caskey, S.J., 1998, Geologic map of the Tule Springs Park Quadrangle, Nevada: Nevada Bureau of Mines and Geology Map 113, 1:24,000.

Bell, J.W., Ramelli, A.R., dePolo, C.M., and Maldonado, F., 1999, Geologic map of the Corn Creek Springs Quadrangle, Nevada: Nevada Bureau of Mines and Geology Map 114, 1:24,000.

Benson, L., and Kleiforth, H., 1989, Stable isotopes in precipitation and groundwater in the Yucca Mountain region, southern Nevada: Paleoclimatic implications, *in* Peterson, D.H., ed., American Geophysical Union Monograph no. 55, p. 41–59.

Benson, L.V., Burdett, J., Lund, S., Kashgarian, M., Lund, S.P., Phillips, F.M., and Rye, R.O., 1996, Climatic and hydrologic oscillations in the Owens Lake Basin and adjacent Sierra Nevada, California: Science, v. 274, p. 746–749.

Benson, L.V., Burdett, J., Lund, S., Kashgarian, M., and Mensing, S., 1997, Nearly synchronous climate change in the northern Hemisphere during the last glacial termination: Nature, v. 388, p. 263–265.

Bischoff, J.L., Stafford, T.W., and Rubin, M., 1997, A time-depth scale for Owens Lake sediments of core OL-92: Radiocarbon dates and constant mass-accumulation rate, *in* Smith, G.I., and Bischoff, J.L., eds., An 800,000-year paleoclimatic record from core OL-92: Boulder, Colorado, Geological Society of America Special Paper 317, p. 91–98.

Brennan, R., and Quade, J., 1997, Reliable late-Pleistocene stratigraphic ages and shorter groundwater travel times from ^{14}C in fossil snails from the southern Great Basin: Quaternary Research, v. 47, p. 329–336.

Coplen, T.B., 1994, Reporting of stable hydrogen, carbon, and oxygen isotopic abundances: Pure Applied Chemistry, v. 66, p. 273–276.

DeDeckker, P., 1981, Ostracods of athalassic saline lakes: A review: Hydrobiologia, v. 81, p. 131–144.

Delorme, L.D., 1969, Ostracodes as Quaternary paleoecological indicators: Canadian Journal of Earth Sciences, v. 6, p. 1471–1476.

Delorme, L.D., 1989, Methods in Quaternary ecology #7: Fresh water ostracodes: Geoscience Canada, v. 16, p. 85–90.

Dettman, D.L., Reische, A.K., and Lohman, K.C., 1999, Controls on the stable isotope composition of seasonal growth bands in aragonitic freshwater bivalves (Unionidae): Geochimica et Cosmochimica Acta, v. 63, p. 1049–1057.

Dohrenwend, J.C., 1984, Nivation landforms in the western Great Basin and their paleoclimatic significance: Quaternary Research, v. 22, p. 275–288.

Enzel, Y., Cayan, D.R., Anderson, R.Y., and Wells, S.G., 1989, Atmospheric circulation during Holocene lake stands in the Mojave Desert: Evidence of regional climate change: Nature, v. 341, p. 44–48.

Epstein, S., and Mayeda, T., 1953, Variation of ^{18}O content of waters from natural sources: Geochimica et Cosmochimica Acta, v. 4, p. 213–224.

Forester, R.M., 1983, Relationship of two lacustrine ostracode species to solute composition and salinity: Implications for paleohydrochemistry: Geology, v. 11, p. 435–438.

Forester, R.M., 1987, Late Quaternary paleoclimate records from lacustrine ostracodes, *in* Ruddiman, W.F., and Wright, H.E., Jr., eds., North America and adjacent oceans during the last deglaciation: Boulder, Colorado, Geological Society of America, The Geology of North America, v. K-3, p. 261–276.

Forester, R.M., 1991a, Ostracode assemblages from springs in the western United States: Implications for paleohydrology, *in* Williams, D.D., and Danks, H.V., eds., Arthropods of springs, with particular reference to Canada: Memoirs of the Entomological Society of Canada, no. 155, p. 181–201.

Forester, R.M., 1991b, Pliocene-climate history of the western United States derived from lacustrine ostracodes: Quaternary Science Reviews, v. 10, p. 133–146.

Haynes, C.V., Jr., 1967, Quaternary geology of the Tule Springs area, Clark County, Nevada, *in* Wormington, H.M., and Ellis, D., eds., Pleistocene studies in southern Nevada: Carson City, Nevada State Museum of Anthropology Paper no. 13, p. 1–128.

Imbrie, J., Hays, J.D., McIntyre, A., Mix, A.C., Morley, J.J., Pisias, N.G., Prell, W.L., and Shackleton, N.G., 1984, *in* Berger, A., et al., eds., The orbital theory of Pleistocene climate: Support from a revised chronology of marine $\delta^{18}O$ record, Milankovitch and climate, part 1: Boston, Riedel, p. 269–305.

Kim, S.-T., and O'Neil, J.R., 1997, Equilibrium and nonequilibrium oxygen isotope effects in synthetic carbonates: Geochimica et Cosmochimica Acta, v. 61, p. 3461–3475.

Kishima, N., and Sakai, H., 1980, Oxygen-18 and deuterium determination on a single water sample of a few milligrams: Analytical Chemistry, v. 52, p. 356–358.

Ku, T.-L., Luo, S., Lowenstein, T.K., Li, J., and Spencer, R.J., 1998, U-series chronology of lacustrine deposits in Death Valley, California: Quaternary Research, v. 50, p. 261–270.

Lin, J.C., Broecker, W.S., Hemming, S.R., Hajdas, I., Anderson, R.F., Smith, G.I., Kelley, M., and Bonani, G., 1997, A reassessment of U-Th and ^{14}C ages for late-glacial high-frequency hydrological events at Searles, Lake, California: Quaternary Research, v. 49, p. 11–23.

Lowenstein, T.K., Li, J., Brow, C., Roberts, S.M., Ku, T., Luo, S., and Yang, W., 1999, 200 k.y. paleoclimate record from Death Valley salt core: Geology, v. 27, p. 3–6.

MacKenzie, J.A., 1985, Carbon isotopes and productivity in the lacustrine and marine environment, *in* Stumm, W., ed., Chemical processes in lakes: New York, Wiley, p. 99–118.

McKinley, P.W., Long, M.P., and Benson, L.V., 1991, Chemical analysis of water from selected wells and springs in the Yucca Mountain area, Nevada and southeastern California: U.S. Geological Survey Open-File Report 90-355, 48 p.

McVickar, J.L., and Spaulding, W.G., 1993, Monitoring and mitigation of paleontological resources final draft technical report upper Las Vegas Wash flood control facility, Clark County, Nevada: Las Vegas, Nevada, Dames and Moore.

Menking, K.M., Bischoff, J.L., Fitzpatrick, J.A., Burdette, J.W., and Rye, R.O., 1997, Climatic/hydrologic oscillations since 155,000 yr B.P., at Owens Lake, California, reflected in abundance and stable isotope composition of sediment carbonate: Quaternary Research, v. 48, p. 58–68.

Mifflin, M.D., 1968, Delineation of groundwater flow systems in Nevada: Desert Research Institute Water Resources Center Report H-W 4, 111 p.

Mifflin, M.D., and Wheat, M.M., 1979, Pluvial lakes and estimated pluvial climates of Nevada: Nevada Bureau of Mines and Geology, Bulletin 94, 57 p.

Ore, H.T., and Warren, C.N., 1971, Late Pleistocene–early Holocene geomorphic history of Lake Mojave, California: Geological Society of America Bulletin, v. 82, p. 2553–2562.

Paces, J.B., Forester, R.M., Whelan, J.F., Mahan, S.A., Bradbury, J.P., Quade, J., Neymark, L.A., and Kwak, L., 1996, Synthesis of groundwater discharge deposits near Yucca Mountain: U.S. Geological Survey Milestone Report 3GQH671M to DOE-YMPSCO, 75 p.

Phillips, F., Campbell, A.R., Smith, G.I., and Bischoff, J.L., 1994, Interstadial climatic cycles: A link between western North America and Greenland?: Geology, v. 22, p. 1115–1118.

Phillips, F.M., Zreda, M.G., Benson, L.V., Plummer, M.A., Elmore, D., and Sharma, P., 1996, Chronology for fluctuations in late Pleistocene Sierra Nevada glaciers: Science, v. 274, p. 749–751.

Quade, J., 1986, Late Quaternary environmental changes in the upper Las Vegas Valley: Quaternary Research, v. 26, p. 340–357.

Quade, J., and Pratt, W.L., 1989, Late Wisconsin groundwater discharge environments of the southwestern Indian Springs Valley, southern Nevada: Quaternary Research, v. 31, p. 351–370.

Quade, J., Cerling, T.E., and Bowman, J.R., 1989, Systematic variation in the carbon and oxygen isotopic composition of Holocene soil carbonate along elevation transects in the southern Great Basin, USA: Geological Society of America Bulletin, v. 101, p. 464–475.

Quade, J., Mifflin, M.D., Pratt, W.L., McCoy, W., and Burckle, L., 1995, Fossil spring deposits in the southern Great Basin and their implications for changes in water-table levels near Yucca Mountain, Nevada, during Quaternary time: Geological Society of America Bulletin, v. 107, p. 213–230.

Quade, J., Forester, R.M., Pratt, W.L., and Carter, C., 1998, Black mats, spring-fed streams, and late-glacial-age recharge in the southern Great Basin: Quaternary Research, v. 49, p. 129–148.

Riggs, A.C., 1984, Major carbon-14 deficiency in modern snail shells from southern Nevada springs: Science, v. 224, p. 58–61.

Roberts, S.M., Spencer, R.J., Yang, W., and Krouse, H.R., 1997, Deciphering some unique paleotemperature indicators in halite-bearing saline lake deposits from Death Valley, California, USA: Journal of Paleolimnology, v. 17, p. 101–130.

Romanek, C., Grossman, E., and Morse, J.W., 1992, Carbon isotopic fractionation in synthetic aragonite and calcite: Effects of temperature and precipitation rate: Geochimica et Cosmochimica Acta, v. 56, p. 419–430.

Sharpe, S.E., Whelan, J.F., Forester, R.M., and McConnaughey, T., 1994, Molluscs as climate indicators: Preliminary stable isotope and community analysis, in International High-Level Radioactive Waste Management Proceedings: American Society of Chemical Engineers and American Nuclear Society, 5th International Conference, 22–26 May 1994, Las Vegas, Nevada, p. 2538–2544.

Smith, G.I., and Bischoff, J.L., 1997, An 800,000-year paleoclimatic record from core OL-92, Owen's Lake, southeast California: Boulder, Colorado, Geological Society of America Special Paper 317, 165 p.

Snyder, C.T., and Langbein, W.B., 1962, The Pleistocene lake in Spring Valley, Nevada, and its climate implications: Journal of Geophysical Research, v. 67, p. 2385–2394.

Spaulding, W.G., 1984, Preliminary assessment of climate change during late Wisconsin time, southern Great Basin and vicinity, Arizona, California, and Nevada: U.S. Geological Survey Water-Resources Investigations Report 84-4328, 40 p.

Spaulding, W.G., 1995, Environmental change, ecosystem responses, and late Quaternary development of Mojave Desert, in Mead, J.I., et al., eds., Late Quaternary environments and deep time: A tribute to Paul S. Martin: Rapid City, South Dakota, Fenske Companies, p. 139–164.

Stute, M., Forster, M., Frischkorn, H., Serejo, A., Clark, J.F., Schlosser, P., Broecker, W.S., and Bonani, G., 1995, Cooling of tropical Brazil (5 °C) during the Last Glacial Maximum: Science, v. 269, p. 379–383.

Szabo, B.J., Kolesar, P.T., Riggs, A.C., Winograd, I.J., and Ludwig, K.R., 1994, Paleoclimatic inferences from a 120,000-yr calcite record of water-table fluctuations in Browns Room of Devils Hole: Quaternary Research, v. 41, p. 59–69.

Taylor, D.W., 1967, Late Pleistocene molluscan shells from the Tule Springs Area, in Wormington, H.M., and Ellis, D., eds., Pleistocene studies in southern Nevada: Carson City, Nevada State Museum of Anthropology Paper no. 13, p. 395–399.

Thomas, J.M., 1996, Geochemical and isotopic interpretation of groundwater flow, geochemical processes, and age dating of groundwater in carbonate-rock aquifers of the southern Basin and Range [Ph.D. thesis]: Reno, University of Nevada, 135 p.

Thomas, J.T., Lyles, B.F., and Carpenter, L.A., 1991, Chemical and isotopic data for water from wells, springs, and streams in carbonate-rock terrane of southern and eastern Nevada and southeastern California, 1985–88: U.S. Geological Survey Open-File Report 89-422, 24 p.

Thompson, R.S., Anderson, K.H., and Bartlein, P.J., 1999, Quantitative paleoclimatic reconstructions from late Pleistocene plant macrofossils of the Yucca Mountain region: U.S. Geological Survey Open-File Report 99-338, 38 p.

Wells, S.G., McFadden, L.D., and Dohrenwend, J.C., 1987, Influence of late Quaternary climatic changes on geomorphic and pedogenic processes on a desert piedmont, eastern Mojave Desert, California: Quaternary Research, v. 27, p. 130–146.

Wells, S.G., Anderson, R.Y., McFadden, L.D., Brown, W.J., Enzel, Y., and Miossec, J-L., 1989, Late Quaternary paleohydrology of the eastern Mojave River drainage, southern California: Quantitative assessment of the late Quaternary hydrologic cycle in large arid watersheds: New Mexico Water Resources Research Institute Report no. 242, 253 p.

Winograd, I.J., and Thordarsen, W., 1975, Hydrogeologic and hydrochemical framework, south-central Great Basin, Nevada-California, with special reference to the Nevada Test Site: U.S. Geological Survey Professional Paper 712-C, 126 p.

Winograd, I.J., Coplen, T.B., Landwehr, J.M., Riggs, A.C., Ludwig, K., Szabo, B.J., Kolesar, P.T., and Revesz, K.M., 1992, Continuous 500,000-year record from vein calcite in Devils Hole, Nevada: Science, v. 258, p. 255–260.

Winograd, I.J., Landwehr, J.M., Ludwig, K., Coplen, T.B., and Riggs, A.C., 1997, Duration and structure of the past four interglaciations: Quaternary Research, v. 48, p. 141–154.

Winograd, I.J., Riggs, A.C., and Coplen, T.B., 1998, The relative contribution of summer and cool-season precipitation to groundwater recharge, Spring Mountains, Nevada, USA: Hydrogeology Journal, v. 6, p. 77–93.

Xia, J., Ito, E., and Engstrom, D.R., 1997, Geochemistry of ostracode calcite, Part 1: An experimental determination of oxygen isotope fractionation: Geochimica et Cosmochimica Acta, v. 61, p. 377–382.

Manuscript Accepted by the Society August 1, 2002

Geological Society of America
Special Paper 368
2003

Regional response of alluvial fans to the Pleistocene-Holocene climatic transition, Mojave Desert, California

Eric V. McDonald
Division of Earth and Ecosystem Science, Desert Research Institute, 2215 Raggio Parkway, Reno, Nevada, 89512, USA

Leslie D. McFadden
Department of Earth and Planetary Sciences, University of New Mexico, Albuquerque, New Mexico, 87131, USA

Stephen G. Wells
Desert Research Institute, 2215 Raggio Parkway, Reno, Nevada, 89512, USA

ABSTRACT

Alluvial fan deposits along the Providence Mountains piedmont in the eastern Mojave Desert that (1) are derived from diverse rock types, (2) are dated with luminescence techniques and soil-stratigraphic correlations to other relatively well dated fan, eolian, and lacustrine deposits, and (3) have some of the highest peaks in the Mojave Desert, provide a unique opportunity to study the influence of Pleistocene-Holocene climatic transition on regional fan deposition across diverse geomorphic settings. Geomorphic and age relations among alluvial and eolian units along the Providence Mountains and Soda Mountains piedmonts indicate that most of the late Quaternary eolian and alluvial fan units were deposited during similar time intervals and represent region-wide changes in geomorphic factors controlling sediment supply, storage, and transport.

Deposition of alluvial fans in the desert southwestern United States during the latest Pleistocene has been largely attributed to (1) a more humid climate and greater channel discharge and (2) time-transgressive changes in climate and an increase in sediment yield. Stratigraphic and age relations among depositional units demonstrate that a regional period of major alluvial fan deposition occurred between ca. 9.4 and 14 ka, corresponding with the timing of the Pleistocene-Holocene climatic transition. This age range indicates that deposition of these fans is not simply a result of greater effective moisture and channel discharge during the last glacial maximum. Increases in sediment yield during the Pleistocene-Holocene transition have been largely attributed to a time-transgressive decrease in vegetative cover with an increase in hillslope erosion. Geomorphic relations along the Providence Mountains, however, suggest that that changes in vegetation cover during the Pleistocene-Holocene climatic transition may have had a limited impact on hillslope instability and sediment yield because of (1) the inherently high infiltration capacity of coarse-textured soils and colluvium, (2) possible strong spatial variations in soil cover across hillslopes, and (3) modern vegetation cover appears to provide enough stability for the buildup of soils and colluvium. An increase in sediment yield may instead be largely due to an increase in extreme storm events, possibly an increase in tropical cyclones. Extreme storms would provide the rainfall intensity and duration to mobilize permeable sediments from mountain catchments and into distal fan areas.

McDonald, E.V. McFadden, L.D. and Wells, S.G., 2003, Regional response of alluvial fans to the Pleistocene-Holocene climatic transition, Mojave Desert, California, *in* Enzel, Y., Wells, S.G., and Lancaster, N., eds., Paleoenvironments and paleohydrology of the Mojave and southern Great Basin Deserts: Boulder, Colorado, Geological Society of America Special Paper 368, p. 189–205. © 2003 Geological Society of America.

INTRODUCTION

The impact of Pleistocene-Holocene climate change on major periods of alluvial fan deposition on desert piedmonts in the southwestern United States has been the focus of many investigations. Two general models describing the linkage between climate change and a major cycle of alluvial fan deposition have evolved from these studies. One model suggests that the formation of alluvial fans occurred largely during the latest Pleistocene when climatic conditions were more humid than at present (Melton, 1965; Lustig, 1965; Ponti, 1985; Christenson and Purcell, 1985; Dorn et al., 1987). This model indicates that a more humid climate led to greater stream discharge and hence an increase in sediment transport resulting in fan deposition. A second model suggests that a cycle of fan aggradation occurred largely as a sequence of geomorphic responses to time-transgressive changes in climate and vegetation during the Pleistocene-Holocene transition (Bull and Schick, 1979; Wells et al., 1987, 1990; Bull, 1991; Harvey and Wells, 1994; Harvey et al., 1999). According to this model, a cycle of fan deposition was triggered by a transition from a wetter to a relatively drier climate when a reduction in effective soil moisture resulted in a loss of vegetative cover on hillslopes that, in turn, resulted in an increase in sediment supply and surface runoff from increasingly barren slopes.

Determining temporal linkages between fluctuations in climate and alluvial fan deposition has been a challenging and controversial problem for a variety of reasons. First, there are few well-dated deposits that can be temporally linked to detailed records of climate change because of the difficulty in dating alluvial fan deposits. This is complicated further by a significant variation in the precision of different methods used in determining relative and numerical ages of alluvial stratigraphic units. The most successful linking of climatic change to alluvial fan deposition have been studies of mountain piedmonts that border pluvial lake basins containing dated lacustrine deposits and shoreline features (Wells et al., 1987; Harvey and Wells, 1994; Reheis et al., 1996; Harvey et al., 1999). Results of these studies have demonstrated that a pronounced period of alluvial fan aggradation in the Mojave Desert is linked to a period of climatic change associated with the Pleistocene-Holocene transition.

A second fundamental problem is determining if the control of geomorphic factors is secondary to the control of climate change in triggering periods of alluvial fan deposition across a region. Factors such as tectonic stability, mountain height, and the size and lithology of the drainage basin may exert a strong influence on alluvial depositional processes (Bull, 1991) and, in turn, the possible timing of fan deposition. If the timing of alluvial fan deposition is strongly dominated by some aspect of climate change, then the temporal relations between fan deposition and variations in climate should occur largely in a similar fashion across a region and across diverse geomorphic settings. Regional evaluation of fan deposition has been difficult because of a lack of adequate regional stratigraphic correlations (reinforced by adequate age control) among fan deposits across diverse geomorphic settings.

The Quaternary piedmont deposits in the Mojave Desert of California (Fig. 1) provide a unique setting in which to examine the timing of late Quaternary climate fluctuations on alluvial fan stratigraphy across diverse geomorphic settings. Stratigraphic and geomorphic relations among alluvial fan, lacustrine, and eolian deposits on the Soda Mountains piedmont (hereafter SMP) along the western margin of the Silver Lake playa (hereafter Silver Lake) in the Mojave Desert provide a well-documented record of late Pleistocene and Holocene climatic change and periods of eolian and alluvial fan deposition (Wells et al., 1987, 1989, this volume; Enzel et al., 1989, 1992; Harvey and Wells, 1994; Harvey et al., 1999). Quaternary alluvial and eolian deposits along the western margin of the Providence Mountains piedmont (hereafter PMP) provide a different geomorphic setting with which to examine the influence of climate change on alluvial fan deposition. First, the diverse rock types of the Providence Mountains (Miller et al., 1985), ranging from coarse-grained plutonic to microcrystalline carbonates, have created four lithologically dissimilar fan chronosequences that are juxtaposed along the mountain front. Second, the drainage basins in the Providence Mountains are considerably largely and at a higher elevation than those of the Soda Mountains (Fig. 2). Soil-stratigraphic correlations between soils formed in alluvial fan deposits on each piedmont, supplemented with numerical age estimations for alluvial and eolian units from the Providence Mountain fans, provide a (1) regional stratigraphic framework of late Quaternary alluvial fan and eolian deposits and (2) adequate age control for examining the impact of changes in climate on fan deposition across diverse geomorphic settings.

The purpose of this paper is to examine the relative influences of Pleistocene-Holocene climate change and diverse geomorphic settings on the timing and mechanisms of alluvial fan deposition. Specifically, this paper (1) presents the late Quaternary stratigraphy and geochronology common to the Providence Mountains piedmont and Soda Mountains piedmont, (2) demonstrates consistency in the timing of key eolian and alluvial depositional events across the east Mojave Desert, and (3) evaluates the role of related changes in climate, vegetation, and geomorphic response associated with alluvial fan deposition during the Pleistocene-Holocene transition.

Geomorphic setting

The Providence Mountains. The Providence Mountains are a prominent feature along the eastern edge of the Mojave Desert (Fig. 1). Elevations range from ~600 m in distal fan areas to >2000 m at the mountain tops (Fig. 2). Fan deposits along the western margin of the Providence Mountains grade to an extensive area of sand sheets associated with the Kelso Dunes. The modern climate of the Providence Mountains is typical of the Mojave Desert with annual rainfall ranging from ~150 mm annually along the alluvial fans to ~250 mm annually at the mountain tops based on regional relations between recorded annual precipitation and elevation (Fig. 3). Most of the annual precipitation in

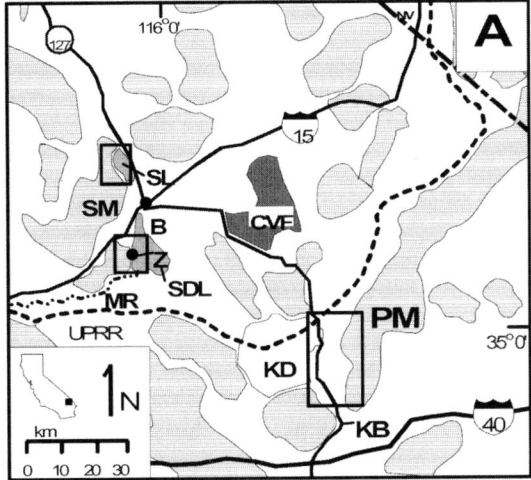
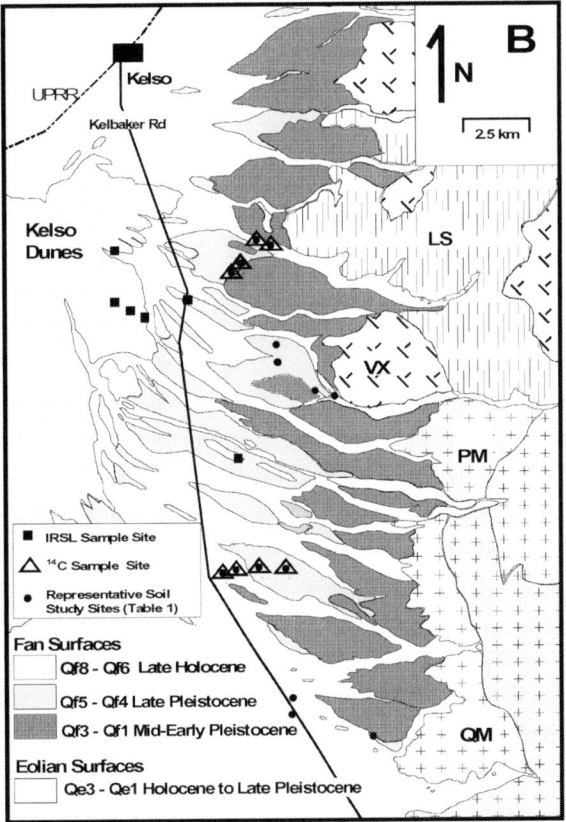

Figure 1. A: Location of study area (open box) along the Providence Mountains (PM) piedmont in relation to study areas along the Soda Mountains (SM) adjacent to the Silver Lake Playa (SL), specifically, Wells et al. (1987), and Harvey and Wells (1994) and Harvey et al. (1999). Other map abbreviations: MR—Mojave River (dash-dot-dot line), B—Baker, UPRR—Union Pacific Railroad (dashed line), CVF—Cima volcanic field, OD—Old Dad Mountain, KD—Kelso Dunes, and KB—the Kelbaker road. B: General distribution of Quaternary depositional units along the Providence Mountains piedmont, sample sites for ^{14}C and infrared stimulated luminescence (IRSL) age control, and locations of representative soil study sites (Table 1). Piedmont deposits further subdivided (dotted line) into sequences based upon the dominant rock types that make up the alluvium: PM—mixed-plutonic rocks, QM quartz monzonite; LS—limestone and marble, and VX—volcanic-mixed. NV—Nevada (long dash–short dash is Nevada-California state line), Z—Zzyzx, SDL—Soda Lake Playa.

the east Mojave Desert is associated with winter Pacific frontal storms. Modern vegetation along the fans is predominantly Mojave Desert Scrub dominated by *Larrea tridentata* (creosotebush) and *Ambrosia dumosa* (white bursage). The vegetation along the mountain hillslopes ranges from blackbrush (*Coleogyne ramosissima*) and Mojave yucca (*Yucca schidigera*) to juniper/pinyon woodland (predominantly on north-facing slopes).

Quaternary depositional units along the PMP were subdivided using relative degrees of development of soils and desert pavements and stratigraphic relations among depositional units (Table 1; Figs. 4, 5). Eight alluvial fan units (Qf1–Qf8, oldest to youngest) and three eolian sand units (Qe1–Qe3, oldest to youngest) have been recognized and are described in detail elsewhere (McDonald, 1994; McDonald and McFadden, 1994).

Late Quaternary alluvial fan units are also separated along the mountain front into four sequences based upon the dominant rock types that make up the alluvium: (1) PM—leucocratic to mesocratic mixed-plutonic rocks (syenite, syenogranite, monzodiorite); (2) QM—quartz monzonite; (3) LS—limestone, marble, and minor amounts of volcanic; and (4) VX—rhyolitic tuff and rhyodacite with lesser amounts of plutonic and limestone. Identification of rock types is based on the mapping of Miller et al. (1985). There is no evidence of tectonic activity along the mountain front during the late Quaternary.

The Soda Mountains. The Soda Mountains lie ~110 km to the northwest of the Providence Mountains (Fig. 1). Elevations range from ~270 m in distal fan areas to nearly 700 m at the mountain tops. Fan deposits along the eastern margin of the Soda Mountains grade to the western margins of the Silver Lake playa and Soda Lake playa basins. These two playas are the remnant of Pleistocene pluvial Lake Mojave. Modern annual rainfall ranges from ~60 mm along the alluvial fans to ~100 mm (estimated) at the mountain tops (Fig. 3). Modern vegetation along the fans and mountain ridges is predominantly Mojave Desert Scrub including

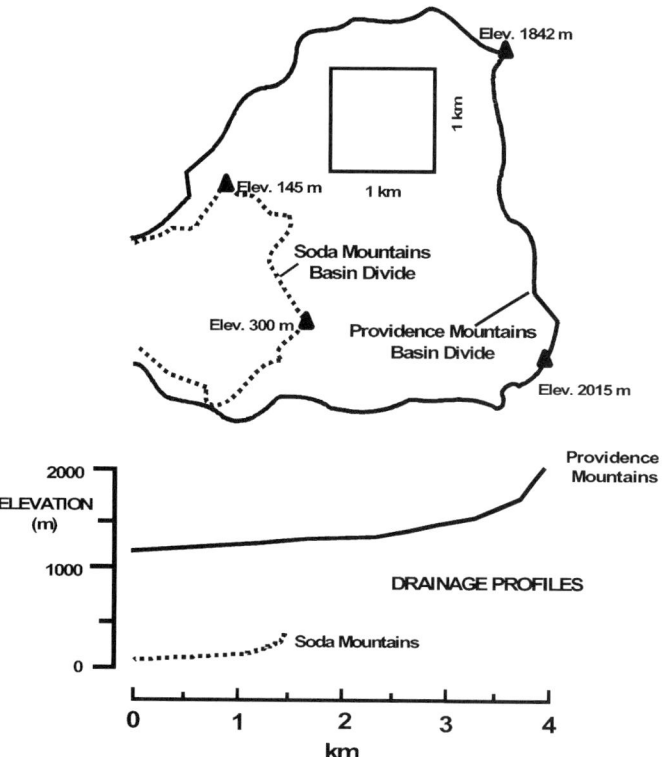

Figure 2. Schematic diagram comparing typical areas and ephemeral stream profiles of drainage basins in the Soda Mountains and Providence Mountains. Elevations along basin perimeter represent two highest elevations along divide. Outline of Soda Mountains drainage basin has been rotated 180°. Drainage basin shown for the Providence Mountains is the headwaters for Winston Wash. The drainage basin shown for the Soda Mountains is the major source area of the alluvial fans studied by Wells et al. (1987).

creosotebush and bursage. Bedrock geology is predominantly metavolcanics and granitics.

Previous studies of the geomorphology and late Quaternary history of the Soda Mountains and pluvial Lake Mojave have focused predominantly along the western and northern margins of Silver Lake playa (Wells et al., 1987, 1989; McFadden et al., 1989; Enzel et al., 1992) and along the western boundary of Soda Lake playa near Zzyzx (Wells et al., 1990; Harvey and Wells, 1994; Harvey et al., 1999; Harvey and Wells, this volume). These studies established detailed chronologies of hillslope response as well as alluvial fan and eolian deposition by comparing the degree of soil formation, desert pavement development, and stratigraphic relations to well-dated pluvial lake shorelines. There is no evidence of tectonic activity along the mountain front during the late Quaternary.

Geochronology

Age control for alluvial and eolian deposits in the Providence Mountains is derived from (1) infrared stimulated luminescence (IRSL) ages of sands associated with alluvial and eolian deposits (Clarke, 1994; Edwards, 1993), (2) soil-stratigraphic correlations to SMP alluvial units using a soil development index (Harden, 1982), and (3) radiocarbon dating of pedogenic carbonate (Wang et al., 1996). All radiometric ages cited and preferred stratigraphic correlations between the Soda and Providence Mountains are shown in Figure 4.

Radiocarbon ages of pedogenic carbonates are reported as measured AMS ^{14}C dates and modeled ^{14}C ages (Fig. 4). Modeled ages are measured ^{14}C ages adjusted using a diffusion/reaction model that attempts to account for the various processes and factors controlling the ^{14}C content of soil carbonate (Amundson et al., 1994; Wang et al., 1994, 1996). This modeling approach was developed because measured ^{14}C ages may underestimate the age of a soil because of (1) a possible time lag between deposition and the onset of soil formation and (2) the time required to form a carbonate coating of adequate thickness for radiocarbon dating.

Unless otherwise noted, all radiocarbon ages cited in this paper are expressed as calibrated ages in order to compare them with the cited IRSL ages. Radiocarbon ages were calibrated using the CALIB v. 3.0 program of Stuiver and Reimer (1993).

Profile development index (PDI) values shown in Figure 4 and values from other studies cited in the text are calculated using methods in Harden (1982) and Harden and Taylor (1983) and all values (this study as well as cited studies) are based on profile maximum values listed in Taylor (1988).

LATEST QUATERNARY STRATIGRAPHY OF THE PROVIDENCE MOUNTAINS PIEDMONT

Holocene deposits

Alluvium of units Qf8 and Qf7 are inset into all other deposits or overlap these deposits, indicating that these are the youngest fan deposits along the piedmont. Unit Qf8 consists of sediments of the active wash and weakly vegetated bars and channels that lack soil development and desert pavements (Figs. 4, 5; Table 1). The next oldest deposit, unit Qf7, is generally within ~1 m of the active wash and is largely confined to narrow terraces along active channels. The unit Qf7 alluvium consists of moderately to poorly stratified sandy-pebble (grus) stream and sheetwash deposits (PM, QM sequences) to well-stratified pebble-gravel braided stream deposits (VX, LS sequences). Soils developed on unit Qf7 are characterized by a thin (0.5–1.0 cm) incipient Av horizon, a lack of B horizons, and weak stage I carbonate morphology (carbonate stage morphology after Gile et al., 1966; Table 1). Incipient pavements have formed in a few places on the VX and LS Qf7 deposits. Nearly all clasts on Qf7 surfaces are unweathered on the upper surfaces, but the undersides may be very weakly oxidized. The oldest radiocarbon dates on pedogenic carbonate range from 4980 cal. yr B.P. (4410 ± 110 ^{14}C yr B.P.) for PM sequence soils to 5900 cal. yr B.P. (5110 ± 60 ^{14}C yr B.P.) for LS sequence soils (Fig. 4). Modeled radiocarbon ages range from ca. 4 to 6 ka for the PM Qf7 soils

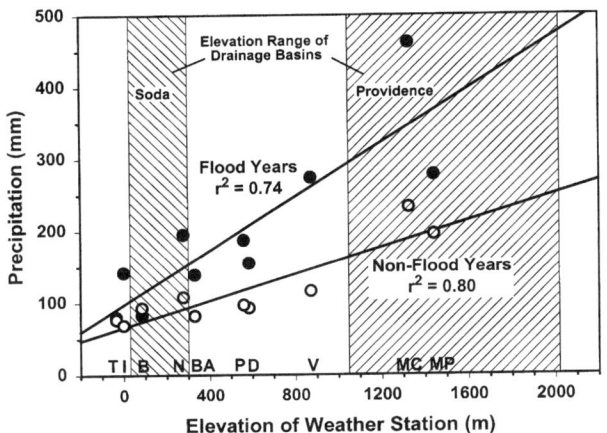

Figure 3. Regional trends in annual precipitation across southern California deserts based on precipitation records for 1948–1999. Flood years (filled circles, upper regression line) represents mean annual precipitation for years when flooding of the Silver Lake playa occurred (1969, 1978, 1980, 1983, 1992, 1998). Nonflood years represents mean annual precipitation for all nonflood years. Weather data from: T—Thermal, I—Indio, B—Blythe, N—Needles, BA—Baker, P—Twentynine Palms, D—Daggett, V—Victorville, MC—Mitchell Caverns, MP—Mountain Pass.

TABLE 1. TYPICAL SOIL CHARACTERISTICS OF THE FOUR FAN SEQUENCES

Fan Unit		PM	QM	VX	LS
Qf8 Late Holocene (active wash)	Horizon: Texture: Stage: Pavement:	No soil Sand 0 None	No soil Sand 0 None	No soil Sand to loamy sand 0 None	No soil Sand to loamy sand 0 None
Qf7 Late Holocene	Horizon: Texture: Stage: Pavement:	Avk-AC-Ck-C Loamy sand I None	Av-A-AC-Ck-C Loamy sand I None	Avk-ACk-Ck-C Loamy sand I None to very weak	Avk-ACk-Ck-C Sandy loam I None to very weak
Qf6 Late Holocene	Horizon: Texture: Stage: SB: Pavement:	Avk-Bwk-Ck-C Sandy loam I Cambic Weak	Av-A-Bw-Ck-C Loamy sand I Cambic Very weak to none	Avk-Bwk-Bk-Ck-C Loamy sand I Cambic Weak to moderate	Avk-Bwk-Bk-Ck-C Sandy loam I–II Cambic Weak to moderate
Qf5 Early Holocene to latest Pleistocene	Horizon: Texture: Stage: SB: Pavement:	Avk-Bwk-Btk-Ck-C Sandy loam I–II Cambic or argillic Moderate to weak	Fan unit not found	Avk-Bwk-Bk-Ck-C Sandy loam I–II Cambic/calcic Moderate to strong	Avk-Btvk-Bk-Ck-C Loam II–III Calcic Moderate to strong
Qf4 Late Pleistocene	Horizon: Texture: Stage: SB: Pavement:	Avk-Btk-Bk-Ck-C Sandy clay loam III Argillic Moderate to strong	Av-ABv-Bt-Bk-Ck-C Clay loam I–II Argillic Weak to moderate	Avk-BAv-Bwk-Bkm-Ck-C Sandy loam III–IV Petrocalcic Strong	Avk-Btvk-Bwk-Bkm-Bk-Ck-C Loam III+ Petrocalcic Strong
Qf3 Middle Pleistocene	Horizon: Texture: Stage: SB: Pavement:	Avk-BAvk-Bt-Bkm-Ck-C Clay loam III–IV Argillic/petrocalcic Strong	Av-BAv-Bt-BCk-Ck-C Loam I Argillic Moderate to weak	Avk-BAv-Btk-Bk-Bkm-Bk-Ck-C Sandy clay loam IV Argillic/petrocalcic Strong	Avk-Btvk-Bwk-Bkm-Bk-Ck-C Loam IV Petrocalcic Strong to very strong
Qf2 Mid–early Pleistocene	Horizon: Texture: Stage: SB: Pavement:	Avk-Btk-Bkm-Ck-C Sandy clay loam IV Petrocalcic Moderate to weak	Av-Bt-Btk-BCk-Ck-C Sandy clay loam I Argillic Weak to moderate	Avk-BAvk-Btk-Bk-Bkm-Ck-C Sandy clay loam IV Argillic/petrocalcic Strong to moderate	Avk-Bkm-Bk-Ck-C Sandy loam IV–V Petrocalcic Moderate to strong

Note: PM—mixed-plutonic; QM—quartz monzonite; VX—mixed-volcanic; LS—limestone. Age estimations of fan units in Table 2. Unit Qf1 not shown because unit is highly eroded. Horizon—typical soil horizon sequence; texture—finest texture of any B or AC horizon; stage—soil carbonate stage; SB—strongest diagnostic B horizon; pavement—general quality of desert pavement where either best developed or best preserved.

Providence Mountains[a]							Soda Mountains[b]					
Geolgic Time	Surface	SDI[c]	IRSL[d] (yrs)	^{14}C[e] (yrs)	Cal-^{14}C[f] (yrs)	^{14}C-Model[g] (ka)		Surface	SDI[h]	SDI[i]	SDI[j]	^{14}C[k] (yrs)	Cal-^{14}C[f] (yrs)
Holocene Late	Qf8	0						Qf6	0	0	0		
								Qe3					
	Qf7	2.1		4410 ± 110 / 5110 ± 60	5900 / 4980	7 - 8 / 4 - 6		Qf5	0.8	0.6	2.3		
	Qe3		4250 ± 290										
Holocene Early/Mid	Qf6	4.2		5380 ± 80 / 4010 ± 70	6190 / 4490	8 - 11 / 7 - 8		Qf4	4.9	3.9	7.9	3400 ± 60	3630
								"Ql5"[l]				3620 ± 60	3910
	Qe2		3500 ± 220 / 3700 ± 425 / 4074 ± 334 / 8420 ± 795					Qf3	7.5	5.3	6.6		
								Qe2					>6820[j]
	Qf5	12.1	10410 ± 890 / 12460 ± 1151	18120 ± 150 / 16310 ± 100	21640 / 19200	29 - 36 / 27 - 33		Qe3	14.3	12.2	19.7	8350 ± 300 / 10333 ± 120	9380 / 12200
								Qf2				11320 ± 120 / 12020 ± 130	13230 / 14020
								Ql2				13670 ± 550	16380
Pleistocene Late	Qe1		16830 ± 1465 / 17300 ± 1935					Qe1/ Qil2				14660 ± 260 / 15350 ± 240	17550 / 18270
								Ql1				16270 ± 310 / ~18000	19160 / 21480
								Qil1				20320 ± 740 / ~22000	22000 / 24500
	Qf4	25.9		28960 ± 440 / 26980 ± 290		48 - 57 / 40 - 48		Qf1	19.3	20.3	46.4		

a = This study
b = Soda Mountains/Silver Lake stratigraphy from Wells et al. (1987; 1990)
c = Soil development index values calculated according to Harden (1982); and Harden and Taylor (1983), values from McDonald (1994)
d = IRSL dates from Clarke (1994)
e = Oldest pedogenic carbonate radiocarbon ages (upper PM soils, lower LS soils) from Wang et al. (1994)
f = Calibrated radiocarbon ages based on Stuiver and Reimer (1993)
g = Maximum surface ages using modeled radiocarbon results from Wang et al. (1994)
h = Values for Soda Mountains piedmont, from McFadden et al. (1989)
i = Values for Salt Springs Hills area, from McFadden et al. (1989)
j = Values for Soda Mountains piedmont and Cima piedmont, from Harden et al. (1991)
k = Radiocarbon ages reported in Wells et al. (1987, 1989, 1990) and Brown (1990)
l = Unlabled late Holocene lake stand, Silver Lake playa, informally defined here for illustration
j = Age estimate from soil stratigraphy, Mcfadden et al. (1992)

Figure 4. Regional correlations and age estimations among Quaternary fan, lacustrine, and eolian deposits in the eastern Mojave Desert. Ages in italics are based on lacustrine sedimentation rates.

and from ca. 7 to 8 ka for the LS Qf7 soils. In contrast, soil stratigraphy suggests that PM unit Qf7 is correlative with unit Qf5 along the Soda Mountains and that the latter was deposited after ca. 3.4 ka (Fig. 4; Wells et al., 1987). Moreover, Bull (1991) notes that incipient varnish on pavement clasts generally indicates a surface age of <1000 yr B.P.

Two eolian units are interstratified with Holocene alluvium (Fig. 4). Unit Qe3 consists of scattered dune and sand sheet deposits that overly unit Qf6 and distal Qf5 fan deposits and is truncated by unit Qf7 deposits. A vegetative cover on much of the unit Qe3 surface indicates that this eolian unit is temporarily stabilized. A single IRSL date of 4250 ± 290 yr B.P. (Clarke, 1994) was obtained from eolian sands just above the contact between Qe3 and the underlying Qf6 deposit. A lack of soil development on this underlying Qf6 deposit indicates that deposition of Qe3 sands must have occurred soon after deposition of the Qf6 gravel. Eolian unit Qe2 consists of multiple remnants of what once had been a widespread sand sheet that covered a large portion of the distal Qf5 surfaces (Figs. 1, 5). Distal Qf7 and Qf6 deposits are deeply inset into and truncate the eastern and southern margins of the Qe2 complex as well as distal Qf5 units. Established vegetation on much of the Qe2 surface remnants indicates that these sand deposits are currently stabilized. Four IRSL dates—3500 ± 220 yr B.P. (0.5 m deep), 3700 ± 425 yr B.P. (8.3 m deep), 8420 ± 795 yr B.P. (2.1 m deep), and 4074 ± 334 yr B.P. (8 m deep; Clarke, 1994; Edwards, 1993)—were obtained from an extensive complex of Qe2 sands that overlies distal Qf5 fans (Figs. 4, 5).

Unit Qf6 is a widespread deposit and represents a significant period of Holocene fan aggradation. The Qf6 unit is an extensive fill terrace deposit along wash channels in proximal fan areas, extends well into mountain front embayments, and extensively overlaps older deposits in distal fan areas. The PM and QM Qf6 deposits consist of poorly to moderately stratified pebble-gravel to well-stratified sandy-pebble braided stream and sheetwash deposits with a subtle bar-and-swale microtopography of low relief. The VX and LS Qf6 deposits consists of well-stratified pebble-gravel braided stream deposits and have a pronounced bar-and-swale microtopography. The PM and QM Qf6 fan deposits have a finer texture than that of the VX and LS Qf6 and Qf7 fan deposits. Soils formed on the Qf6 generally have thin (1–2 cm) Av horizons and weak Bwk horizons with stage I to II carbonate morphology (Table 1). Desert pavement development is very

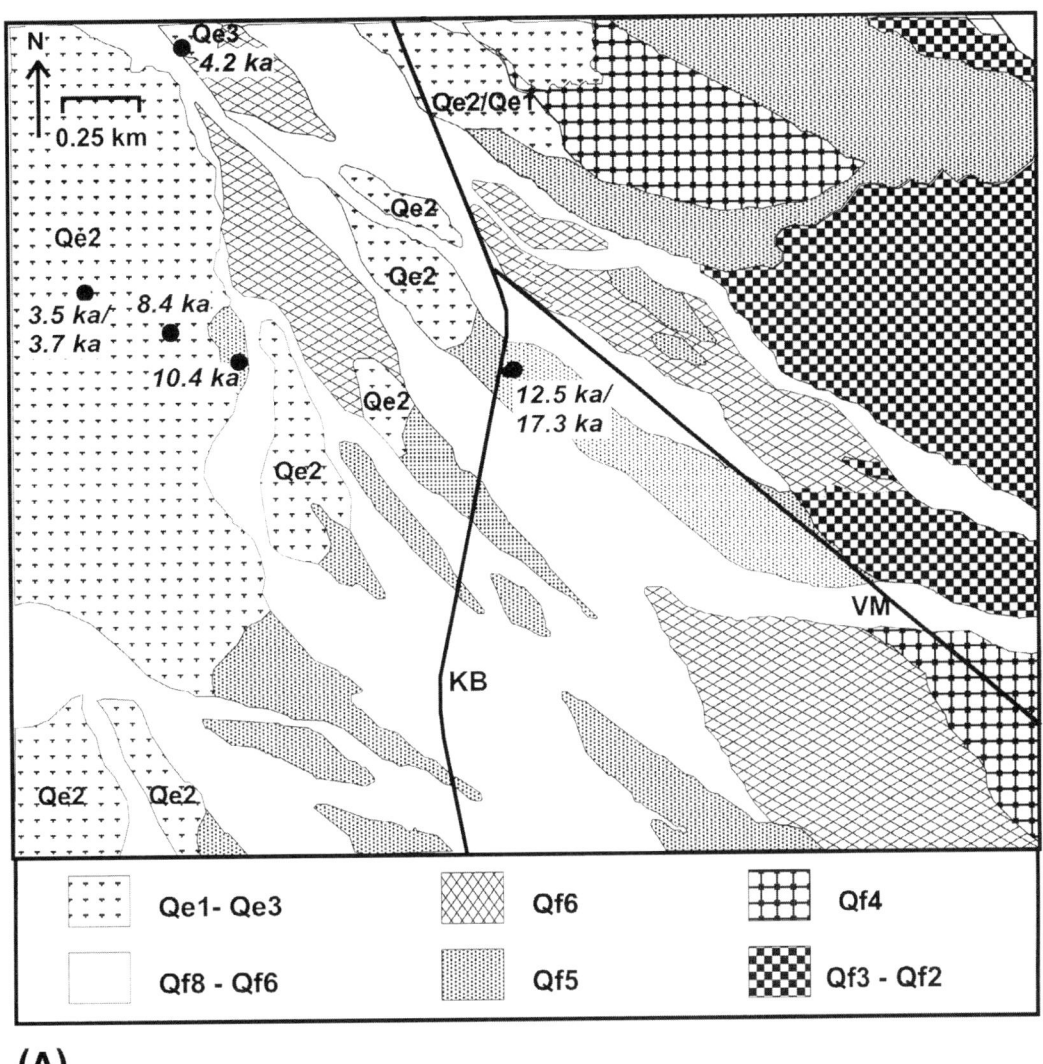

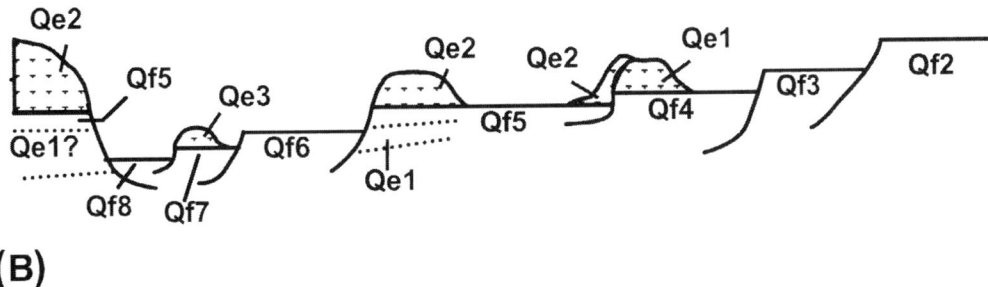

Figure 5. A: Map of Quaternary depositional units along part of the LS fan sequence. Map units defined in Table 1. Italic-labeled dots show location and age of infrared stimulated luminescence sample sites shown in Figure 4. Abbreviations: KB—Kelbaker road and VM—Vulcan Mine access road. B: Generalized schematic cross section of alluvial fan and eolian units shown in A.

weak to moderate depending on the rock type (Table 1). Unit Qf6 can be distinguished readily from unit Qf7 by the accumulation of incipient varnish on siliceous pavement clasts and the weak degree of dissolution of limestone clasts on Qf6 surfaces and the development of a weak B horizon. The oldest radiocarbon dates on pedogenic carbonate range from 6190 cal. yr B.P. (5380 ± 80 ^{14}C yr B.P.) for soils of the PM sequence to 4490 cal. yr B.P. (4010 ± 70 ^{14}C yr B.P.) for soils of the LS sequence (Fig. 4). Modeled radiocarbon ages range from ca. 8 to 11 ka for the PM Qf6 soils and from 7 to 8 ka for the LS Qf6 soils. Geomorphic relations and IRSL ages for units Qe3 and Qe2 suggest that deposition of unit Qf6 may have begun just before 4200 yr B.P. and continued as late as ~3500 yr B.P. Soil stratigraphy suggests that PMP unit Qf6 is correlative with SMP unit Qf4. Radiocarbon ages suggest that SMP unit Qf4 was deposited after ca. 3600 ka (Fig. 4; Wells et al., 1987). Soils developed in all Qf6 deposits along the Providence Mountains piedmont, as well as soils formed in the late Holocene Qf4 deposits along the Silver Lake playa, have only weakly developed Bw horizons that overlie weak Bk or Ck horizons with stage I or II carbonate morphology.

Latest Pleistocene to early Holocene deposits

Alluvial fan unit Qf5 is a widespread deposit that extends far into distal fan areas (Figs. 1, 5). The PM Qf5 deposit is a noticeably coarse-textured alluvial unit that generally consists of pebble-gravel to pebble-cobble, poorly to moderately stratified sheetflood, sheetwash, and debris flow deposits. The surface of the PM Qf5 deposits generally consists of (1) large, elongated lobes or fields of cobbles and boulders (boulders up to 2–3 m in length are not uncommon in some of the larger depositional lobes), (2) remnants of debris flow levees, and (3) bouldery lobes that are 5–8 km downstream of the fan apex. These features suggest that at least the surface sediments of unit Qf5 were the result of frequent debris flow deposition. The VX and LS Qf5 deposits are moderately to well stratified and have surfaces with pronounced planar and trough crossbedding indicating that deposition was largely by ephemeral braided streams. Deposits of Qf5 age were not recognized within the QM fan sequence and may have been buried or stripped by younger deposits or may have never been deposited. Evidence of the possible widespread burial of QM Qf5 deposits is that the surface of extensive QM Qf6 deposits is only 1 or 2 m below the surface of the Pleistocene (pre-Qf5) deposits in QM fan environments. Bull (1991) notes that drainage basins of quartz monzonites appear to have produced a large pulse of sediment that overwhelmed valley floors resulting in the burial of Pleistocene and early Holocene deposits by nearly continuous deposition into the middle and late Holocene.

Soils on Qf5 deposits are moderately developed with thick (3–5 cm) Av horizons, either Bw or weak Bt horizons, and underlying Bk horizons with stage I to III carbonate morphology (Table 1). Development of desert pavement ranges from moderate to strong, with clasts that are moderately to strongly varnished; and the soil has a noticeable reduction in the depositional bar-and-swale microtopography. Unit Qf5 can be distinguished readily from younger units based upon soil and pavement development.

Distal Qf5 deposits that are marginal to the Kelso Dunes are elongated and linear and form distinct, ramp-like surfaces that are several meters above younger fan deposits (Figs. 1, 5). These ramp-like features are capped by a 0.5–1 m thick layer of moderately stratified gravel and cobbles that overlies interstratified eolian sands, sandy-pebble alluvium, and lenses of gravel-cobble alluvium. These ramps are interpreted to be the remnants of Qf5 axial washes that dissected large Qe1 distal sand sheets or dunes that prograded eastward, away from the Kelso Dunes. Subsequent erosion of the Qe1 sediments created inverted topography with axial Qf5 channel deposits forming elongated fan surfaces elevated above the modern washes.

Sand sampled from within Qf5 alluvium yielded an IRSL age of 10,410 ± 890 yr B.P., and sand sampled from within the surface layer of well-stratified cobbles and gravel yielded an IRSL age of 12,460 ± 1151 yr B.P. (Fig. 4). Soil stratigraphy suggests that unit Qf5 is correlative with unit Qf2 along the Soda Mountains and that Qf2 was deposited between ~9380 cal. yr B.P. (8350 ± 300 ^{14}C yr B.P.) and 14,020 cal. yr B.P. (12,020 ± 130 ^{14}C yr B.P.) (Fig. 4; Wells et al., 1987). The oldest radiocarbon dates on pedogenic carbonate range from 21,640 cal. yr B.P. (18,120 ± 150 ^{14}C yr B.P.) for soils of the PM sequence to 19,200 cal. yr B.P. (16,310 ± 60 ^{14}C yr B.P.) for soils of the LS sequence (Fig. 4). Modeled radiocarbon ages range from ca. 29 to 36 ka for the PM Qf7 soils and from 27 to 33 ka for the LS Qf7 soils.

Eolian deposit Qe1 occurs as a thin, discontinuous sand sheet that overlies Pleistocene fan deposits (pre-Qf5 deposits) at several locations on the piedmont (Figs. 1, 5) and underlies Qf5 deposits. As discussed above, the lateral distribution of Qe1 sands and the elevated, ramp-like features of distal Qf5 deposits indicate that unit Qe1 may have covered extensive portions of the lower piedmont at one time. An IRSL date of 16,830 ± 1465 yr B.P. was obtained from the base of a unit Qe1 remnant that overlies pre-Qf5 fan deposits. The presence of a well-developed soil, which has strongly developed Bwk horizons with 7.5 YR hues, indicates the long-term stability of the sand sheet where this IRSL date was obtained. Sand collected from a 1.5 m thick layer of eolian sand stratified with scattered lenses of gravel and cobbles and that lies below the Qf5 alluvium yielded an IRSL age of 17,300 ± 1935 yr B.P. (Fig. 5; Clarke, 1994).

Evaluation of ages for late Quaternary deposits

Although they are in correct relative order and generally in the same age range (i.e., Holocene and latest Pleistocene), radiocarbon dates of pedogenic carbonate do not agree with (1) age estimates for alluvial and eolian units based on the IRSL dates discussed above (2) current knowledge of soil-forming rates in the Mojave Desert (McFadden et al., 1989, 1992; Reheis et al., 1989, 1991; Harden et al., 1991), and (3) stratigraphic correlations with nearby and relatively well dated piedmont deposits (Wells et al., 1987). The IRSL technique used has been validated with accurate ages derived for eolian sands bracketing Mount Mazama tephra in Oregon (Clarke, 1994) and has provided stratigraphically reasonable dates elsewhere in the Mojave Desert (Rendell and Sheffer, 1996; Clarke et al., 1996).

Modeled radiocarbon ages for pedogenic carbonate may be older than the actual age of the soil if (1) noncontinuous deposition of pedogenic carbonate occurred or (2) radiometrically dead carbon from limestone were incorporated (Amundson et al., 1994; Wang et al., 1996). Both modeled and measured radiocarbon ages would overestimate the actual age of the soil and alluvium if the original alluvium contained old organic matter (Wang et al., 1996). Possible contamination with radiometrically dead carbonate from limestone was shown not to be a problem in the dates reported in this study (Wang et al., 1996). The other two potential sources of error yielding ages that are too old, however, are possible.

Noncontinuous deposition of pedogenic carbonate occurred in these soils. Pedogenic carbonates below 75 cm in the Qf5 and Qf4 soils, where the oldest radiocarbon ages occur, would have experienced the greatest degree of noncontinuous deposition because of changes in the flux of water and carbonate. Numerical modeling of soil-water balance indicates that most of the pedogenic carbonate below ~75 cm would have accumulated early in the development of soils on the Qf5 surfaces. This is because the flux of water and carbonate below 75 cm would have significantly decreased during the Holocene because of a drier climate coupled with textural development of B and Av horizons (McDonald, 1994; McDonald et al., 1996). Carbonate accumulation that predominantly occurred early in the period of soil development would result in a strong overestimation of modeled radiocarbon ages (Wang et al., 1996).

Alluvium in which the dated soils formed probably contained old, preexisting organic matter, resulting in model and measured radiocarbon dates that overestimate the true age of the Qf7 through Qf5 deposits. Alluvium would have been largely derived from deposits where accumulation of soil organic matter would have occurred, including soils formed in hillslope colluvium and floodplain sediments (Wells et al., 1987; Bull, 1991; Harvey et al., 1999). Radiocarbon dates of ca. 4410 and 5110 ^{14}C yr B.P. (4980–5900 cal. yr B.P.) on pedogenic carbonates in Qf7 fan deposits (Fig. 4) support this interpretation. Soil and pavement of the Qf7 fans have only minimal development with weak Av horizons and small discontinuous patches of incipient pavement (McDonald, 1994; McDonald and McFadden, 1994). Further, limestone surface clasts show no signs of surface dissolution or pitting; and most of the Qf7 deposits lie within a few decimeters of active wash channels. Radiocarbon age estimates for Qf6 are similar to the radiocarbon ages for Qf7; however, the stronger degree of soil and pavement development of Qf6 relative to Qf7 clearly indicates that Qf6 is at least a few thousand years older. Together, these soil and geomorphic features do not support an age of 4–5 ka (model ages of 4–8 ka) for the Qf7 soil and surfaces but are consistent with radiocarbon ages that are too old if the decomposition of older, preexisting organic carbon is incorporated into the pedogenic carbonates.

The radiocarbon content of a sediment from a pair of modern washes in the Providence Mountains yielded modern ages (Wang et al., 1996); however, these samples were derived from washes in the distal fan environment. It is unlikely that the organic carbon content of these washes accurately represents the organic carbon content that would have been associated with alluvium derived from mountain basins during deposition of older alluvial units. Radiocarbon dating of pedogenic carbonate in soils along the White Mountains in Nevada and California also resulted in an overestimation of the age of deposits independently dated by conventional radiocarbon methods due to the inclusion of older organic carbon (Pendall et al., 1994). In contrast, radiocarbon ages of pedogenic carbonate for soils formed in lacustrine beach gravel along the margin of Silver Lake playa yielded generally similar age estimates as radiocarbon dates obtained by conventional methods (McFadden et al., 1992). The closer agreement between these dating techniques probably reflects the fact that deposits of lacustrine beach gravel are not likely to contain abundant detrital organic matter that can be incorporated into pedogenic carbonates.

Because at least two of the major assumptions governing the diffusion-reaction model for radiocarbon dating of pedogenic carbonates are not validated, we do not view the modeled radiocarbon ages as reliable. These dates significantly overestimate the ages of the soil and alluvium. It is also likely that the measured radiocarbon ages on pedogenic carbonates overestimate the true age of Qf7 through Qf4 soils due to contamination by older organic carbon.

REGIONAL STRATIGRAPHIC CORRELATIONS AND GEOMORPHIC HISTORY AMONG LATE QUATERNARY DEPOSITS

Stratigraphic and age relations among late Quaternary alluvial and eolian units along the Providence Mountains piedmont indicate that most of these units are correlative with alluvial and eolian units along the Soda Mountains piedmont (Fig. 4). Luminescence dates from the Providence Mountains piedmont and conventional radiocarbon dates from the Soda Mountains piedmont and nearby playas provide adequate age control linking general periods of deposition between these piedmonts as well as with late Quaternary climatic events. Stratigraphic correlations shown in Figure 4 are strengthened by the strong similarity of PDI values and overall degree of soil and desert pavement development in fan deposits on both piedmonts. The similarity in PDI values is significant given that the PDI values from outlying areas are derived from soils described by several different scientists and that variability among PDI values can occur as a result of differences in individual methods (Reheis et al., 1989). Similar PDI values among different study areas indicates in part that progressive changes in soil morphology, such as development of structure and texture in B horizons, result in similar, but specific changes in soil development that can be readily recognized across the region by different scientists. In other words, the strongly similar degrees of soil morphology as represented by PDI values supports that these soils represent similar lengths of soil formation and provide a strong foundation for establishing correlation among alluvial units across the region.

Deciphering the late Quaternary geomorphic record in the east Mojave is rooted strongly in the record of pluvial Lake Mojave (Ore and Warren, 1971; Wells et al., 1987; Brown et al., 1990; Harvey and Wells, 1994, this volume; Harvey et al., 1999; Wells et al., this volume). A detailed history of pluvial Lake Mojave is provided by a wide range of paleoenvironmental data and age estimates on lacustrine and eolian deposits (Wells et al., 1987, 1989; Brown et al., 1990; Wells et al., this volume). Fluctuating lake levels resulted in two high lake stands bracketed by three periods of intermittent lake activity (Figs. 4, 6). Calibrated ^{14}C ages indicate that pluvial lake activity began ca. 24.5 ka (Intermittent Lake I) and ended by ca. 9.4 ka (Intermittent Lake III) with high lake stands occurring between ca. 21.5 and 19.2 ka (Intermittent Lake I) and between ca. 16.4 and 13.2 ka (Intermittent Lake II). The following discussion relates regional eolian and alluvial events to the paleoenvironmental record based on the history of pluvial Lake Mojave.

Eolian events

Deposition of unit Qe1 along both mountain piedmonts appears to have occurred during approximately the same time interval of Intermittent Lake II (Figs. 4, 6). Stratigraphic relations along the Soda Mountain piedmont indicate that deposition of unit Qe1 occurred before deposition of unit Qf2 (Wells et al., 1987). Moreover, the absence of buried soil features in the top of unit Qe1 suggest that deposition of unit Qe1 may have continued nearly until deposition of unit Qf2. A period of eolian activity coinciding with Intermittent Lake II is recorded within lacustrine deposits that just underlie sediments dated at ca. 17.6 ka (Cores SIL-I, SIL-M, Wells et al., 1989; Brown et al., 1990; Wells et al., this volume). Together, these features suggest that the onset of unit Qe1 deposition coincided with fluctuating lake levels associated with Intermittent Lake II between ca. 18.3 and 17.6 ka and that eolian deposition probably continued to some degree during the Intermittent Lake II phase (Wells et al., 1987, 1989, this volume). Fluctuating lake levels and occasional periods of drying during lake recession would enhance deflation of sand and silt from distal fluvial deposits and lake basins, increasing widespread eolian deposition along mountain piedmonts around Lake Mojave (Wells et al., 1987; Lancaster and Tchakerian, 1996, this volume).

Luminescence dating and stratigraphic relations indicate that deposition of PMP unit Qe1 began before 17.3 ka and ended before deposition of PMP unit Qf5 dated at 12.5 ka. Deposition of PMP unit Qe1 would have responded to the same increase in the supply of eolian sediment from the Lake Mojave basin that caused SMP unit Qe1. The Providence Mountains unit Qe1 is a result of an eastward expansion onto the piedmont of sand that forms the present-day Kelso Dunes and Devils Playground. Activity of both the Kelso Dunes and Devils Playground is linked

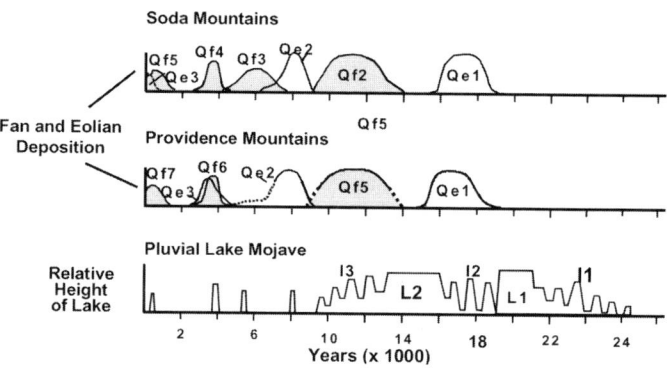

Figure 6. Summary of periods of fan and eolian deposition along the Providence Mountains piedmont and near Silver Lake–Soda Mountains and fluctuations of pluvial Lake Mojave during the last 23,000 yr. Plots of geomorphic events imply general interval of time and levels of activity but do not imply relative volumes of sediment.

to the supply of sediment from the pluvial Lake Mojave basin (Lancaster, 1994; Lancaster and Tchakerian, this volume).

Buried soils have not been identified between units Qe1 and Qf5 along the Providence Mountain piedmont, suggesting that deposition of unit Qe1 may have continued until deposition of unit Qf5. This lack of a buried soil, however, could also be due to erosional truncation of unit Qe1 during deposition of unit Qf5.

Deposition of unit Qe2 along both mountain piedmonts appears to have begun about the same time and appears to be related to an increase in the supply of eolian sediment associated with lake recession at the end of pluvial Lake Mojave (Figs. 4, 6). Stratigraphic relations among dated shoreline features indicate that deposition of unit Qe2 along the Soda Mountain piedmont began after ca. 9.4 ka and ended before 3.9 ka (Wells et al., 1987, 1989). Soil stratigraphy of shoreline and middle Holocene fan deposits suggests that deposition of unit Qe2 probably ended before ca. 6.8 ka (McFadden et al., 1992). IRSL dates and stratigraphic relations indicate that deposition of unit Qe2 along the Providence Mountains piedmont began after 10.4 ka and before 8.4 ka and was stabilized ca. 3.5 ka. An alternative explanation is that PMP unit Qe2 was deposited closer to ca. 8.4 ka and that the surface of the dated Qe2 dune complex was reactivated during the late Holocene resulting in the deposition of PMP unit Qe3 beginning at ca. 4.2 ka. The three youngest IRSL ages may instead date a layer of PMP unit Qe3 sands that overlie and truncate older Qe2 sands. Late Holocene reactivation of PMP unit Qe2 is supported by luminescence ages for sand ramps along nearby Old Dad Mountain (Fig. 1) that indicate periods of increased eolian activity between ca. 6.7 and 3.3 ka (Rendell and Sheffer, 1996; Lancaster and Tchakerian, 1996).

Alluvial fan events

Alluvial fan deposits that predate the latest Pleistocene occur along both the Soda Mountains and Providence Mountains piedmonts. A well-developed soil and desert pavement on SMP unit Qf1 is truncated by shoreline features created during highstands of pluvial Lake Mojave, indicating that SMP unit Qf1 predates pluvial lake activity during the latest Pleistocene and is older than ca. 24.5 ka. Soil stratigraphy suggests that SMP unit Qf1 is correlative with PMP unit Qf4, which was deposited before PMP unit Qe1 and before ca. 17.3 ka.

Stratigraphic and age relations among eolian and alluvial units indicate that the deposition of alluvial fan unit PMP Qf5 also appears to have coincided with deposition of SMP unit Qf2 (Figs. 4, 6). Stratigraphic relations indicate that deposition of both unit SMP Qf2 and PMP unit Qf5 occurred after deposition of unit Qe1 and before deposition of unit Qe2 (Figs. 4, 6). Geomorphic relations near Silver Lake playa suggest that deposition of unit SMP Qf2 grades to shoreline B or that deposition of SMP unit Qf2 began shortly after creation of shoreline B (Wells et al., 1987). Shoreline B formation occurred during Intermittent Lake II, and the oldest radiocarbon date for shoreline B deposits is ca. 14.0 ka. (Fig. 4; Wells et al., 1989). Stratigraphic relations and radiocarbon dating of shoreline features indicate that deposition of SMP unit Qf2 ended ca. 9.4 ka (Wells et al., 1989; Brown et al., 1990). Stratigraphic relations and IRSL dates discussed above indicate that deposition of PMP unit Qf5 occurred during a similar time interval with deposition beginning after 16.8 ka and ending before 8.4 ka (Fig. 4). Strongly similar degrees of soil development between SMP unit Qf2 and PMP unit Qf5 further support that these alluvial deposits are correlative (Fig. 4).

Stratigraphic and age relations among eolian and alluvial units suggest that the deposition of late Holocene alluvial fan units along both the Providence Mountains and Soda Mountains piedmont also may have occurred during similar time intervals (Figs. 4, 6). Deposition of PMP unit Qf6 occurred between ca. 4.2 and 3.5 ka. Deposition of SMP unit Qf3 occurred between ca. 9.4 ka and 3.6 ka, and deposition of SMP unit Qf4 occurred after ca. 3.6 ka. Geomorphic relations suggest that deposition of unit SMP Qf3 may have overlapped with deposition of unit SMP Qe2, which would place deposition of unit SMP Qf3 closer to ca. 6 ka (Brown et al., 1990; McFadden et al., 1992). Soil development reflected by PDI values suggests that SMP unit Qf4 is correlative with PMP unit Qf6 (Fig. 4). Both PMP unit Qf7 and SMP unit Qf5 have minimal soil development and lack desert pavement with interlocking clasts, suggesting that these alluvial deposits may be correlative. Lack of adequate age control, however, prevents confirming that these units may have been deposited during similar time intervals.

REGIONAL DEPOSITION OF ALLUVIAL FANS IN RESPONSE TO THE PLEISTOCENE-HOLOCENE CLIMATIC TRANSITION

Deposition of PMP unit Qf5 and SMP unit Qf2 between ca. 14 and 9.4 ka (Figs. 4, 6) indicates that this major period of fan deposition across the east Mojave was a region-wide event. A regional period of deposition is further supported by stratigraphic correlations and general age estimates (based upon the degree of soil and desert pavement development) for other fan deposits across the Mojave (McFadden et al., 1989; Bull, 1991; Wells et al., 1990; Harden et al., 1991). Region-wide fan deposition implies that this activity was a result of some aspect of climatic variation rather than active tectonics or a complex response to crossing intrinsic geomorphic conditions (Schumm, 1977; Blair and McPherson, 1994).

Relation between regional fan deposition and late Quaternary climate

Age relations of PMP unit Qf5 and SMP unit Qf2 indicate that this pronounced episode of fan deposition was not simply a result of greater effective moisture during the latest Pleistocene, as suggested by several studies (Melton, 1965; Lustig, 1965; Ponti, 1985; Christenson and Purcell, 1985; Dorn et al., 1987; Dorn, 1994; Blair and McPherson, 1994). First, fan deposition began near the end of a period of greater effective moisture, coinciding with a gradual decrease in effective moisture and pluvial lake levels during the Pleistocene-Holocene transition (Figs. 4, 6). Sedimentologic records from pluvial Lake Mojave indicate that lake levels began to significantly fluctuate at ca. 13.2 ka, with the nearly complete disappearance of a long-standing pluvial lake by ca. 9.4 ka (Wells et al., 1989, 1994, this volume; Enzel et al., 1989, 1992). Other pluvial lakes in the vicinity of the Mojave Desert also show gradual drying in conjunction with a transition to a relatively drier climate during this time interval (Quade, 1986; Smith and Street-Perrott, 1983). Paleobotanical studies for several mountains in the Mojave Desert also indicate that a gradual decrease in vegetation and effective soil moisture occurred during this time interval (Van Devender et al., 1987; Harvey et al., 1999).

Second, fan deposition appears to have been minimal between ca. 24 and 14 ka when paleoenvironmental data also indicate that there was a considerable increase in effective moisture. Paleobotanical and soil hydrology data suggest that precipitation in the Sonoran and Mojave Deserts may have increased by at least 40–150% above current mean annual rainfall during the latest Pleistocene (Spaulding, 1985; Van Devender et al., 1987; Phillips, 1994; McDonald et al., 1996). Evidence that both the Providence and Soda Mountains were subjected to greater effective moisture during the latest Pleistocene is provided by linkages among anomalous, historic atmospheric circulation patterns over the north Pacific Ocean, flooding of the Mojave River, and development of ephemeral lakes that occupied the nearby Silver Lake playa (Enzel et al., 1989, 1992). Historical flooding and ephemeral lakes correlate with years of regional changes in the oceanic-atmospheric circulation patterns. These changes generally consisted of a southern shift in the Aleutian low and subtropical jet stream. The southern shift of the jet stream resulted in a considerable increase in the frequency of frontal storms sweeping in from the Pacific across southern California. Similar but more persistent atmospheric circulation patterns appear to have resulted in the formation of pluvial Lake Mojave during the latest Pleistocene (Wells et al., 1989; Enzel et al., 1989, 1992; Enzel and Wells, 1997). Although formation of these perennial lakes was due primarily to increased rainfall and flooding in the Transverse Ranges (~250 km to the west of the Providence Mountains), a considerable increase in annual precipitation probably also occurred across the Mojave Desert. Evidence for this is the considerable increase in regional rainfall associated with years of historic flooding of the Silver Lake playa, indicating that sufficient moisture was delivered across the coastal mountains and into the Mojave Desert (Fig. 3). Similar increases in annual rainfall due to increases in frontal storm activity likely occurred during the latest Pleistocene as well.

An increase in ephemeral channel flow most likely would have coincided with regional increases in effective moisture. For example, there was a slight increase in ephemeral stream activity along the Providence Mountains during the winters of 1992–1993 and 1997–1998 when there was a considerable increase in regional rainfall associated with anomalous storm events discussed in the preceding paragraph. There appears to have been minimal sediment transport associated with these years with above average rainfall, however. Increased channel flow throughout most of the latest Pleistocene without a corresponding increase in alluvial sedimentation implies a high water-to-sediment ratio. The predominance of fan aggradation coinciding with the Pleistocene-Holocene transition indicates a regional geomorphic response leading to a significant increase in the flux of sediment into ephemeral drainages across the region discussed in more detail below (Wells et al., 1987; Harvey and Wells, 1994; Harvey et al., 1999).

The relative influences of drainage basin morphometry and bedrock lithology on fan deposition

Perhaps as significant as alluvial fan deposition coinciding with the timing of the Pleistocene-Holocene climatic transition is the fact that regional fan aggradation occurred despite considerable geomorphic differences among drainage basins in the Providence and Soda Mountains. Generally, drainage basins in the Providence Mountains are considerably larger and at higher elevations than drainage basins in the Soda Mountains. For example, one of the largest drainage basins along the western flank of the Providence Mountains has a basin area of ~8.8 km^2 and ranges in elevation from ~1200 to 2000 m (Fig. 2). In contrast, one of the largest basins along the Soda Mountains has an area of ~1.9 km^2 and ranges in elevation from ~100 to 300 m. The catchments within the Soda Mountains are also generally steeper, and the main trunk streams have much smaller valley width-to-height ratios compared to catchments and trunk streams in the Providence Mountains. Another key geomorphic contrast is the wide variation in rock types that make up the drainage basins along the Providence Mountains. Stratigraphic sequences of fan deposits along the Providence Mountains piedmont that are physically correlative have been derived from drainage-basin source areas consisting of diverse rock types such as coarse-grained plutonic, microcrystalline siliceous, and massive carbonate rocks. The one exception is the absence or lack of exposure of QM unit Qf5 (Table 1), which is derived from quartz monzonite.

The geomorphic diversity among drainage basins results in an equally wide range of key variables controlling sediment supply. These include sediment production, sediment size, channel discharge, water-to-sediment ratio, and sediment storage within each basin. For example, sediment production and hillslope hydrology can strongly vary among drainage basins composed of either coarse-grained plutonic and carbonate rock types (Bull, 1991). If sediment supply to alluvial fans were primarily due to an increase in effective moisture and stream activity during the latest Pleistocene, it is reasonable to expect that the timing of fan aggradation would have varied considerably among the different basin rock types across the Providence Mountains and between the diverse basin morphometry of the Providence and Soda Mountains. This is because varied responses in sediment production, supply, and transport could be expected from such a diversity in drainage basin geomorphology. Our conclusion that the timing of fan aggradation along these two mountain piedmonts was similar and occurred between 14 and 9 ka indicates that alluvial fan deposition, at least in the east Mojave, Desert was not due primarily to an increase in effective moisture and concomitant increase in channel discharge associated with a period of increase pluvial activity between ca. 24 and 9 ka.

REMAINING QUESTIONS: THE INTERRELATED ROLES OF CLIMATE CHANGE, VEGETATION RESPONSE, AND SEDIMENT SUPPLY IN REGIONAL FAN AGGRADATION

Pronounced aggradation of alluvial fans across the desert southwestern U.S. between ca. 14 and 9.4 ka, therefore, has been attributed to an increase in sediment yield corresponding with time-transgressive changes in climate and vegetation during the Pleistocene-Holocene transition (Wells et al., 1987, 1990; Bull, 1991; Harvey and Wells, 1994; McDonald, 1994; McDonald and McFadden, 1994; Reheis et al., 1996; Harvey et al., 1999). Specifically, a cycle of fan deposition was triggered as a result of the transition from a wetter to a relatively drier climate when a reduction in effective soil moisture resulted in a reduction in the vegetative cover on hillslopes that, in turn, resulted in an increase in soil erosion and a concomitant increase in sediment supply. Numerous paleobotanical studies indicate that throughout the Mojave Desert there was a gradual upward shift in the elevation of vegetation coinciding with a decrease in effective soil moisture through the end of the latest Pleistocene and into the early Holocene, between ca. 14 and 8 ka (e.g., Spaulding, 1985; Van Devender et al., 1987; Harvey et al., 1999; Wigand and Rhode, 2002). Fan entrenchment and stabilization of the alluvial fan surface followed when a decrease in sediment supply from increasingly barren hillslopes resulted in an increase in steam power leading to incision of the alluvium.

There are two critical factors among all the studies of alluvial fans cited above. First is that an increase in sediment yield resulted from time-transgressive changes in vegetation and a concomitant decrease in the stability of soils along drainage basin hillslopes. Second is that there was a considerable increase in sediment stored on drainage basin hillslopes as soil and colluvium as a result of greater effective moisture that enhanced weathering of basin bedrock during the latest Pleistocene. An assessment of vegetation, geomorphic, and climatic attributes in the Providence Mountains provides an opportunity to conduct a generalized evaluation of hillslope dynamics during the Pleistocene-Holocene climatic transition. Results of this evaluation, discussed in more detail below, suggest that increased sediment yields in the Providence Mountains cannot be attributed primarily to time-transgressive changes in vegetation and increasing erosion of soils along basin hillslopes.

Vegetation change and hillslope erosion

The role of time-transgressive changes in vegetation along basin hillslopes in triggering an increase in sediment yield and fan aggradation during the Pleistocene and Holocene transition along the Providence Mountain piedmont may have been limited compared to hillslope response to vegetation change cited in other studies. Field observations of current soil and colluvial cover across hillslopes along several PM drainage basins (lying between ~1100 and 1700 m) indicate that most of the bedrock along these slopes is covered by between 10 and 50 cm of soil and colluvium. Vegetation ranges from that dominated by creosotebush and white bursage to blackbrush and Mojave yucca (*Yucca schidigera*) at higher hillslope elevations (juniper/pinyon woodland is largely along north-facing slopes). Further, plant density is high, with much of the soil surface protected by shrub canopies (Fig. 7). The fact that an appreciable cover of soil and colluvium are present indicates that soil is probably stable under a temperate desert scrub. In other words, hillslope areas that probably underwent a change in vegetation from predominantly juniper/pinyon woodland to temperate desert scrub can still develop and maintain an appreciable cover of soil and colluvium. Thus, modern soil and vegetation relations suggest that it is unlikely that changes in vegetation cover alone would have been a primarily responsible for an increase in soil erosion and sediment yield.

Another key relationship is that a reduction in plant cover across desert hillslopes may not result necessarily in an increase in soil erosion. First, soils along basin hillslopes during the late Pleistocene probably contained abundant clasts, including the soil surface, from local weathering of the bedrock. Although the effect of surface stones on surface runoff is complex and depends on such factors as stone size, percent cover, and the degree to which a stone is imbedded into the soil surface, surface stone cover has been shown to increase infiltration and decrease surface runoff by providing surface roughness limiting overland flow and increasing surface microtopography, which enhances infiltration (Yair, 1987; Poesen, 1992; Abrahams et al., 1994). Stone cover can also protect exposed soil by limiting erosion from raindrop impact. Second, hillslope runoff derived from colluvium

Figure 7. Photograph of hillslope at ~1500 m elevation showing typical vegetation cover for PM basin hillslopes. The majority of vegetation in view consists of blackbrush, Mojave yucca, and galleta grass (*Hilaria jamesii*). Hillslope covered with ~1 m of colluvium and soil.

has been shown to be highly limited except under exceptionally high rainfall intensities because of the overall high rate of surface infiltration relative to exposed bedrock areas. Moreover, surface runoff generated by exposed bedrock along hillslope ridgelines can be absorbed entirely by colluvium downslope (Yair, 1987, 1992; Yair and Enzel, 1987). The potentially high infiltration capacity of hillslope soil and colluvium suggests that changes in either vegetation type and/or density may have had a limited impact on the stability and erosion of soils along basin hillslopes and promoting an increase in sediment yields.

Soil cover and sediment storage along basin hillslopes

Geomorphic factors governing soil formation and the storage of sediment along basin hillslopes would have varied considerably among the different rock types that make up the Providence Mountains suggesting that removal of sediment stored in the form of an extensive soil and colluvial cover due to climatic transition may have had a limited role. Based on current knowledge of weathering processes in desert environments, it is reasonable to conclude that an stable, appreciable cover of soil and colluvium may have developed along many of the basin hillslopes underlain by coarse- to medium-grained plutonic rocks found in the southern part of the Providence Mountains (PM and QM sequences). These rock types are conducive to enhanced weathering, especially grussification, during prolonged periods of greater effective soil moisture (Birkeland, 1984; Bull, 1991). In contrast, the formation of an extensive cover of soils and colluvium stabilized along basin hillslopes underlain by limestone, marble, and volcanic rocks along the northern portion of the Providence Mountains

(VX and LS sequences) was probably limited. Soil formation along these hillslopes would be limited relative to that of the plutonic rock types because of the resistance of microcrystalline to massive carbonate and siliceous rocks to disintegrate into fine-textured colluvium that would enhance soil formation. Although the addition of dust would enhance the accumulation of a fine-textured matrix in soils derived from carbonate and volcanic materials, especially in conjunction with increased eolian activity associated with deposition of unit Qe1, it is unknown if the rate of accumulation along basin slopes would have been sufficient to dramatically impact the character of these soils.

Development of an extensive cover of stable soil and colluvium would be restricted further on LS and VX basin hillslopes compared to soils formed on the PM basin hillslopes due to considerable differences in basin topographic relief. The mean hillslope angle for LS and VX basins exceeds 40° and is ~10°–20° steeper than PM hillslopes (Fig. 8). We are not implying that soil cover would be absent on LS and VX hillslopes of the Providence Mountains but that the soil thickness and lateral extent would probably be substantially less than that found on hillslopes underlain by plutonic rocks. Field observations of the modern soil cover suggest that the overall degree of soil cover among the LS drainage basins is considerably less than that of the PM drainage basins. We hypothesize that sediment derived from weathering of LS and VX basin hillslopes was largely stored as talus or colluvial wedges along valley bottom channels rather than as an extensive cover of soil along hillslopes.

Extreme storm events in sediment mobilization

An apparent increase in summer monsoon activity during the early Holocene has been associated primarily with promoting channel incision and fanhead trenching and secondarily with enhanced mobilization of hillslope sediments (Wells et al., 1987, 1990; Bull, 1991; Harvey and Wells, 1994; McDonald, 1994; McDonald and McFadden, 1994; Reheis et al., 1996; Harvey et al., 1999). Increased summer storm activity has been interpreted in nearly all geomorphic and paleobotanical studies in the southwestern U.S. as an increase in summer convective storms. This interpretation is based on an early Holocene increase in the abundance of succulents and C-4 grasses preserved within packrat middens across the Mojave and Sonoran deserts (Spaulding, 1985; Spaulding and Graumilch, 1986; Bull, 1991).

Alternate interpretations are that this increase in succulents and grasses resulted from an increase in (1) late season Pacific frontal storms during April through early June when soil temperatures are warm or (2) an increase in tropical cyclones that occur during late August through early October (Van Devender et al., 1987). The potentially important role tropical cyclones may play has been largely overlooked in paleobotanical reconstructions of the desert southwest (McAuliffe and Van Devender, 1998). Tropical cyclones form in the Pacific off the coast of southern Mexico and occasionally migrate northward away for their normal paths over the Pacific and cross Baja California

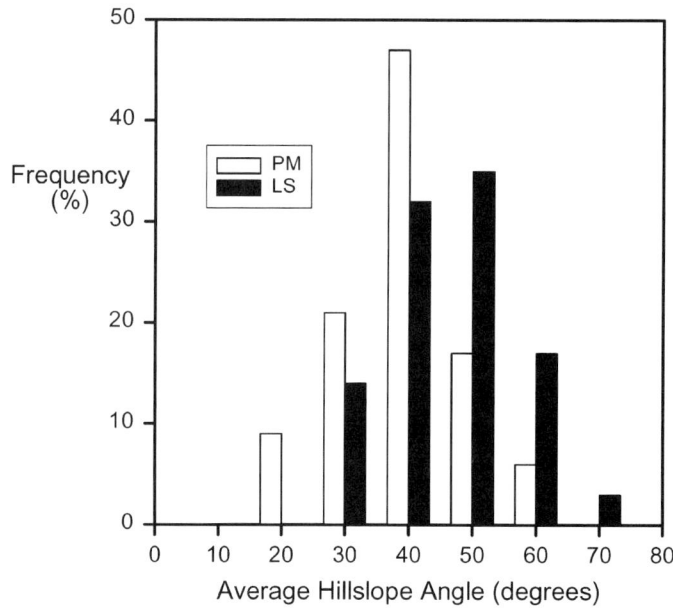

Figure 8. Frequency of hillslope angles for PM and LS drainage basins. Frequency based on measurements of 50 hillslopes for each drainage.

and Sonora Mexico (Ely, 1997). Historic tropical cyclones share strongly similar atmospheric circulation anomalies with those that produce extreme winter storms in southern California and have been responsible for flooding the Mojave River and the Silver Lake playa (Ely et al., 1994). These storms can yield several days of heavy rainfall across wide regions as they dissipate over land. Tropical cyclones have been associated with some of the largest flood events in southern Utah and Arizona (Ely, 1997). Tropical cyclones could have provided a significant source of late summer precipitation to vegetation in the Mojave during the early Holocene for similar reasons as suggested for the Arizona uplands of the Sonoran Desert by McAuliffe and Van Devender (1998). An increase in tropical cyclones, therefore, may have enhanced early Holocene increases in vegetation associated with warm-season rainfall.

Given the characteristics of tropical cyclones discussed in the preceding paragraph, an interesting question is raised: Was deposition of alluvial fans along the Providence Mountains during the Pleistocene-Holocene transition enhanced by an increase in tropical cyclone activity? Postulated changes in atmospheric circulation in the Mojave Desert may have enhanced penetration of tropical air into the desert southwest, which would have driven both convective storms and tropical cyclones (Harvey et al., 1999). A key aspect of tropical cyclones and fan aggradation is that tropical cyclones provide precipitation high in intensity, long in duration, and covering a large area. Long duration storms would provide more sustained channel flow for transporting and depositing alluvium in distal fan areas. As discussed above, distal deposits of the PMP Qf5 that are up to about eight kilometers from the mountain front are largely well bedded alluvium, complete with planar and trough crossbedding, suggesting deposition by sustained braided stream processes rather than short-lived, flashy runoff commonly associated with summer convective storms. Further, precipitation that is of high intensity and long duration would be considerably more effective in mobilizing colluvium that has a high infiltration rate.

We do not imply that summer convective storms were not an important factor in mobilizing sediment but rather raise two critical points. First, mobilization of sediment and transport into distal fan areas probably required an increase in the size and frequency of extreme storm events compared to today's climate. The regional occurrence of large-scale, sheetflood bedforms on pre-Holocene alluvial fans in the Mojave indicate that extreme storms and channel discharge events occurred during the early Holocene (Wells and Dohrenwend, 1985; Wells et al., 1987). Second, tropical cyclones may have provided a generally underrecognized source of precipitation that could have had a profound impact on regional fan aggradation. Determining the magnitude of enhanced tropical cyclones in the Mojave and their potential impact on regional fan deposition, if any, will require further investigations, especially involving paleoecology, surface-water hydrology of soils on basin hillslopes, paleoclimatology and climate modeling.

SUMMARY AND CONCLUSIONS

Geomorphic and age relations among alluvial and eolian units along the Providence Mountains and Soda Mountains piedmonts indicate that most of the late Quaternary eolian and alluvial fan units are stratigraphically correlative, were probably deposited during similar time intervals, and represent region-wide changes in geomorphic factors controlling sediment supply, storage, and transport. Similar timing of depositional events among the Providence Mountains and Soda Mountains piedmonts is significant because there is considerable variation in the geomorphology and ecology of the mountain catchments. Drainage basins along the Providence Mountains are considerably higher and larger than basins in the Soda Mountains. Further, there exists strong variation in basin morphology and rock type among the basins in the Providence Mountains. Regional fan deposition occurred regardless of lithology, size, and elevation or topographic relief of drainage basins. Wide variations in sediment production, storage, and transport are likely to be associated with diverse geomorphic settings; however, the best available geomorphic and age relations indicate that major intervals of aggradation transcended all geomorphic variation. Similar deposition intervals across the region imply that the mechanisms driving deposition are extrinsic factors related to climatic variation and not intrinsic factors such as complex response related to crossing inherent geomorphic thresholds or the instability of basin sediments due to tectonic activity.

Stratigraphic and age relations among alluvial and eolian units in the eastern Mojave Desert demonstrate that a regional period of major alluvial fan deposition occurred between ca. 9.4 and 14 ka and that correspond with the timing of the Pleistocene-Holocene climatic transition. This age range indicates that deposition of these fans is not simply a result of greater effective moisture and channel discharge during the last glacial maximum. The regional extent of fan deposition across diverse geomorphic settings strongly supports the models of Bull (1991), Wells et al. (1987), and Harvey et al. (1999) that proposed that a significant period of fan aggradation in the Mojave Desert occurred during the Pleistocene-Holocene transition. The results of this study differ from those cited above; however, in linking climate change to increases in sediment yield and fan aggradation with a concomitant increase in hillslope instability. Geomorphic relations along the Providence Mountains suggest that that changes in vegetation cover during the Pleistocene-Holocene climatic transition along the Providence Mountains may have had a limited impact on hillslope instability and sediment yield because of the (1) inherently high infiltration capacity of coarse-textured soils and colluvium, (2) possible strong spatial variations in the nature of soil cover across hillslopes, and (3) modern vegetation cover appears to provide enough stability for the buildup of soils and colluvium. An increase in sediment yield may instead be largely due to an increase in extreme storm events, perhaps an increase in tropical cyclones. Extreme storms would provide the rainfall intensity and duration to mobilize permeable sediments from mountain catchments and into distal fan areas.

Remaining problems

Although analysis of the timing of alluvial fan deposition along the Providence Mountains raises interesting questions regarding how climatic variation may drive increases in sediment yield and alluvial aggradation, several key problems remain and will require further investigation. Few studies of climatic and geomorphic factors controlling alluvial fan processes have incorporated a geomorphic analysis of hillslope environments similar to that of Harvey and Wells (1994). Further investigations are needed that integrate the spatial and temporal distribution of soils, colluvium, and vegetation across basin slopes and the hydrologic response of these slopes to variations in climate. Such investigations also need to consider if and how interrelated changes in climatic variation, sediment yield, and fan aggradation may have varied fans of Holocene and prelatest Pleistocene age relative to fans deposited during the Pleistocene-Holocene transition. The role of extreme storm events in generating discrete periods of alluvial fan deposition, especially the role of tropical cyclones, also requires further analysis.

Although IRSL ages, reinforced with stratigraphic correlation to alluvial units along the Soda Mountain piedmont the based on soil stratigraphy, provide strong local age control, the radiocarbon ages on pedogenic carbonates remain problematic. A project to date unit Qf5 and older alluvial fan deposits using in situ cosmogenic nuclides in currently under way. We hope that results will clarify age relation of unit Qf5 as well as serve as a tool for testing regional correlation of alluvial fan units.

ACKNOWLEDGMENTS

We thank Nick Lancaster, Michele Clarke, and Ann Wintle for collaboration on the IRSL dates, and Ron Amundson and Yang Wang for collaboration on pedogenic ^{14}C dates. Many thanks to the staff at UCR Sweeney Granite Mountains Reserve and the CSU Fullerton Desert Studies Center for their friendship and logistical support. We also thank NASA (NAG 5 1828), NSF (EAR 9118335), and scholarships to McDonald from the University of New Mexico, Sigma Xi, and the Geological Society of America for funding for this work. Thanks to Steven Reneau for thoughtful comments on an earlier version of this paper. The senior author is especially grateful to Steven Reneau and Arron Yair for many discussions regarding the relationship between climate change, extreme storm events, and alluvial fan response.

REFERENCES CITED

Abrahams, A.D., Howard, A.D., and Parsons, A.J., 1994, Rock-mantled-slopes, *in* Abrahams, A.D., and Parsons, A.J., eds., Geomorphology of desert environments: London, Chapman and Hall, p. 173–212.

Amundson, R., Wang, Y., Chadwick, O., Trumbore, S., McFadden, L., McDonald, E., Wells, S., and Deniro, M., 1994, Factors and processes governing the carbon-14 content of carbonate in desert soils: Earth and Planetary Science Letters, v. 51, p. 761–767.

Birkeland, P.W., 1984, Soils and geomorphology: New York, Oxford University Press, 372 p.

Blair, T.C., and McPherson, J.G., 1994, Alluvial fan processes and forms, *in* Abrahams, A.D., and Parsons, A.J., eds., Geomorphology of desert environments: London, Chapman and Hall, p. 354–402

Brown, W.J., Wells, S.G., Enzel, Y., Anderson, R.Y., and McFadden, L.D., 1990, The late Quaternary history of pluvial lake Mojave: Silver Lake and Soda Lake basins, California, *in* Reynolds, R.E., et al., eds., At the end of the Mojave: Quaternary studies in the eastern Mojave Desert: Redlands, California, Special Publication of the San Bernardino County Museum Association, 1990 Mojave Desert Quaternary Research Center Symposium, May 18–21, 1990, p. 55–72.

Bull, W.B., 1991, Geomorphic responses to climatic change: New York, Oxford University Press, 326 p.

Bull, W.B., and Schick, A.P., 1979, Impact of climatic change on an arid watershed, Nahal Yael, southern Israel: Quaternary Research, v. 11, p. 153–171.

Christenson, G.E., and Purcell, C., 1985, Correlation and age of Quaternary-fan sequences, Basin and Range province, southwestern United States, *in* Weide, D.L., ed., Soils and Quaternary geology of the southwestern United States: Boulder, Colorado, Geological Society of America Special Paper 203, p. 115–122.

Clarke, M.L., 1994, Infrared stimulated luminescence ages from aeolian sand and alluvial fan deposits from the eastern Mojave Desert, California: Quaternary Geochronology, v. 13, p. 533–538.

Clarke, M.L., Wintle, A.G., and Lancaster, N., 1996, Infrared stimulated luminescence dating of sands from the Cronese basins, Mojave Desert: Geomorphology, v. 17, p. 199–206.

Dorn, R.I., 1994, The role of climatic change in alluvial fan development, *in* Abrahams, A.D., and Parsons, A.J., eds., Geomorphology of desert environments: London, Chapman and Hall, p. 354–402.

Dorn, R.I., DeNiro, M.J., and Aije, H.O., 1987, Isotopic evidence for climatic influence on alluvial-fan development in Death Valley: Geology, v. 15, p. 108–110.

Edwards, S.R., 1993, Luminescence dating of sand from the Kelso Dunes, California, *in* Pye, K., ed., The dynamics and environmental context of aeolian sedimentary systems: Geological Society of London Special Publication No. 72, p. 59–68.

Ely, L.L., 1997, Response of extreme floods in the southwestern United States to climatic variations in the late Holocene: Geomorphology, v. 19, p. 175–202.

Ely, L.L., Enzel, Y., and Cayan, D.R., 1994, Anomalous North Pacific atmospheric circulation and large winter floods in the southwestern United States: Journal of Climate, v. 7, p. 977–987.

Enzel, Y., and Wells, S.G., 1997, Extracting Holocene paleohydrology and paleoclimatology information from modern extreme flood events: An example from southern California: Geomorphology, v. 19, p. 203–226.

Enzel, Y., Cayan, D.R., Anderson, R.Y., and Wells, S.G., 1989, Atmospheric circulation during Holocene lake stands in the Mojave Desert: Evidence of regional climate change: Nature, v. 341, p. 44–48.

Enzel, Y., Brown, W.J., Anderson, R.Y., McFadden, L.D., and Wells, S.G., 1992, Short-duration Holocene lakes in the Mojave River drainage basin, southern California: Quaternary Research, v. 38, p. 60–73.

Gile, L., Peterson, F.F., Grossman, R.B., 1966, Morphologic and genetic sequences of carbonate accumulation in desert soils: Soil Science, v. 101, p. 347–360.

Harden, J.W., 1982, A quantitative index of soil development from field descriptions: Examples from a soil chronosequence in central California: Geoderma, v. 28, p. 1–28.

Harden, J.W., and Taylor, E.M., 1983, A quantitative comparison of soil development in four climatic regimes: Quaternary Research, v. 20, p. 342–359.

Harden, J.W., Taylor, E.M., Hill, C., Mark, R.L., McFadden, L.D., Reheis, M.C., Sowers, J.M., and Wells, S.G., 1991, Rates of soil development from four soil chronosequences in the southern Great Basin: Quaternary Research, v. 35, p. 383–399.

Harvey, A.M., and Wells, S.G., 1994, Late Pleistocene and Holocene changes in hillslope sediment supply to alluvial fan systems: Zzyzx, California, *in* Millington, A.C., and Pye, K., eds., Environmental change in drylands: Biogeographical and geomorphological perspectives: Chichester, Wiley, p. 67–84.

Harvey, A.M., Wigand, P.E., and Wells, S.G., 1999, Response of alluvial fan systems to late Pleistocene to Holocene climatic transition: Contrasts between the margins of pluvial Lakes Lahontan and Mojave, Nevada and California, USA: Catena, v. 36, p. 255–281.

Lancaster, N., 1994, Controls on aeolian activity: New perspectives from the Kelso Dunes, Mojave Desert, California: Journal of Arid Environments, v. 27, p. 113–124.

Lancaster, N., and Tchakerian, V.P., 1996, Geomorphology and sediments of sand ramps in the Mojave Desert: Geomorphology, v. 17, p. 151–166.

Lustig, L.K., 1965, Clastic sedimentation in Deep Springs Valley, California: U.S. Geological Survey Professional Paper 352F, p. 131–92.

McAuliffe, J.R., and Van Devender, T.R., 1998, A 22,000-year record of vegetation change in the north-central Sonoran Desert: Paleogeography, paleoclimatology, and paleoecology, v. 141, p. 253–275.

McDonald, E.V., 1994, The relative influences of climatic change, desert dust, and lithologic control on soil-geomorphic processes and hydrology of calcic soils formed on Quaternary alluvial-fan deposits in the Mojave Desert, California [Ph.D. thesis]: Albuquerque, University of New Mexico, 383 p.

McDonald, E.V., and McFadden, L.D., 1994, Quaternary stratigraphy of the Providence Mountains piedmont and preliminary age estimates and regional stratigraphic correlations of Quaternary deposits in the eastern Mojave Desert, California, *in* Wells, S.G., et al., eds., Quaternary stratigraphy and dating methods: Understanding geologic processes and landscape evolution in southern California, *in* McGill, S.F., and Ross, T.M., eds., Geological investigations of an active margin: Geological Society of America, Cordilleran Section Guidebook, p. 205–210.

McDonald, E.V., Pierson, F.B., Flerchinger, G.N., and McFadden, L.D., 1996, Application of a process-based soil-water balance model to evaluate the influence of late Quaternary climate change on soil-water movement in calcic soils: Geoderma, v. 74, p. 167–192.

McFadden, L.D., Ritter, J.B., and Wells, S.G., 1989, Use of multiparameter relative-age methods for age estimations and correlation of alluvial fan surfaces on a desert piedmont, eastern Mojave Desert, California: Quaternary Research, v. 32, p. 276–290.

McFadden, L.D., Wells, S.G., Brown, W.J., and Enzel, Y., 1992, Soil genesis on beach ridges of pluvial lake Mojave: Implications for Holocene lacustrine and eolian events in the Mojave Desert, southern California: Catena, v. 19, p. 77–97.

Melton, M.A., 1965, The geomorphic and paleoclimatic significance of alluvial deposits in southern Arizona: Journal of Geology, v. 73, p. 1–38.

Miller, D.M., Glick, L.L., Goldfarb, R., Simpson, R.W., Hoover, D.B., Detra, D.E., Dohrenwend, J.C., and Munts, S.R., 1985, Mineral resources and resource potential map of the south Providence Mountains Wilderness Study Area, San Bernardino County, California: U.S. Geological Survey Miscellaneous Field Studies Map MF-1780-A, scale 1:62,500.

Ore, H.T., and Warren, C.N., 1971, Late Pleistocene–early Holocene geomorphic history of Lake Mojave, California: Geological Society of America Bulletin, v. 82, p. 2553–2562.

Pendall, E.G., Harden, J.W., Trumbore, S.E., and Chadwick, O.A., 1994, Isotopic approach to soil carbonate dynamics and implications for paleoclimatologic interpretations: Quaternary Research, v. 42, p. 60–71.

Phillips, F.M., 1994, Environmental tracers for water movement in desert soils in the American southwest: Soil Science Society of America Journal, v. 58, p. 15–24.

Poesen, J.W.A., 1992, Mechanisms of overland flow generation and sediment production on loamy and sandy soils with and without rock fragments, in Parsons, A.J., and Abrahams, A.D., eds., Overland flow: Hydraulics and erosion mechanics: New York, Chapman and Hall, p. 275–306.

Ponti, D.J., 1985, The Quaternary alluvial sequence of the Antelope Valley, California, in Weide, D.L., ed., Soils and quaternary geology of the southwestern United States: Boulder, Colorado, Geological Society of America Special Paper 203, p. 79–96.

Quade, J., 1986, Late Quaternary environmental changes in the upper Las Vegas Valley, Nevada: Quaternary Research, v. 26, p. 340–357.

Reheis, M.C., Harden, J.W., McFadden, L.D., and Shroba, R.R., 1989, Development rates of late Quaternary soils, Silver Lake playa, California: Soil Science Society of America Journal, v. 53, p. 1127–1140.

Reheis, M.C., Sowers, J.M., Taylor, E.M., McFadden, L.D., and Harden, J.W., 1991, Morphology and genesis of carbonate soils on the Kyle Canyon fan, Nevada, USA: Geoderma, v. 52, p. 303–342.

Reheis, M.C., Slate, J.L., Throckmorton, C.K., McGeehin, J.P., Sarna-Wojcicki, A.M., and Dengler, L., 1996, Late Quaternary sedimentation on the Leidy Creek fan, Nevada-California: Geomorphic responses to climate change: Basin Research, v. 12, p. 279–299.

Rendell, H.M., and Sheffer, N.L., 1996, Luminescence dating of sand ramps in the eastern Mojave: Geomorphology, v. 17, p. 187–198.

Schumm, S.A., 1977, The fluvial system: Wiley, New York, 338 p.

Smith, G.I., and Street-Perrott, A.F., 1983, Pluvial lakes of the western United States, in Porter, S.C., ed., The late Pleistocene: Minneapolis, Minnesota, University of Minnesota Press, p. 190–212.

Spaulding, W.G., 1985, Vegetation and climates of the last 45,000 yr in the vicinity of the Nevada Test Site, south-central Nevada: U.S. Geological Survey Professional Paper 1329, 83 p.

Spaulding, W.G., and Graumlich, L.J., 1986, The last pluvial episodes in the deserts of southwestern North America: Nature, v. 320, p. 441–444.

Stuiver, M., and Reimer, P.J., 1993, Extended ^{14}C database and revised CALIB 3.0 14C program: Radiocarbon, v. 35, p. 215–230.

Taylor, E.M., 1988, Instructions for the soil development index template—Lotus 1-2-3: U.S. Geological Survey Open-File Report 233A.

Van Devender, T.R., Thompson, R.S., and Betancourt, J.L., 1987, Vegetation history of the deserts of southwestern North America; The nature and timing of the late Wisconsin-Holocene transition, in Ruddiman, W.F., and Wright, H.E., Jr., eds., North America and adjacent oceans during the last deglaciation: Boulder, Colorado, Geological Society of America, The Geology of North America, v. K-3, p. 323–352.

Wang, Y., Amundson, R., and Trumbore, S., 1994, A model for soil $^{14}CO_2$ and its implications for using ^{14}C to date pedogenic carbonate: Geochimica et Cosmochimica Acta, v. 58, p. 393–399.

Wang, Y., McDonald, E.V., Amundson, R., McFadden, L.D., and Chadwick, O., 1996, Paleoenvironmental implications of carbon and oxygen isotopes in alluvial soils, Providence Mountains, California: Geological Society of America Bulletin, v. 108, p. 379–391.

Wells, S.G., and Dohrenwend, J.C., 1985, Relict sheetflood bedforms on late Quaternary alluvial-fan surfaces in the southwestern United States: Geology, v. 13, p. 512–516.

Wells, S.G., McFadden, L.D., and Dohrenwend, J.C., 1987, Influence of late Quaternary climatic changes on geomorphic and pedogenic processes on a desert piedmont, eastern Mojave Desert, California: Quaternary Research, v. 27, p. 130–146.

Wells, S.G., McFadden, L.D., Anderson, R.Y., Brown, W.J., Enzel, Y., and Miossec, J-L., 1989, Late Quaternary paleohydrology of the eastern Mojave River drainage, southern California: Quantitative assessment of the late Quaternary hydrologic cycle in large arid watersheds: New Mexico Water Resources Research Institute Report No. 242.

Wells, S.G., McFadden, L.D., and Harden, J., 1990, Preliminary results of age estimations and regional correlations of Quaternary alluvial fans within the Mojave Desert of southern California, in Reynolds, R.E., et al., eds., At the end of the Mojave: Quaternary studies in the eastern Mojave Desert: Redlands, California, Special Publication of the San Bernardino County Museum Association, 1990 Mojave Desert Quaternary Research Center Symposium, May 18–21, 1990, p. 45–54.

Wells, S.G., Brown, W.J., Enzel, Y., Anderson, A., and McFadden, L.D., 1994, A brief summary of the late Quaternary history of pluvial Lake Mojave, eastern California, in Wells, S.G., et al., eds., Quaternary stratigraphy and dating methods: Understanding geologic processes and landscape evolution in southern California, in McGill, S.F., and Ross, T.M., eds., Geological investigations of an active margin: Geological Society of America, Cordilleran Section Guidebook, p. 182–188.

Wigand, P.E., and Rhode, D., 2002, Great Basin vegetation history and aquatic systems: The last 150,000 yr: Smithsonian Contributions to Earth Sciences, no. 33, p. 309–368.

Yair, A., 1987, Environmental effects of loess penetration into the northern Negev Desert: Journal of Arid Environments, v. 13, p. 9–24.

Yair, A., 1992, Headwater control on channel runoff, in Parsons, A.J., and Abrahams, A.D., eds., Overland flow: Hydraulics and erosion mechanics: New York, Chapman and Hall, p. 53–68.

Yair, A., and Enzel, Y., 1987, The relationship between annual rainfall and sediment yield in arid and semiarid areas: The case of the northern Negev: Catena Supplement 10, p. 121–135.

MANUSCRIPT ACCEPTED BY THE SOCIETY AUGUST 1, 2002

Late Quaternary variations in alluvial fan sedimentologic and geomorphic processes, Soda Lake basin, eastern Mojave Desert, California

Adrian M. Harvey
Department of Geography, University of Liverpool, PO Box 147, Liverpool L69 3BX, England

Stephen G. Wells
Desert Research Institute, 2215 Raggio Parkway, Reno, Nevada 89506-0220, USA

ABSTRACT

Alluvial fans at the front of the Soda Mountains at Zzyzx, in the Mojave Desert, California, have responded differently to late Pleistocene to Holocene climatic changes. The alluvial fans have been mapped in the field and the depositional facies interpreted as debris-flow and fluvial channel and sheetflood sediments. The relative age relationships of the fan segments have been determined primarily on the basis of soil development. The overall sequence has been established in relation to dated shorelines of late Pleistocene pluvial Lake Mojave, and the ages suggested by regional correlations. Six sets of alluvial fan deposits have been identified and labeled, oldest to youngest, Qf0–Qf5. Qf0 and Qf1 sediments predate pluvial Lake Mojave I (18.5–16.5 ka). Qf0 sediments are seen only in sections in fanhead trenches. Soil characteristics and regional correlations suggest an age for Qf0 much greater than for Qf1. Qf1 appears to date from the late Pleistocene, but prior to the Lake Mojave I highstand. Qf2 dates from the period following the Lake Mojave I and II highstands (18.4–16.6–13.7–11.4 ka) but prior to the youngest dated shoreline of pluvial Lake Mojave (10–9.3 ka). Fan depositional phases Qf3–Qf5 postdate the youngest lake shoreline, and are therefore Holocene in age.

From the late Pleistocene to the Holocene there was a switch from deposition dominantly by debris-flow to fluvial channel and sheetflood processes, which was accompanied by changes in fan style from fan aggradation to progradation and dissection. However, during the mid Holocene (ca. 4.3–3.5 ka) the Qf4 sediments suggest a short-lived reversal of this trend with a local increase in sedimentation and a short-lived reversion to debris-flow deposition on some fans. Different fans along the mountain front responded differentially to climatic change over the period since the late Pleistocene, with the largest fans switching from debris-flow to fluvial processes first, and some of the smallest fans becoming inactive during the Holocene. The results indicate that the fan processes are controlled by water and sediment supply from the hillslopes, switching as these processes changed in response to climatic changes. There is no evidence for tectonically induced change over this period, and changes in fan geomorphology induced by base-level change are restricted to the toe areas of some fans. At the local level, topographic catchment thresholds control the response of individual fans to climatically induced changes in runoff and sediment supply.

Harvey, A.M., and Wells, S.G., 2003, Late Quaternary variations in alluvial fan sedimentologic and geomorphic processes, Soda Lake basin, eastern Mojave Desert, California, *in* Enzel, Y., Wells, S.G., and Lancaster, N., eds., Paleoenvironments and paleohydrology of the Mojave and southern Great Basin Deserts: Boulder, Colorado, Geological Society of America Special Paper 368, p. 207–230. © 2003 Geological Society of America.

INTRODUCTION

Geomorphologists and sedimentologists are confronted with the challenge to elucidate the relative influence of tectonic activity, climatic change, and intrinsic geomorphic conditions on alluvial fan processes (Bull, 1977; Nilson, 1982; Frostick and Reid, 1989; Blair and McPherson, 1994; Ritter et al., 1995; Harvey, 1997). Wells et al. (1997) have argued that the identification of regional patterns of age relationships of alluvial fan surfaces provide an indication of the underlying controls of alluvial fan depositional sequences, whereby:

1. geologically instantaneous and contemporaneous deposition over a series of fans within a restricted area would imply control by intrinsic thresholds related to individual storm events;

2. temporally limited but regionally correlative fan deposition would imply response to regional climatic change; and

3. long-term, regionally noncorrelative fan deposition would imply tectonic control.

Previously we have demonstrated that geologically instantaneous storm-generated deposition yielded significant spatial and temporal variations in alluvial fan facies within northwest England (Wells and Harvey, 1987). Those variations were a result of intrinsic thresholds influenced by the geomorphology of individual watersheds, which in turn controlled water-to-sediment ratios during the storm event. We hypothesized that, for those small alluvial fans, intrinsic thresholds serve as a primary control on the depositional and geomorphic processes, whereas, fluctuations in climatic and tectonic regimes are secondary controls operating over larger regions and over longer periods of time. In this paper, we test this hypothesis and elucidate the spatial and temporal variations in sedimentologic and geomorphic processes operating on alluvial fans in the Mojave Desert, California. These fans are in a very different environment, and include fans that are that are significantly larger and that formed over longer periods of time.

Age control for sequences of alluvial fan deposition in arid continental environments is uncommon. However, recent work within the eastern Mojave Desert over the past decade has established a regional chronology for geomorphic and environmental change spanning the last 100 ka. This has been based on: (1) studies of soil-profile development (Wells et al., 1987; Reheis et al., 1989; McFadden et al., 1989); (2) radiocarbon age estimation of pedogenic carbonate (McDonald, 1994; McDonald and McFadden, 1994; Wang et al., 1994, 1996; McDonald et al., 1996); (3) cosmogenic dating of surface exposure and K-Ar and Ar-Ar dating of basalt lava flows (Turrin et al., 1984; Wells et al., 1995); (4) radiocarbon-dated geomorphic and sedimentologic features associated with pluvial lakes (Wells et al., 1987, this volume; McDonald et al., this volume); and (5) luminescence-dated eolian deposits (Clarke, 1994; Amundson et al., 1994; Lancaster and Tchakerian, this volume).

In this paper, we capitalize on this setting to establish a stratigraphy for 15 alluvial fan complexes at Zzyzx, situated along the mountain front of the southern Soda Mountains (Fig. 1). We identify changes in the sedimentology and geomorphology of the stratigraphic sequences that occurred during fan development. The Zzyzx area provides an opportunity to document differential fan behavior since the late Pleistocene. The fans are fed by a range of different-sized catchments on a variety of bedrock geologies, and extend to the margins of Soda Lake, a modern playa occupying the floor of the lake basin of Pleistocene pluvial Lake Mojave (Fig. 1). The high lake shorelines have been dated (Wells et al., 1987, 1990a; Enzel et al., 1989a), and the morpho-stratigraphic relationships between them and fan segments allow a broad chronology of fan development to be identified.

First we establish the stratigraphy of the alluvial fan surfaces in relation to the framework provided by the paleo-shorelines. We characterize the fan surfaces by their soil profile development, and by correlation, we set the sequence within its regional context. In the second part of the paper we describe the geomorphic development of the fan complexes within the timeframe established from the soil chronosequences. We identify different patterns of morphological development, related to threshold-controlled responses to climatic change over the last 30 ka.

Late Pleistocene to Holocene environmental change in the Mojave Desert

Evidence for previous climates in the Mojave Desert comes primarily from two sources: paleo lake levels (Wells et al., 1998, this volume) and vegetation reconstructions based on pollen and macro plant remains preserved in packrat (*neotoma*) middens (Van Devender, 1977; Spaulding, 1985, 1990; Wells and Woodcock, 1985; Woodcock, 1986; Grayson, 1993). Lake levels in pluvial Lake Mojave were high during the late Pleistocene, reaching maxima at 18.5–16.5 ka and 13.7–11.4 ka, in response to strong zonal atmospheric circulation (Knox, 1984) and a southward shift in the midlatitude westerlies bringing high precipitation to the Transverse Ranges (Wells et al., 1990a). This circulation broke down during the Pleistocene-to-Holocene climatic transition (Bull, 1991), and meridional circulation allowed the penetration of subtropical "monsoonal" air into the Mojave Desert (Bryson and Lowry, 1955; Bryson, 1957), bringing summer convectional rainstorms. Lake Mojave desiccated after ca 9.3 ka, but shallow lakes have formed episodically during the Holocene in response to northern Pacific air masses bringing high precipitation to the Transverse Ranges in winter (Enzel et al., 1989a, 1989b; Wells et al., 1990a, 1998; Enzel and Wells, 1997). Today, several weather types cause flooding in the desert southwest (Ely, 1997), of which monsoonal conditions are the most likely to cause geomorphic activity in small desert mountain catchments.

The late Quaternary vegetation sequence in the Mojave region reflects the climatic sequence, with a general lowering of vegetation zones during the Pleistocene, but an increase in elevation during the Holocene. However, altitudinal vegetation reconstructions for late Pleistocene in the Mojave suggest that the limited elevation of the Soda Mountains was insufficient for the development of Juniper woodland (Harvey et al., 1999). Throughout the late Pleistocene and early Holocene the vegetation of the

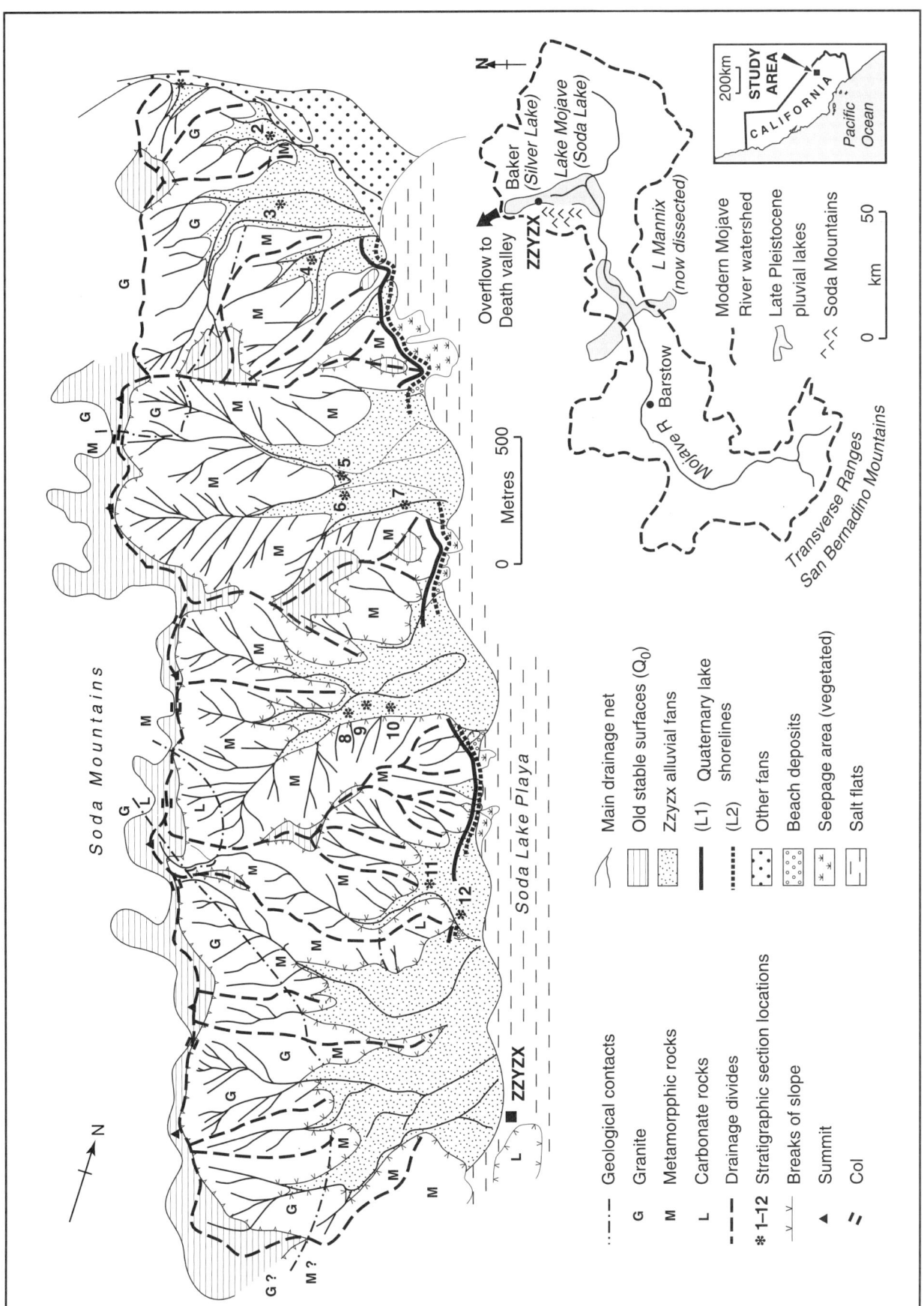

Figure 1. General and location maps of the southern Soda Mountains and Zzyzx fans, showing Geology and locations of shorelines of pluvial Lake Mojave. Also shown are locations of described sections (see Fig. 6). Insets show location within California and within the Mojave River system.

Soda Mountains and the Zzyzx fans would have been temperate desert scrub. The modern thermophilous desert vegetation developed following the mid-Holocene (Grayson, 1993).

Previous work on alluvial fans in the Mojave Desert and adjacent areas (Bull, 1991) indicates that fans responded to climatically induced pulses of sediment, supplied from the hillslopes of the catchments or from dissection of older fan deposits. Major late Pleistocene fan accumulation from active hillslope sources has been recognized in Death Valley (Denny, 1965; Hunt and Mabey, 1966; Bull, 1991; Blair and McPherson, 1994), and within the Mojave basin at Silver Lake (Wells et al., 1987, 1990b; McFadden et al., 1989) and in the Providence Mountains (McDonald and McFadden, 1994; McDonald et al., 1996; McDonald et al., this volume). A major pulse of activity has been identified during the Pleistocene to Holocene climatic transition (Bull, 1991; Harvey et al., 1999), coincident with major piedmont flooding (Wells and Dohrenwend, 1985). In these areas pulsed phases of fan activity occurred at various times during the Holocene. Previous studies of the Zzyzx fans include a general description (Wells et al., 1990c), general morphometric summaries (Harvey, 1992a, 1992b), a consideration of contrasts in hillslope processes within the fan catchments between the late Pleistocene and the Holocene (Harvey and Wells, 1994), and a consideration of the effects of the late Pleistocene to Holocene climatic transition on fan processes (Harvey et al., 1999). In this paper we present a detailed study of the fan geomorphic and sedimentologic sequences within the context of the regional alluvial fan sequences, and consider how the morphology of the Zzyzx fans responded to climatically induced changes in fan processes since the late Pleistocene.

RESEARCH AREA AND METHODOLOGY

The Zzyzx fans are located at the mountain front along the eastern margin of the southern Soda Mountains in the eastern Mojave Desert (Fig. 1). The mountain front is tectonically stable, according to the criteria of Bull (1978), and the fans have backfilled into the mountain catchments. These catchments are developed on three major Mesozoic bedrock lithologies; granitic rocks, metavolcanics, and limestones (Fig. 1).

Two high lake shorelines of Pleistocene pluvial Lake Mojave are preserved as a zone of erosional notches on the bedrock interfluves between the fans, at ~10–12 m above the modern playa floor (elevation of the modern playa floor: ~275–280 m). These shorelines were formed during highstands, coincident with perennial lake stages Mojave I and II, between ca. 18.4–16.6 and 13.7–11.4 ka (Enzel et al., 1989a; Wells et al., 1990a, 1998, this volume; Brown et al., 1990), when the lake reached 287 m and 285 m altitudes, respectively. On the fans one composite shoreline zone can be identified at about this altitude (L1 on Figure 1). A lower shoreline is evident ~2–3 m above the margins of the modern playa, this shoreline, at ~280 m, is probably the equivalent of the lowest ancient shoreline identified at Silver Lake, the northern extension of pluvial Lake Mojave (Fig. 1), and dated to ca. 10–9.3 ka (Wells et al., 1987). That shoreline (L2 on Figure 1) can be traced around the bedrock interfluves and across the fans at Zzyzx.

Geomorphic and sedimentologic mapping

Fifteen alluvial fan complexes were distinguished on the basis of geographic location, including from south to north: Southern (SOS, SOC, SON), Springer/Zzyzx (SPR, ZZX), Josh/Gate (JSG, GTE), Palm Cones (PC), Camino Viejo (CV), Johnny (JNF), Steve (STV), Mesquite (MSQ), Vulture (VLT), Solitary (SOL), and Northern (NOR) fan complexes (see Figure 2 for locations). They range from relatively large fluvially dominant fan complexes to small steep debris cones (Fig. 3), comprising stratified fluvial and sheetflood and massive debris-flow deposits respectively (Fig. 4). Within these 15 complexes, 32 individual alluvial fan apices have been identified (Fig. 2). The geomorphology and sedimentology of the fans were mapped in the field at a scale of 1:6000 onto base maps constructed from enlargements of the U.S. Geological Survey 1:24,000 topographic maps, with some details added from vertical black and white aerial photography. Morphology, sedimentology, and age-related properties of the fan surfaces were recorded on the base maps from observations in the field. Mapping of the fan morphology involved delineation of the major geomorphic units within each fan. In the fan-toe areas special attention was paid to the field relationships between these units and the shoreline remnants. The depositional characteristics of the fans were mapped on the basis of surface form and exposure of the sedimentary features in section, using a modification of the facies scheme previously developed (Table 1; Wells and Harvey, 1987).

Alluvial fan stratigraphy

Age relationships of the alluvial fan deposits were established by (1) observing topographic positions in relation to the pluvial lake shorelines, (2) observing and measuring postdepositional fan-surface modification, (3) recording basic stratigraphic relations (e.g., inset and overlapping), and (4) observing soil-bounded unconformities in vertical sections. On the basis of field relations with the shorelines and stratigraphic position, three groups of fan segments and associated stratigraphic units could be identified:

(i) Alluvial fan surfaces and their underlying deposits that are truncated by and therefore older than the oldest and highest shoreline (L1 on Figure 1) are designated as unit Qf1.

(ii) Alluvial fan surfaces and their underlying deposits that truncate the L1 shoreline zone, but which are truncated by the younger shoreline (L2 on Figure 1) are designated as unit Qf2. These were laid down during the period between the two shorelines.

(iii) Alluvial fan surfaces that truncate and therefore postdate the younger shoreline (L2 on Figure 1) are designated as units Qf3, Qf4, and Qf5 (oldest to youngest).

Differentiation between these three groups and between the three units within the youngest group was achieved on the basis

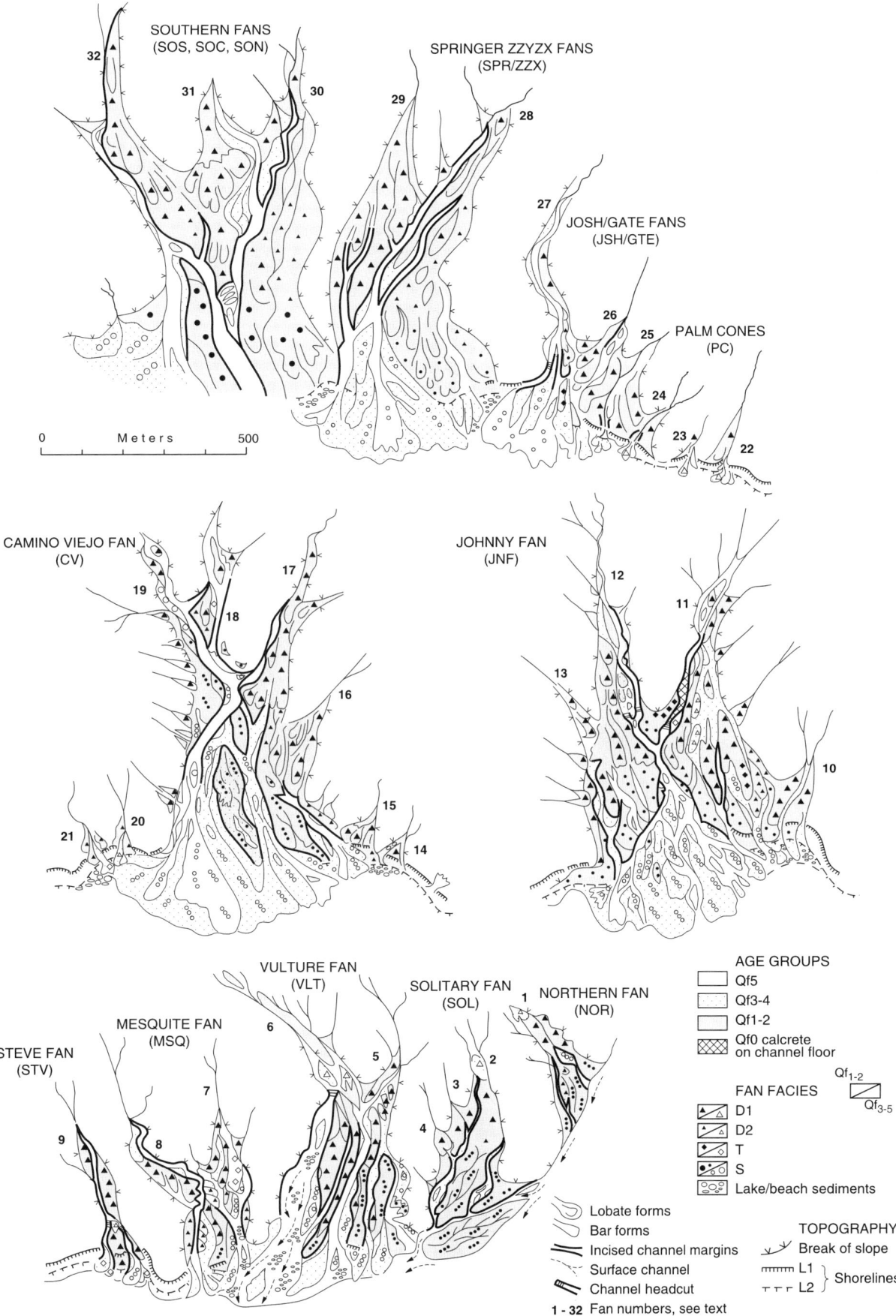

Figure 2. Summary map of the sedimentary units on the Zzyzx alluvial fans. Abbreviations in capitals refer to fan complexes, numbers refer to fan apices feeding discrete fan segments (see text). For the sedimentary units, solid symbols relate to Qf1–2 and open symbols to Qf3–5 age units.

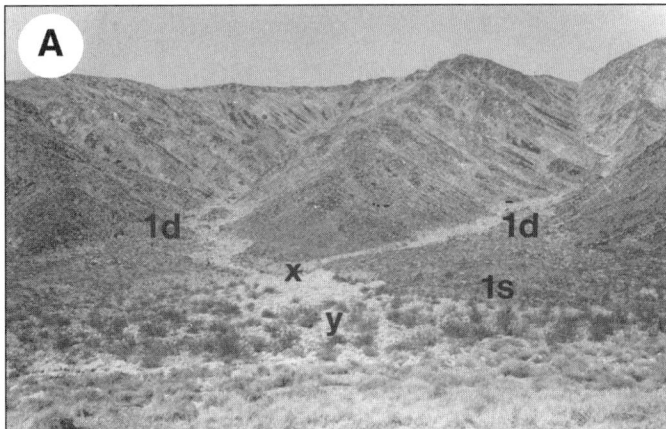

Figure 3. General views of selected Zzyzx fans. A: Johnny Fan, a large fluvially dominant fan. B: Steve Fan, a steep debris-flow dominated fan (for locations see Fig. 2). 1d—Qf1 debris-flow deposits; 1s—Qf1 fluvial deposits; 2d—Qf2 debris-flow deposits; x—location of Qf0 deposits exposed in fanhead trench; y—younger (Qf3–5) fan deposits; L1—upper shoreline; L2—lower shoreline.

Figure 4. Examples of fan sediments. A: Coarse bouldery debris-flow deposits, Johnny Fan apex zone. B: Qf1 stratified, fluvial/ sheetflood deposits; CV Fan midfan zone; arrows indicate erosional horizons.

of postdepositional modification of the fan surfaces, using criteria similar to those developed by McFadden et al. (1989), including rock weathering and varnish characteristics, and stone-pavement and soil-profile characteristics (summarized on Table 2). As observed throughout the Mojave Desert, rock weathering and varnish properties on alluvial fan surfaces show progressive changes with increasing age of deposits (Table 2). These surface properties vary with clast lithology (McDonald and McFadden, 1994). The clasts derived from carbonate source areas (Fig. 1) rarely develop varnish coatings, but do exhibit surface weathering features such as solution pits and microrills (cf. Amit and Gerson, 1986). The clasts derived from metavolcanic and mixed granitic source areas (Fig. 1) develop more extensive and thicker rock varnish coatings on progressively older fan surfaces. Granitic clasts on the older fan (Qf1) surfaces show both development of varnish coatings on some clast faces, and chemical and mechanical weathering by granular disintegration on others.

Stone pavement shows progressive development from no pavement on young (Qf5) fan surfaces to moderately developed pavements (terminology after Al-Farraj and Harvey, 2000) on the oldest (Qf1) fan surfaces (Table 2). As recognized on fan surfaces throughout this region (Wells et al., 1987; McFadden et al., 1989), particle size decreases and clast angularity increases with increasing pavement development, resulting in a progressive obliteration of any original depositional fabric. There are, however, variations in pavement properties related to original particle size and lithology of the alluvial fan deposits, whereby larger clasts take longer to fracture and form uniform pavement surfaces. The physical and chemical weathering of carbonate and metavolcanic clasts is dominated by fracture processes, typically yielding angular and interlocking pavement clasts. Diminution of clasts derived from granitic sources appears to result from fracturing and granular disintegration, ultimately producing a fine gritty-textured pavement surface. The most mature pavements, however, do not occur on the alluvial fan surfaces, but on older planated bedrock surfaces

	TABLE 1. SEDIMENTARY FACIES ON THE ZZYZX FANS
D1:	Bouldery debris flows; matrix-supported large cobbles and boulders, which in section and on the surface show pronounced basal shear and frontal compressional fabrics. Surface form is of pronounced lobe and levee topography.
D2:	Fluid, muddy debris flows; matrix-supported mud-rich debris flows. Higher matrix content and smaller clasts than D1. Less obvious compressional fabric. More subdued surface relief.
T:	Stony matrix-poor debris flows and transitional deposits; loose poorly sorted cobbles and small boulders with little or no matrix, little fabric other than packing. Surface form is lobate but with less relief than on D1.
S1:	Fluvial boulder bars; large imbricated boulders—clast-supported, usually showing nose to tail sorting in bar forms.
S2:	Fluvial cobble bars; large imbricated cobbles—clast-supported, usually showing nose to tail sorting in bar forms, more common than S1. In sections show erosional bases and moderate stratification and sorting. Surface form: lobate or bar-shaped.
S3:	Fluvial gravel sheets; gravel and cobble sheets—clast-supported, locally showing imbrication, usually well sorted, and in section good stratification. Subdued sheetforms locally showing bar and swale topography.

Note: Based on Wells and Harvey, 1987.

	TABLE 2. SUMMARY OF AGE-RELATED FAN SURFACE CHARACTERISTICS
Qf1:	Dark, well-developed varnish on metavolcanic or granite clasts largely obscures rock texture. Granite clasts may also show granular disintegration. On smaller sediments, well-developed pavements of fractured angular clasts occur. The larger clasts are still intact, though they may show evidence of fracture. Original depositional fabrics are obliterated. Soils show 7.5YR hue and stage II $CaCO_3$.
Qf2:	Dark varnish is present on metavolcanic and granite clasts, but rock texture is still visible through the varnish. Moderate pavement development occurs on smaller sediments, with minor fracturing of larger clasts. Original depositional fabrics are faint. Soils are thin, with A/B horizon differentiation, 10YR hue, and stage I + $CaCO_3$.
Qf3:	Spotty, pale reddish varnish occurs. There is incipient pavement development, with little or no fracturing but original depositional fabrics are only weakly preserved. Soils (10YR hue) have little or no horizon differentiation.
Qf4:	No varnish occurs, but clasts appear darkened. There is no pavement development; nor is fracturing evident, and original depositional fabrics are unaltered. There is incipient soil accumulation.
Qf5:	This unit comprises unaltered fresh sediment.

such as spur tops and summits (designated Q0, Harvey and Wells, 1994). On these low-relief surfaces the pavement is composed of relatively uniform, small, angular clasts derived from the underlying bedrock, together with fragments of fractured petrocalcic layers (caliche rubble, Lattman, 1973), derived from pedogenic K horizons capping the bedrock surfaces.

Soil morphology and stratigraphy

Many previous studies have used soil chronosequences to aid the correlation and relative dating of dry-region alluvial depositional surfaces (e.g., McFadden and Weldon, 1987; McFadden et al., 1989; Harden, 1990; Bull, 1990, 1991; Harvey et al., 1995) based primarily on field descriptions of soil horizon development (Harden, 1982, 1990; Birkeland, 1985, 1990; Harden et al., 1991) and on soil $CaCO_3$ status (Gile et al., 1966, 1981; Machette, 1985). We described soil profiles at 7 sites covering the relative age range of the Zzyzx fan surfaces. All sites were on fans GTE and JNF (Fig. 2), fed by the dominant metavolcanic lithology. Stable sites were selected (i.e., surfaces displaying minimum evidence of vertical erosion or deposition) in order to assess the maximum degree of soil development associated with a particular fan surface. Soil profiles were described in the field using the terminology of the Soil Survey Staff (1975). Twelve main soil properties were recorded at each site (Table 3). In addition, special features, such as the degree of clast weathering, were noted for each horizon where appropriate. The vertical arrangement of the soil horizons and their properties were described from the land surface down to the parent material and/or bottom of the cut. In addition, the properties of the land surface features (i.e., topography, vegetation, slope aspect) near each profile site were recorded.

SOIL CHRONOSEQUENCES AND SOIL STRATIGRAPHY

The soil profiles on the Zzyzx fans (Table 3) show comparable patterns of profile development to those described by Wells et al. (1987) and McFadden et al. (1989) on alluvial fans along the northern Soda Mountains in the Silver Lake basin (Fig. 1). A common horizon in both areas is the vesicular A (Av or Avk) horizon, ranging in thickness from 0.5 to 4 cm (McFadden, 1988; McFadden et al., 1986, 1998). These vesicular horizons

TABLE 3. FIVE PEDONS (A–E) SUMMARIZING MORPHOLOGICAL AND PHYSICAL CHARACTERISTICS OF SOILS DEVELOPED ON ALLUVIAL FAN DEPOSITS ALONG THE SODA MOUNTAINS NEAR ZZYZX, CALIFORNIA

Horizon	Thickness (cm)	Color Dry	Texture %G <2mm	Structure	Consistency Dry–wet	CaCO₃	Clay films	Pores	Roots	Salts	Comments
Pedon A. Geomorphic position: alluvial fan (Qf1) surface. Topographic position: cut in wall of fanhead trench. Slope: ~8°. Parent material: debris flow deposits, metavolcanic clasts. Vegetation: creosote, cactus, woody species <1%. Pavement: moderate, partially interlocked, some fracturing, metavolcanic clasts. Varnish: well developed; top 5YR 2.5/2, base 5YR 6/4, max 5YR 5/4.											
Avk	0–4	10YR 6/4, 4/4	35–40LfS granules pebbles	2 fpl 1 fsbk	so–so, po sh	Very e→n.o.	n.o.	vf–vf fdv	3f–vf	n.o.	
Btk	4–15	10YR 5/6, 8.75YR 5/4	60% SiL granules to boulders	3 cgr 1 msbk	shs, p	ev, d & s Coats: Stage I very discontinuous Top: carbonate as a result of silt (red) imput	Co 1 nbr	3 ft	3 vf: f	n.o.	Buried varnished stone
Btk₂	15–30	8.75YR 7/4; 10YR 5/6(u)	75% fSL cgr-boulder	Sg 1 fgr	loss, sp	Carbonate coats; clay film with thin discontinuos carbonate (sides almost continuous); highly variable, 3% to 100%	Clay films on clast tops Clay is n.o. effervescent m = es,d	inter	3 f & fv	n.o.	Salty taste throughout
Bwk	30–54	8.75YR 7/4; 10YR 5/6	>75%fSL 80–90	Sg lfgr	loss, ps	Nearly continuous; weakly laminar	n.o.	inter	3 vf	n.o.	Carbonate: top—red silt; e, d; bottom—Stage I+; (≈1mm)
Bwk2	54–85	8.75YR 7/3; 10YR 5/6	>90%LfS	Sg	loso, po	es, d; thick coats on bottom; some continuous; red silt coats tops of clast; carbonate bottom, <1 mm	n.o.	Inter	1 vf & f	n.o.	Carbonate rind different from above, more grainy, less thick
Btkb	85+	clay: 7.5YR 6/4 7.5YR 5/6 carb: 10YR 8/3		locally matrix gr→fsbh	H	Stage III: much carbonate that impregnates this horizon may be flushed through from above. Evidence: carbonate impregnated Bt matrix	co, & 1 nbr	n.o.		n.o.	
Pedon B. Geomorphic position: alluvial fan (Qf2) surface. Topographic position: channel cut into terrace in Qf2 gravels, 10 m up fan from burrow pit. Slope: ~5°. Parent material: alluvial fan gravels, metavolcanic clasts. Vegetation: creosote, saltbrush. Pavement: large clasts, weak. Varnish: moderate.											
Avk	0–3	10YR 6/4, 5/4	>80%LS	2msbk	Sono, po	Noneffer	n.o.	2f & m v	3f & vf	n.o.	
Bwk	3–20	8.75 7/4(mixed) ped surface: 7.5YR 7/4	>70% SL	f & m/lo gr	soss, so lo	Matrix: noneff Very rare, very thin coating on sides, bottom of gravel	Co	Inter	2vf & f	n.o.	
Bk	20–35	8.75YR 7/4; 5/4 mixed	>90%SL (mostly pebbles)	Sg	lono, po	Matrix: es Gravel: 10%–80%; bottom coated with thin (<0.5m) CaCO₃; highly variable	n.o.	Inter	3f	n.o.	
Bky	35–60	10YR 6/4, 5/4	90% SL	Sg	loss, sp	Matrix: e Gravel: 10%–30% of bottom covered by very thin coating	n.o.	Inter	2vf & f	Gypsum crystal on one clast base	
Bk	60–70+	10YR 7/3,5/4	79%SL	Sg	loss, sp	Matrix: es Gravel: discontinuous, thin (<0.5mm on 20%–50% of clast bottom)	n.o.	Inter	n.o.	n.o.	

TABLE 3. FIVE PEDONS (A–E) SUMMARIZING MORPHOLOGICAL AND PHYSICAL CHARACTERISTICS OF SOILS DEVELOPED ON ALLUVIAL FAN DEPOSITS ALONG THE SODA MOUNTAINS NEAR ZZYZX, CALIFORNIA (continued)

Horizon	Thickness (cm)	Color Dry	Texture %G <2mm	Structure	Consistency Dry-wet	CaCO$_3$	Clay films	Pores	Roots	Salts	Comments
Pedon C. Geomorphic position: alluvial fan (Qf2) surface and on depositional bar. Topographic position: excavation into stream-cut in bar. Slope: 5–8°. Parent material: alluvial fan gravels. Vegetation: saltbrush, creosote, cactus, 5% cover. Pavement: moderate. Varnish: moderate to dark. Profile exhibits evidence of bioturbation (infilled burrow); strongly developed Bt associated with burrow.											
Av	0–3	10YR 6/4, 4/4	>80SL	2fpl	Soft vss, po	Top = n.o. n.o.	n.o.	3f & vf vesic.	2f	n.o.	
B	3–16	7.5YR 6/4, 4/6	>90SL	Sg	loss, po	Red silty caps clast; minimal amount on bottom; n.o.	n.o. coats on stone	inter	2vf f	n.o.	Bt is not observed.
Bt	16–35	Ped 7.5YR 5.5/6 5YR 5/4 H: 7.5YR 5/6; (w) 5/4	75ScL	Fsbk gr	hs, p	n.o.	2nbr films on stones 2n films on grain	inter	1vf	n.o.	
C (burrow)	35–45	10YR 7/3; 5/4	65 granule pebbles	S Sg	loso, po	n.o.	n.o.	inter	2f	n.o.	
Btkb	45–63	6.75YR 6/4; 10YR 5/6	>90SL	M	hs, p	Stage I: discontinuous coats, sides and bottom Weakly–slightly e, d	1nbr 1n film grain	inter	3f	n.o.	
Bk1b	63–104	10YR 7/4; 4/4	>80LfS	sg	lono, po	Coating thickness = 0.1–0.5 mm Matrix: strongly effective carbonate on side (thin) and discontinuous on bottoms	n.o.	inter	2f 1m	n.o.	Clear, smooth?
Bk2	104–128+	10YR 6/4, 4/4	>80LS	sg	loso, po	Eff, d; red fine sand and silt in clast top; thin discontinuous coats on bottom; maximum 20% cover, most less	n.o.	inter	1f	n.o.	
Pedon D. Geomorphic position: alluvial fan (Qf3) surface. Topographic position: inset terrace, within fan. Slope: ~5° Parent material: alluvial fans gravels, metavolcanic clasts. Vegetation: creosote, grasses. Pavement: weak. Varnish: slight to moderate.											
Av	0–2.5	10YR 6/4; 5/4	30LfS	cpl	so–so,po sh	e, d	n.o.	vf–f dv & t	3vf–f	n.o.	Locally few vesicles
Bwk	2.5–30	10YR 6/6; 4/5	>90LS	m; lmgr	so–vss,po lo	es, d; fine red silt in top; bottom– very rare thin coat but on few clast	n.o.	inter	2f–vf	n.o.	
Bk	30–70	10YR 6/4; 5/6	>80S	sg	loso,po	Thin discontinuous on bottoms: 40% to 50% cover; es, d; % decrease with depth	n.o.	inter	n.o.	n.o.	
Bk2	70–80+	10YR 6/4; 4/6	>90LS	sg	loso,po	Very thin discontinuous coat that thins with depth; es, d	n.o.	inter	n.o.	n.o.	

TABLE 3. FIVE PEDONS (A–E) SUMMARIZING MORPHOLOGICAL AND PHYSICAL CHARACTERISTICS OF SOILS DEVELOPED ON ALLUVIAL FAN DEPOSITS ALONG THE SODA MOUNTAINS NEAR ZZYZX, CALIFORNIA (continued)

Horizon	Thickness (cm)	Color Dry	Texture %G <2mm	Structure	Consistency Dry–wet	CaCO$_3$	Clay films	Pores	Roots	Salts	Comments
Av	0–2	10YR 7/3; 4/3	>80LS	2msbk	soso,po	e, d	n.o.	3vf & fdv	3f	n.o.	
Bk	2–18	10YR 7/4; 4/4	>80LS	sg	loso,po	e, d Very rare on side and clast bottom (<1%)	n.o	inter		n.o.	
Bk2	18–43	10YR 6/4; 5/4	>90LS	sg	loso,po	es, d Discontinuous on sides; <1–22 at bottom silt cap	n.o.	inter	2f–vf	n.o.	
Bk3	43–69+	10YR 7/4; 4/4	>90S	>sg	loso,po	Very thin (<0.1 mm); side/bottom; <1% of clast surface; w/r, e, d	n.o.	inter	n.o.	n.o.	Silt (red) on clast surface. Reworked carbonate coated pebbles.

Pedon E. Geomorphic position: alluvial fan (Qf4) surface, depositional bar. Topographic position: stream cut into bar on fan surface. Slope: ~5°. Parent material: Alluvial fan gravels, metavolcanic clasts. Vegetation: creosote <5%, saltbrush, cactus. Pavement: none. Varnish: none to very slight. Varnish on some clasts, interpreted as reworked from older fan deposits surface clasts bases 7.5 YR 6/6 exhibit salt weathering cryptogamic surface crust.

Note: Terminology (abbreviations) follows Soil Survey Staff (1975).

commonly occur beneath very weakly to strongly developed pavements. Stone pavements and Av horizons are genetically linked, forming by aeolian accretion, dilation, and vertical lifting processes (Wells et al., 1985; McFadden et al., 1986, 1998; McDonald, 1994; McDonald et al., 1996). Within the southern Soda Mountains the Av horizons are typically thicker on the relatively older fan surfaces, a trend observed by McDonald (1994) on the Providence Mountains fans. Such observations support the use of stone pavement and soil properties for estimating relative ages of the underlying alluvial fan deposits. In both regions of the Soda Mountains, the vesicular horizons developed on younger (e.g., units Qf3 and Qf4) and on the oldest (Qf1) alluvial fans contain disseminated pedogenic calcium carbonate (Avk horizons). Pedogenic calcium carbonate was not observed in the Av horizons developed on alluvial fan unit Qf2 within the Zzyzx area.

Fan surfaces and soils postdating the youngest Lake Mojave shorelines

The youngest alluvial fan surface (unit Qf5) shows no soil or pavement development, reflecting recent depositional processes on these fans. Alluvial fan unit Qf4 has relatively thick but weakly developed calcic (Bk) horizons (< stage I) with very thin, discontinuous carbonate coatings on the sides and bottoms of clasts. The broad distribution of pedogenic carbonate through the profile may result from the coarse texture (>80% gravel) and relatively high permeability of these deposits. Clasts within the soil horizons of unit Qf4 show signs of being reworked from older alluvial fan deposits. The top horizons contain buried clasts with degraded varnish coatings. The lower horizons include carbonate-coated pebbles in which the carbonate coatings are thick and degraded. On such clasts the thickest carbonate coatings do not occur systematically along the bottom or along the sides of the clasts. These data indicate that coarse sediment within alluvial fan unit Qf4 is derived, in part, from older fans that have better-developed soils with thicker accumulations of pedogenic calcium carbonate. Reworking of older soils into unit Qf4 occurs in part because these deposits are inset within and topographically below the older fan deposits. The maximum horizon dry color is 10YR 6/4, indicating almost no reddening.

The soil formed in the next oldest fan, unit Qf3, is similar to unit Qf4 in that it is coarse grained with thick but weakly developed calcic horizons. The profile of unit Qf3 is different, however, in that it contains a weakly developed 27-cm thick Bwk horizon and stage I carbonate horizon (Table 3). Fine pink to very pale brown (maximum color 8.75YR 7/4) silt can be observed capping coarser clasts.

Fan surfaces and soils predating the youngest Lake Mojave shorelines

Alluvial fan unit Qf2 truncates the highest shoreline of pluvial Lake Mojave (L1), but is truncated by the youngest Lake Mojave shoreline (L2). Soil profiles developed on alluvial fan unit

Qf2 show significant spatial variations on some alluvial fans (Table 3). Pedon B is characterized by Av-Bwk-Bk-Bky-Bk horizons, whereas Pedon C is characterized by Av-Bw-Bt-C horizons. Maximum dry colors for Pedon B are typically 8.75YR 7/4 both on ped faces and on disaggregated samples with loose to granular structure in the Bwk horizon. Pedon C displays maximum dry colors of 5YR 5/4 and is subangular blocky in the Bt horizon. Spatial variations in profile development have been observed by McDonald and McFadden (1994) on fan surfaces in the Providence Mountains, and they have correlated these fans to unit Qf2 in our study area. However, in the southern Soda Mountains, the spatial variability in profile development appears to reflect spatial variation in bioturbation associated with burrowing. Higher permeability along these macropores may favor enhanced translocation of clays. Thus, the soil profile properties of Pedon B more accurately reflect the nature of soils developed on unit Qf2. This profile has pedogenic calcium carbonate accumulation stage I+.

The oldest stratigraphic unit that occurs on the alluvial fan surfaces (Qf1) is truncated by the highest shoreline (L1). The surfaces are characterized by the moderate to well-developed pavement with numerous overturned stones displaying 5YR 5/4 undersides. Many of the surface clasts have deeply weathered fractures, which penetrate at least 15% of the clast diameter. The soil profile developed in this unit is characterized by a 8.5YR 7/4 (maximum dry color), 26 cm Bt horizon. The degree of calcium carbonate accumulation varies between stage II+ with local zones of weak laminar layers (stage III). Where observed, the base of this unit lies stratigraphically over a truncated, very well developed soil in older alluvial fan sediments.

The oldest alluvial fan unit (Qf0) observed in the study region has no geomorphic expression at the land surface and occurs only where exposed in sections, buried by younger alluvial fan deposits. Unit Qf0 is distinguished from overlying alluvial fan deposits by a soil-bounded unconformity. The soil profile associated with the unconformity is always eroded such that the entire soil profile was never observed. Field morphologic descriptions show that the soil profile, prior to its truncation, was well developed. In places, the profile is characterized by a 7.5YR 5/6, subangular blocky Btkb horizon with stage III pedogenic calcium carbonate accumulation.

LOCAL AND REGIONAL CORRELATIONS AND AGE ESTIMATES FOR ALLUVIAL FAN DEPOSITS

Systematic description of soil and stone pavement properties remains one of most fundamental and consistent methods for establishing stratigraphic relations and for providing local and regional correlations of Quaternary deposits in arid environments. The soils within both the Soda and Silver Lake basins exhibit progressive changes in the morphological properties on successively higher and stratigraphically older alluvial fan deposits. In both areas, these changes can be used to establish a chronosequence because of similar parent material, elevation and regional topography, flora, and climatic history. However within the Soda Lake basin study area, the soils lack the direct isotopic age control determined for alluvial fans within the Silver Lake basin (Wells et al., 1998). Within the Soda Lake basin, age estimates of the soils and their underlying deposits are based on (1) the geomorphic relation of the fan deposits to the shorelines and (2) on correlations with a nearby soil chronosequence that has isotopic age constraints. Based upon comparisons between the soil profiles described above and the profiles of alluvial fans in the Silver Lake basin (Wells et al., 1987; Reheis et al., 1989) as well as those methods discussed above, we infer the following local correlations:

Those alluvial fans postdating the youngest shoreline (L2), units Qf5, Qf4, and Qf3 of the Zzyzx fans, correlate with the late Holocene Qf5–6, middle Holocene Qf4, and early Holocene Qf3, respectively, in the Silver Lake area.

Those alluvial fans forming between shorelines L1 and L2, unit Qf2, correlate with the latest Pleistocene to early Holocene Qf2 observed in the Silver Lake basin.

Those alluvial fans predating the highest shoreline L1, units Qf1 and Qf0, correlate with late Pleistocene Qf1 and middle (?) Pleistocene Qf0 alluvial fans of the northern Soda Mountains area.

A concern over the use of soil-geomorphic data for providing age estimations and for establishing regional soil-geomorphic correlations is that much remains unknown about how rates of pavement evolution and soil development vary with lithology. Many chronosequence studies have been limited to deposits composed of similar lithologies, e.g., igneous and metamorphic siliceous lithologies (Birkeland, 1990), or calcareous lithologies (Lattman, 1973; Gerson et al., 1985; Amit and Gerson, 1986; Harden et al., 1991; Amit et al., 1993; Al-Farraj and Harvey, 2000). Direct comparison among chronosequences composed of contrasting lithologies has been difficult because of a lack of adequate age control. Recent advances in Quaternary numerical dating techniques (luminescence and cosmogenic surface exposure dating), however, have enhanced the use of Quaternary alluvial units and geomorphic surfaces for interpreting local and regional geologic and climatic events within the Mojave Desert (Wells et al., 1990b, 1995, 1998; McDonald, 1994; McDonald and McFadden, 1994; McDonald et al., 1996). Using the Harden (1982) soil development indices (SDI), McDonald (1994), McDonald and McFadden (1994) and McDonald et al. (1996) have demonstrated that correlations yield reasonably consistent age ranges for alluvial deposits across the Mojave. Using the regional correlations of McDonald and McFadden (1994) and McDonald et al. (1996) and new dating results (Wells et al., 1995; Anderson and Wells, this volume, Chapter 6), we have refined the age estimates for the deposition of the Zzyzx fan sequences in the southern Soda Lake basin (Fig. 5).

The minimum age for our oldest fan unit (Qf0) is suggested by comparisons with soils in the Cima volcanic field (Wells et al., 1995) where stratigraphic relations between alluvial fans and basaltic lava flows are well defined and the lava flow surfaces have multiple cosmogenic surface-exposure dates. Given the minimal degree of Bt horizon preservation but the stage III pedogenic

	S Soda Mts (Zzyzx) This study	Estimated age (ka) a	Pluvial Lake Mojave b	Silver Lake and vicinity c	Providence Mountains d	Cima Piedmont e
HOLOCENE Late	Modern sediments			Qf_5	Qf_9	Qf_9
HOLOCENE Late	Qf_5	< 1.8		Qf_5 [Qe3]	Qf_8	Qf_8
HOLOCENE Middle	Qf_4	4.3–3.5		Qf_4 (3.4)	Qf_7 (4.4–2.2)	Qf_7
HOLOCENE Early	Qf_3	9.3–5.2		Qf_3 [Qe2]	Qf_6 (5.4–2.8)	Qf_6
HOLOCENE Early	L_2	10–9.3	Youngest shoreline	Ql_2 (8.4)		
PLEISTOCENE Late	Qf_2	11.5–9.3	Intermittent lake	Qf_2 (10.3–9.2) Qe_1	Qf_5 (18.1–10.4)	Qf_5
PLEISTOCENE Late	L_1	14–11.5	Lake Mojave II (14–11.5)	Ql_1 (20.3–14.7)		
PLEISTOCENE Late	L_1	18.4–16.6	Lake Mojave I (18.4–16.6)	Ql_1 (20.3–14.7)		
PLEISTOCENE Late	Qf_1	c 34–18.4		Qf_1	Qf_4 (28.7–26.9)	Qf_4
PLEISTOCENE Late						Qv_5 (72–65)
PLEISTOCENE Middle	Qf_0	> 68			Qf_3	
PLEISTOCENE Middle						Qv_4 (130)
PLEISTOCENE Middle						Qv_3 (170–150)
PLEISTOCENE Middle				Qf_2 (=Bishop Ash 780)		

Figure 5. Regional correlation between the late Quaternary sedimentary units recognized on the Zzyzx fans and those recognized on other fan groups within the eastern Mojave Desert. a—date ka, adopted in this study; b—from Brown et al. (1990), Wells et al. (1990a, 1997, this volume); c—from Reheis et al. (1989), Wells et al. (1997, 1998), Anderson and Wells (this volume, Chapter 6); d—from McDonald (1994), McDonald and McFadden (1994), McDonald et al. (1996), Wang et al. (1996); e—from McDonald and McFadden (1994), Wells et al. (1995).

$CaCO_3$ accumulation associated with Qf0, these fan deposits may be correlated with Cima alluvial fan unit Qf3 in the Cima volcanic field (McDonald and McFadden, 1994). The fan deposits in the Cima area are overlain by a basaltic lava flow (unit Qv4 of Wells et al., 1995), which has surface exposure ages of 65 ± 9 ka and 72 ± 7 ka. The lack of a well-developed Bkm (pedogenic petrocalcic and capping laminar layer) on Qf0 soils would suggest an age less than that of the Providence Mountains alluvial fans interbedded with the 0.74 Ma Bishop ash (unit Qf2 of McDonald and McFadden, 1994, and McDonald et al., 1996).

Alluvial fan unit Qf1 predates the high shoreline of Lake Mojave (L1, 18.4–16.6 ka) and postdates the buried sediments of unit Qf0. The suggested correlation of our Qf1 with alluvial fan unit Qf4 in the Cima volcanic field (McDonald and McFadden, 1994) suggests that Qf1 was deposited after ca. 34 ka, the average age of two cosmogenic surface exposure ages derived from a basaltic flow underlying that deposit. These bracketing ages are supported by recent work of Anderson and Wells (this volume, Chapter 6), in which Qf1 is correlated with alluvial fan deposits in the Dumont basin, north of Silver Lake, which have been radiocarbon-dated to between 30 and 18 ka. We favor an age closer to 30 ka because the relative degree of stone pavement and soil development appear to require at least >10,000 yr to form (Machette, 1985; McFadden et al., 1989; Amit and Gerson, 1986; Amit et al., 1993).

Alluvial fan unit Qf2 postdates the high shoreline zone (L1, 18.4–11.5) but predates the low shoreline (L2, 10–9.3 ka), thus falling within the late Pleistocene to early Holocene transition period (Wells et al., 1987, 1994, this volume; Harvey et al., 1999). Alluvial fan units Qf3, Qf4, and Qf5 postdate the low shoreline (L2, ca. 9.3 ka), and are therefore are all of Holocene age. Correlations between the Soda Lake basin fans and the IRSL- (infrared stimulated luminescence) dated alluvial fans of the Providence area and the radiocarbon-dated sequences of Silver Lake basin (Wells et al., 1987) and the Dumont basin (Anderson and Wells, this volume, Chapter 6) yield specific age estimates for the deposition of Qf3 between 9.3 and 5.2 ka, Qf4 between 4.3 and 3.5 ka, and Qf5 to post-1.8 ka.

LATE QUATERNARY GEOMORPHIC DEVELOPMENT OF THE ALLUVIAL FANS

The alluvial fans along the southern Soda Mountain front differ from one another in their age relations, morphology, and sedimentology, reflecting variations in their source-area geomorphology (Harvey and Wells, 1994; Fig. 1). The distal reaches of the all fans, excluding SOL and NOR (Fig. 2) toe out upon and interfinger with the fine-grained fill of the Soda Lake playa. Thus, the distal reaches of the alluvial fans have responded to a progressive basin-filling (Wells et al., 1989, this volume). Since the local lowering of base level through the fall in lake level to ca. 9.3 ka, the playa floor has been accumulating sediment. Over the past 10 ka, sedimentation rates on playa floors within this region range up to ~1.2 m/1000 yr (Wells et al., this volume). Thus, the distal geometry of the fan complexes may in part reflect a rise in base level of up to ~10 m during the Holocene.

The spatial distribution of sedimentary facies (Table 1) and stratigraphic units within the 15 alluvial fan complexes is illustrated in Figure 2. Not every alluvial fan complex exhibits the complete sequence of stratigraphic units; rather, some fan complexes are dominated by older stratigraphic units (i.e., SOL, NOR, SON on Figure 2) and others have extensive areas dominated by younger units (i.e., CV, VLT on Figure 2). On a limited number of fans sediments of the younger stratigraphic units are present near the fan apex (e.g., MSQ, VLT on Figure 2), whereas, on the majority of fans the younger stratigraphic units dominate the distal fan zones.

Sedimentology, morphology, and chronology of the alluvial fan complexes

At the northern end of the mountain front, two small fans (NOR, SOL, Figure 2) issue from relatively small drainage basins with moderate relief developed on granitic bedrock (Fig. 1). The distal fans are topographically above the former lake levels but have been trimmed by a channel of a large alluvial fan that wraps around the mountain margin. Both fan complexes are dominated by late Pleistocene sediments (unit Qf1, Figure 2). Proximal debris flows (D1 facies grading downfan to D2; Table 1) give way downfan to alternating debris flows and fluvial sheet gravels (D2 and S2 facies on Table 1; Fig. 6, sections 1, 2). The Qf1 fan surfaces are incised by ephemeral channels throughout their lengths, and Holocene fan deposits (primarily Qf4) occur within the wider reaches of these fan trenches. In the apices of these complexes, two generations of hillslope debris flows were deposited directly on the older fan surfaces. The older set of hillslope deposits (D1 facies, Table 1) are either late unit Qf1 or Qf2 age, and the younger hillslope deposits are Qf4 transitional deposits (T facies, Table 1).

The next set of alluvial fan complexes to the south are VLT, MSQ, and STV fans (Fig. 2), whose source areas are primarily in metavolcanic rocks, but with granitic terrain in the upper part of the catchment feeding VLT fan (Fig. 1). The distal portions of these complexes either are trimmed by the shorelines or prograde across the playa floor. On the VLT complex only the upper shoreline (L1, Figure 1) is clearly observed, but both upper and lower shorelines (L1, L2, Figure 1) are present on MSQ and STV fan complexes. These fan complexes are dominated by late Pleistocene (unit Qf1) sediments, debris flows (D1, D2 facies, Table 1) on STV fan and proximal debris flows, grading distally to fluvial sediments on VLT and MSQ fans (Fig. 2). Field relations provided evidence for a late phase of Qf1 sedimentation of significantly coarser bouldery D1 facies than in the earlier depositional events, which were dominated by more fluid debris flows (D2 facies; Figure 6, sections 3, 4). The late Pleistocene fan sediments are trimmed by shoreline L1 (Fig. 1), indicating that primary fan sedimentation had ceased in these reaches before 18.6 ka. Fan trenching followed Qf1 deposition, eroding through the L1 shoreline. Small volumes of unit Qf2 were deposited as insets within the trench and in the distal zones of VLT and MSQ fans topographically below the upper shoreline. During the Holocene, VLT and MSQ fan complexes have undergone proximal trenching and distal progradation of younger (mostly Qf4) fluvial sediments (S2, S3, Table 1). During the middle Holocene, young debris flows (Qf4) were deposited proximally, especially on MSQ fan complex (Fig. 2). On STV fan complex the L1 shoreline truncates Qf1 debris flows. Latest Pleistocene–early Holocene (Qf2) debris flows were deposited across the L1 shoreline, and prograded to beyond that shoreline, where these deposits were then trimmed by the L2 shoreline. Holocene Qf3, Qf4, and Qf5 transitional and fluvial deposits are inset below the L2 shoreline, prograding onto the Soda Lake playa.

The two largest fan complexes (JNF, CV, Figure 2) issue from catchments dominated by metavolcanic rocks, but with small exposures of granitic rocks in the northern subcatchment of JNF fan and limestone in the southern subcatchment of CV fan (Fig. 1). Both fan complexes are dominated by late Pleistocene (Qf1) sediments composed of proximal debris flows grading to stratified fluvial sediments in midfan (Fig. 6, sections 5–10). On the JNF complex, middle Pleistocene Qf0 sediments and an associated petrocalcic horizon are exposed by fan trenching in midfan and underlie the Qf1 sediments (Fig. 6, sections 5, 6). On both fan complexes major hillslope debris flows of late Qf1 or Qf2 age toe out on the Qf1 fan surfaces (Fig. 6, section 10). On both fan complexes, the distal portions of Qf1 segments are trimmed by shoreline L1 (Fig. 2). The Qf1 deposits were trenched proximally prior to the deposition of unit Qf2, which forms inset terraces within the fanhead trenches (Fig. 6, section 8), and then prograde from former intersection points to extensive distal fan surfaces of fluvial sediments. They are cut by shoreline L2 (Fig. 2). Further incision then occurred in the fanhead trenches of both fan complexes and Holocene (Qf3, Qf4, and Qf5) fluvial sediments were deposited as inset terraces within the fanhead trenches and prograded distally (Fig. 2).

South of CV fan complex there are several small debris cones (referred to as PC on Figure 2), and larger debris cones (GTE) of the composite JSH/GTE fan complexes (Fig. 2) that have source areas in metavolcanics. Both shorelines are well preserved along the mountain front in this area. The small PC cones and GTE fan complex are characterized by late Pleistocene (Qf1) debris flows (D1, Table 1) which have been eroded by the L1 shoreline to form a steep wavecut scarp. Debris flows of unit Qf2 age are deposited in trenches that cut shoreline L1, but these deposits are trimmed by shoreline L2, forming a smaller wavecut scarp. Holocene (units Qf3, Qf4, and Qf5) transitional deposits (T facies, Table 1) are set in trenches cut through both shorelines and prograde short distances away from the wavecut scarps. The JSH fan is more complex with proximal Qf1 debris flows (D1 facies, Table 1) changing distally to Qf1 or younger Qf2 fluid debris flows (D2 facies, Table 1) and fluvial sediments (S2 facies, Table 1). As elsewhere, exposures in latest Pleistocene sediments show the youngest parts of unit Qf1 to be coarse bouldery debris flows (D1 facies, Table 1) over older finer debris flows (D2 facies, Table 1; Figure 7, section 11). The shorelines are not observed on JSH complex because of

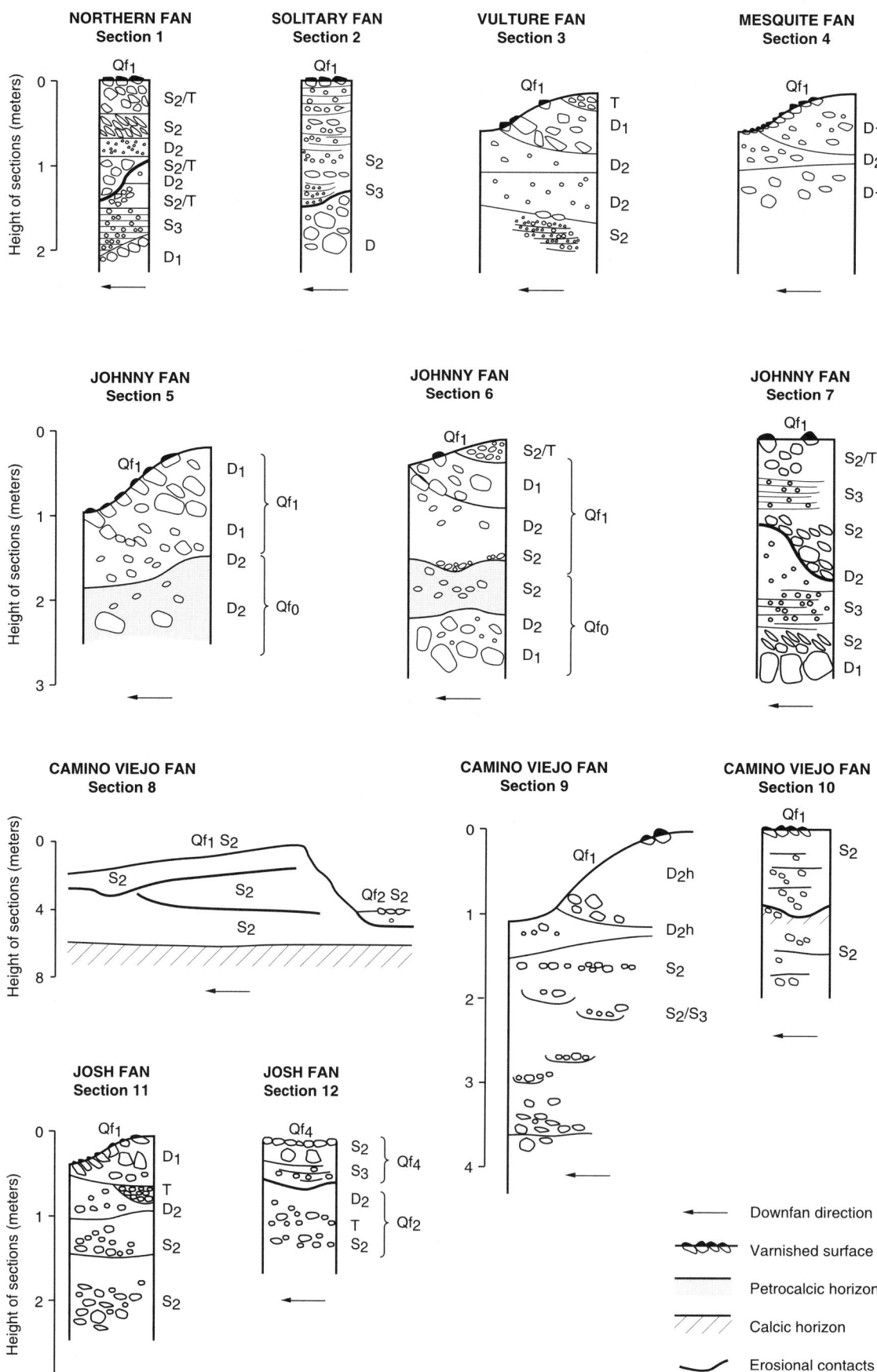

Figure 6. Schematic stratigraphic sections (see Fig. 1 for locations).

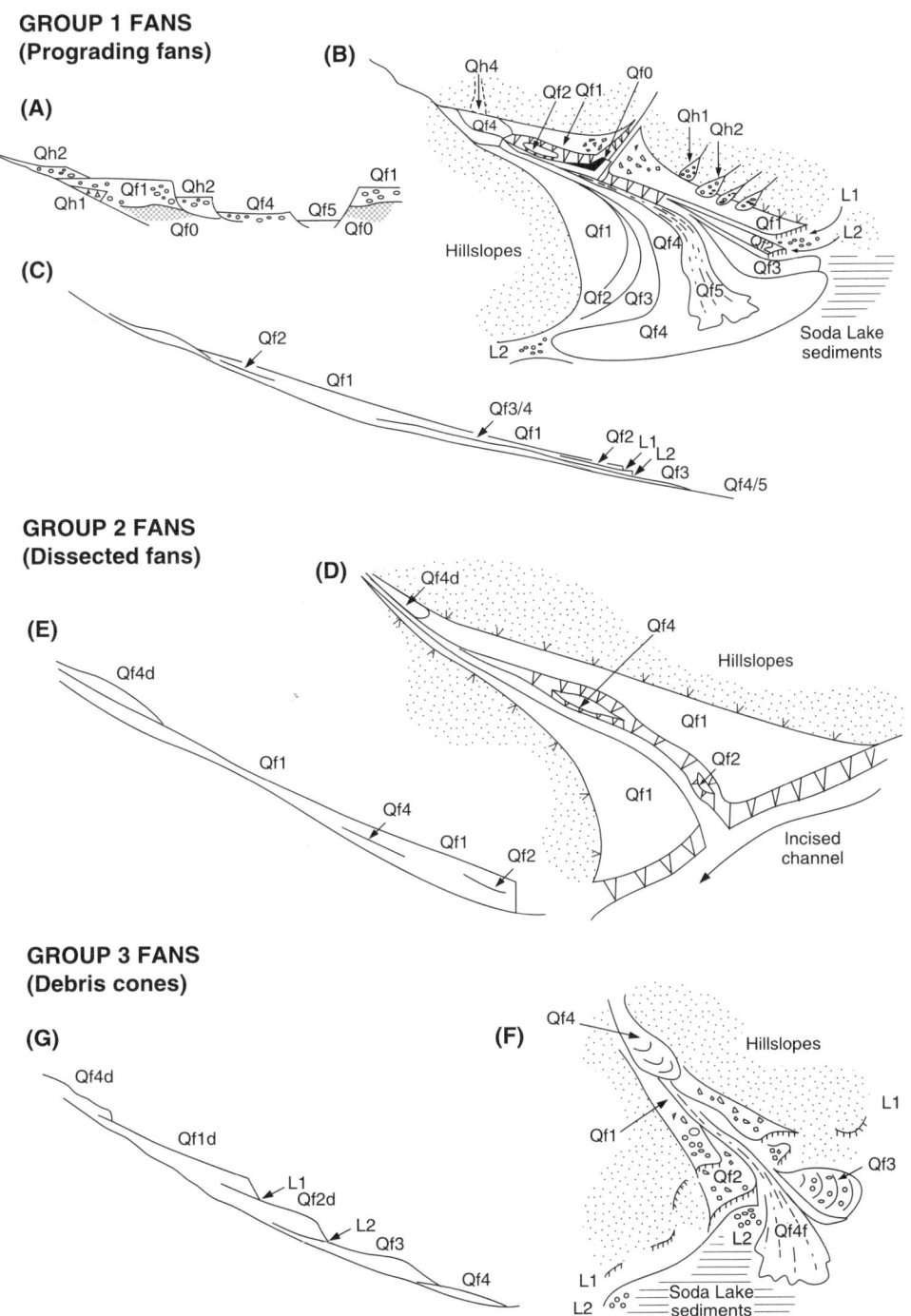

Fig. 7. Schematic models of fan development for the fan groups identified on Table 5, showing (A) schematic cross-fan profile for Group 1 fans, (B) schematic model and (C) summary downfan profile for Group 1 fans, (D) schematic model and (E) summary downfan profile for Group 2 fans, and (F) schematic model and (G) summary downfan profile for Group 3 fans.

extensive middle Holocene (Qf4) deposition of fluvial sheet gravels (Fig. 6, section 12) in the distal zone.

The southernmost fan complexes (SPR, ZZX, and SON, SOC, and SOS fans, Figure 2) are fed by steep drainage basins developed dominantly on granitic rocks (Fig. 1) with smaller areas on metavolcanics and limestone. Both shorelines are visible on ZZX and SON fans but have been obscured in the distal areas of SOS (Fig. 2) by human alteration of the fan surface. The apices of the larger fan complexes (ZZX, SON, SOS, Figure 2) show late Pleistocene massive bouldery debris-flow deposits (D1 facies, Table 1), grading distally into finer, more fluid debris flows or fluvial deposits (D2 or S2 facies, Table 1). On each fan complex the surfaces of the late Pleistocene fans are trenched, and on the larger fans younger, dominantly fluvial deposits, prograde across the shorelines (Qf2 on ZZX and SON fans, and Qf3–5 throughout). The distal end of SPR fan complex has been truncated by ZZX fan, and the distal deposits on SOC fan complex are middle Holocene (Qf4) debris flows.

LATE QUATERNARY VARIATIONS IN ALLUVIAL FAN PROCESSES AND DEVELOPMENTAL STYLES

Classification of fan evolutionary style

During fan development, changes in the supply of water and sediment to a fan may cause changes in the relative importance and spatial distribution of erosion and deposition. This will result in changes in the morphological style of the fan. We have recognized five process-based morphological styles in the Zzyzx fan complexes (Table 4; see also Harvey, 2003), and have grouped the fans on the basis of similar evolutionary sequences of fan style over the period of fan development. From the field evidence mapped and summarized on Figure 5, and with the fan surfaces being assigned to age groups according to their surface and soil properties (Table 2; Fig. 2), it is possible to classify the fans at two scales.

Classification at the scale of the fan complex

At the scale of the fan complex, we identify three main styles of fan evolution that have produced the present-day plan-view assemblages of stratigraphic and sedimentologic units and characteristic longitudinal profiles (Fig. 7). The larger, fluvially dominated fan complexes (VLT, MSQ, JNF, CV, JSH, ZZX, SON, SOS on Figure 2) show plan-forms of typical telescopic fans (Harvey, 1997), and long-profiles with successive intersection points (Fig. 7A–C). Several fan complexes show styles dominated by dissection (Fig. 7D–E). The two northern fans (NOR, SOL, Figure 2) show some dissection and relatively little Holocene depositional activity. The SPR fan complex shows plan and long-profile characteristics dominated by local base level–induced dissection and is tributary to ZZX fan, whose incising channel provides a local base level for SPR fan. The final major group of fan complexes comprises all the smaller fans and debris cones (STV, CV and PC cones, GTE fans, and SOC fan, Figure 2). These steep fans and cones all show very minor trenching with spatially limited progradation that is restricted to distal lobes (Fig. 7F–G).

Classification at the scale of the individual fan apex/segment

At the scale of the 32 discrete fan apices recognized in the field (Fig. 2), it is possible to identify the fan segments fed by each apex, and to classify each by the dominant process style at each of the stages of fan evolution recognized (Table 5). For each time period represented by an alluvial fan stratigraphic unit, the dominant geomorphic style for each apex-fed fan segment has been assessed and the fan segment classified into one of 3 major groups (Table 5—note that two groups include subgroups). This classification allows trends in the geomorphic evolution of the fan complexes to be identified (Fig. 8).

Group 1 fan segments (Table 5; Fig. 8) are all large alluvial fans. Group 1a (VLT 5, 6, JNF 12, CV 19, JSH 27, ZZX 28, SON 30, SOS 32; for locations see Figure 2), show a characteristic aggradation to progradation trend with a progressive shift from debris flow to fluvial processes from the late Pleistocene into the Holocene. On Group 1b fans (MSQ 7, 8), the mid-Holocene (Qf4) sequences are complicated by significant debris-flow activity.

Group 2 fan segments are dissected fans and comprise two subgroups: Group 2a (NOR 1 and SOL 2) dissected in relation to the channel of the neighboring large fan complex, and Group 2b (JNF 11, CV 17, 18, SPR 29), tributary fans, where dissection is related to local base-level lowering, following the incision of the main fan channels.

TABLE 4. PROCESS-BASED FAN MORPHOLOGICAL STYLES, RECOGNIZED IN THE ALLUVIAL FANS OF THE SOUTHERN SODA MOUNTAINS

1. Aggradation	On nontrenched fan complexes, defined as deposition that takes place near fan apices and that may extend distally over much of the fan surfaces. On fan surfaces that had been previously trenched within the fanhead area, younger fan deposits that may occur within the trench are also a form of aggradation.
2. Progradation	This is the case of proximal trenching by the trunk channel and deposition on the distal fan surfaces. Distal deposition may extend the distal limit of the fan complex.
3. Dissection	With little addition of sediment to the fan complex, processes of dissection may occur. These include trenching of the fan surface by the trunk channel or incision by channels heading on the fan surface. The main zone of dissection may be at the fan apex, midfan, or in distal fan areas.
4. Complex behavior	Alluvial fans may show combinations of erosion and deposition unrelated to each other, such as proximal aggradation concurrent with distal dissection.
5. Stabilization and/or passive or limited activity	This is the case where under current conditions there is little or no sediment provided to, or erosion from, the fan complex. There are no fresh erosional forms, with the whole fan surface showing evidence of stability, through stone pavement and soil development. The implication is that there has been minimal erosion or deposition for a period longer than that required for the initiation of identifiable soil and pavement formation to take place.

Note: See also Harvey (2003).

TABLE 5. CHANGES IN FAN STYLE THROUGH THE LATE PLEISTOCENE AND HOLOCENE FOR THE FAN SEGMENTS FED BY THE 32 FAN APICES IDENTIFIED IN FIGURE 2

Fan	D	B	Pleistocene			Holocene	
			Qf1	Qf2	Qf3	Qf4	Qf5
Group 1a							
VULT-5	.056	.34	Ac	Pf	Pf	Ac	Pf
VULT-6	.237	.38	Ac	Pf	Pf	Ac	Pf
JN-12	.349	.30	Ac	Pf	Pf	E(Pf)	Pf
CV-19	.455	.25	Af	Pf	Pf	Pf	Pf
JSH-27	.205	.28	Ac	Ad	Pf	Af	Pf
ZZX-28	.256	.30	Ac	Pf	Af	Pf	Pf
SON-30	.187	.34	Ac	Pf	Af	Af	Pf
SOS-32	.262	.35	Ac	Pf	Pf	Pf	Pf
Group 1b							
MSQ-7	.062	.43	Ac	Pf	X	Ad	E(Pf)
MSQ-8	.097	.36	Ad	Pf	X	Ad	Pf
Group 2a							
NOR-1	.037	.33	Ac	E(Ad)	Z	E(Ad)	X
SOL-2	.065	.34	Ac	Z	E(X)	Ac	X
Group 2b							
JN-11	.293	.41	Ac	Pf	Af	Af	X
CV-17	.224	.33	Ac	Pf	X	Pf	X
CV-18	.108	.39	Ac	Pf	X	Pf	X
SPR-29	.104	.35	Ad	X	X	X	X
Group 3							
SOL-3	.012	.44	Ad	Z	X	Z	Z
SOL-4	.006	.33	Ad	Z	X	Z	Z
STV-9	.072	.40	Ad	Pd	Z	E(Pd)	X
JN-10	.039	.41	Ad	Pd	Z	X	Z
JN-13	.059	.46	Ad	Z	Z	Z	Z
CV-14	.019	.38	Ad	Pd	Z	Z	Z
CV-15	.062	.32	Ad	Pd	Z	Pd	Z
CV-16	.037	.36	Ad	Ad	Z	Pd	Z
CV-20	.010	.44	Ad	Pd	Pd	Pd	Pd
CV-21	.012	.40	Ad	Pd	Pd	Pd	X
PC-22	.033	.41	Ad	Pd	Z	Pd	Z
PC-23	.010	.42	Ad	Pd	Z	Pd	Z
GT-24	.020	.45	Ad	Pd	X	Pd	Z
GT-25	.062	.45	Ad	Pd	Z	E(Pd)	Z
GT-26	.037	.44	Ad	Ad	Z	Ad	Z
SOC-31	.076	.44	Ad	Ad	X	Pd	Pd

Note: D—Drainage area to fan (km^2); B—Basin slope (bedrock relief/bedrock basin length). Fan styles: A—aggradation, P—progradation, f—fluvial, c—composite, d—debris flow, X—dissection, Z—passive and/or limited activity, E—complex behavior (dominant behavior shown in parentheses).

Group 3 fan segments are all relatively small fans or cones (Table 5). This group shows a sequence of latest Pleistocene debris-flow aggradation followed by latest Pleistocene to early Holocene progradation by debris-flow processes. There is a trend during the Holocene toward limited activity and stabilization. Only the larger fans in this grouping (STV 9; CV cones 20, 21; SOC 31) show continued progradational activity during the Holocene.

Trends in late Quaternary evolution of the fan groups

The morphological trends over the late Quaternary are summarized by the schematic models shown on Figure 7, and by the summary Table 6 and Figure 8. Several overall trends can be identified. (1) There is a trend in sedimentary processes over the past 30 ka from debris-flow activity toward fluvial processes, especially for the larger fans (Table 6; Fig. 8). During the late Pleistocene, 56% of the fan segments were aggrading by debris-flow processes, and by mid Holocene only 13% were experiencing debris-flow aggradation. (2) There is a trend in depositional style from aggradation toward progradation in both large and small fans (Table 6; Figure 8). During the late Pleistocene, all of the fan segments were aggrading, but by late Holocene no fans had an aggradational style. (3) In addition, there is a trend in the geomorphic style toward dissection, which is most apparent on

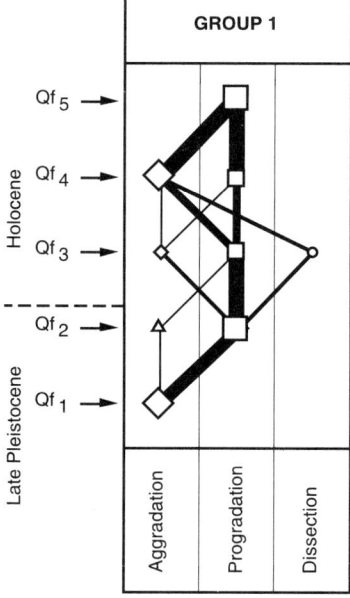
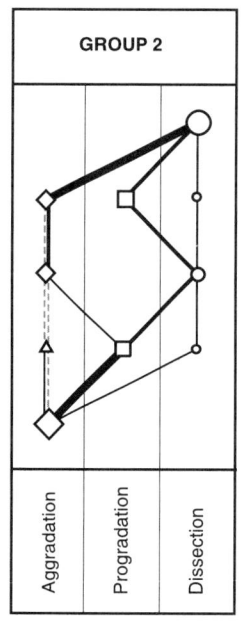
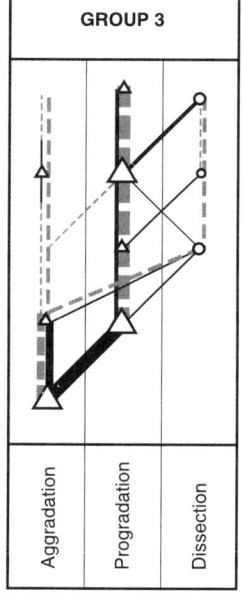

Figure 8. Schematic diagrams to illustrate the changes in fan style (for definitions see Table 4) through late Quaternary time on the three groups of fans (see Table 5). Symbol sizes and line thicknesses are proportional to the number of fans showing specific processes at each time stage, or specific changes in processes between successive time stages, respectively. Dashed lines indicate inactivity, i.e., where there is no evidence of erosional or depositional change between successive time stages, and the morphology is inherited.

tributary fans influenced by falling local base levels caused by incising main channels. During the latest Pleistocene to early Holocene only 3% of the fan segments experienced dissection, but by early Holocene 32% were being dissected (Table 6; Figure 8). (4) Finally, there is a trend in the geomorphic processes toward stabilization of the fan surfaces, especially on the smaller cones (Table 6, Figure 8). During the latest Pleistocene to early Holocene, only 9% of the sampled fans were inactive, however, by late Holocene 38% had become inactive.

Summary of the late Quaternary fan sequences

During the late Pleistocene (between 30 and 18.6 ka, unit Qf1) all fans were aggrading. The hillslopes within the source areas were actively supplying sediment to the fans by a variety of processes (Harvey and Wells, 1994), including rockfall processes forming talus slopes below rock outcrops, slopewash processes forming colluvial mantles on the footslopes, and hillslope debris-flow processes forming debris-flow lobes. Alluvial fan aggradation ranged from debris flows on the smaller fans and near fan apices to fluvial and sheetflood processes on the larger fans. The field evidence suggests that the last phases of hillslope debris-flow activity continued after the cessation of late Pleistocene fan sedimentation and may have continued during the transitional period from latest Pleistocene to the Holocene (Qf2).

During the latest Pleistocene–early Holocene transitional period (ca. 11.5–9.3 ka, unit Qf2), 18% of the sampled fans were aggrading by debris-flow process, 32% were prograding by debris-flow processes, 38% were prograding by fluvial processes, 3% were being dissected, and 9% were relatively inactive (Table 6). The larger fans (Group 1) developed fanhead trenches and prograded distally by fluvial processes. On the largest fans of this group some fluvial deposition occurred within the fanhead trenches (Qf2), later to be dissected to form terraces (Fig. 5). On the smaller fans and cones (Group 3) some dissection within the distal areas occurred following the fall in pluvial Lake Mojave from the higher shoreline. During this transitional period, debris-flow activity was such that depositional lobes prograded beyond shoreline L1. On many hillslopes there was renewed hillslope destabilization very late during Qf1 time or into Qf2 time, resulting in debris-flow lobes extending onto Qf1 fan surfaces (Figs. 5, 7; Harvey and Wells, 1994; Harvey et al., 1999).

Between 8.5 ka and 5.2 ka (Qf3) only 9% of the fan segments were aggrading by composite and fluvial processes; 6% and 18% were prograding by debris-flow and fluvial processes, respectively; 32% were undergoing dissection, and 35% were stabilized. During the early Holocene, hillslope debris-flow activity ceased, and the hillslopes either stabilized or underwent a switch to fluvial erosion focused along the linear headwater channels (Harvey and Wells, 1994). There is little evidence of deposition

TABLE 6. CHANGES IN FAN STYLE: SUMMARY

Category	Pleistocene Qf1 (%)	Qf2 (%)	Qf3 (%)	Holocene Qf4 (%)	Qf5 (%)
Aggradation					
Debris flow	18 (56)	6 (19)	– –	4 (13)	– –
Composite (+Fluv)	14 (44)	– –	3 (9)	6 (19)	– –
Progradation					
Debris flow	– –	10 (31)	2 (6)	10 (31)	1 (3)
Fluvial	– –	12 (38)	6 (19)	6 (19)	11 (34)
Dissection	– –	1 (3)	10 (31)	2 (31)	8 (25)
Passive and/or limited activity	– –	3 (9)	11 (34)	4 (13)	12 (38)

Note: Number of fan apices (and percentages) in each category total 32. On this table, complex behavior has been reclassified according to dominant behavior.

within the fan apices during the early Holocene. However, on the larger fans (Group 1) continued fanhead entrenchment occurred with distal progradation by fluvial or sheetflood deposition. On many of the smaller fans and cones (Group 3) there is little evidence of erosion or deposition during the early Holocene, and these fans appear to have stabilized (Table 6). On a limited number of small fans and cones, distal fan dissection and continued lobe progradation by debris-flow or transitional deposition occurred along the former shoreline zone.

During the middle Holocene (Qf4), 31% of the fans were aggrading again by debris-flow and fluvial processes, 50% were prograding by debris flow and fluvial processes, and 6% were being dissected. Minor renewal of hillslope debris-flow activity between 4.3 ka and 3.5 ka renewed the supply of sediment to the apices of some fans. Otherwise, the middle Holocene was dominated by progressive channel incision in midfan areas and extensive distal progradation by fluvial processes on the larger fans. Distal lobe progradation occurred on many of the smaller fans and cones. There was a general trend for deposition on most fans at this time, ~13% of the fan segments were inactive or stabilized.

In the past 1.8 ka, none of the fan segments have undergone an aggradational style, 37% have prograded by debris-flow and fluvial processes, 25% have been dissected, and 38% are stable. There is little field evidence for hillslope instability by mass movement. Rather, active channel incision occurs in the steeper headwater areas (Harvey and Wells, 1994). The larger fans appear to have been and are currently prograding. The late Holocene fan deposits (Qf5) are less extensive than the middle Holocene deposits. Most of the smaller cones appear to be inactive.

DISCUSSION

Factors controlling alluvial fan processes

The changes that have occurred since the late Pleistocene in the geomorphology of the Zzyzx fans are responses to climatic change. The Soda Mountain front is tectonically stable, and tectonics cannot have had any influence on late Quaternary fan dynamics. Base level initially fell with the fall in lake level, then would have risen with sedimentation on the playa floor during the Holocene. However, base-level change has had little effect on fan dynamics, and that effect has been restricted to the shoreline zone, primarily on the smaller fans. Fanhead trenching has occurred well above the shoreline zone, especially on the larger fan complexes, and is clearly related to changes in hydrology and sediment supply. Moreover, similar late Quaternary sequences of erosion and deposition can be seen throughout the Zzyzx fans, and can be broadly correlated with similar sequences throughout the region (Fig. 6). This indicates regional climatically led changes in hydrology and sediment supply as the primary causes of the observed changes in fan dynamics. The major change in fan style occurred between the Pleistocene and the Holocene, as an overall trend toward greater fluvial activity. This was not the result of exhaustion of the sediment reservoir in the catchment (cf. Bull, 1991); there is ample hillslope sediment availability, but was the result of climatically induced changes in hillslope processes (Harvey and Wells, 1994).

Although we can identify the overall trends, they were not uniform from group to group over the last 30 ka. Within this overall climatically led context, variations between the fan complexes must relate to the operation of geomorphic thresholds (Schumm, 1979). Three important thresholds may be operating:

1. between debris-flow and fluvial or sheetflood processes, controlled by water:sediment ratios supplied to the fan (Wells and Harvey, 1987);

2. between erosional and depositional processes, controlled by the threshold of critical stream power (Bull, 1979), defined as the relationship between actual stream power and critical power, i.e., the power required to transport the sediment supplied;

3. between activity and inactivity, which may simply be a reflection of runoff power, under conditions of very limited sediment supply.

All three thresholds are controlled by water and sediment supply from the catchments, and hence by drainage area and relief characteristics. For catchments of similar climate and geology they can be defined by measures of drainage basin area and relief (Wells and Harvey, 1987). Larger catchments have greater runoff and their fans would tend to be fluvially dominant, and steeper catchments would have higher rates of sediment supply and therefore be dominated by debris flows (Harvey, 1997). Fans from small, perhaps less steep catchments would be more likely to stabilize.

Generalized catchment thresholds between debris-flow and fluvially dominant deposition have previously been defined for the Zzyzx fans (Harvey, 1992a). Now we are able to refine that definition by considering the thresholds for the catchments of the 32 fan segments, and how the thresholds change over the Qf1–5 time scale. On Figure 9 fan processes on the 32 fan segments have been plotted in relation to drainage basin area and basin slope (Table 5), for the five time-periods (Qf1–5). On each plot, the fan styles tend to cluster, suggesting threshold conditions separating the processes.

At any one time, not all of the three threshold types defined above may operate. During the late Pleistocene (Qf1) a threshold distinguishes between fluvial and debris-flow aggradation. Then during Qf2, after fanhead trenching on most fans, a similar threshold distinguishes progradation by fluvial and sheetflood processes from progradation by debris-flow lobes. During the Holocene there is a reduction in debris-flow activity and more fans undergo dissection.

During the mid Holocene (Qf4), when hillslope debris flows were again active, a threshold similar to those during Qf1 and Qf2 times distinguishes debris-flow and fluvial-dominant fans. During the intervening Qf3 and then during Qf5, when hillslope debris flows were not active a rather different threshold is apparent, based more on drainage area than on basin slope, distinguishing continued fluvial progradation on the larger fans from inactivity on most of the smaller fans.

Throughout, fluvial processes are associated with the larger catchments and debris flows, during periods of hillslope instability, with the smaller steeper catchments. There are changes through the time sequence; during the Holocene the fluvial realm extends into smaller and steeper catchments than during the late Pleistocene. More important perhaps, is the general consistency of the threshold separating fluvial processes on the larger fans from either debris flows (during Qf1, 2, 4 periods of hillslope activity) or fan stability (during Qf3, 5 periods of hillslope inactivity). Despite the differences in bedrock geology, little distinction can be made between fan catchments on granitic and metavolcanic bedrock. The larger fans (from groups 1a, 1b on Table 5) switch from late Pleistocene composite (proximal debris flows, distal fluvial and sheetflood deposits) aggradational fans to Holocene prograding fluvially dominant fans. The smaller fans (largely composing Group 3 on Table 5), switch from late Pleistocene debris-flow activity to Holocene stability, with occasional reactivation by debris flows particularly in mid Holocene (Qf4 time). Only a few fans lie close to the thresholds between these two groups (fans 1, 2, 5, 8, 9, 16, 18, 29). They are of intermediate size, and some do show switches between modes. Many are in groups 2a and 2b (Table 5), and have been subject to local base level–induced dissection. Also, 4 of the 8 are granite-fed fans, though whether this is significant is not certain.

In summary, the changes in fan style are broadly synchronous along the mountain front of the southern Soda Mountains, and indicate response to climatically induced changes in water and sediment supply to the fans. The ways in which individual fans responded to these changes are conditioned by catchment-related thresholds.

Regional climatic implications

The variations in fan behavior at Zzyzx reflect temporal variations in runoff power and sediment supply from the catchment hillslopes during the late Quaternary, and may be partly related to vegetation cover on the hillslopes and partly to climatic and hillslope hydrological conditions. Previous climatic reconstructions, based on paleovegetation reconstructions from data derived from packrat middens, or hydrological reconstructions based on pluvial lake levels (see above), allow distinction between regional and local paleoclimates. Paleovegetation data (see above) suggest that even during the late Pleistocene the hillslopes of the southern Soda Mountains supported temperate desert scrub vegetation, rather than Juniper woodland as would have been the case in higher mountain areas (Harvey et al., 1999). Lake-level data express the regional climatic and hydrological characteristics, but not necessarily those directly related to local geomorphic controls. Fluctuations in the levels of pluvial Lake Mojave reflect conditions in the headwaters of the Mojave River system in the Transverse Ranges of southern California, rather than climatic conditions in the Mojave Desert itself (Wells et al., 1990a, 1998), but the geomorphic processes on the hillslopes of the southern Soda Mountains must reflect local climatic and hydrological conditions.

The late Pleistocene (Qf1 time) was a period of excess sediment supply from the hillslopes and aggradation on the fans. The high sediment availability and the evidence for widespread mass movement on the hillslopes (Harvey and Wells, 1994) suggests a climate perhaps with higher bedrock weathering rates and certainly much higher soil moistures than at present. This implies a climate probably colder than at present and certainly much wetter. The Pleistocene to Holocene transition (Qf2 time) was the last time of major hillslope debris-flow activity, again implying hillslope instability through slope failure and mass movement, with higher soil moistures and a wetter climate than today. However, fanhead trenching on the larger fans suggests a change in critical power relationships by an increase in runoff power in relation to Qf1 times. This supports the regional evidence for the incursions of tropical air under "monsoonal" conditions, bringing high storm rainfall intensities (Wells and Dohrenwend, 1985; Harvey et al., 1999).

During the Holocene (Qf3–5 time) a change in geomorphic regime on the hillslopes is evident, whereby widespread mass movement dominantly by debris flows in the late Pleistocene to early Holocene (Qf1–2), gave way to largely stable hillslopes affected primarily by linear fluvial incision in the headwater channels during the Holocene (Harvey and Wells, 1994). This implies a decrease in soil moistures but an increase in the effec-

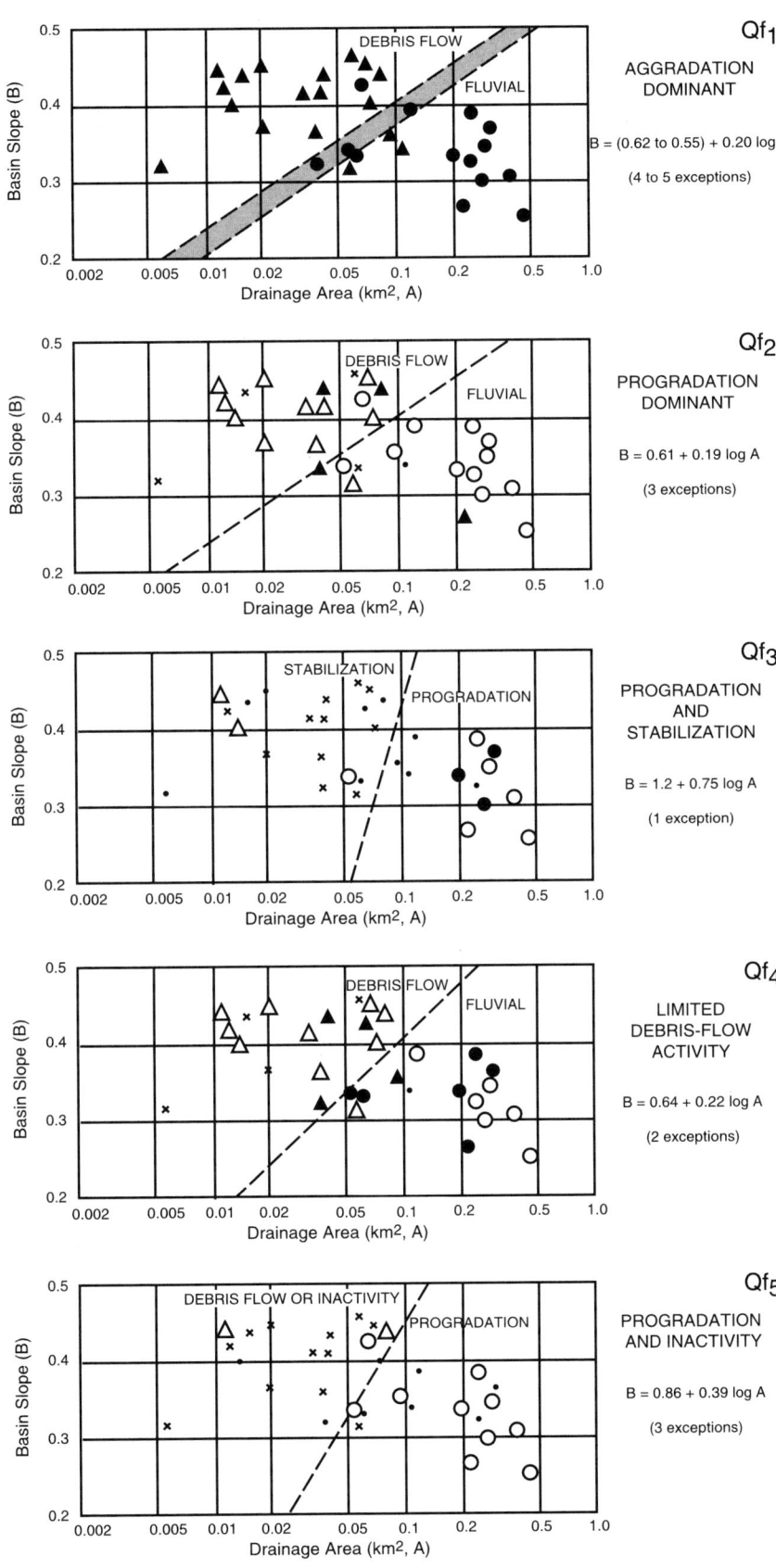

Figure 9. Dominant fan processes at times Qf1–Qf5, for the 32 fan segments, plotted against bedrock drainage area and bedrock basin slope (for definitions see Table 4). Dashed lines, and equations quoted, are for suggested threshold conditions (see text for explanation); exceptions relate to the number of anomalous cases, dissecting (local base level-controlled) fans excluded.

tiveness of storm runoff power (Bull and Schick, 1979), in keeping with a Holocene trend toward greater aridity (Baker, 1977).

The Qf4 period (ca. 4.4–3.5 ka) differs from Qf3 and Qf5 periods, in that there is evidence for localized hillslope debris-flow activity, and limited debris-flow supply to fan proximal locations. There is also evidence for continued fanhead trenching and widespread distal progradation. These trends suggest perhaps a mid-Holocene wetter period. The hillslope debris-flow activity, and especially the fanhead incision and distal progradation suggest a period of greater storm magnitude and greater runoff power.

CONCLUSIONS

1. The sedimentary and morphological sequences identified on the Zzyzx alluvial fans at the mountain front of the southern Soda Mountains accord with the emerging regional picture of late Quaternary alluvial fan sequences in the eastern Mojave Desert. The sequences identified at Zzyzx accord with those identified throughout the eastern Mojave Desert. The evidence suggests response to late Quaternary regional climatic changes.

2. In the absence of tectonic activity, or of base-level changes effective in causing major changes in fan morphology, the overriding control of alluvial fan dynamics is climatic change. Climatic change governs the generation of water and sediment from the hillslopes of the alluvial fan catchments. Of critical importance appear to be soil moisture and weathering characteristics, influencing sediment generation by mass movement processes on the hillslopes especially during the late Pleistocene, together with the incidence of intense storm rainfalls, affecting runoff power since the late Pleistocene.

3. Alluvial fan response to regional patterns of climatic change is conditioned by local intrinsic thresholds, relating to the generation of water and sediment from the fan catchment hillslopes. The threshold conditions are subject to change as the climate itself changes.

ACKNOWLEDGMENTS

We are grateful to the graphics section of the Department of Geography, University of Liverpool, for producing the diagrams. Wells acknowledges support of his field work by the New Mexico Water Resource Research Institute, the U.S. Geological Survey, and NASA (via a subcontract from the University of Washington). Harvey is grateful to the Royal Society for a grant toward the costs of field work at Zzyzx and to the Fulbright Foundation for the award of a Fulbright Scholarship, to enable him to work at the Desert Research Institute at Reno and to carry out further field work at Zzyzx. We thank our coworkers on Mojave Desert soils (L. McFadden, Y. Enzel, T. Royek, T. Skirvin, C. Renault) for field assistance and for discussions on the implications of the soil properties. The late Jean-Luc Moissec initiated the work at Zzyzx. He was tragically killed in a road accident while returning from his field season at Zzyzx. This study represents the culmination of work that he initiated. We acknowledge the support of the Rob Fulton, manager of the Desert Studies Center at Zzyzx. We also thank Josh Van Tassel for his valuable field assistance.

REFERENCES CITED

Al-Farraj, A., and Harvey, A.M., 2000, Desert pavement characteristics on wadi terrace and alluvial fan surfaces: Geomorphology, v. 35, p. 279–297.

Amit, R., and Gerson, R., 1986, The evolution of Holocene Reg (gravelly) soils in deserts: An example from the Dead Sea region: Catena, v. 13, p. 59–79.

Amit, R., Gerson, R., and Yaalon, D.H., 1993, Stages and rate of gravel shattering processes by salts in desert reg soil: Geoderma, v. 57, p. 295–324.

Amundson, R., Wang, Y., Chadwick, O., Trumbore, S., McFadden, L., McDonald, E., Wells, S., and De Niro, M., 1994, Factors and processes governing the ^{14}C content of carbonate in desert soils: Earth and Planetary Science Letters, v. 125, p. 385–405.

Baker, V.R., 1977, Stream channel response to floods, with examples from central Texas: Geological Society of America Bulletin, v. 88, p. 1057–1071.

Berger, R., and Libby, W.F., 1967, UCLA radiocarbon dates VI: Radiocarbon, v. 9, p. 477–504.

Birkeland, P.W., 1985, Quaternary soils of the western United States, in Boardman, J., ed., Soils and Quaternary landscape evolution: Chichester, Wiley, p. 303–324.

Birkeland, P.W., 1990, Soil-geomorphic research: A selective overview, in Knuepfer, P.L.K., and McFadden, L.D., eds., Soils and landscape evolution: Geomorphology, v. 3, p. 207–224.

Blair, T.C., and McPherson, J.G., 1994, Alluvial fan processes and forms, in Abrahams, A.D., and Parsons, A.J., eds., Geomorphology of desert environments: London, Chapman and Hall, p. 354–402.

Brown, W.J., Wells, S.G., Enzel, Y., Anderson, L.D., and McFadden, L.D., 1990, The late Quaternary history of pluvial Lake Mojave: Silver Lake and Soda Lake basins, California, in Reynolds, R.E., et al., eds., At the end of the Mojave: Quaternary studies in the eastern Mojave Desert: Redlands, California, Special Publication of the San Bernardino County Museum Association, 1990 Mojave Desert Quaternary Research Center Symposium, May 18–21, 1990, p. 55–72.

Bryson, R.A., 1957, The annual march of precipitation in Arizona, New Mexico and northwestern Mexico: Technical report on the meteorology and climatology of arid regions, no. 6: Tucson, University of Arizona, Institute of Atmospheric Physics, 24 p.

Bryson, R.A., and Lowry, W.P., 1955, Synoptic climatology of the Arizona summer precipitation singularity: American Meteorological Society Bulletin, v. 36, p. 329–339.

Bull, W.B., 1977, The alluvial fan environment: Progress in Physical Geography, v. 1, p. 222–270.

Bull, W.B., 1978, Geomorphic tectonic activity classes of the south front of the San Gabriel Mountains, California: U.S. Geological Survey, Contract Report 14-08-001-G-394, Office of Earthquakes, Volcanoes, and Engineering, Menlo Park, California, 59 p.

Bull, W.B., 1979, Threshold of critical power in streams: Geological Society of America Bulletin, v. 90, p. 453–464.

Bull, W.B., 1990, Stream-terrace genesis: Implications for soil development, in Knuepfer, P.L.K., and McFadden, L.D., eds., Soils and landscape evolution: Geomorphology, v. 3, p. 351–367.

Bull, W.B., 1991, Geomorphic response to climatic change: Oxford, Oxford University Press, 326 p.

Bull, W.B., and Schick, A.P., 1979, Impact of climatic change on an arid watershed, Nahal Yael, southern Israel: Quaternary Research, v. 11, p. 153–171.

Clarke, M.L., 1994, Infrared stimulated luminescence ages from aeolian sand and alluvial fan deposits from the eastern Mojave Desert, California: Quaternary Science Reviews, v. 13, p. 533–538.

Denny, C.S., 1965, Alluvial fans in Death Valley region, California and Nevada: U.S. Geological Survey Professional Paper 466, 59 p.

Ely, L.L., 1997, Response of extreme floods in the southwestern United States to climatic variations in the late Holocene, in Kochel, R.C., and Miller, J.R., eds., Geomorphic responses to short-term climatic change: Geomorphology, v. 19, p. 175–201.

Enzel, Y., and Wells, S.G., 1997, Extracting Holocene paleohydrology and paleoclimatology information from modern extreme flood events: An example from southern California, in Kochel, R.C., and Miller, J.R., eds., Geomorphic responses to short-term climatic change: Geomorphology, v. 19, p. 203–226.

Enzel, Y., Anderson, R.Y., Brown, W.J., Cayan, D.R., and Wells, S.G., 1989a, Tropical and subtropical moisture and southerly displaced North Pacific storm track: Factors in the growth of late Quaternary lakes in the Mojave Desert, in Betancourt, J.L., and MacKay, A.M., eds., Proceedings of the Sixth Annual Pacific Climate (PACLIM) Workshop, Technical Report 23 of the Interagency Ecological Studies Program for the Sacramento–San Joachim Estuary, U.S. Geological Survey, Tucson, Arizona, p. 135–139.

Enzel, Y., Cayan, D.R., Anderson, R.Y., and Wells, S.G., 1989b, Atmospheric circulation during Holocene lake stands in the Mojave Desert: Evidence for regional climatic change: Nature, v. 341, p. 44–47.

Frostick, L.E., and Reid, I., 1989, Climatic versus tectonic controls of fan sequences: Lessons from the Dead Sea, Israel: Journal of the Geological Society, London, v. 146, p. 527–538.

Gerson, R., Amit, R., and Grossman, S., 1985, Dust availability in deserts: A study in the deserts of Israel and the Sinai: Institute of Earth Sciences, Hebrew University of Jerusalem, Contract Report DAJA-830C00041 for United States Army Research Development and Standardization Group, 190 p.

Gile, L.H., Peterson, F.F., and Grossman, R.B., 1966, Morphological and genetic sequence of carbonate accumulation in desert soils: Soil Science, v. 101, p. 347–360.

Gile, L.H., Hawley, J.W., and Grossman, R.B., 1981, Soils and geomorphology in the Basin and Range area of southern New Mexico: Guidebook to the Desert Project: New Mexico Bureau of Mines and Mineral Resources, Memoir 39.

Grayson, D.K., 1993, The desert's past: A natural prehistory of the Great Basin: Washington, D.C., Smithsonian Institution, 356 p.

Harden, J.W., 1982, A quantitative index of soil development from field descriptions: Examples from a chronosequence in central California: Geoderma, v. 28, p. 1–28.

Harden, J.W., 1990, Soil development on stable landforms and implications for landscape studies, in Knuepfer, P.L.K., and McFadden, L.D., eds., Soils and landscape evolution: Geomorphology, v. 3, p. 399–416.

Harden, J.W., Taylor, E.M., McFadden, L.D., and Reheis, M.C., 1991, Calcic, gypsic and siliceous soil chronosequences in arid and semiarid environments, in Nettleton, W.D., ed., Occurrence, characteristics and genesis of carbonate, gypsum and silica accumulations in soils: Soil Science Society of America Special Publication, v. 26, p. 1–16.

Harvey, A.M., 1992a, Controls of sedimentary style on alluvial fans, in Billi, P., et al., eds., Dynamics of gravel bed rivers: Chichester, Wiley, p. 519–535.

Harvey, A.M., 1992b, The influence of sedimentary style on the morphology and development of alluvial fans: Israel Journal of Earth Sciences, v. 41, p. 123–137.

Harvey, A.M., 1997, The role of alluvial fans in arid zone fluvial systems, in Thomas, D.S.G., ed., Arid zone geomorphology: Process, form and change in drylands (2nd edition): Chichester, Wiley, p. 231–259.

Harvey, A.M., 2003, The response of dry-region fans to Quaternary climatic change, in Alsharhan, A.S., et al., eds., Desertification in the Third Millenium: Rotterdam, Balkema, p. 83–98.

Harvey, A.M., and Wells, S.G., 1994, Late Pleistocene and Holocene changes in hillslope sediment supply to alluvial fan systems: Zzyzx, California, in Millington, A.C., and Pye, K., eds., Environmental change in drylands: Biogeographical and geomorphological perspectives: Chichester, Wiley, p. 67–84.

Harvey, A.M., Miller, S.Y., and Wells, S.G., 1995, Quaternary soil and river terrace sequences in the Aguas/Feos river systems: Sorbas basin, southeast Spain, in Lewin, J., et al., eds., Mediterranean Quaternary river environments: Rotterdam, Balkema, p. 263–281.

Harvey, A.M., Wigand, P.E., and Wells, S.G., 1999, Response of alluvial fan systems to the late Pleistocene to Holocene climatic transition: Contrasts between the margins of pluvial Lakes Lahontan and Mojave, Nevada and California, USA: Catena, v. 36, p. 255–281.

Hunt, C.B., and Mabey, D.R., 1966, Stratigraphy and structure, Death Valley, California: U.S. Geological Survey Professional Paper 494A, 162 p.

Knox, J.C., 1984, Responses of river systems to Holocene climates, in Wright, H.E., ed., Late Quaternary environments of the United States, Volume 2: The Holocene: London, Longman, p. 26–41.

Lattman, L.H., 1973, Calcium carbonate cementation of alluvial fans in southern Nevada: Geological Society of America Bulletin, v. 84, p. 3013–3028.

Machette, M.N., 1985, Calcic soils of the southwestern United States, in Weide, D.L., ed., Soils and Quaternary geology of southwestern United States: Boulder, Colorado, Geological Society of America Special Paper 203, p. 1–21.

McDonald, E.V., 1994, The relative influence of climatic change, desert dust and lithological control on soil-geomorphic processes and hydrology of calcic soils found on quaternary alluvial fan deposits in the Mojave Desert, California [Ph.D. thesis]: Albuquerque, University of New Mexico, 383 p.

McDonald, E.V., and McFadden, L.D., 1994, Quaternary stratigraphy of the Providence Mountains piedmont and preliminary age estimates and regional stratigraphic correlations of Quaternary deposits in the eastern Mojave Desert, California, in McGill, S.F., and Ross, T.M., eds., Geological investigations of an active margin: Geological Society of America, Cordilleran Section Field Trip Guidebook, San Bernardino, California, San Bernardino County Museum, p. 205–210.

McDonald, E.V., Pierson, F.B., Flerchinger, G.N., and McFadden, L.D., 1996, Application of a process-based soil-water balance model to evaluate the influence of late Quaternary climate change on soil water movement: Geoderma, v. 74, p. 167–192.

McFadden, L.D., 1988, Climatic influences on rates and processes of soil development in Quaternary deposits of southern California, in Reinhardt, J., and Sigleo, W.R., eds., Palaeosols and weathering through geologic time: Principles and applications: Boulder, Colorado, Geological Society of America Special Paper 206, p. 153–177.

McFadden, L.D., and Weldon, R.J., 1987, Rates and processes of soil development on Quaternary terraces in Cajon Pass, southern California: Geological Society of America Bulletin, v. 98, p. 280–293.

McFadden, L.D., Wells, S.G., and Dohrenwend, J.C., 1986, Influences of Quaternary change on processes of soil development on desert loess deposits of the Cima volcanic field, California: Catena, v. 13, p. 361–389.

McFadden, L.D., Ritter, J.B., and Wells, S.G., 1989, Use of multiparameter relative-age methods for age estimation and correlation of alluvial fan surfaces on a desert piedmont, eastern Mojave Desert, California: Quaternary Research, v. 32, p. 276–290.

McFadden, L.D., McDonald, E.V., Wells, S.G., Anderson, K., Quade, J., and Foreman, S.L., 1998, The vesicular layer and carbonate collars of desert soils and pavements: Formation, age, and relation to climatic change: Geomorphology, v. 24, p. 101–145.

Nilson, T.H., 1982, Alluvial fan deposits, in Scholle, P., and Spearing, D., eds., Sandstone depositional environments: American Association of Petroleum Geologists, Tulsa, Oklahoma, Memoir v. 31, p. 49–86.

Reheis, M.C., Harden, J.W., McFadden, L.D., and Shroba, R.R., 1989, Development rates of late Quaternary soils: Silver Lake Playa, California: Soil Science Society of America Journal, v. 53, p. 1127–1140.

Ritter, J.B., Miller, J.R., Enzel, Y., and Wells, S.G., 1995, Reconciling the roles of tectonism and climate in Quaternary alluvial fan evolution: Geology, v. 23, p. 245–248.

Schumm, S.A., 1979, Geomorphic thresholds: The concept and its application: Institute of British Geographers, Transactions, New Series, v. 4, p. 485–515.

Soil Survey Staff, 1975, Soil taxonomy: A basic system of soil classification for making and interpretation of soil surveys: USDA/SCS Agricultural Handbook 436, United States Government Printing Office, Washington, D.C., 754 p.

Spaulding, W.G., 1985, Vegetation and climates of the last 45,000 yr in the vicinity of the Nevada Test Site, south-central Nevada: U.S. Geological Survey Professional Paper 1329, 83 p.

Spaulding, W.G., 1990, Vegetation dynamics during the last deglaciation, southeastern Great Basin, USA: Quaternary Research, v. 33, p. 188–203.

Turrin, B., Dohrenwend, J.C., Wells, S.G., and McFadden, L.D., 1984, Geochronology and eruptive history of the Cima volcanic field, eastern Mojave Desert, California, in Dohrenwend, J.C., ed., Surficial geology of the eastern Mojave Desert, California: Geological Society of America 1984 Annual Meeting Field Trip Guidebook, p. 393–399.

Van Devender, T.R., 1977, Holocene woodlands in the southwestern deserts: Science, v. 198, p. 189–192.

Wang, Y., Amundson, R., and Trumbore, S., 1994, A model for $^{14}CO_2$ and its implications for using ^{14}C to date pedogenic carbonate: Geochimica et Cosmochimica Acta, v. 58, p. 393–399.

Wang, Y., McDonald, E.V., Amundson, R., McFadden, L.D., and Chadwick, O., 1996, An isotopic study of soils in chronological sequences in alluvial deposits, Providence Mountains, California: Geological Society of America Bulletin, v. 108, p. 379–391.

Wells, P.V., and Woodcock, D., 1985, Full-glacial vegetation of Death Valley, California: Juniper woodland opening to Yucca semidesert: Madrono, v. 32, p. 11–23.

Wells, S.G., and Dohrenwend, J.C., 1985, Relict sheetflood bedforms on late Quaternary alluvial-fan surfaces in the southwestern United States: Geology, v. 13, p. 512–516.

Wells, S.G., and Harvey, A.M., 1987, Sedimentologic and geomorphic variations in storm-generated alluvial fans, Howgill Fells, northwest England: Geological Society of America Bulletin, v. 98, p. 182–198.

Wells, S.G., McFadden, L.D., Dohrenwend, J.C., Turrin, B.D., and Mahrer, K.D., 1985, Late Cenozoic landscape evolution of lava flow surfaces of the Cima volcanic field, Mojave Desert, California: Geological Society of America Bulletin, v. 96, p. 1518–1529.

Wells, S.G., McFadden, L.D., and Dohrenwend, J.C., 1987, Influence of late Quaternary climatic change on geomorphic and pedogenic processes on a desert piedmont, eastern Mojave Desert, California: Quaternary Research, v. 27, p. 130–146.

Wells, S.G., Anderson, R.Y., Enzel, Y., and Brown, W.J., 1990a, An overview of floods and lakes within the Mojave River drainage basin: Implications for latest Quaternary paleoenvironments in southern California, in Reynolds, R.E., et al., eds., At the end of the Mojave: Quaternary studies in the eastern Mojave Desert: Redlands, California, Special Publication of the San Bernardino County Museum Association, 1990 Mojave Desert Quaternary Research Center Symposium, May 18–21, 1990, p. 31–37.

Wells, S.G., McFadden, L.D., and Harden, J., 1990b, Preliminary results of age estimations and regional correlations of Quaternary alluvial fans within the Mojave Desert of southern California, in Reynolds, R.E., et al., eds., At the end of the Mojave: Quaternary studies in the eastern Mojave Desert: Redlands, California, Special Publication of the San Bernardino County Museum Association, 1990 Mojave Desert Quaternary Research Center Symposium, May 18–21, 1990, p. 45–53.

Wells, S.G., Moissec, J-L., and Harvey, A.M., 1990c, Sedimentary processes during Holocene and Pleistocene alluvial fan deposition along the southern Soda Mountains, California: A summary, in Reynolds, R.E., et al., eds., At the end of the Mojave: Quaternary studies in the eastern Mojave Desert: Redlands, California, Special Publication of the San Bernardino County Museum Association, 1990 Mojave Desert Quaternary Research Center Symposium, May 18–21, 1990, p. 39–44.

Wells, S.G., Anderson, D.E., Williamson, T.N., and Enzel, Y., 1994, Hydrogeomorphic and precipitation regimes during the Holocene: Reconstructions from fluvial and lacustrine sediments in southern California's deserts: Eos (Transactions, American Geophysical Union) 1994 Fall Meeting, v. 75, p. 371.

Wells, S.G., McFadden, L.D., Poths, J., and Olinger, C.T., 1995, Cosmogenic ^{3}He surface-exposure dating of stone pavements: Implications for landscape evolution in deserts: Geology, v. 23, p. 613–616.

Wells, S.G., McDonald, E.V., Harvey, A., Ritter, J., and Knott, J., 1997, Influence of climatic regimes and their variations on the deposition of alluvial fans in Quaternary extensional basins: Geological Society of America Abstracts with Programs, v. 29, no. 6, p. 240.

Wells, S.G., Anderson, D.E., Anderson, K.C., and McDonald, E.V., 1998, Reconstruction of late Quaternary hydrologic regimes along the southern Great Basin and Mojave Desert; California and Nevada, USA, in Grossman, M., et al., eds., Palaeohydrology with an emphasis on humid temperate and tectonically active zones: Abstracts Volume of the Third International Meeting on Global Continental Palaeohydrology (GLOCOPH), Sept 4–6, 1998, Rissco University, Kumagaya Campus, Japan, p. 32–33.

Woodcock, D., 1986, The late Pleistocene of Death Valley: A climatic reconstruction based on microfossil data: Palaeogeography, Palaeoclimatology, Palaeoecology, v. 57, p. 272–283.

MANUSCRIPT ACCEPTED BY THE SOCIETY AUGUST 1, 2002

Late Quaternary eolian dynamics, Mojave Desert, California

Nicholas Lancaster
Desert Research Institute, 2215 Raggio Parkway, Reno, Nevada 89512, USA

Vatche P. Tchakerian
Department of Geography, Texas A&M University, College Station, Texas 77843, USA

ABSTRACT

Changes in climate may impact eolian sediment transport systems via changes in sediment supply, availability, and mobility. Relations between periods of eolian accumulation and stabilization and the proxy record of paleohydrologic changes in the east-central Mojave Desert suggest that climatically driven variations in sediment supply from fluvial and/or lacustrine sources played a major role in determining the duration and timing of periods of late Pleistocene and Holocene eolian activity. Changes in sediment availability and mobility resulting from climatic changes that affected the region played a less important role.

INTRODUCTION

Accumulation of eolian deposits requires (1) a source of available sediment, (2) sufficient wind energy to transport that sediment, and (3) conditions that promote accumulation in the depositional zone. Most eolian sand in arid regions occurs adjacent to well-defined regional- and local-scale sediment transport systems in which sand is moved by wind from source areas (e.g., distal fluvial deposits) via transport pathways to depositional sinks (dune fields, sand seas) (e.g., Fryberger and Ahlbrandt, 1979; Mainguet, 1983; Wilson, 1971).

The dynamics of eolian sediment transport systems on any time scale are determined by the relations between the supply, availability, and mobility of sediment of a size suitable for transport by wind (Kocurek and Lancaster, 1999). In turn, sediment supply, availability, and mobility are determined in large part by regional and local climate and vegetation cover. Sediment supply is the emplacement of sediment that serves as a source of material for the eolian sediment transport system. It may be affected by variations in flood magnitude and frequency, river sediment load, lake and sea levels, and rates of bed rock weathering that affect sediment source areas. Climatic changes impact sediment availability (the susceptibility of a sediment surface to entrainment of material by wind) and mobility (transport rates) via changes in the magnitude and frequency of winds capable of transporting sediment, vegetation cover, and soil moisture.

Eolian sediments and landforms are widespread throughout the Mojave Desert of southern California and southern Nevada, although they are not a dominant component of the landscape in most areas (Tchakerian, 1997). In addition, eolian dust forms an important component of many soils in the region (McDonald et al., 1995, this volume; McFadden et al., 1987). The majority of the sand deposits occur along well-defined regional- and local-scale sand transport corridors that extend from source areas in the western and central Mojave toward depositional sinks that lie to the east and south (Zimbelman et al., 1995). Large parts of these systems are stabilized by vegetation and experience little or no eolian activity in present climatic regimes (Bach, 1995). However, it has long been recognized that eolian activity has been more extensive and more intense than it is at present during periods of the late Quaternary (e.g., Smith, 1967). Other evidence for past eolian activity includes eolian sand mantles on beach ridges at Silver Lake (Wells et al., 1987), eolian silt mantles on the lava flows of the Cima volcanic field (McFadden et al., 1987), and ventifacts covered by desert varnish (Laity, 1992). Subsequently, numerous detailed stratigraphic and geomorphic investigations of eolian sand deposits (e.g., Tchakerian, 1991; Lancaster and Tchakerian, 1996) together with application of luminescence dating techniques (Clarke and Rendell, 1998; Clarke et al., 1996a, 1996b; Rendell et al., 1994; Rendell and Sheffer, 1996; Wintle et al., 1994) have provided an increasingly clear picture of the history of the major eolian systems in the east-central Mojave Desert.

The close proximity of source areas and depositional sinks, together with the large amount of research that has been carried out in the Mojave, make this an ideal area in which to assess the response of eolian sediment transport systems to the direct and indirect effects of late Quaternary climate change and to develop general process-response models. In this paper, we summarize the results of this work over the past decade, and discuss its significance in relation to evidence for climatic changes in the region derived from other sources.

EOLIAN GEOMORPHIC SYSTEMS IN THE MOJAVE DESERT

The Mojave Desert is characterized by broad valleys separated by fault-bounded mountain ranges. In the central and eastern part of the desert, most of these ranges trend north-northwest to south-southeast (Fig. 1). Playas fed by ephemeral streams occur in many, but not all, of the intermontane basins. Prevailing winds in the Mojave Desert are from the west and south, but the structurally controlled topography exerts a significant influence on regional wind patterns and winds tend to blow parallel to the valley axes (Laity, 1987). Throughout the central and eastern Mojave Desert, topographic control of winds and sand transport results in well-defined aeolian sediment transport corridors (Fig. 1) that are characterized by areas of active and inactive dunes and sand sheets, together with sand ramps (topographically controlled accumulations of eolian, fluvial, and slope deposits) (Zimbelman et al., 1995).

Eastern Mojave Desert

There are three main eolian sand transport corridors in the eastern Mojave Desert (Clarke and Rendell, 1998; Zimbelman et al., 1995) (Figs. 1 and 2). The most extensive of these systems is called the Bristol Trough, and extends 150 km southeast from Bristol Playa to the Colorado River. Eolian sand deposits are concentrated around the three large playas in this area (Bristol, Cadiz, and Danby) and include active dune fields (Cadiz Dunes, parts of the Rice Valley dune field), vegetation-stabilized linear and crescentic dunes (Rice Valley dune field), many stabilized sand ramps (Iron Mountain, Arica Mountains, and Big Maria Mountains) as

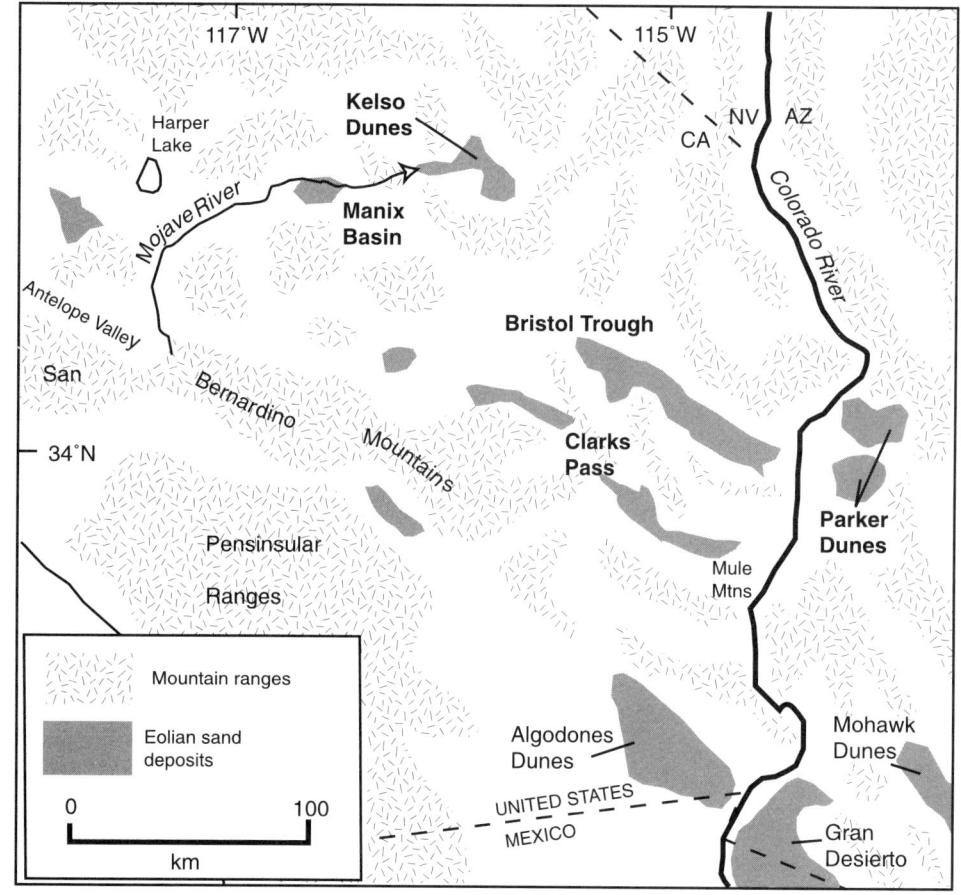

Figure 1. The Mojave Desert and adjacent areas, showing major areas of eolian sand, together with location of the principal sand transport pathways (bold).

well as extensive vegetated sand sheets that surround and link the areas of sand dunes and sand ramps.

The second major sand transport corridor is called the Clarks Pass system (Zimbelman et al., 1995), which extends along the northern flank of the Transverse Ranges (Bach, 1995). The transport corridor continues via Dale Lake and Clarks Pass, terminating at the Mule Mountains just west of the Colorado River (Fig. 1). Sand ramps are prominent east of Johnson Lake (Bach, 1995) and on the western flanks of the Sheephole and Mule Mountains. Elsewhere, vegetated sand sheets dominate this system. The third system lies east of the Colorado River and consists of large areas of vegetation-stabilized linear and crescentic dunes of the Parker Dunes in Arizona (Fig. 1). It was originally considered to be part of the Bristol Trough system (Zimbelman et al., 1995), but recent work has shown it to be mineralogically distinct from the sand transport systems that lie west of the Colorado River, indicating that these dunes are derived from Colorado River sediments (Zimbelman and Williams, 2002; Muhs et al., 2003).

Central Mojave Desert

The major eolian system in this area extends for 60 km eastwards from the fan delta of the Mojave River as it exits the Afton Canyon to the Kelso Dunes (Fig. 1) (Ramsey et al., 1999; Smith, 1984; Zimbelman et al., 1995). Included in this system are the sand ramps and small dunes around the Cronese basins (see Clarke et al., 1996b). This eolian system consists of active crescentic dunes and sand sheets that extend east from the Mojave River Sink south of Soda Lake to the vicinity of Balch, stabilized sand sheets and relict sand ramps in the area of the Devils Playground, and active and dormant dunes of the Kelso dune field (Lancaster, 1993; Sharp, 1966). East of Kelso dune field, sand sheets mantle alluvial fans at the base of the Providence Mountains and eolian sand forms an important part of many distal alluvial fans in this area (Clarke, 1994; McDonald et al., this volume).

A second system in this area extends east from the Manix basin (Fig. 1) and consists of sand ramps (e.g., Soldier Mountain) and sand sheets that mantle the western parts of the Cady Mountains (Fig. 2). Eolian deposits and small dunes are widespread in the Manix basin itself, east of Barstow (Bach, 1995). Smaller areas of sand sheets and dunes extend east from Coyote Lake and Harper Lake playas (Figs. 1, 2) (Bach, 1995; Dean, 1978). Although not discussed in this paper, an additional area of extensive active sand sheets, small dune fields, and sand ramps extends south from Death Valley along the lower Amargosa River valley to Dumont Dunes (Dean, 1978; Eymann, 1953) and forms a further, entirely separate, eolian sediment transport system.

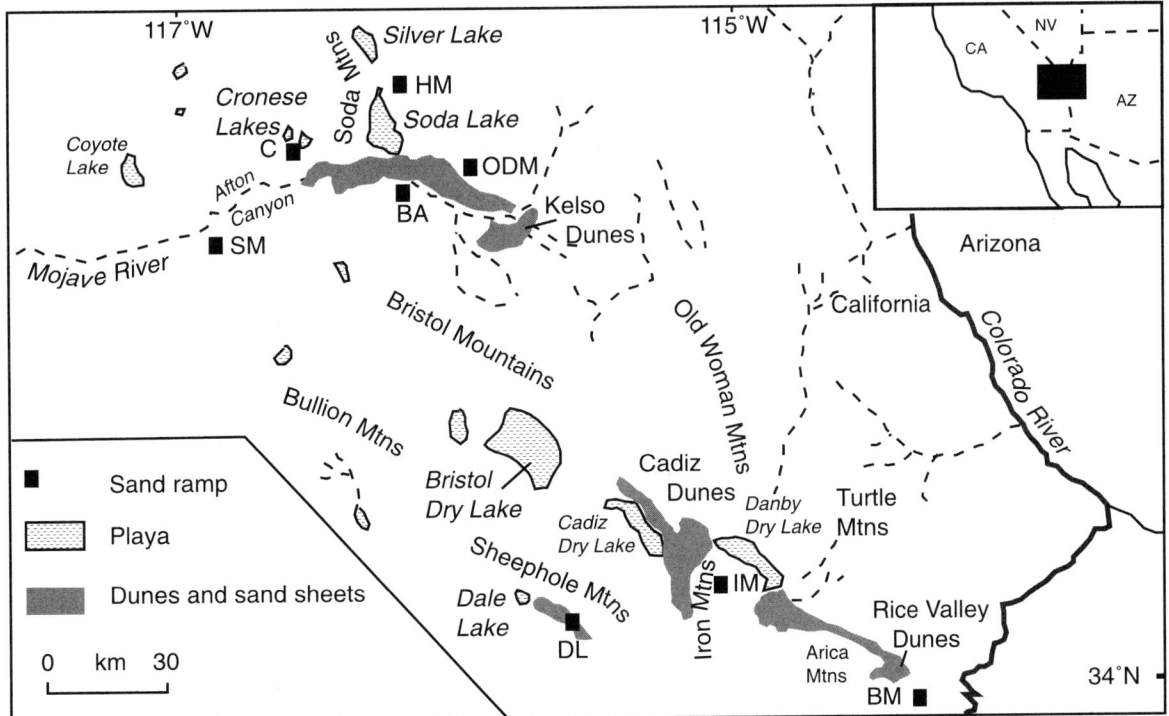

Figure 2. Location of sand ramp study sites in the central and eastern Mojave Desert (after Rendell and Sheffer, 1996). SM—Soldier Mountain; C—Cronese basin; HM—Hanks Mountain; BA—Balch; ODM—Old Dad Mountain; DL—Dale Lake; IM—Iron Mountain; BM—Big Maria.

Western Mojave Desert

Eolian deposits are widespread in the area of the western Mojave Desert, but have not been investigated in detail. Dean (1978) and Bach (1995) describe extensive, but thin, eolian sand sheets in the area of Antelope Valley (Fig. 1), with thicker sand accumulations including small dunes and sand ramps east of major playas (Rosamond, Buckhorn, Edwards dry lakes) (Dibblee, 1967).

HISTORY OF EOLIAN ACCUMULATION

Information on the history of eolian accumulation, stabilization, and reworking in this region is provided mainly by sediments of the sand ramps that occur along the sand transport pathways. Because the sand ramps represent a long-term net accumulation of eolian, alluvial, and slope deposits, with little reworking, they provide a long history of eolian processes in the Mojave Desert (Lancaster and Tchakerian, 1996). By contrast, areas of dunes may be partially or completely reworked by each period of eolian activity and therefore preserve a relatively limited record of past eolian dynamics. In this section, we review the evidence from stratigraphic, sedimentologic, and luminescence dating studies on each of the major sand transport corridors. Full details of the luminescence dating procedures and data are found in Clarke (1994); Clarke et al. (1996a, 1996b); Rendell et al. (1994); Rendell and Sheffer (1996); and Wintle et al. (1994). Sand ramp stratigraphy and sedimentology are discussed by Tchakerian (1991) and Lancaster and Tchakerian (1996).

Bristol Trough

Evidence for the history of eolian activity in this corridor is provided by the deposits of sand ramps at Iron Mountain and Big Maria Mountain (Figs. 2, 3). Iron Mountain sand ramp lies in the middle part of the system, south of Danby playa. Big Maria sand ramp lies southeast of the Rice Valley dune field and represents the far distal portion of the sand transport corridor. At Iron Mountain, sand ramp deposits (Fig. 3) consist of a medium-fine sand with a significant component of weathered granite (grus) of coarse sand size and prominent alluvial/debris flow units (Units 2 and 4). The lowermost unit (Unit 1) is composed of compact red sand and is capped by a soil with a well-developed Bt horizon. The soil is covered by a 20 cm-thick deposit of angular gravel-sized granitic talus interpreted as a local debris flow (Unit 2). Unit 3 consists of fine gray-brown sand with sparse grus and is capped by a thin, but well-developed, soil. The soil is covered by a 50 cm fine fluvial gravel unit (Unit 4) with 50–80 cm deep channels incised into the soil zone. The uppermost unit (Unit 5) consists of 100 cm of indurated red-brown eolian sand with as much as 15% grus.

Sand ramp deposits at Big Maria (Fig. 3) consist of a lowermost 1.5–2 m thick unit of gray-brown fine eolian sand (Unit 1) overlain by a 2.2 m-thick buried soil (Unit 2) with carbonate rhizoliths and prominent reddening throughout. The upper part of the soil is truncated by erosion. The uppermost part of the ramp consists of two eolian sand units (3 and 4) with a total thickness of 3.1 m. These sands are capped by a weak paleosol and an angular talus lag, which is in turn covered by as much as 1 m of modern active eolian sand (not shown in Figure 3).

Feldspar thermoluminescence (FTL) and infrared stimulated luminescence (IRSL) ages reported by Rendell and Sheffer (1996) indicate that there are two major and one minor periods of late Pleistocene and Holocene sand accumulation at Iron Mountain and Big Maria sand ramps. The oldest period of sand accumulation at Iron Mountain occurred prior to 20 ka. It is separated from the younger period of accumulation by a period of soil formation indicating stabilization of the sand ramp. The second major period of eolian deposition took place from ca. 15 to 9 ka. A third period of eolian stabilization, soil formation, and local fluvial activity occurred sometime after 9 ka. The history of eolian accumulation at Big Maria is similar to Iron Mountain. The older sand deposits were deposited prior to 22 ka, with a second period of eolian activity in the period 15–6 ka.

Clarks Pass system

The history of this system is documented by the thick sand ramp deposits on the western flanks of the Sheephole Mountains near Dale Lake. The stratigraphy and sediments of the Dale Lake sand ramp have been studied in detail by Tchakerian (1991, 1992). Luminescence dating information is contained in Rendell et al. (1994). Sediments of the Dale Lake sand ramp consist of at least 18 m thickness of eolian and fluvial sands, with abundant talus and grus in some sedimentary units. A total of six sedimentary units are recognized (Rendell et al., 1994; Tchakerian, 1991) (Fig. 4). Unit 1 consists of indurated fine to medium red eolian sand and is capped by a buried soil with a nodular stage II calcic horizon, while Unit 2 consists of medium sand with calcrete nodules and some grus, together with minor gravel-filled fluvial channels. A prominent paleosol with a stage II–III calcic horizon caps this unit. Unit 3 consists of coarse reddish yellow eolian sands with grus and small gravel-filled channels. Unit 4 is a larger gravel filled channel. The uppermost units (5–7) are composed of fine, moderately well sorted eolian sand, with some grus in Unit 5. Although there are significant disparities between the results of quartz thermoluminescence (QTL) and FTL age determinations on the sands of the Dale Lake sand ramp, the QTL and IRSL results indicate the existence of two main periods of aeolian deposition, between >35 and 25 ka, and 15–10 ka (Rendell et al., 1994). The intervening time was apparently one of eolian stability and soil formation in this area.

Kelso Dunes system

The Kelso Dunes eolian sediment transport system (Fig. 5) extends for 60 km eastwards from the fan delta of the Mojave River as it exits Afton Canyon to the Kelso dune field. The Mojave River is considered to be the principal source for sediment for

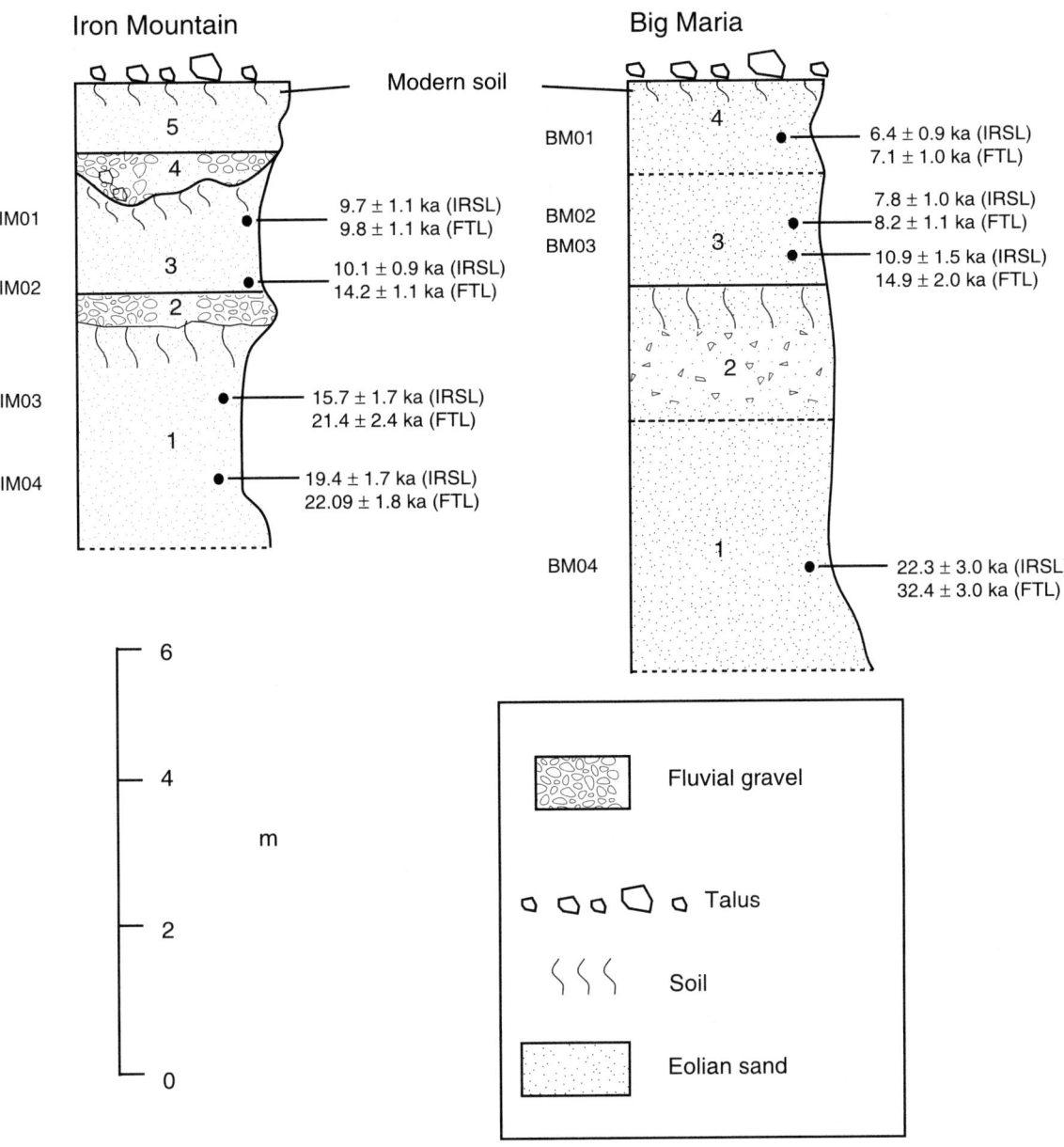

Figure 3. Stratigraphic columns of sand ramps at Iron Mountain and Big Maria. Stratigraphy modified from Lancaster and Tchakerian (1996). Sample numbers and luminescence ages from Rendell and Sheffer (1996). See text for a description of stratigraphic units. IRSL—infrared stimulated luminescence. FTL—feldspar thermoluminescence.

Kelso Dunes (Ramsey et al., 1999; Sharp, 1966), although local sources (e.g., the washes that drain the Granite Mountains and Cima Dome) are also important contributors to the dune field (Ramsey et al., 1999).

The Mojave River heads in the San Bernardino Mountains, a tectonically active area of granitic rocks. Following heavy winter rainfall, the Mojave River may flow through the Afton Canyon to the Mojave River Sink and the adjacent East Cronese Lake or as far as Silver Lake (Enzel et al., 1992; Wells et al., this volume). During intervals of the late Pleistocene and early Holocene, the area of Soda Lake and Silver Lake was occupied by a shallow (~50 m) lake called Lake Mojave (Brown et al., 1990; Wells et al., this volume). The area of the Mojave River Sink is today characterized by gravel-size deflation lags, coppice dunes, and small climbing and falling dunes. To the east of this area lie a series of 2–5 m-high active crescentic dunes and undulating sand sheets.

Sand is transported east from the Mojave River Sink source zone through Devils Playground, a 10 km-wide corridor between the Old Dad and Bristol Mountains, to the depositional sink of the Kelso dune field (Figs. 2, 5). In the western part of Devils Play-

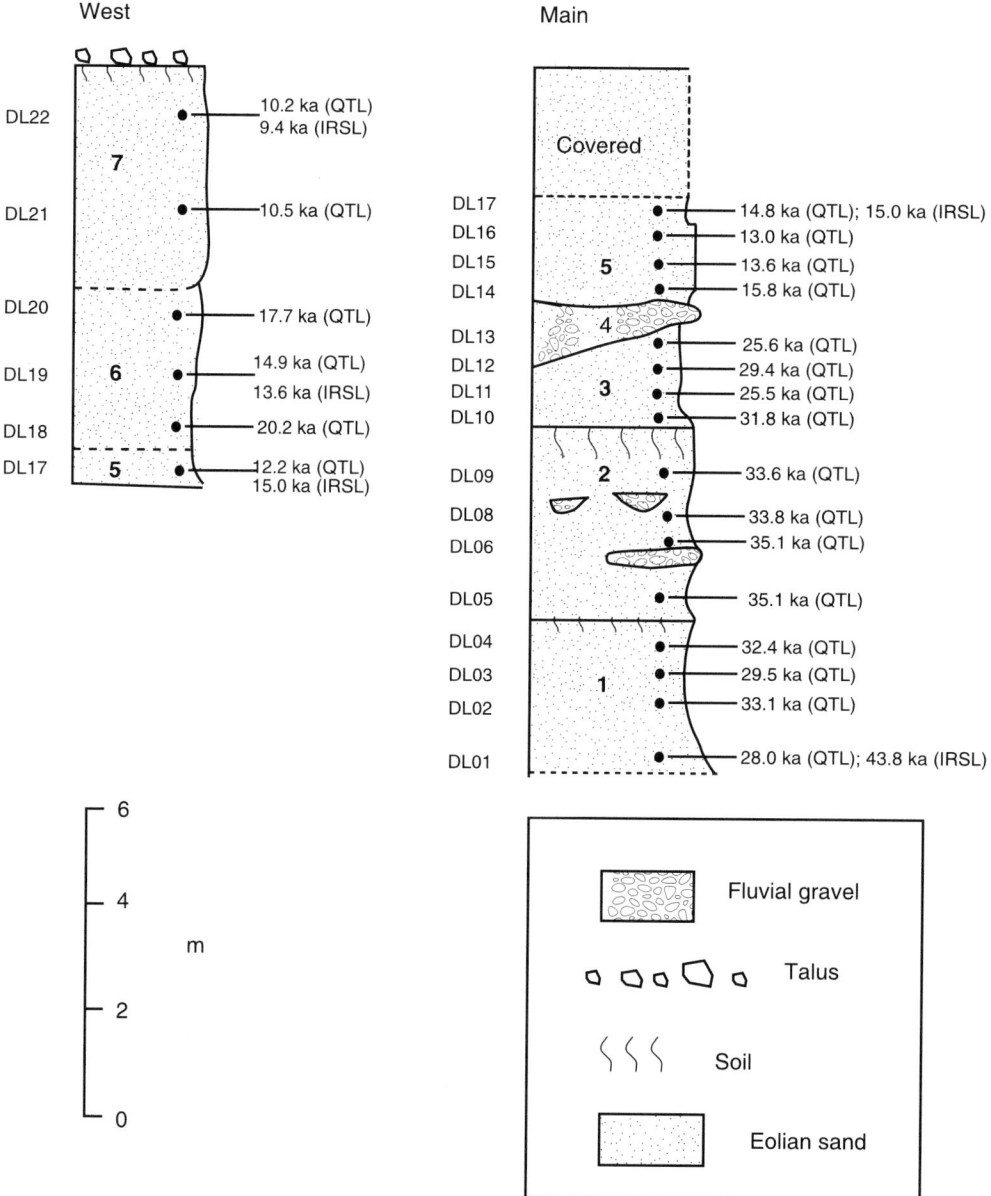

Figure 4. Stratigraphic column of sand ramp at Dale Lake. Stratigraphy after Tchakerian (1991) and Lancaster and Tchakerian (1996). Luminescence ages and sample numbers from Rendell et al. (1994). QTL—quartz thermoluminescence. IRSL—infrared stimulated luminescence.

ground are active crescentic dunes up to 5 m high as well as climbing and falling dunes draped over the hills north of Balch (Fig. 5). The eastern part of Devils Playground is, by contrast, characterized by sand sheets that are currently stabilized by vegetation. A series of largely vegetation-stabilized sand ramps mantle the Old Dad Mountains on the northeast side of Devils Playground (Fig. 5).

Kelso Dunes is the major depositional sink for this system. The dune field occupies an area of 100 km² on the piedmont slopes of the Granite and Providence Mountains and represents a sand accumulation of ~1 km³ (Lancaster, 1993; Sharp, 1966). The dune field consists of a 40 km² area of active dunes surrounded on their west, north, and east sides by a 60 km² area of low vegetation-stabilized dunes (Fig. 5). In addition, eolian sand is interbedded with alluvial fan sequences east of Kelso Dunes (Clarke, 1994; McDonald and McFadden, 1994).

The history and characteristics of eolian deposits in this system have been intensively studied in recent years. Investigations have concentrated on the Cronese basins, representing conditions adjacent to the source zone (Clarke and Rendell, 1998; Clarke et al., 1996b; Rendell et al., 1994). Devils Playground (the transport corridor), and (3) Kelso Dunes and adjacent areas (the sediment sink) (Clarke, 1994; Lancaster, 1993; Wintle et al., 1994).

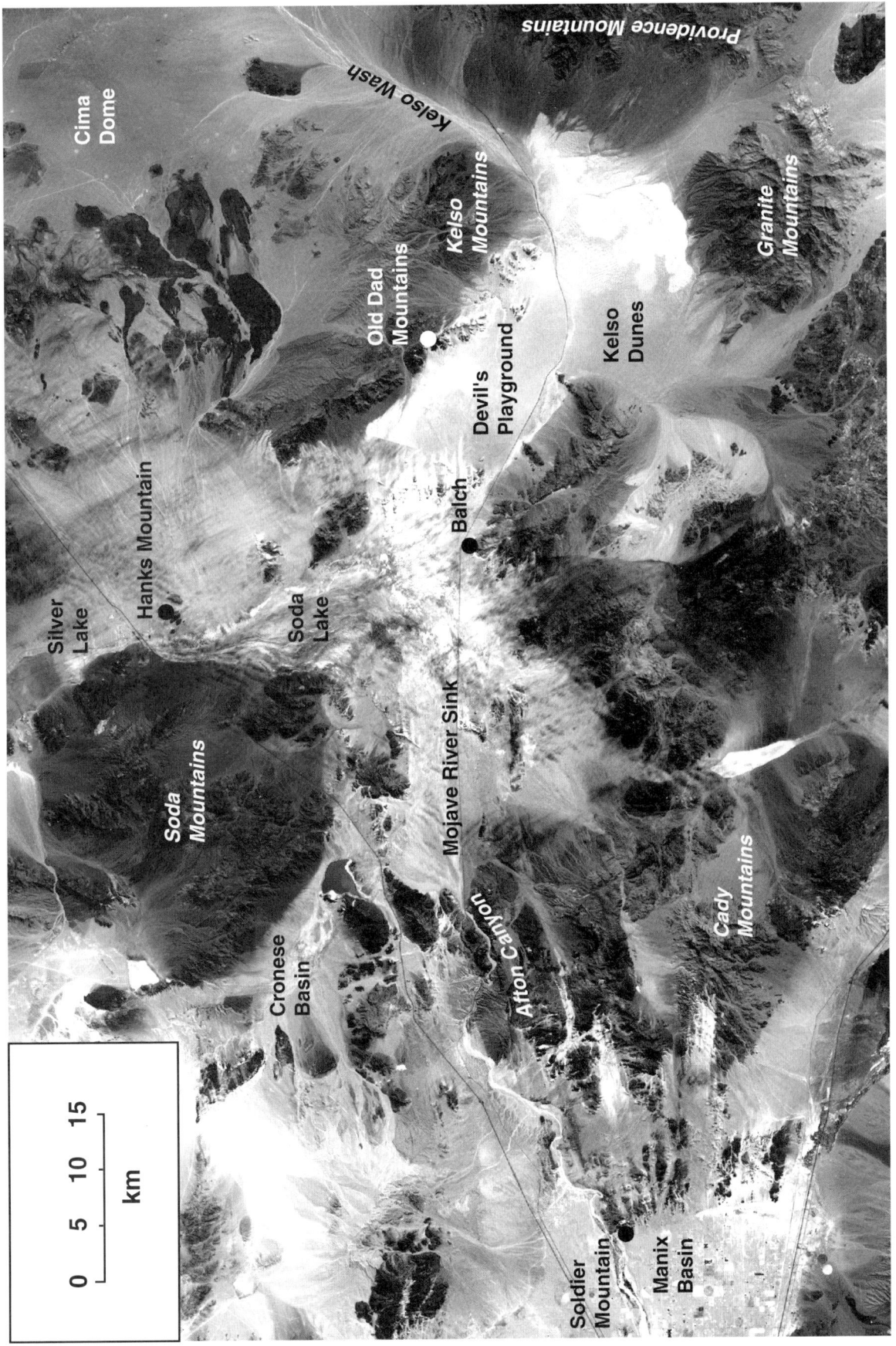

Figure 5. The Kelso Dunes eolian sediment transport system (Landsat TM, band 4). Note: prominent sand streaks and sheets extending east from the Manix basin; extensive areas of small active dunes in the Mojave River Sink and areas immediately to the east. When this image was acquired (summer 1993), East Cronese basin held a small lake resulting from inflow from the Mojave River in the winter of this year.

The Cronese basins. The Cronese basins (Fig. 6) are located downstream of Afton Canyon, north of the Mojave River Wash. Floodwaters reach the East Cronese basin directly from the Mojave River and a narrow spillway connects the East Cronese basin to the West Cronese basin (Thompson, 1929). Water flows between the two basins once the lake level in the East Cronese basin has reached a minimum depth of two meters (Brown, 1989). Both basins also receive some runoff from adjacent alluvial fans. It should be noted that, prior to the deposition of the Mojave River fan delta east of Afton Canyon and the consequent filling of the basin, the direction of flow in the Cronese lake basins was likely reversed, so that they did not receive water from the Mojave River at this time (Brown, 1989).

Eolian deposits in the Cronese basins consist of active climbing and falling dunes, together with stabilized sand ramps and low dune fields (Fig. 6). The most prominent of the dunes is the Cat Dune, a falling dune that fills a ravine between elevations of 370 and 670 m on the east side of the Cronese Mountains (Evans, 1961). The dune is indurated and its surface is covered by a thin angular gravel talus. At the base of the main dune is a smaller talus-covered falling dune informally named the "Kitten Dune." In addition, extensive sand ramps mantle the western slopes of the Cronese Mountains, where there are also a series of low dune ridges.

Sediments of Cat Dune (Fig. 7) consist of medium to fine, moderately to well-sorted eolian sand with an admixture of grus and angular gravel- to cobble-sized talus derived from adjacent hillslopes. In the lower part of the section (Unit A1), the talus clasts are slightly to moderately weathered and have caliche rinds on their lower faces. Talus in the upper part of the section is angular and unweathered (Unit B1). A loose pebble- to small boulder-sized talus covers the indurated surface of the dune. The composition of the Cat Dune sediments is very similar to those of Unit Qe2 of Brown (1989), which forms a mantle at the eastern base of the Cronese Mountains. The equivalent of Unit Qe2 at Silver Lake is cut by the C and D shorelines of Lake Mojave with inferred Holocene ages (Wells et al., 1987).

Kitten Dune comprises two main eolian units (Fig. 7). Unit A2 consists of moderately well sorted medium sand with 5–10% highly weathered grus and weak calcic horizon development. It is capped by a layer of 0.05–0.20 m-diameter boulders of highly weathered metamorphic rocks. Unit B2 overlies these boulders and consists of massive to friable well-sorted red-brown medium eolian sand with 5–10% granule-sized grus. This sand ramp is covered by a moderately developed desert pavement of angular, weakly varnished cobbles and small boulders. Based on IRSL ages, the Cat and Kitten Dunes apparently accumulated between 24 and 8 ka (Clarke et al., 1996b; Rendell et al., 1994), while adjacent stratigraphically equivalent eolian sediments (Unit Qe2 of Brown, 1989) were deposited between 24 and 18 ka (Clarke et al., 1996b).

Eolian landforms in the West Cronese basin include yardangs, dune ridges, and sand ramps (Fig. 6). The sand ramps mantle the west- and northwest-facing slopes of the northern part of the

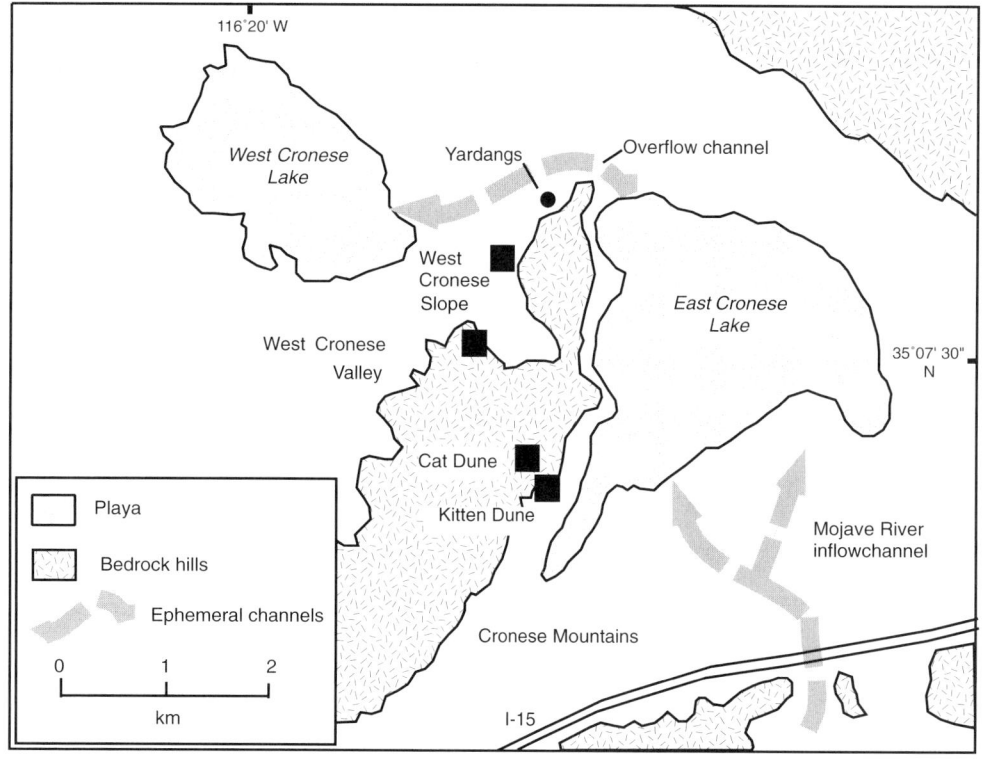

Figure 6. Location of major sand ramps and other eolian deposits studied in the Cronese basins. Based on 7.5′ topographic quads for the area plus mapping from aerial photographs (NHAP, 1983). Boxes locate sections illustrated in Fig. 7 and described in the text.

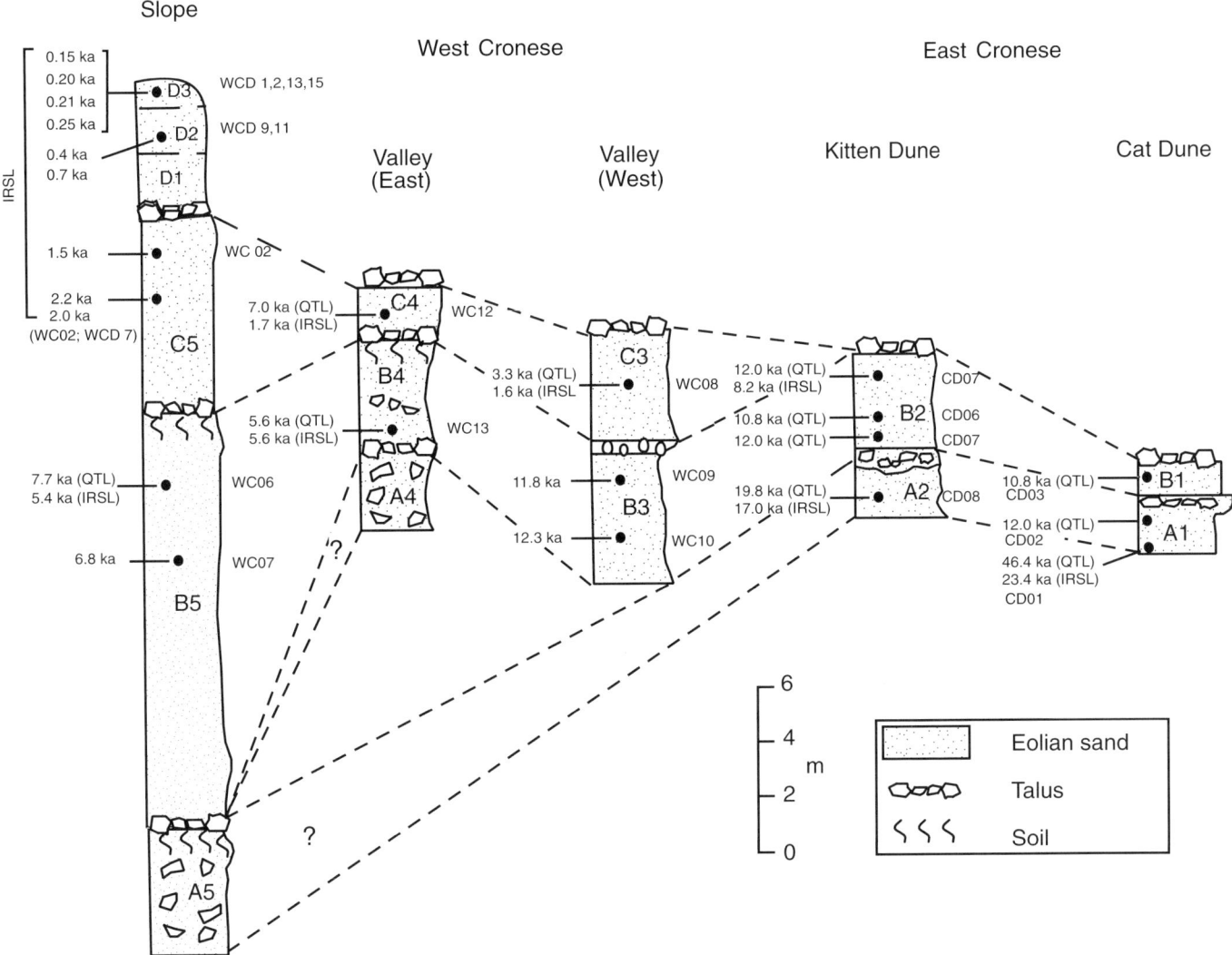

Figure 7. Stratigraphic columns of sand ramps in the Cronese basin. Stratigraphy modified from Lancaster and Tchakerian (1996) and Clarke et al. (1996b). Luminescence ages from Rendell et al. (1994), Clarke et al. (1996b), and Rendell and Sheffer (1996). For location of sections, see Figure 6. QTL—quartz thermoluminescence. IRSL—infrared stimulated luminescence.

Cronese Mountains. Those on the southern part of the range are deeply dissected. All ramps are covered by cobble- to boulder-sized angular talus. The northern sand ramps (locations West Cronese Slope and Valley) consist of four main stratigraphic units (Fig. 7). At the base of the main exposed section, the oldest unit (A5) consists of well-indurated structureless to very weakly horizontally laminated red-brown sand with a carbonate cemented zone and abundant rhizoliths in the upper 0.1–0.2 m. The upper surface of this unit is in erosional contact with the overlying units and is covered in many places by an angular cobble- to small boulder-sized colluvium. The next younger deposit (Unit B5) consists of up to 8 m of horizontally bedded and laminated coarse to medium buff sand. At the top of this unit, there is slight to moderate reddening and abundant calcic laminae, indicative of a weakly developed buried soil. The still-younger Unit C5 is composed of 2 m of friable, slightly indurated pale brown massive sand with a thin cover of angular gravel-sized colluvium. Lapping onto the exposed surfaces of this ramp are three crescentic dune ridges (Units D1–D3) that are composed of slightly indurated, friable, very well laminated, pale brown medium and coarse sand with beds dipping at ±10° to the southeast. At the northwest tip of the Cronese Mountains, deposits of a similar composition are carved into a series of yardangs ~1 m high with long axes oriented southwest-northeast. The yardangs are formed in horizontally stratified coarse yellow-red eolian sands that lie on a surface that is cut into well-indurated red-brown sand with abundant calcic nodules and blebs. This sand appears to be equivalent to Unit A5.

IRSL ages for the West Cronese Slope sand ramp and dunes (Clarke et al., 1996b) indicate that Unit C5 was formed 1.4–2.2 ka, with unit B5 being deposited between 5 and 7 ka. The younger dune ridges (D1–D3) and the deposit in which the yardangs are carved were formed during the late Holocene between 0.15 and 2 ka (Clarke et al., 1996b). A radiocarbon date (Beta-84635) from charcoal exposed at the base of one of the yardangs gave a calibrated age of 1310–1385 A.D.

To the south, a series of dissected sand ramps fill a south-north–trending valley in the main Cronese Mountains. Two main eolian units are identified at this site. Unit B3 is composed of slightly indurated pale brown sand with rare angular gravel-sized grus. The upper surface of this unit is marked by a weak paleosol and locally by a 0.5–1 m-deep gravel-filled channel. Unit B4 consists of a 5 m thickness of massive, friable reddish brown to buff medium sand with common grus and concentrations of angular cobble to boulder clasts at depths of 2 m and 6 m. IRSL ages (Rendell and Sheffer, 1996; Clarke et al., 1996b) indicate that unit B3 was deposited ca. 12 ka; Unit B4 formed in the period 5–6 ka; and Units C3 and C4 ca. 1.6 ka.

Eolian deposits from sand ramps at Hanks Mountain, southeast of Baker and at Balch (Figs. 2, 8, 9), provide a record of eolian deposition in the Soda Lake–Silver Lake source area. The Hanks Mountain sand ramp (Fig. 9) comprises 1 m of medium to fine sand (Unit C), a 2 m-thick unit of calcrete-cemented colluvium (Unit B), underlain by medium reddish-brown sand (Unit A). Luminescence ages from Hanks Mountain sand ramp indicate that it was formed between 20 and 30 ka (Rendell and Sheffer, 1996), equivalent to the proposed age for the pre–Lake Mojave eolian unit in the Silver Lake area (Wells et al., this volume).

The sand ramp at Balch is a composite of eolian and fluvial deposits (Fig. 9). The upper part of the ramp consists of 5 m of gray-brown sand (Unit A), overlain by 8 m of fluvial sand with gravel stringers and 1–1.5 m deep gravel-filled channels (Unit B) and 2 m of fine gray-brown eolian sand (Units C and D) with small fluvial gravel lenses, capped by a well-developed calcrete paleosol. Luminescence ages for the upper part of the Balch sand ramp (Rendell and Sheffer, 1996) indicate that it accumulated in the period 20–25 ka. The topographically lower, but stratigraphically higher, part of the ramp (Unit E) is composed of 4 m of fine gray-brown sand that contains varying amounts of grus. It overlies a unit of angular granitic pebbles and cobbles and is also capped by a angular pebble stratum without any soil development. This part of the ramp accumulated in the early to mid-Holocene (3–9 ka) (Rendell and Sheffer, 1996).

The Devils Playground. Eolian landforms in the Devils Playground area include active crescentic dunes and falling dunes to the west and vegetation-stabilized sand sheets and sand ramps to the east (Fig. 9). Active dunes consist of pale brown to buff moderately well sorted medium sand. The sand sheets typically consist of 1.5–2 m (locally as much as 4 m) of gray-brown fine sand that

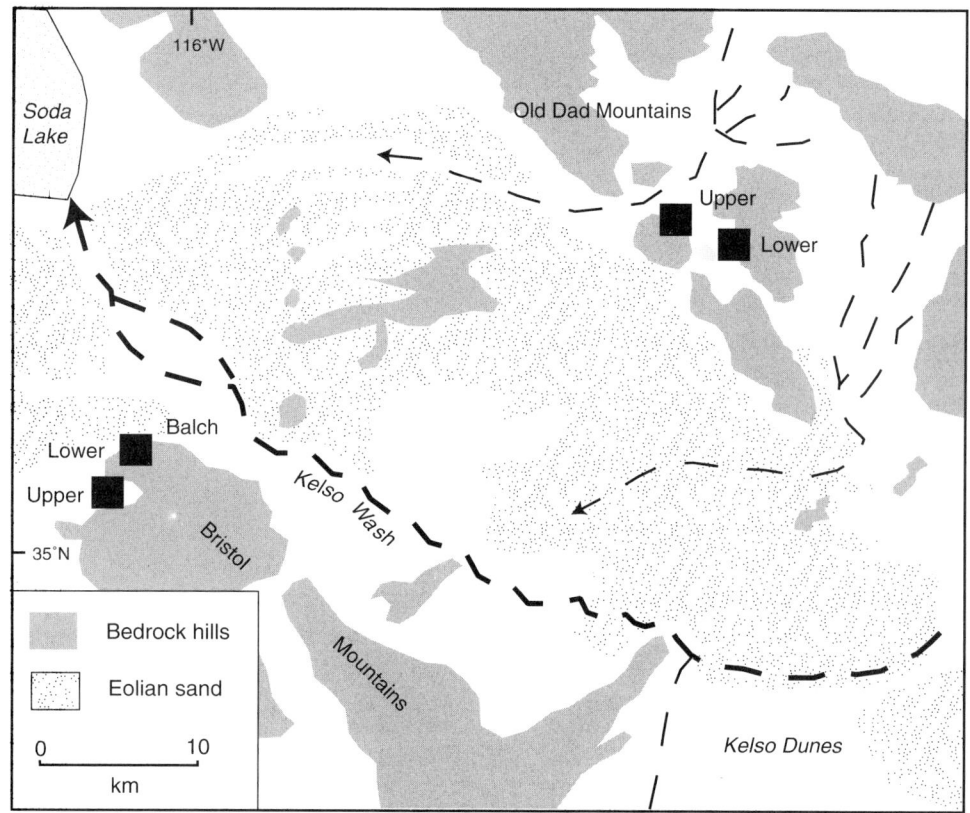

Figure 8. The Devils Playground area, showing location of study sites described in text.

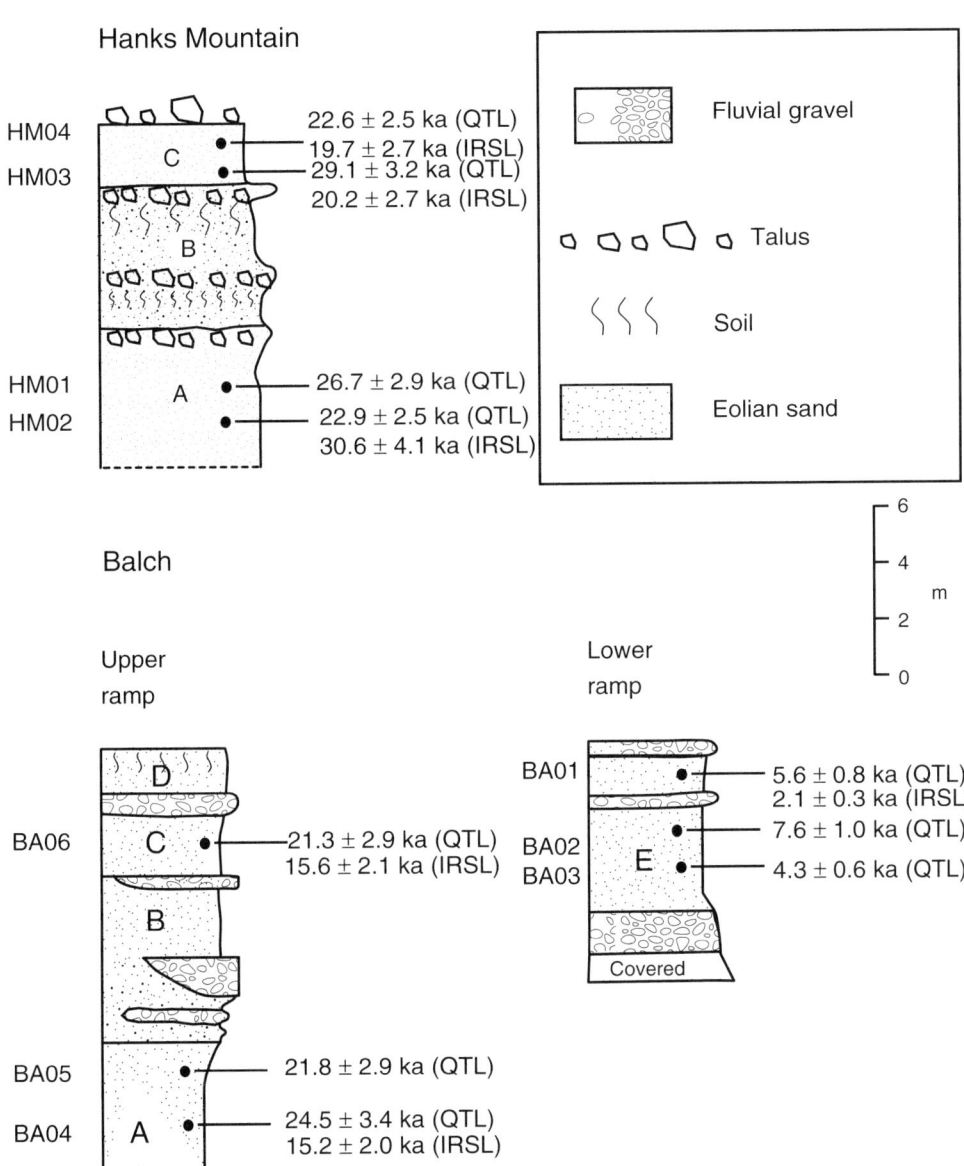

Figure 9. Stratigraphic columns of sand ramps at Hanks Mountain and Balch. Stratigraphy modified from Lancaster and Tchakerian (1996). Luminescence ages from Rendell and Sheffer (1996). QTL—quartz thermoluminescence. IRSL—infrared stimulated luminescence.

overlies fine gravel to granules that is partially cemented by a powdery carbonate. Sand ramps mantle the Old Dad Mountains on the northeast side of the valley as well as the northwest parts of the Bristol Mountains on the southern side.

The history of eolian activity in this transport corridor is documented by sand ramps in the Old Dad Mountains on the northern side of the corridor. Sand ramps in this area consist of both falling (lee slope) ramps on the northeast slopes of the mountain and climbing (windward slope) ramps on the west and southwest sides (Fig. 10). The best sections are provided by the falling ramps. Their upper parts (Unit D2) consist of friable medium to fine moderately-well to well-sorted sand with a cover of boulder- to cobble-sized talus. At a depth of 1 m is a layer of 15–20 cm clasts, which are underlain by massive red-brown sand that contains abundant grus (Unit A2). The lower part of the ramp is exposed in a stream section. From the base, it comprises 1.5 m of pale brown sand with sparse grus (Unit A1) capped by a very weakly developed soil; 2 m of mixed fluvial sands, gravel channels, and stringers (Unit B), capped by a well-developed argillic, and in places calcic, buried soil; and 2 m of fine red-brown sands with rare coarse sand grus (Unit D1). Luminescence ages for the Old Dad Mountains ramps (Rendell and Sheffer, 1996) indicate that they accumulated during the period from ca. 14 to 3 ka.

Kelso Dunes. Kelso dune field (Figs. 5, 11) consists of a "core" of 3 large WSW-ENE–trending complex linear ridges up to 160 m high and 1900–2000 m apart. On the west and northwest sides of the core of the dune field lie degraded, vegetated, straight-crested and barchanoid crescentic ridges up to 15 m high

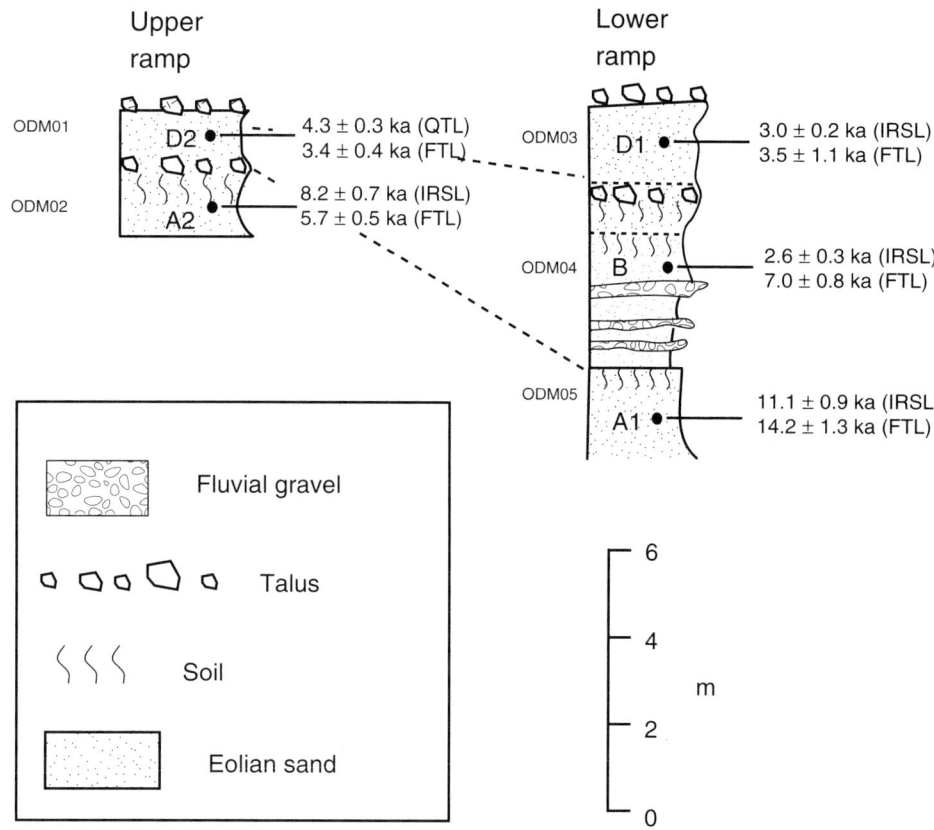

Figure 10. Stratigraphic columns of sand ramps in the Old Dad Mountains. Stratigraphy modified from Lancaster and Tchakerian (1996). Luminescence ages from Rendell and Sheffer (1996). For location of sections see Figure 8. QTL—quartz thermoluminescence. FTL—feldspar thermoluminescence. IRSL—infrared stimulated luminescence.

and linear dunes up to 5 m high. The eastern section of the dune field consists of a 1–5 m thickness of sand formed into five smaller areas of dunes, each cut by washes and separated from the "core" area by Cottonwood Wash, which is incised into eolian and fluvial deposits by as much as 20 m. To the north is a smaller version of the main area of active dunes that consists of three linear ridges up to 50 m high with superimposed 2–4-m-high crescentic dunes. Areas of low, partially active, linear and crescentic ridges, as well as extensive areas of sand sheets occur on the northeast and southern edges of this part the dune field. East of Winston Wash are areas of vegetated and degraded crescentic ridges with a height of 8–10 m.

A characteristic feature of Kelso Dunes is the juxtaposition of areas of dunes of distinctly different morphological type, size, and spacing and alignment. A total of 14 dune units can be identified on aerial photographs (Fig. 11). The dune field appears to have developed by the coalescence of a series of smaller genetically independent dune fields, each of which represents an episode of sediment input or reworking of existing dunes (Lancaster, 1993).

Estimates of the age of Kelso Dunes have varied widely between "several thousand years and possibly 10,000–20,000 [years]" (Sharp, 1966) to "very likely >100,000 yr, and quite possibly more than a million years" (Yeend et al., 1984; Smith, 1984). Because the Kelso dune field overlies, and in some cases is intercalated with, alluvial fan deposits derived from the Granite and Providence Mountains, it is possible to use the ages of these alluvial fans (Wells et al., 1990) to constrain the age of the dunes. The western part of the dune field rests on fans of early or possibly middle Pleistocene age. Late Pleistocene fan units underlie the southern margins of this part of the dune field. A well-developed soil is developed on eolian sand that lies on the Granite Mountains alluvial fans south of the main dune field. This soil appears similar to those developed on sand ramps elsewhere in the region with an age of >20 ka (Rendell and Sheffer, 1996).

Dunes east of Cottonwood Wash appear to be much younger and mostly lie on fan surfaces of Holocene and latest Pleistocene age (Clarke, 1994; McDonald and McFadden, 1994). Unit Qe1 of McDonald and McFadden (1994) consists of thin discontinuous sand sheets that overlie alluvial fan unit Qf4, which is of late Pleistocene age. Eolian Unit Qe2 includes much of the eastern part of the dune field (geomorphic units I through V on Figure 11) and overlies fan unit Qf5 (early Holocene to latest Pleistocene). In turn, Qe2 is truncated by the late Holocene alluvial fan unit (Qf6).

Several key dune geomorphic units (Fig. 11) have been the subject of luminescence dating using IRSL techniques (Clarke, 1994; Wintle et al., 1994). These included the core of active dunes (unit VI), vegetation stabilized dunes and sand sheets on the eastern margin of the dune field (Unit II), vegetation-stabilized crescentic ridges on the northern margin of the dune field

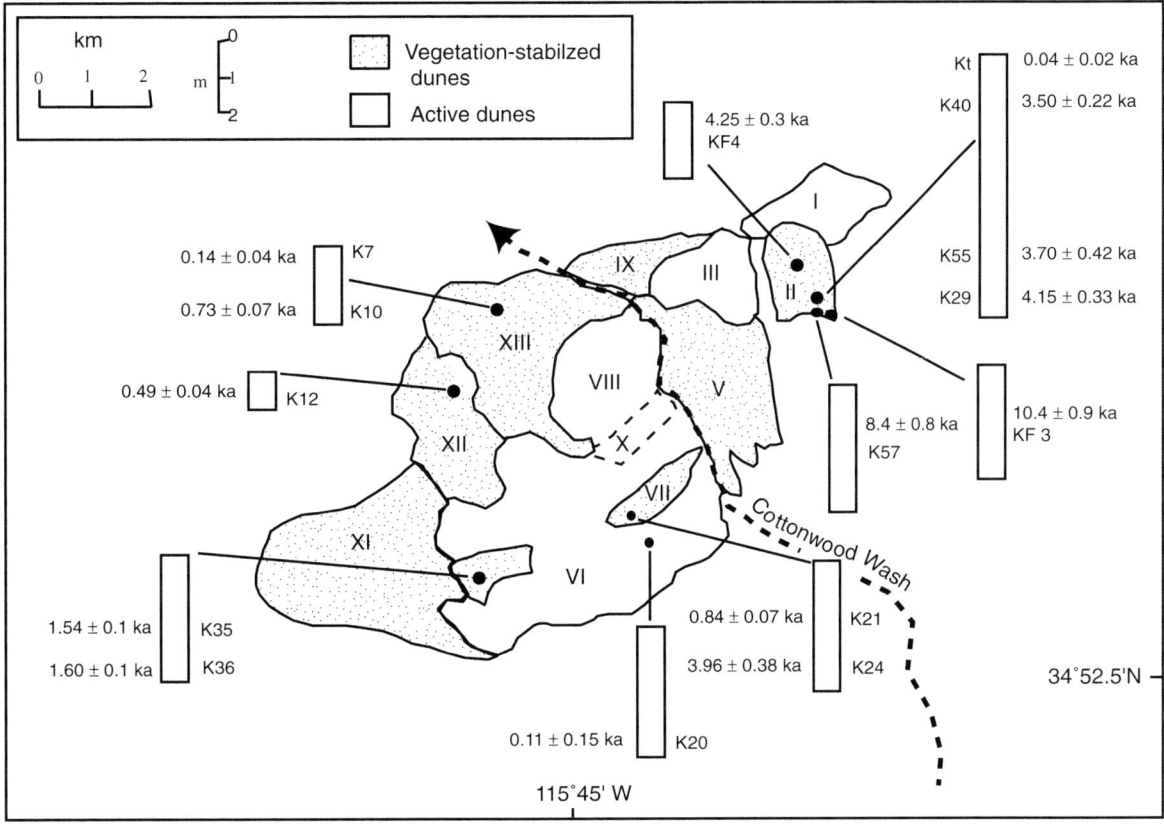

Figure 11. Map of the Kelso dune field showing location of luminescence samples. Luminescence ages from Wintle et al. (1994) and Clarke (1994).

(units IX and XII), and vegetated crescentic ridges on the southwest side of the dunes (Unit XIV). IRSL dates from Unit II indicate that it accumulated between 10.4 and 3.5 ka. Unit VI was apparently in place by 4 ka. The crescentic dunes of unit XIV were formed ca. 1.5 ka and dunes on the north side of the dune field were formed, or reworked, between 0.8 and 0.5 ka. The oldest sands known are those which were deposited as sand sheets on alluvial fan surfaces as much as 5 km southeast of the present dune margins between 16 and 18 ka (Clarke, 1994). The pattern of luminescence ages suggests that there is an increase in the minimum ages of dune and sand sheet units from northwest to southeast, supporting the hypothesis that the dune field accumulated by stacking or shingling of successive generations of eolian units on the piedmont of the Providence and Granite Mountains (Lancaster, 1994).

The Manix basin

The eastern margin of the Manix basin is covered by extensive sand sheets and sand ramps against the western flanks of the Cady Mountains. The eolian history of this area is documented by the sand ramp at Soldier Mountain (Fig. 12). This sand ramp has a maximum thickness of 25–30 m and is cut by gullies where runoff from adjacent slopes has concentrated at the bedrock-sand interface on its northern and southern edges. The surface of the ramp is covered by a mantle of angular small boulder- to cobble-sized clasts composed of schist derived from Soldier Mountain. In places, the truncated Bt horizon of a buried soil developed in the uppermost eolian unit is exposed. The highest parts of the ramp adjacent to and east of the peak of Soldier Mountain are composed of active sand.

The stratigraphy and sediments of the Soldier Mountain sand ramp have been exposed by mining operations on its northern part. Six main sedimentary units can be identified, separated by truncated buried soils and/or thin talus layers similar to those found on the surface of the ramp today (Fig. 12). The buried soils and talus layers slope gently up to the north and converge toward the southern edge of the exposure.

The sediments of the sand ramp are mostly massive, moderately to poorly sorted, fine to coarse angular to subangular quartz sands with a small admixture (<10%) of lithic fragments. Units 1, 2, and 6 are weakly to moderately horizontally laminated, with alternating finer and coarser laminae typical of the deposits of eolian sand sheets, and are weakly cemented by calcium carbonate. There are occasional stringers of angular schist bedrock talus within the lower units. Unit 5 exhibits well-developed

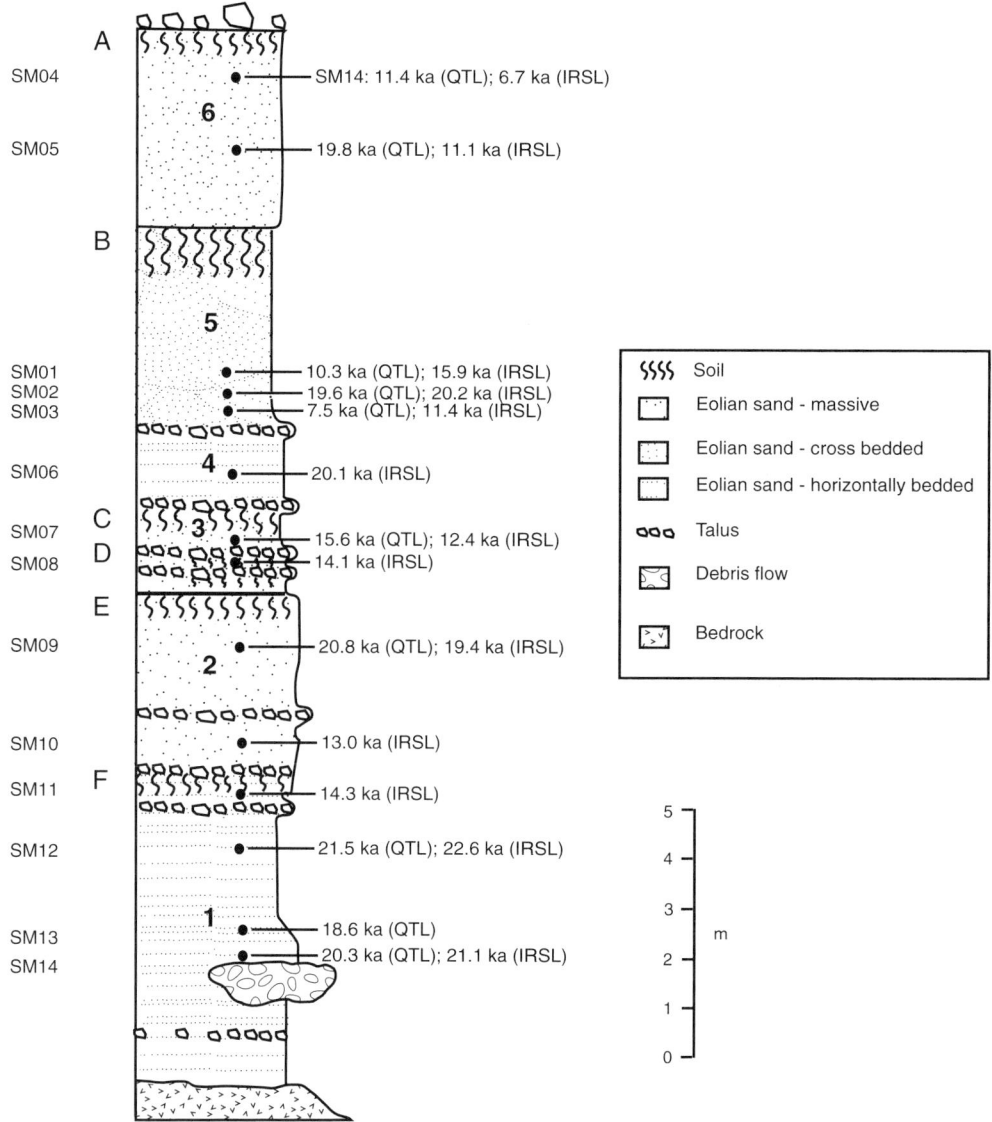

Figure 12. Stratigraphic column of sand ramp at Soldier Mountain. Stratigraphy modified from Lancaster and Tchakerian (1996). Luminescence ages from Rendell and Sheffer (1996). QTL—quartz thermoluminescence. IRSL—infrared stimulated luminescence.

cross-stratification dipping 28°–30° to the northeast. Unit 4 is strongly horizontally laminated and probably represents reworking of eolian sands by local fluvial action. Unit 4 is characterized by closely spaced talus and buried soils. Unit 1 (the lowermost unit) contains debris flow deposits of large and small boulders.

Buried soils are variably developed and preserved at Soldier Mountain, indicating that eolian deposition was episodic. In most cases, only the Bt and the Bk horizons remain. The thickest soils (0.3–0.8 m) are those that cap the uppermost eolian sand (Unit 6) and Unit 5 (A and B on Figure 12). The best-developed soils are E and F, with stage I Bk horizons.

Luminescence ages for the Soldier Mountain sand ramp (Rendell and Sheffer, 1996) indicate that Soldier Mountain Sand ramp accumulated between 20–25 ka and 7–8 ka, but there are a number of inconsistencies in the ages obtained. A moderately developed buried soil complex (soils C to E, Figure 12) appears to have developed from 11 or 12 to 14 ka based on IRSL ages in Rendell and Sheffer (1996).

DISCUSSION

Geomorphic and stratigraphic relations, together with the luminescence ages obtained from different eolian units enable several episodes of late Pleistocene and Holocene eolian deposition and/or dune remobilization to be identified (Fig. 13). Following the interpretations of Rendell and Sheffer (1996), we

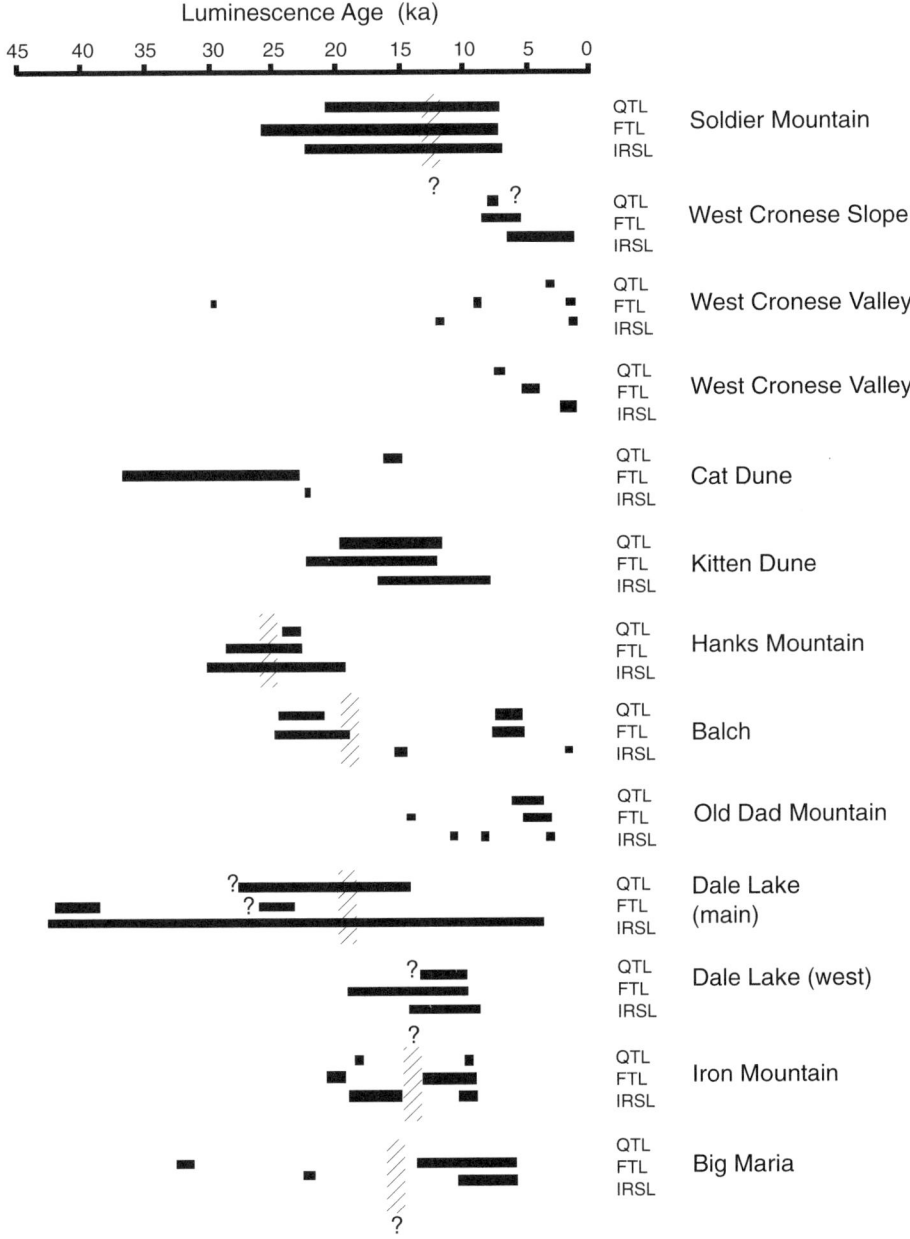

Figure 13. Summary diagram of luminescence-dated periods of eolian deposition in the Mojave Desert. After Rendell and Sheffer, 1996. QTL—quartz thermoluminescence. FTL—feldspar thermoluminescence. IRSL—infrared stimulated luminescence.

consider the earliest phase of eolian activity to have spanned the period 20–30 ka, with deposits at Cat Dune, Hanks Mountain, Balch, Dale Lake, Iron Mountain, and Big Maria. This period is likely equivalent to that which deposited unit Qe1 at Silver Lake (Wells et al., this volume). With the exception of Cat Dune and Hanks Mountain, a prominent and well-developed buried soil follows this period of accumulation, indicating a long period of stability. A later phase of eolian accumulation is identified at the Cat Dune, West Cronese, Dale Lake, Soldier Mountain, Iron Mountain, and the Big Maria sand ramps. This period extends from 15 ka (20 ka at Soldier Mountain) to 7 ka. Middle (4–2 ka) and late Holocene (<2 ka) eolian deposits occur in the Cronese basin, Balch, the Old Dad Mountains, and at Kelso Dunes.

Rendell and Sheffer (1996) suggest that the ages of late Pleistocene eolian accumulation on sand ramps overlap with those for pluvial periods and higher levels of lakes in the Mojave

River system (Fig. 14) and argue for the influence of sediment supply as a major control on periods of eolian deposition. We agree with this interpretation, but, as discussed below, the regional picture is more complex, and likely reflects changes in sediment supply to eolian processes as a result of the interplay between deltaic, lacustrine, and fluvial processes in different basins of the Mojave River system (e.g., Enzel et al., this volume; Wells et al., this volume). In the case of sand ramps in the eastern Mojave Desert (Dale Lake, Iron Mountain, and Big Maria), the source of sediment in these systems is not the Mojave River system, and periods of eolian deposition may reflect enhanced sediment supply from alluvial fan systems that aggraded during periods transitional between wetter and drier climates (Harvey and Wells, this volume).

Central to the understanding of links between eolian geomorphic systems and the sources of their sediments is a precise comparison of the chronologies of the eolian and fluvial-lacustrine systems. In order to compare luminescence and radiocarbon chronologies, it is necessary to calibrate the latter using Mazaud et al. (1991) for radiocarbon ages >18,360 yr B.P. and Bard et al. (1993) and the CALIB v3.0 program (Stuiver and Remer, 1993) for radiocarbon ages younger than 18,360 yr B.P. All radiocarbon ages below are given as calibrated ages in ka.

In the Manix basin, the Soldier Mountain sand ramp accumulated episodically between 20–25 and 7–8 ka. The sand ramp is located downwind (east) of the extensive sandy delta developed by the Mojave River in the western and central parts of this basin, as well as the area of Lake Manix. The eolian sediments of the Soldier Mountain ramp therefore were likely derived from the deltaic and lacustrine sediments to the west. The Afton basin of Lake Manix was occupied by lakes at 35–32.6, 23.8–22.7, and ca. 20.4 ka (Meek, 1989; Enzel et al., this volume). Ages determined from shore features show that the Coyote Lake basin was occupied by a lake between 19.3 and 17 ka, which may have continued to exist until 14.8–12.5 ka. This lake history is interpreted to be the result of diversion of the Mojave River to the Coyote basin, as the delta distributary channel shifted to the north (Enzel et al., this volume). These data suggest that Soldier Mountain sand ramp likely accumulated as a result of the deflation of sand from the desiccated Troy and Afton basins of Lake Manix following diversion of the main flow of the Mojave River to Coyote Lake after 20 ka.

The large complex of sand ramps in the Cronese basin has been regarded by Tchakerian (1989) and Clarke et al. (1996b) as the product of deflation of sediments of the adjacent East and West Cronese lake basins. Although this is likely true of the Holocene-age sediments in this area, the volume of older sediment that comprises the climbing and falling dune complexes of Cat Dune and adjacent areas to the west suggests a much more extensive sediment source. It is possible that these sediments were derived from the drying of the Afton basin and later the Coyote basin of Lake Manix to the west. Numerous small sand ramps between the Cronese basin and these potential source areas indicate the existence of eolian sediment transport from more westerly sources. Similarly, older sand ramp sediments in the West Cronese basin may have been derived from the Coyote basin directly to the west, following its desiccation in the period 13–11 ka.

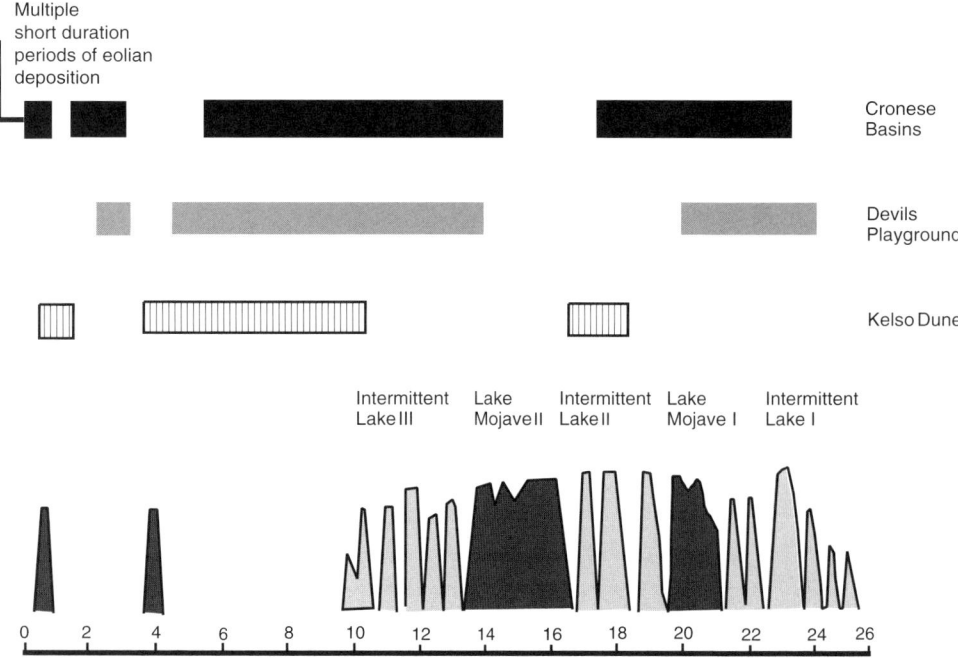

Figure 14. Comparison of periods of eolian construction in the Kelso Dunes system and the level of lake Mojave. Data on lake levels from Brown (1989) and Wells et al., this volume; luminescence ages from data in Rendell and Sheffer (1996), Wintle et al. (1994), and Clarke and Rendell (1998). Note that lacustrine time scale has been converted to calibrated years (ka) to facilitate comparison with the luminescence ages (as in text).

Sand ramps at Balch and Hanks Mountain, as well as sands in the Old Dad Mountains and Kelso Dunes are likely derived from sediments deposited in the Silver Lake–Soda Lake basin. Wells et al. (this volume) show that lakes existed in the Silver Lake basin from 24.5 to 9.7 ka with two persistent highstands from 21.9 to 19.4 ka (Lake Mojave I) and 16–13.3 ka (Lake Mojave II), with final desiccation ca. 10 ka. It is not clear if the Mojave River fed Soda Lake prior to 25 ka. Sand ramp sediments at Hanks Mountain and the upper ramp at Balch largely predate the existence of Lake Mojave I, suggesting that they were derived from an ancestral distal fluvial system in the Soda Lake–Silver Lake basin, and may be equivalent to the basal eolian deposits identified in cores in these basins (Wells et al., this volume). Latest Pleistocene and Holocene eolian sand in the Kelso Dunes system was likely derived from the southern part of the Lake Mojave basin, together with local inputs (Ramsey et al., 1999). The oldest sands at Kelso Dunes, with IRSL ages ca. 17 ka (Clarke, 1994), appear to post date Lake Mojave I, and were deposited during times of intermittent lake conditions (Intermittent Lake II), while the remainder of the dated dune sediments were likely deposited during the desiccation of Lake Mojave II after 13 ka and the existence of Intermittent Lake III (Fig. 14). Sediments of the lower sand ramp at Balch and at Old Dad Mountains were also deposited during this period, with a possible later input 3–4 ka, or following the lacustrine event in the Silver Lake basin 3.8–3.9 ka (Clarke and Rendell, 1998; Enzel et al., 1992). The sedimentary evidence from the Silver Lake basin of Lake Mojave indicates that intermittent lake periods were characterized by a series of drying events during which lake levels fell rapidly and remained low for a number of years (Wells et al., this volume). During one such event, eolian sand was deposited on the lake basin floor (Wells et al., this volume). It is likely that these short-duration drying events exposed large areas of the Mojave River fan delta to eolian processes, resulting in input of sediment to the eolian system. The resolution of the luminescence dates obtained does not, however, permit more precise correlations, for example with the prominent drying event in the Silver Lake basin at 18 ka (Wells et al., this volume).

Late Holocene eolian deposits occur close to modern sand source areas in the west Cronese, Balch, and Old Dad Mountain areas. Multiple, short-duration episodes of eolian accumulation are recorded at these sites (Clarke and Rendell, 1998; Rendell and Sheffer, 1996). IRSL ages for sediments in the Kelso sand transport system (including the Cronese basins) appear to cluster between 0.14 and 0.20, 0.4 and 0.5, 0.7 and 0.85, 1.4 and 1.7, 2 and 3, 3.5 and 4.2, and 5.4 and 6.7 ka (Clarke and Rendell, 1998). The ages appear to correlate well with the record of extreme floods in rivers in Arizona and Utah (Ely, 1997), as well as with the record of major floods in the Mojave River system (Enzel et al., 1992; Enzel and Wells, 1997). We agree with Clarke and Rendell (1998) that episodes of eolian deposition at this temporal scale reflect enhanced sediment supply from the Mojave River system, similar to modern analogues of flooding and supply of eolian sediment (e.g., in 1993). In addition, multiple periods of eolian activity and reworking of dune sediments likely occurred throughout the late Holocene, most notably ca. 1.5 ka and 0.4–0.7 ka (Wintle et al., 1994).

The record of periods of nondeposition and/or soil formation on the sand ramps throughout the Mojave suggests one major and a number of minor periods of regional-scale eolian stability. The first well-dated period occurred 20–15 ka, with shorter periods of soil formation ca. 14 and 4 ka. The first period coincides with the existence of Lake Mojave I (21.9–19.4 ka) and Intermittent Lake II (19.4–16 ka). The second period of eolian stability and soil formation ca. 14 ka coincides with the existence of Lake Mojave II (16–13.3 ka). Although the main cause of high levels of Lake Mojave was an order of magnitude increase in discharge of the Mojave River, hydrological models (Wells et al., this volume) indicate an increase of 50–200% in local precipitation during high lake stands. Although not significant to the hydrological budget of the area, it probably had a major impact on dune mobility. Therefore, periods of eolian stability and soil formation likely represent times of increased precipitation and vegetation cover in the Mojave Desert that reduced sediment availability and mobility. This scenario also applies to the Holocene, when the period 4.5–3 ka appears to have been cooler and wetter in many parts of the southwestern United States. with lowering of the woodland–desert scrub boundary (Spaulding, 1985; Spaulding et al., 1994), increased groundwater recharge in southern Nevada, and shallow lakes in Death Valley, Searles Lake, and the Silver Lake basin (Enzel et al., 1992).

CONCLUSIONS

The late Pleistocene and Holocene record of eolian deposition in the Mojave Desert suggests strongly that sediment supply from linked fluvial and lacustrine sources is a major control of eolian dynamics in the region. These changes can be evaluated using the model of Kocurek and Lancaster (1999) which recognizes a matrix of eolian system states determined by the supply, availability, and mobility of sediment. Interactions between these factors result in episodic input of sediment to the eolian system. In the Mojave Desert, sediment supply was high during most of the latest Pleistocene and early Holocene. Sediment was stored in the subbasins of the Mojave River system during periods of highest lake levels, with contemporaneous input of sand to the eolian system during periods of lower and/or fluctuating lakes (e.g., Intermittent Lakes I–III in the Lake Mojave basin). Changes in the position of terminal basins of the Mojave River gave rise to periods of localized eolian deposition. During the periods of greatest precipitation (inferred from the paleohydrology of the periods of highest lakes), sediment availability was restricted by the existence of permanent water bodies in the principal sediment source areas and mobility was reduced by increased vegetation cover along transport pathways. The eolian sediment transport system was therefore shut down, leading to the observed periods of stability of eolian deposits and soil formation. During the terminal Pleistocene and early Holocene, sediment supply declined abruptly as flow in the Mojave River decreased and the lake system desiccated (Wells et al., this volume). However, sediment availability increased because vegetation cover was reduced, and rose to levels that were limited only by the transport capacity of the local winds in a climate somewhat drier than today (Spaulding et al., 1994). In this period of the Pleistocene-Holocene transition, sediment stored in the latest Pleistocene period of high sediment supply was mobilized as lagged input to the system, giving rise to a major episode of eolian construction. Eolian activity was enhanced by generally dry conditions during the later part of this time. During the Holocene, the system was controlled by sediment supply. Episodes of short-lived lakes and/or increased flood frequency in the Mojave River system were marked by pulses of enhanced fluvial sediment supply followed by eolian deposition as the fluvial sediments were reworked by the wind. A mid Holocene episode of higher regional rainfall and therefore increased vegetation cover limited sediment availability and caused the observed period of eolian stability in some areas from ca. 2 to 4 ka. A return to drier conditions after 2 ka resulted in renewed eolian activity and reworking of sediment produced and stored in the previous wetter periods. The current sediment configuration of the eolian system appears to be limited by low sediment supply from the Mojave River, although field observations (see Clarke and Rendell, 1998) suggest increased eolian sediment input following large modern flood events (e.g., 1993).

ACKNOWLEDGMENTS

Research on which this paper is based was supported in part by the National Science Foundation (EAR 9204648), The National Geographic Society, and NATO Collaborative Research Grant 900151.

REFERENCES CITED

Bach, A.J., 1995, Climatic controls on aeolian activity in the Mojave and Colorado Deserts, California [Ph.D. thesis]: Tempe, Arizona State University, 308 p.

Bard, E., Arnold, M., Fairbanks, R.G., and Hamelin, B., 1993, ^{230}Th, ^{234}U and ^{14}C ages obtained by mass spectrometry on corals: Radiocarbon, v. 35, p. 191–201.

Brown, W.J., 1989, Late Quaternary stratigraphy, paleohydrology, and geomorphology of pluvial Lake Mojave, Silver Lake and Soda Lake basins, southern California [M.S. thesis]: Albuquerque, University of New Mexico, 266 p.

Brown, W.J., Wells, S.G., Enzel, Y., Anderson, R.Y., and McFadden, L.D., 1990, The late Quaternary history of pluvial Lake Mojave: Silver Lake and Soda Lake basins, California, in Reynolds, R.E., et al., eds., At the end of the Mojave: Quaternary studies in the eastern Mojave Desert: Redlands, California, Special Publication of the San Bernardino County Museum Association, 1990 Mojave Desert Quaternary Research Center Symposium, May 18–21, 1990, p. 55–72.

Clarke, M.L., 1994, Infrared stimulated luminescence ages from aeolian sand and alluvial fan deposits from the eastern Mojave Desert, California: Quaternary Science Reviews, v. 13, p. 533–538.

Clarke, M.L., and Rendell, H.M., 1998, Climatic change impacts on sand supply and the formation of desert sand dunes in the southwest USA: Journal of Arid Environments, v. 39, no. 3, p. 517–532.

Clarke, M.L., Richardson, C.A., and Rendell, H.M., 1996a, Luminescence dating of Mojave Desert sands: Quaternary Science Reviews, v. 14, no. 7–8, p. 783–790.

Clarke, M.L., Wintle, A.G., and Lancaster, N., 1996b, Infrared stimulated luminescence dating of sands from the Cronese basins, Mojave Desert: Geomorphology, v. 17, no. 1–3, p. 199–206.

Dean, L.E., 1978, The California desert sand dunes: Department of Earth Sciences, Riverside, University of California, 72p.

Dibblee, T.W., Jr., 1967, Areal geology of the western Mojave Desert, California: U.S. Geological Survey Professional Paper 522, p. 152.

Ely, L.L., 1997, Response of extreme floods in the southwestern United States to climatic variations in the late Holocene: Geomorphology, v. 19, no. 3–4, p. 175–202.

Enzel, Y., and Wells, S.G., 1997, Extracting Holocene paleohydrology and paleoclimatology information from modern extreme flood events: An example from southern California: Geomorphology, v. 19, no. 3–4, p. 203–226.

Enzel, Y., Brown, W.J., Anderson, R.Y., McFadden, L.D., and Wells, S.G., 1992, Short-duration Holocene lakes in the Mojave River drainage basin, southern California: Quaternary Research, v. 38, no. 1, p. 60–73.

Evans, J.R., 1961, Falling and climbing sand dunes in the Cronese ("Cat") Mountain area, San Bernardino County, California: Journal of Geology, v. 70, p. 107–113.

Eymann, J.L., 1953, A study of sand dunes in the Colorado and Mojave Deserts [M.S. thesis]: Los Angeles, University of Southern California, 91 p.

Fryberger, S.G., and Ahlbrandt, T.S., 1979, Mechanisms for the formation of aeolian sand seas: Zeitschrift für Geomorphologie, v. 23, p. 440–460.

Kocurek, G., and Lancaster, N., 1999, Aeolian system states: Theory and Mojave Desert Kelso Dunefield example: Sedimentology, v. 46, no. 3, p. 505–516.

Laity, J.E., 1987, Topographic effects on ventifact development, Mojave Desert, California: Physical Geography, v. 8, p. 113–132.

Laity, J.E., 1992, Ventifact evidence for Holocene wind patterns in the east-central Mojave Desert: Zeitschrift für Geomorphologie, Suppl-Bd, v. 84, p. 73–88.

Lancaster, N., 1993, Development of Kelso Dunes, Mojave Desert, California: National Geographic Research and Exploration, v. 9, no. 4, p. 444–459.

Lancaster, N., 1994, Studies of Quaternary eolian deposits of the Mojave Desert, California, in McGill, S.F., and Ross, T.M., eds., Geological investigations of an active margin: Geological Society of America, Cordilleran Section Guidebook, San Bernardino, San Bernardino County Museum Association, p. 172–175.

Lancaster, N., and Tchakerian, V.P., 1996, Geomorphology and sediments of sand ramps in the Mojave Desert: Geomorphology, v. 17, no. 1–3, p. 151–166.

Mainguet, M., 1983, Tentative mega-morphological study of the Sahara, in Gardner, R., and Scoging, H., eds., Mega geomorphology: Oxford, Clarendon Press, p. 113–133.

Mazaud, A., Laj, C., Bard, E., Arnold, M., and Tric, E., 1991, Geomagnetic field control of ^{14}C production over the last 80 k.y.: Implications for the radiocarbon time scale: Geophysical Research Letters, v. 18, p. 1885–1888.

McDonald, E., and McFadden, L.D., 1994, Quaternary stratigraphy of the Providence Mountains piedmont and preliminary age estimates and regional stratigraphic correlations of Quaternary deposits in the eastern Mojave Desert, California, in McGill, S.F., and Ross, T.M., eds., Geological investigations of an active margin: Geological Society of America, Cordilleran Section Guidebook, San Bernardino, California, San Bernardino County Museum, p. 205–210.

McDonald, E., McFadden, L.D., and Wells, S.G., 1995, The relative influences of climate change, desert dust and lithologic control on soil-geomorphic processes on alluvial fans, Mojave Desert, California, in Reynolds, R.E., and Reynolds, J., eds., Ancient Surfaces of the east Mojave Desert: San Bernardino, California, San Bernardino County Museum Association, p. 35–42.

McFadden, L.D., Wells, S.G., and Jercinovich, M.J., 1987, Influences of eolian and pedogenic processes on the origin and evolution of desert pavements: Geology, v. 15, p. 504–508.

Meek, N., 1989, Geomorphic and hydrologic implications of the rapid incision of Afton Canyon, Mojave Desert, California: Geology, v. 17, p. 7–10.

Tchakerian, V.P., 1992, Aeolian geomorphology of the Dale Lake sand sheet: San Bernardino County Museum Special Publication, v. 92-2, p. 46-49.

Tchakerian, V.P., 1997, North America, in Thomas, D.S.G., ed., Arid zone geomorphology: Process, form, and change in drylands: New York, John Wiley and Sons, p. 523-541.

Thompson, D.G., 1929, The Mojave Desert region, California: U.S. Geological Survey Water-Supply Paper, v. 578, p. 759 p.

Wells, S.G., McFadden, L.D., and Dohrenwend, J.C., 1987, Influence of late Quaternary climatic changes on geomorphic processes on a desert piedmont, eastern Mojave Desert, California: Quaternary Research, v. 27, p. 130-146.

Wells, S.G., McFadden, L.D., and Harden, J., 1990, Preliminary results of age estimations and regional correlations of Quaternary alluvial fans within the Mojave Desert region of southern California, in Reynolds, R.E., et al., eds., At the end of the Mojave: Quaternary studies in the eastern Mojave Desert: Redlands, California, Special Publication of the San Bernardino County Museum Association, 1990 Mojave Desert Quaternary Research Center Symposium, May 18-21, 1990, p. 45-54.

Wilson, I.G., 1971, Desert sandflow basins and a model for the development of ergs: Geographical Journal, v. 137, no. 2, p. 180-199.

Wintle, A.G., Lancaster, N., and Edwards, S.R., 1994, Infrared stimulated luminescence (IRSL) dating of late-Holocene aeolian sands in the Mojave Desert, California, USA: The Holocene, v. 4, no. 1, p. 74-78.

Yeend, W., Dohrenwend, J.C., Smith, R.S.U., Goldfarb, R., Simpson, R.W.J., and Munts, S.R., 1984, Mineral resources and mineral resource potential of the Kelso Dunes wilderness study area (CDCA-250), San Bernardino County, California: U.S. Geological Survey Open-File Report 84-647, 19 p.

Zimbelman, J.R., and Williams, S.H., 2002, Geochemical indicators of separate sources for eolian sands in the eastern Mojave Desert, California, and western Arizona: Geological Society of America Bulletin, v. 114, p. 490-496.

Zimbelman, J.R., Williams, S.H., and Tchakerian, V.P., 1995, Sand transport paths in the Mojave Desert, southwestern United States, in Tchakerian, V.P., ed., Desert aeolian processes: New York, Chapman and Hall, p. 101-130.

Muhs, D.R., Reynolds, R.R., Been, J., and Skipp, G., 2003, Eolian sand transport pathways in the southwestern United States: Importance of the Colorado: Quaternary International, v. 104, p. 3-18.

Ramsey, M.S., Christensen, P.R., Lancaster, N., and Howard, D.A., 1999, Identification of sand sources and transport pathways at the Kelso Dunes, California, using thermal infrared remote sensing: Geological Society of America Bulletin, v. 111, p. 646-662.

Rendell, H.M. and Sheffer, N.L., 1996, Luminescence dating of sand ramps in the eastern Mojave Desert: Geomorphology, v. 17, no. 1-3, p. 187-198.

Rendell, H., Lancaster, N., and Tchakerian, V.P., 1994, Luminescence dating of late Pleistocene aeolian deposits at Dale Lake and Cronese Mountains, Mojave Desert, California: Quaternary Science Reviews, v. 13, p. 417-422.

Sharp, R.P., 1966, Kelso Dunes, Mojave Desert, California: Geological Society of America Bulletin, v. 77, p. 1045-1074.

Smith, H.T.U., 1967, Past versus present wind action in the Mojave Desert region, California: U.S. Army Cambridge Research Laboratory Report, AFCRL-67-0683, 26 p.

Smith, R.S.U., 1984, Eolian geomorphology of the Devil's Playground, Kelso Dunes and Silurian Valley, California, in Dohrenwend, J.C., ed., Surficial geology of the eastern Mojave Desert, California: Geological Society of America 1984 Annual Meeting Field Trip 14 Guidebook, p. 162-173.

Spaulding, W.G., 1985, Vegetation and climates of the last 40,000 yr in the vicinity of the Nevada Test Site, south-central Nevada: U.S. Geological Survey Professional Paper 1329, 83 p.

Spaulding, W.G., Koehler, P.A., and Anderson, R.S., 1994, A late Quaternary paleoenvironmental record from the central Mojave Desert, in Reynolds, R.E., ed., Off Limits in the Mojave Desert: Field trip guidebook to the 1994 Mojave Desert Quaternary Research Center field trip to Fort Irwin and surrounding areas: San Bernardino, California, San Bernardino County Museum Association, p. 53-55.

Stuiver, M., and Reimer, P.J., 1993, Extended 14C database and revised CALIB 3.0 14C age program: Radiocarbon, v. 35, p. 215-230.

Tchakerian, V.P., 1989, Late Quaternary aeolian geomorphology, east-central Mojave Desert, California [Ph.D. thesis]: Los Angeles, University of California, 174 p.

Tchakerian, V.P., 1991, Late Quaternary aeolian geomorphology of the Dale Lake sand sheet, southern Mojave Desert, California: Physical Geography, v. 12, no. 4, p. 347-369.

MANUSCRIPT ACCEPTED BY THE SOCIETY AUGUST 1, 2002

Printed in the USA